Springer Proceedings in Physics **418**

Indexed by Scopus

The series Springer Proceedings in Physics, founded in 1984, is devoted to timely reports of state-of-the-art developments in physics and related sciences. Typically based on material presented at conferences, workshops and similar scientific meetings, volumes published in this series will constitute a comprehensive up to date source of reference on a field or subfield of relevance in contemporary physics. Proposals must include the following:

- Name, place and date of the scientific meeting
- A link to the committees (local organization, international advisors etc.)
- Scientific description of the meeting
- List of invited/plenary speakers
- An estimate of the planned proceedings book parameters (number of pages/articles, requested number of bulk copies, submission deadline).

Please contact:

For Americas and Europe: Dr. Zachary Evenson; zachary.evenson@springer.com
For Asia, Australia and New Zealand: Dr. Loyola DSilva; loyola.dsilva@springer.com

Nurul Razliana Abdul Razak ·
Mohd Arif Anuar Mohd Salleh ·
Dewi Suriyani Che Halin ·
Kamrosni Abdul Razak · Flora Somidin ·
Muhammad Fadlin Hazim Baser
Editors

Proceedings of the Green Materials and Electronic Packaging Interconnect Technology Symposium

EPITS 2024, 26–27 August, Ho Chi Minh City, Vietnam

Springer

Editors
Nurul Razliana Abdul Razak
Faculty of Mechanical Engineering and Technology
Universiti Malaysia Perlis
Arau, Perlis, Malaysia

Dewi Suriyani Che Halin
Faculty of Chemical Engineering and Technology
Universiti Malaysia Perlis
Arau Perlis, Perlis, Malaysia

Flora Somidin
Faculty of Chemical Engineering and Technology
Universiti Malaysia Perlis
Arau Perlis, Perlis, Malaysia

Mohd Arif Anuar Mohd Salleh
Faculty of Chemical Engineering and Technology
Universiti Malaysia Perlis
Arau Perlis, Perlis, Malaysia

Kamrosni Abdul Razak
Faculty of Chemical Engineering and Technology
Universiti Malaysia Perlis
Arau Perlis, Perlis, Malaysia

Muhammad Fadlin Hazim Baser
Faculty of Chemical Engineering and Technology
Universiti Malaysia Perlis
Arau Perlis, Perlis, Malaysia

ISSN 0930-8989 ISSN 1867-4941 (electronic)
Springer Proceedings in Physics
ISBN 978-981-96-2870-4 ISBN 978-981-96-2871-1 (eBook)
https://doi.org/10.1007/978-981-96-2871-1

This Springer imprint is published by the registered company Springer Nature Singapore Pte Ltd.
The registered company address is: 152 Beach Road, #21-01/04 Gateway East, Singapore 189721, Singapore

Preface

This proceedings volume of the Electronic Packaging Interconnect Technology Symposium (EPITS) 2024 offers a comprehensive overview of recent advancements and emerging technologies across a range of fields integral to electronic packaging and materials science. EPITS 2024 continues its tradition of convening a diverse group of international scientists, engineers, researchers, and students, facilitating a dynamic exchange of breakthrough ideas, knowledge, and insights. The symposium emphasizes cutting-edge developments in electronic packaging while expanding its focus to include pioneering research in green materials, advanced materials, and surface coating technologies, reflecting the evolving needs and interdisciplinary nature of modern engineering and materials science.

This volume includes papers addressing critical issues and innovations within electronic packaging, as well as contributions from the broader fields of advanced and green materials also surface engineering. EPITS 2024 not only provides an exceptional platform for sharing empirical research findings and technological innovations but also fosters collaborations among academicians, industry practitioners, and policymakers. Through this proceeding, readers gain access to diverse research perspectives, which not only advance scientific understanding in electronic and materials engineering but also align with global sustainability goals and the push for greener, more efficient technologies.

Preface

Contents

Advanced Materials

Electronic Packaging

A Comparative Study of Mechanical Properties in Gold with Different Purity Wire Bonding

Ahmad Zarif Azahar[1,2], Maria Abu Bakar[2(✉)], and Azman Jalar[2,3]

[1] Western Digital, SanDisk Storage Malaysia Sdn. Bhd., Plot 301A, Persiaran Cassia Selatan 1, Taman Perindustrian Batu Kawan, MK13, Batu Kawan, 14100 Seberang Perai Selatan, Penang, Malaysia

[2] Institute of Microengineering and Nanoelectronics, Universiti Kebangsaan Malaysia, 43600 Bangi, Selangor, Malaysia

maria@ukm.edu.my

[3] Department of Applied Physics, Faculty of Science and Technology, Universiti Kebangsaan Malaysia, 43600 Bangi, Selangor, Malaysia

Abstract. One way to establish interconnection between electronic chips in the semiconductor packaging technology. The process uses metal wires, typically gold (Au), copper (Cu) and silver (Ag) to establish these interconnections. Semiconductor industries often use Au as the medium of interconnection in high end product as its excellent electrical conductivity ensures efficient signal transmission. Additionally, gold's high resistance to corrosion compared to copper and silver, prevents connection degradation over time, a vital feature for reliable electronics. Moreover Aus wire materials are further segregated into the gold purity, which measured by the number of nines, N, in the industry. This paper studies the bonding quality of different Au purity in wire bonding, 2N, 3N and 4N. The experimental work focuses on the wire bonding process of 30 Au (gold) wires bonding on 30 Al (aluminum) bonding pad with variation of the Au purity. Analysis of the wire bonding quality is focused on the destructive and non destructive critical to quality (CTQ) items which includes ball size, ball thickness, ball shear and intermetallic coverage (IMC). The findings 2N Au wire deformation is lower compared to 4N which showing lower ball size, 43.0µm for 2N and 44.3µm for 4N. However, in ball thickness, 2N wire has higher thickness value compared to 4N with 19.4 µm and 17.2 µm respectively. Results also showed that the ball shear result for 2N wire are higher in gram force with 19.4g compared to the 3N and 4N purity wire of 17.4g and 17.2g respectively. IMC coverage on 4N purity show more coverage compared to the 2N purity with critical bonding parameters programmed are remained as constants. The study also analyse that all Au purity sample are able to meet the industrial spec JEDEC JESD47J of shear per area value above 0.0062 g force/μm^2.

Keywords: intermetallic compound coverage (IMC) · Number of nines (N) · Critical to quality (CTQ) Shear per area

N. R. Abdul Razak et al. (Eds.): ECOOP 1996, SPPHY 418, pp. 3–9, 2025.
https://doi.org/10.1007/978-981-96-2871-1_1

1 Introduction

Wire bonding is a fundamental technique in microelectronic, enabling electrical connection between integrated circuits (ICs) and their packaging. The process utilize thin metallic wires, typically gold, aluminum or copper, to establish crucial link between the internal circuitry of a chip and its external pins. While other metals mentioned can be used for wire bonding, gold reigns supreme for several reason [1–3]. Gold boost exceptional conductivity, nearly ten times greater than some solders material. This translate to minimal signal loss and efficient current flow within the bonded connections which is crucial for high performance electronic devices. Moreover, gold does not readily form oxides that hinder electrical conductivity unlike other material used for wire bonding. This inherent resistance to oxidation ensures long lasting and reliable connection over time [4, 5]. In addition, gold's softness is a major advantage. During wire bonding, the wire needs to be malleable enough to form connection without damaging the delicate surfaces of the microelectronic components. Gold bonding allows for gentle bonding without compromising the integrity of the bonded parts. In industry, the gold's purity plays a role when implies to different application.

The purities are often differentiate by the number of nines, N in gold wire. The number N representing a "nine" in gold's percentage purity. Thus, 2N gold wire indicates 99% gold purity, while 3N signifies 99.9% purity and 4N represent the highest purity at 99.99%. Purities variations influence the wire properties, affecting its suitability in specific bonding application [6]. 4N offers the highest electrical conductivity and the best resistance to oxidation, making 4 N wire a good choice for application demanding top notch conductivity and where thicker wires are used. While 2N exhibits the highest tensile strength among the three, making it suitable for application that require robust connection that can withstand stress. High-performance applications demanding superior conductivity might prioritize 4N wires. Conversely, applications requiring strong and reliable connections that can handle stress might favor 2N wires.

2 Methodology

Three different purity Au wire, Sample A, Sample B, and Sample C were prepared with all of the wire thickness of 17.78μm. Sample A has the lowest Au purity of 99.00%, 2N, followed by Sample B with 99.90%, 3N, and lastly Sample C with the highest Au purity of 99.99%, 4N. As for the bonding sample, silicon (Si) die with dimension of 8787 μm (w) × 31 μm (t) × 12219 μm (l), contains of more than 30 Al bond pad aligning at the edge of the Si die was used in this study. Standard die attach film (DAF) of 8 μm thickness was attached underneath the Si die. The Si die with DAF was later attached to substrate and go through baking process.

The wire was installed onto the wire bonder machine (KNS IConn) to perform ball bonding process via thermosonic bonding with capillary dimension of 32 μm chamfer diameter. The bonding program was program with wire attached to each bonding pad available on the Si die. The bonding parameter was set was loaded with Sample A wire and proceed with wire bonding. Upon completion, the Sample A wire was uninstalled and replaced with Sample B. Without changing the parameter, wire bonding process was performed with Sample B. Upon completion, the activity was repeated with Sample C. The parameter used for the study is listed in Table 1.

Table 1. Wire bonding parameter

Ultrasonic Generator	85 mAmps
Bonding Force	16 g
Bonding Time	10 ms

After completion, the samples were then taken for critical to quality (CTQ) measurement which includes non-destructive and destructive test. For non-destructive test, bonded ball size and ball thickness were taken. For destructive test, ball shear and intermetallic coverage (IMC) were taken into data collection as the qualitative measurement.

For ball size and ball thickness CTQ, optical microscope (Nikon MM-800) with a magnification of 50 X was used to measure 10 samples of bonded ball for each Sample A, B and C. Figure 1(a) illustrate the location of ball size and ball thickness measured on the bonded ball.

Destructive test of Ball Shear assesses the strength of the bonded ball. A dage machine (Steller 4000) with pre-installed tool for shearing on the cartridge was used to measure the ball shear. A chisel-shaped shearing tool was carefully aligned above the specific bond or ball that needs to be tested. The tool will apply a shearing force to the ball which will eventually break as the force increases as per shown in Fig. 1(b). The force required to break the bond are the key measurement, which was measured in gram force, g. A total of 30 bonded ball were sheared and measured.

In order to analyze the intermetallic compound (IMC) formed between the gold wire and aluminum bond pad, the samples were characterized. This involved etching the bond with a sodium hydroxide solution rinsing, and then using a needle to separate the gold wire from the pad. An optical microscope captured the IMC coverage on the bond ball using 100X magnification, which was then converted to grayscale for measurement using Olympus Stream Essential software as per shown in Fig. 1(c). This software allows measurement of the areas of remaining gold and IMC on the separated bond ball, in which measuring the percentage of area cover by the green and red colors.

3 Results & Discussion

Histogram for all sample Ball Size measurement is shown in Fig. 2 for each wire purity sample. It is observed that 4N wire sample has the highest ball size value with 44.27 μm average reading. Followed by 3N wire sample with 43.73 μm and 2N has the lowest ball size with value of 43.02 μm. The standard deviation of 4N is the lowest with 0.14 followed by 3N and 2N with 0.23 and 0.43 respectively. From the standard deviation, it is observed that the 2N wire has less consistent ball size compared to 3N and 4N.

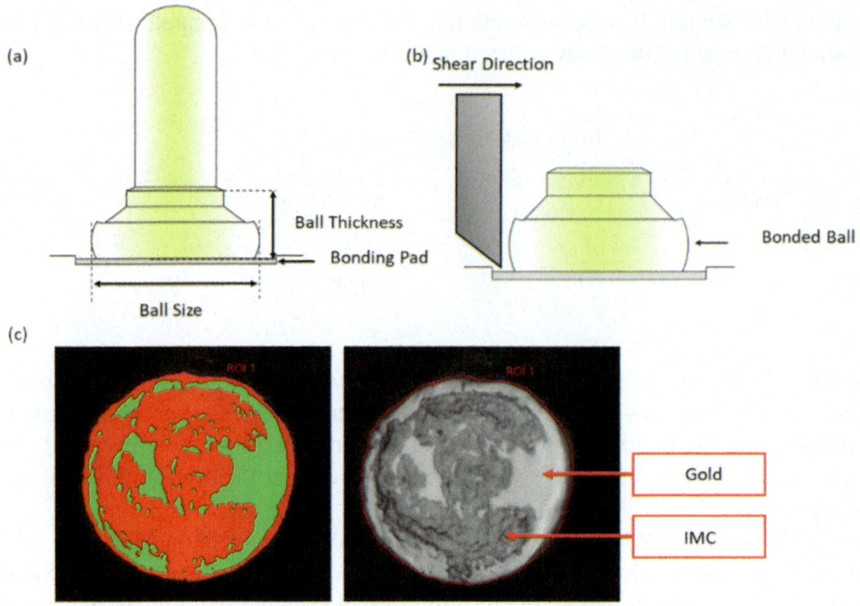

Fig. 1. (a) Illustration of ball size and ball thickness measurement, (b) Ball shear illustration, (c) Example of IMC coverage at the ball bonded

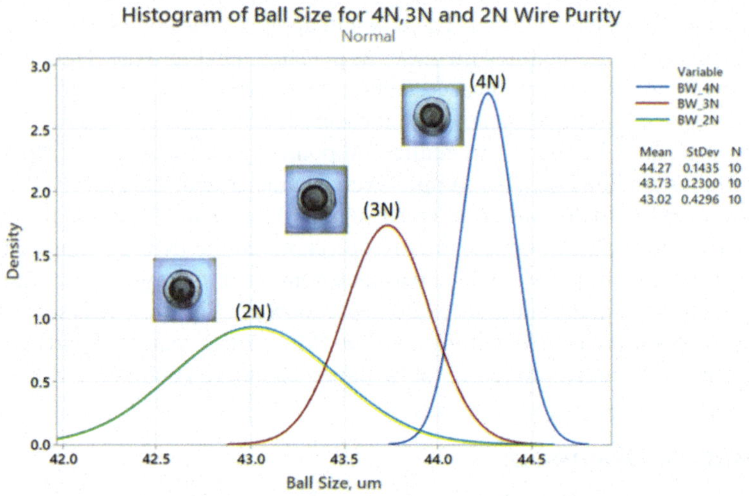

Fig. 2. Ball size trend of 2N, 3N, and 4N Au Wire

Another non-destructive test was done in all sample and histogram was plotted as per shown in Fig. 3(a). 2N wire measured the thickest ball thickness reading with 12.92µm followed by 3N with 10.45µm and 4N with 10.18µm. Standard deviation of 2N has the least deviation with 0.25 compared to the other sample of 3N with 0.59 and 4N with 0.42. The deviation on 2N is lesser compared to the 3N and 4N due to the deformed balls observed for 2N are more rounded as shown inset of Fig. 2. Thus, the measurement of the diameter of the ball is more accurate compared to 3N and 4N.

The deformation observed on the sample may be influenced by the physical properties of the wire. Zulkifli MN et al. reported that an increase in bonding power will increase the deformation in a non-uniform fashion. Increasing bonding force tend to further flatten the grains [7]. Whereby the higher the percentage of Au purity, the lower the physical properties which enable the wire to be deformed more easily. In the study, it is observed that 2N has the highest physical properties as it has the lowest physical deformation, highest ball thickness and lowest ball size, compared to the other sample 3N and 4N although all samples are bonded with the same parameter.

Figure 3(b) shows the histogram of destructive test exercised on the sample. Ball shear reading observed that 2N has the highest ball shear reading with 19.42g followed by 3N and 4N with the reading of 17.43g and 17.20g respectively. It is also observed that the standard deviation of 2N is the highest with 1.08 compared to the other sample with 0.68 for 4N and 0.77 for 3N.

According to the JDEC JESD47K qualification of integrated circuits, states that the shear of Au bonding on Al pad should be greater than 0.0062 g force/μm^2 [8]. Table 2 tabulate the shear per area values for all the sample bonded. All four samples were able to meet the shear per area requirement, thus the quality of wire bonding is considered as acceptable according to the industrial JDEC JESD47K [8] specification. It is observed that the shear per area for 2N wire has the highest value of 0.013368 g force/μm^2 followed by 3N with 0.011613 g force/μm^2 and lastly 4N with 0.011185 g force/μm^2 for this study.

Table 2. Shear per area value for all sample measured

No	Shear Value (g)	Ball Size (μm)	Area (μm^2)	Shear per Area (g/μm^2)
4N	17.2	44.3	1538.1	0.011185
3N	17.4	43.7	1501.2	0.011613
2N	19.4	43.0	1452.8	0.013368

Lastly, Fig. 3(c) shows the IMC coverage observed from the study. The coverage of IMC is measured by the percentage of the surface area of the flipped bonded ball. IMC coverage should be above 60% of the bonded ball area consist of Au-Al intermetallic compound, which was also reported by few other researchers to eliminate reliability failure risk [9–12]. From the study, it is observed that 4N wire has the highest IMC coverage with 76.27% followed by 3N with 72.79% and lastly 2N with 70.42%.

From the findings it can be deduce that with same parameter, different Au purity wire will have different response due to its physical properties. As such, 2N wire has thicker ball thickness and smaller ball size compared to higher purity Au wire as the

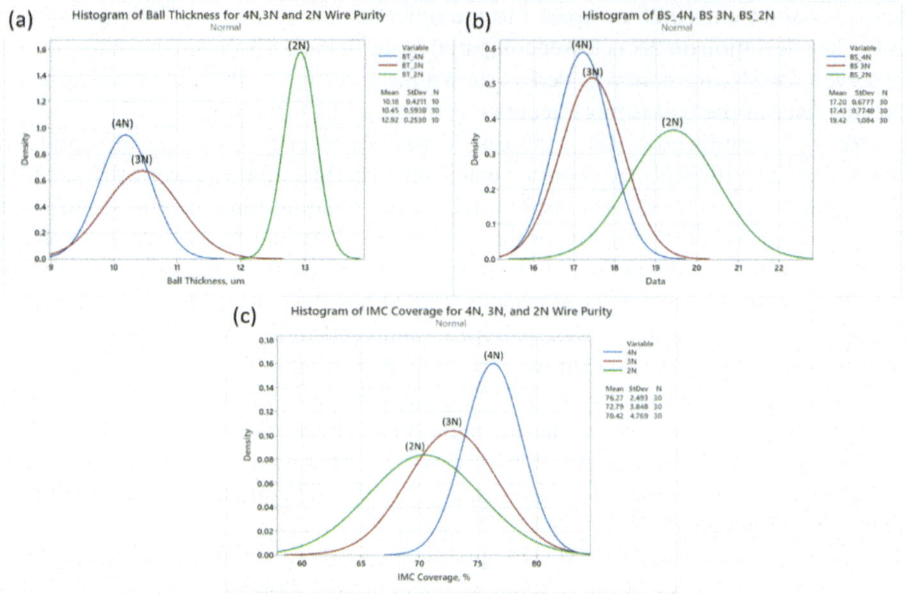

Fig. 3. (a) Ball thickness trend of 2N, 3N, and 4N Au wire (b) Ball shear trend of 2N, 3N, and 4N Au wire, (c) IMC trend of 2N, 3N, and 4N Au wire

physical properties of 2N is higher compared to the other samples. Ultrasonic generator, USG, helps in promoting the formation of IMC during bonding given that suitable force is used for better USG transfer [13]. This also observed in ball shear result as such, more force needed to remove the bonded ball for 2N Au wire compared to 3N and 4N wires. However, in IMC coverage, it is observed that 4N wire has larger quantitative data compared to the lower purity Au wire which has higher IMC coverage surface area. Higher purity wire may have better diffusion rate to form Au-Al intermetallic compound [14].

4 Conclusion

In this paper, we evaluated 17.78μm Au wire size wire bonded on aluminum pads on a Si die. These 17.78μm wire samples consist of different Au purity, 2N with 99.00%, 3N with 99.90% and 4N with 99.99%, whereby all the sample are bonded onto the aluminum pads with the same constant parameter. The increasing of Au purity 2N to 4N resulted in more deformed ball formation, higher ball size and lower ball thickness. It was observed that the difference of ball size between 4N and 2N wire is at 1 μm difference. However, it is more significant comparison on ball thickness, the difference of value obtained is more than 2μm between 2N and 4N wire. On the ball shear result, it is observed that 2N wire will have higher ball shear result. 2N wire measured 19.42 g force for ball shear while 4N wire measured at 17.20 g. Calculating the shear per area of the measured sample for this study, all sample are able to meet the industrial JEDEC JESD47J specification of shear per area must be above 0.0062 g force/μm^2 for the wire bonding to be considered as quality bonding. In terms of IMC coverage, it is observed that the lower N wire will have lower IMC coverage area, 2N wire IMC covered 70.42% while 3N and 4N measured at 72.79% and 76.27% respectively. It is believed that the IMC coverage result could be better if the parameter is adjusted and optimized according to each wire suitability.

Acknowledgments. The authors would like to acknowledge the financial support and research facilities provided by Western Digital via SanDisk Storage Malaysia Sdn. Bhd (research grant: RR-2023-007) through collaboration activities with Universiti Kebangsaan Malaysia (research grant: TAP-K022551).

References

1. Supramaniam, S., Periasamy, S., Nurhisham, M., Bakar, M.A., Jalar, A.: Measurement of aluminium remnant thickness on copper wire bonding using 3D laser scanning microscope. In: 2023 IEEE International Symposium on the Physical and Failure Analysis of Integrated Circuits (IPFA), Pulau Pinang, Malaysia, pp. 1–6 (2023)
2. Omar, A.A., Jalar, A., Hamid, K.A.: Evolution of discrete semiconductor copper wedge bond in biased temperature humidity chamber. Jurnal Teknologi **84**(6–2), 129–135 (2022)
3. Azahar, A.Z., Bakar, M.A., Jalar, A., et al.: Effect of die deflection during Au wire bonding process on bonding quality in overhang semiconductor package. J. Mater. Eng. Perform. **33**, 5836–5845 (2024)
4. Omar, A.A., Jalar, A., Bakar, M.A., Hamid, K.A.: Sulfur ingression on encapsulated leadframe of discrete semiconductor package. IEEE Trans. Compon. Packag. Manuf. Technol. **12**(3), 544–550 (March 2022)
5. Qin, I., Shah, A., Xu, H., Chylak, B., Wong, N.: Advances in wire bonding technology for different bonding wire material. Int. Symp. Microelectron. **2015**(1), 000406–000412 (2015)
6. Gao, H., Lu, J., Lu, R., Xin, W., Xu, X., Yilmaz, H.: Reliability study of silver, copper, and gold wire bonding on IC device. Int. Symp. Microelectron. **2014**(1), 000850–000855 (2014)
7. Zulkifli, M.N., Abdullah, S., Othman, N.K., Jalar, A.: Some thoughts on bondability and strength of gold wire bonding. Gold Bull. **45**, 115–125 (Sept. 2012)
8. JEDEC Solid State Technology Association: Stress-test-driven qualification of integrated circuits, Standard No. JESD47K [Online]. Available: https://www.jedec.org (Dec. 2021)
9. Chen, Y.S., Lee, M.T., Liu, N., Liu, K.: A study for the effectiveness of wire bond process parameters on Al-Cu intermetallic compound distributions and the correlation between bonded ball adhesions and IMC coverage. In: 2018 IEEE International Symposium on the Physical and Failure Analysis of Integrated Circuits (IPFA), pp. 1–5 (2018)
10. Yap, T.J., Lin, Y.K., Au, Y.K., Eu, P.L.: Importance of Cu/Al intermetallic coverage in copper wire bonding with sensitive pad structure. In: 2012 14th International Conference on Electronic Materials and Packaging (EMAP), pp. 1–5 (2012)
11. Xi, J., et al.: Evaluation of Ag wire reliability on fine pitch wire bonding. In: 2015 IEEE 65th Electronic Components and Technology Conference (ECTC), pp. 1392–1395 (2015)
12. Kim, G.R., Ha, S.S., Park, C.H., Pae, S., Choi, B.: The bondability and reliability of a ternary Ag alloy wire on an Al bonding pad under N2-free conditions. Mater. Trans. **59**(9), 1487–1492 (2018)
13. Xie, S., Lin, P., Yao, Q.: Overview of Au-Al bond interface. In: 2021 International Conference on Electronics, Circuits and Information Engineering (ECIE), IEEE, pp. 250–253 (2021)
14. Zhou, H., et al.: Research progress on bonding wire for microelectronic packaging. Micromachines **14**(2), 432 (2023)

Artificial Neural Networks Analysis for Predicting Gold Diffusion in Solder Joints Under Various Rework and Thermal Cycling Conditions

Adlil Aizat Ismail[1,2], Maria Abu Bakar[2(✉)], Azman Jalar[2,3], Mohd Ridzwan Yaakub[4], Muhammad Iqbal Abu Latiffi[4], Erwan Basiron[1,2], and Muhammad Nizam Ilias[1,2]

[1] Western Digital, SanDisk Storage Malaysia Sdn. Bhd., Plot 301A, Persiaran Cassia Selatan 1, Taman Perindustrian Batu Kawan, MK13, Batu Kawan, Seberang Perai Selatan, 14100 Penang, Malaysia

[2] Institute of Microengineering and Nanoelectronics, Universiti Kebangsaan Malaysia, 43600 Bangi, Selangor, Malaysia
maria@ukm.edu.my

[3] Department of Applied Physics, Faculty of Science and Technology, Universiti Kebangsaan Malaysia, 43600 Bangi, Selangor, Malaysia

[4] Center for Artificial Intelligence Technology, Faculty of Information Science and Technology, Universiti Kebangsaan Malaysia, 43600 Bangi, Selangor, Malaysia

Abstract. This research utilises artificial neural networks (ANNs) analysis to investigate gold diffusion based on gold content weight percentage (wt%) in the intermetallic compound (IMC) layer of the ball grid array (BGA) component solder joints under various conditions. The presence of a gold element in the IMC layer on the BGA component side significantly affects the reliability of SnAgCu (SAC) solder joints. Using no rework as a control sample, the ANNs model evaluated and predicted the effects of hot air and laser rework on the gold content wt% in the BGA component side's IMC layer. Energy-dispersive X-ray spectroscopy (EDX) measurements of the corresponding gold content wt%, soldering process parameters, and IMC layer height were included in the datasets. The conditions tested were thermal cycle test (TCT) for 0 and 900 cycles, no rework, hot air rework, and laser rework. The network design consists of three levels: an input layer, one hidden layer with different activation functions, and one output layer. With an R-Square value of 0.880 for the validation set, the ANNs model demonstrated good prediction of gold diffusion in solder joints and closely matched the actual values. By leveraging advanced computational techniques, this research provides insights into optimizing the BGA rework process and enhancing solder joint reliability in electronic manufacturing and repair applications.

Keywords: artificial intelligence · artificial neural network · solder joint reliability · laser rework · thermal cycle test

N. R. Abdul Razak et al. (Eds.): ECOOP 1996, SPPHY 418, pp. 10–17, 2025.
https://doi.org/10.1007/978-981-96-2871-1_2

1 Introduction

The structural reliability and efficiency of solder joints, which connect the ball grid array (BGA) components to the printed circuit board (PCB), can be impacted by several factors. The diffusion of gold into the solder matrix is one of the major factors influencing solder joint reliability [1]. The intermetallic compound (IMC) layers on the BGA component side of the solder joints often contain gold, which substantially affects the mechanical characteristics and long-term stability of the joints [2].

Gold diffusion can cause brittle phases to form in the solder joints, which can weaken the joint's mechanical strength and increase the risk of failure particularly when subjected to thermal cycle test (TCT) and rework conditions. Therefore, it is crucial to understand and predict how gold diffusion will behave under different conditions to optimize solder joint reliability and improve the general durability of electronic devices [3].

Artificial neural networks (ANNs) are used in this work to evaluate and predict gold diffusion in solder joints under different conditions. ANNs are forms of artificial intelligence (AI) that replicate the human brain's neural network and provide powerful features for modelling complex correlations in data. Considering their ability to adapt and predictive ability, ANNs are particularly suitable for the analysis of complex interactions such as gold diffusion in solder joints [4].

The research objective is to develop a prediction model that provides valuable information about how different rework and TCT conditions affect the diffusion of gold in solder joint's reliability. Using advanced computational methods, this research aims to improve solder joint's reliability in electronic manufacturing and repair applications and optimize BGA rework processes.

2 Methodology

The sample preparation involved using six test vehicles, each consisting of two BGA components assembled on an organic solderability preservative (OSP) PCBs. The specific solder joint type used in this research was SnAgCu (SAC), known for its prevalent application in electronic packaging due to its favourable mechanical properties and thermal fatigue resistance. Each BGA component has 170 ball-type solder joints with a diameter of 0.49 ± 0.5 mm with an unpopulated column in the middle. The BGA component's substrate was an electroless nickel immersion gold (ENIG) based. BGA component U123 was the reworked component whilst the U129 was the adjacent component of the reworked component. The information on the sample size and condition is tabulated in Table 1.

Two different rework processes were performed: hot air rework and laser rework. The hot air rework method used a system that provided controlled convection heat to melt solder joints, allowing components to be removed and replaced without inflicting severe thermal damage to surrounding surfaces [5]. The laser rework method uses a laser beam to apply radiation heat to melt solder joints precisely which reduces thermal stress on nearby components [6]. The samples were subjected to TCT based on the JESD22-A104F Temperature Cycling requirement to simulate the thermal stresses encountered during the operational life of electronic devices [7]. The TCT profile involved cycling

Table 1. Details on the samples and quantity

Sample Type	Details	BGA Quantity	Solder Joints Quantity Observed	TCT Cycles
P0	No rework (Control)	2	20	0
P900	No rework (Control)	2	20	900
H0	Hot air rework	2	20	0
H900	Hot air rework	2	20	900
L0	Laser rework	2	20	0
L900	Laser rework	2	20	900

the samples between 0 °C and 100 °C temperature ranges with hot and soak duration at 15 min each. Two conditions were tested; TCT0: no thermal cycling and TCT900: 900 thermal cycles.

All samples were cross-sectioned to expose the solder joints for analysis. The cross-sectioning process involved mounting the samples in epoxy resin, followed by grinding and polishing to obtain a smooth surface for defect inspection using an optical microscope [8]. The cross-sectioned BGA component samples were inspected for potential solder joint defects using an optical microscope to ensure the integrity and reliability of the solder joints, as the IMC layer plays a crucial role in the mechanical strength and electrical conductivity of the connections [8]. Scanning electron microscopy (SEM) combined with energy-dispersive X-ray spectroscopy (EDX) was employed to analyze the microstructure and elemental composition of the solder joints. SEM provided high-resolution images of the cross-sectioned samples, while EDX facilitated the quantification of gold content on the BGA side [7].

Each BGA cross-section contained ten solder joints, which were individually identified for analysis. The IMC layer height at each solder joint were measured at the ten highest points using an SEM and the gold content wt% along with the absolute error % within each solder joint was determined using EDX at ten spots on the BGA sides of the solder joint as shown in Fig. 1(a). These measurements were recorded for each solder joint. The average of the ten points measurements for IMC layer height and gold content wt% were then calculated. This method provided a reliable measure of the IMC layer height & gold content wt% for further analysis [9].

The ANNs model was built using JMP Pro statistical software and consisted of one input layer, one hidden layer, and an output layer. Five features were uploaded to the input layer: solder joint condition (no rework, hot air rework, laser rework), TCT condition (TCT0, TCT900), component (U123 (reworked), U129 (adjacent)), BGA side IMC layer height, and BGA side absolute (abs.) error (%). Each hidden layer consisted of fifteen nodes with three types of activation functions: hyperbolic tangent (tanh), linear, and Gaussian [10]. The output layer consisted of one node indicating the BGA side gold content wt%. Figure 1(b) displays the ANNs model's diagram.

The model used three-fold cross-validation and transformed covariates when fitting [11]. The equations describe how inputs were processed through the network layers into

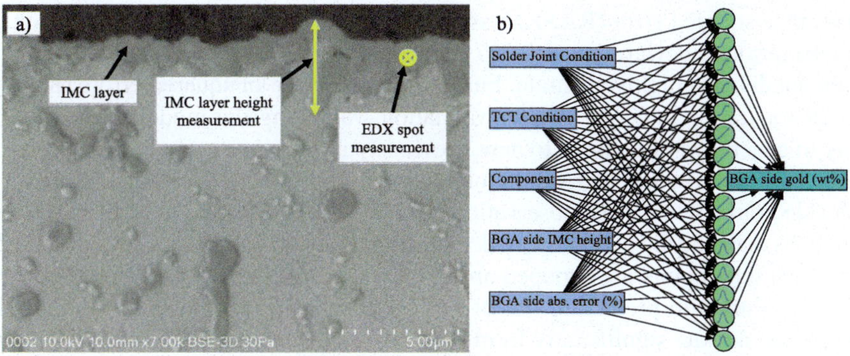

Fig. 1. (a) IMC height & EDX spot measurements on the BGA solder joint's cross section at the BGA side; (b) The ANNs model diagram

final outputs. The input layer consists of five nodes, each representing one of the input features. The input vector x is expressed as follows:

$$x = [x_1, x_2, x_3, x_4, x_5] \tag{1}$$

where x represents each input feature: x_1: solder joint condition; x_2: TCT condition; x_3: component; x_4: BGA side IMC layer height; x_5: BGA side abs. error (%).

The hidden layer applies three activation functions: tanH, linear, and Gaussian, each with five nodes [12–14]. The hidden layer computes a weighted sum of the inputs and applies activation functions. The equation for the j-th hidden node:

$$a_j = f_j(\Sigma_{i=j}^{5} w_{ij}^{(1)} x_i + b_j^{(1)}) \tag{2}$$

where a_j is the activation (output) of the j-th hidden node, f_j is the activation function for the j-th hidden node (tanH, linear, and Gaussian), $w_{ij}^{(1)}$ is the weight from the i-th input to the j-th hidden node, and $b_j^{(1)}$ is the bias term for the j-th hidden node.

The output layer sums the activations from the hidden layer and produces the final prediction. The equation for the output node is:

$$y = \Sigma_{j=1}^{15} w_j^{(2)} a_j + b^{(2)} \tag{3}$$

where y is the predicted BGA side gold content wt%, $w_j^{(2)}$ is the weight from the j-th hidden node to the output node, and $b^{(2)}$ is the bias term for the output node.

3 Results & Discussion

Figure 2(a) shows the data from the training and validation phases of an ANNs model for BGA side gold content wt%. The ANNs model accurately predicted BGA side gold content wt%, with R-Square values of 0.782 for the training dataset and 0.880 for the validation dataset. This explains 78.2% and 88.0% of the variation, respectively. The model's root average squared error (RASE) was 1.290 for training and 0.932 for validation, and the mean absolute deviation was 0.727 for training and 0.603 for validation, indicating that the model was highly accurate. Furthermore, the sum of squared errors (SSE) was 133.029 for training and 34.791 for validation, validating the model's reliability and ability to generalize effectively to new, previously unseen data.

For both the training and validation datasets, the actual by predicted plots for the BGA side gold content wt% observations in Fig. 2(b) and (c) are used to compare the predictions made by the neural network model to the actual observed values. In Fig. 2(b), most of the data points are grouped around the diagonal line, showing that the ANNs model is accurate in predicting the BGA side gold content wt% in the training data. Some points deviate significantly from the line, indicating that the model's predictions are less reliable. However, there are very few outliers. The validation plot in Fig. 2(c), similar to the training plot, most of the data points cluster around the diagonal line, showing excellent prediction accuracy on the validation data. A few deviations from the line indicate some forecast errors. However, the general pattern indicates that the model performs well with additional data. The validation plot exhibits a similar pattern to the training plot, indicating that the model generalizes effectively to new data. This is supported by the high R-Square value and low error metrics discussed previously. The model demonstrates strong performance and good generalization, making it a valuable tool for understanding and predicting the outcomes of different soldering conditions and physical characteristics of solder joints.

The ANNs prediction profiler plot highlights the influence of various input variables on the predicted BGA side gold content wt%. The plot in Fig. 2(d) shows that using laser rework compared to hot air rework leads to a slight decrease in the predicted gold content in the solder joint. This suggests that the solder joint condition significantly affects the model's predictions. Additionally, in Fig. 2(e) and (f) respectively, the TCT condition from 0 to 900 cycles and component type, U123 and. U129 have minimal but noticeable decreasing effects. Within the observed range in Fig. 2(g), the BGA side IMC layer height has no effect, however, the BGA side absolute error (%) in Fig. 2(h) has a positive influence on the estimated gold content, implying that higher abs. Errors during EDX measurements correspond to increasing expected gold content wt%. Overall, the profiler plot emphasizes the relevance of solder joint condition in forecasting BGA side gold concentration, while also providing insights on the influence of other variables.

The actual SEM image results from the reworked U123 BGA component solder joints post TCT900 in Fig. 3 were used to benchmark the ANNs model. Figure 3(a) shows sample L900 reworked with laser rework process has no crack whilst the sample H900 reworked using hot air rework process in Fig. 3(b) has a crack. According to Ismail et al., there were significant amounts of gold residual in the solder joints IMC layer and particles because the saturation of gold content wt% in the melted tin-copper solder restricted the amount of gold that could dissolve. The continued development of IMC particles caused their proportion to rise as TCT cycles increased; as a result, the $AuSn_4$ IMC grains experienced dominant fracture mode cracks [15].

This research showed that variations in the gold content impact the mechanical integrity and endurance of the solder joints. Specifically, the presence of gold content in the IMC layer was found to have a substantial effect on solder joint reliability. These

(a)

Training	
BGA side gold (wt%)	
Measures	**Value**
RSquare	0.782091
RASE	1.2895206
Mean Abs Dev	0.727025
-LogLikelihood	133.85672
SSE	133.02906
Sum Freq	80

Validation	
BGA side gold (wt%)	
Measures	**Value**
RSquare	0.87994
RASE	0.9326161
Mean Abs Dev	0.6034419
-LogLikelihood	53.967078
SSE	34.790915
Sum Freq	40

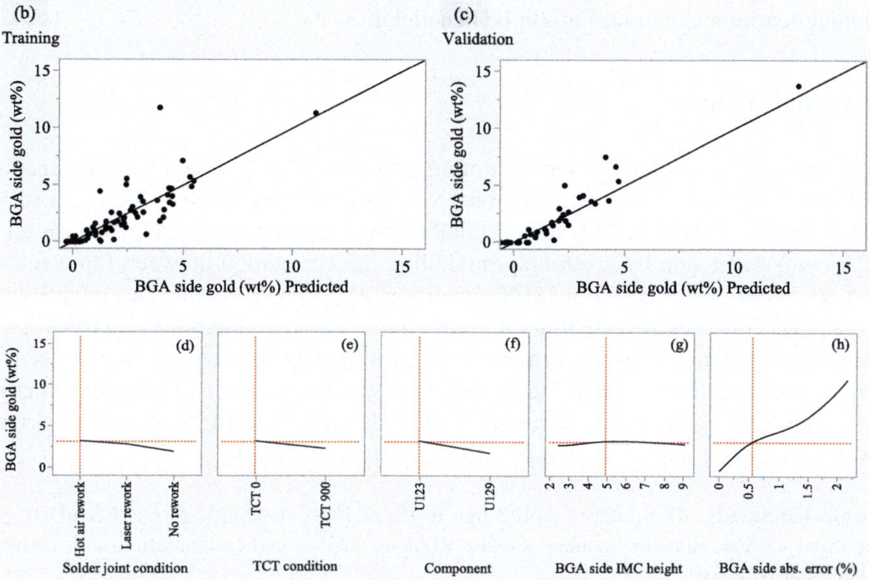

Fig. 2. (a) Statistics from the training and validation phases of an artificial neural network model; Actual by predicted plots for the BGA side gold content wt% results: (b) training plot; (c) validation plot; BGA side gold content wt% prediction profiler: (d) solder joint condition; (e) TCT condition; (f) component; (g) BGA side IMC height; (h) BGA side absolute error (%)

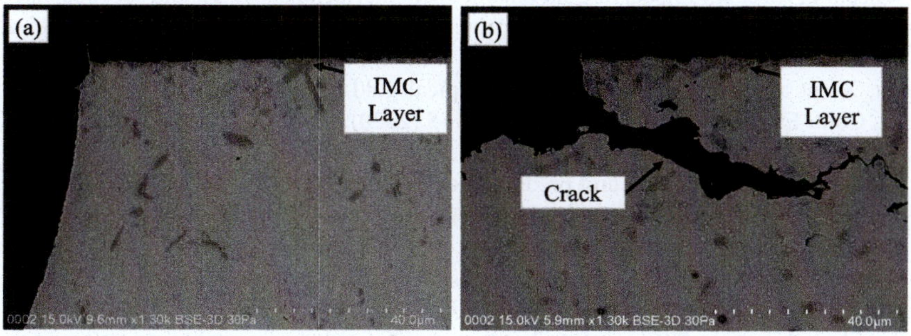

Fig. 3. SEM image of the reworked U123 BGA solder joints (a) sample L900 (b) sample H900

variations were well represented by the ANNs model, which offered a reliable method for predicting gold diffusion in various types of reworks and TCT conditions. This research has demonstrated the potential of ANNs to improve solder joint reliability in electronic manufacturing and repair applications and optimize BGA component rework processes. The insight acquired through this research can help create solder joints that are more robust, improving the lifespan and overall performance of electrical devices. Additional factors, like various solder compositions or TCT patterns, could be incorporated into this research in future studies to improve the ANNs model further and further aid in understanding and predicting the complex behaviour of solder joints, exploring other machine-learning techniques might be beneficial.

4 Conclusion

This research concludes by demonstrating how well ANNs predict the diffusion of gold elements in BGA solder joints under various conditions, such as TCT and various rework methods. The ANNs created reliable predictions of gold content wt% on the BGA component side by accurately modelling the correlation between input features such as solder joint condition, TCT condition, IMC layer height, and measurement errors. These findings are essential for improving solder joint reliability and BGA rework process optimization, which in turn improves the longevity and performance of electronic devices. The effectiveness of this technique shows how advanced computational methods can be applied to electronic manufacturing and repair, opening possibilities for new developments and improvements in solder joint reliability.

Acknowledgments. The authors would like to thank Western Digital through SanDisk Storage Malaysia Sdn. Bhd for financial support (TAP-K022551) and collaboration with Universiti Kebangsaan Malaysia.

References

1. Wheeling, R., et al.: Impact of gold thickness on interfacial evolution and subsequent embrittlement of tin-lead solder joints. J. Electron. Mater. **51**, 7337–7352 (2022). https://doi.org/10.1007/s11664-022-09891-2
2. Du, X., et al.: Reliability evaluation of typical gold-containing solder joints for products. Int. Conf. Electron. Packag. Technol., pp. 1–5 (2020). https://doi.org/10.1109/ICEPT50128.2020.9202536
3. Cong, S., et al.: Reliability and prediction of Sn36Pb2Ag solder joints under thermal aging test. Mater. Res. Exp. **10**(4), 046301 (2023). https://doi.org/10.1088/2053-1591/acca6d
4. Zippelius, A., et al.: Reliability analysis and condition monitoring of SAC+ solder joints under high thermomechanical stress conditions using neuronal networks. Microelectron. Rel. **129**, 114461 (2022). https://doi.org/10.1016/j.microrel.2021.114461
5. Ismail, A.A., et al.: Effects of underfill materials on behavior of intermetallic compound thickness of ball grid array solder joints using ANOVA. Int. Electron. Manuf. Technol. Conf., pp. 1–5 (2022). https://doi.org/10.1109/IEMT55343.2022.9969487

6. Ismail, A.A., et al.: Laser rework process for efficient lead-free solder joints Ball Grid Array (BGA) component rework. IEEE Trans. Compon. Packag. Manuf. Technol. **14**(4), 735–742 (2024). https://doi.org/10.1109/TCPMT.2024.3379219
7. Ismail, A.A., et al.: Cardinal and ordinal directions approach in investigating arrayed solder joints crack propagation of ball grid array semiconductor packages. IEEE Trans. Compon. Packag. Manuf. Technol. **12**(9), 1492–1501 (2022). https://doi.org/10.1109/TCPMT.2022.3198406
8. Ismail, A.A., et al.: Intermetallic compound thickness of ball grid array solder joints under thermal cycling test using ANOVA. Int. Electron. Manuf. Technol. Conf., pp. 1–4 (2022). https://doi.org/10.1109/IEMT55343.2022.9969477
9. Samavatian, V., et al.: Correlation-driven machine learning for accelerated reliability assessment of solder joints in electronics. Sci. Rep. **10**, 14821 (2020). https://doi.org/10.1038/s41598-020-71926-7
10. Pan, L., et al.: Methodology of artificial intelligence aided hybrid modeling for predicting solder joint reliability of BGA package. Electron. Compon. Technol. Conf., pp. 1200–1204 (2020). https://doi.org/10.1109/ECTC51909.2023.00205
11. Xiong, Z., et al.: Evaluating explorative prediction power of machine learning algorithms for materials discovery using k-fold forward cross-validation. Comput. Mater. Sci. **171**, 109203 (2020). https://doi.org/10.1016/j.commatsci.2019.109203
12. Rasamoelina, A.D., Adjailia, F., Sinčák, P.: A review of activation function for artificial neural network. IEEE Wld. Symp. Appl. Mach. Intell. Inform., pp. 281–286 (2020). https://doi.org/10.1109/SAMI48414.2020.9108717
13. Koksal, E.S., Aydin, E.: Physics informed piecewise linear neural networks for process optimization. Comput. Chem. Eng. **174**, 108244 (2023). https://doi.org/10.1016/j.compchemeng.2023.108244
14. Zhang, Y., et al.: Gaussian-type activation function with learnable parameters in complex-valued convolutional neural network and its application for PolSAR classification. Neurocomputing **518**, 95–110 (2023). https://doi.org/10.1016/j.neucom.2022.10.082
15. Ismail, A.A., et al.: Undissolved gold in fine-pitch BGA solder joint under thermal cycling test. J. Mater. Sci. Mater. Electron. **35**, 866 (2024). https://doi.org/10.1007/s10854-024-12595-4

Effect of Indium Addition on the Microstructure Formation and the Wettability of Sn10Cu Lead-Free Solder Alloys

Muhammad Fadlin Hazim Baser, Mohd Arif Anuar Mohd Salleh(✉), Rita Mohd Said, and Nur Syahirah Mohamad Zaimi

Centre of Excellence Geopolymer and Green Technology (CEGeoGTech), Faculty of Chemical Engineering Technology, Universiti Malaysia Perlis (UniMAP), Taman Muhibbah, 02600 Arau, Perlis, Malaysia

arifanuar@unimap.edu.my

Abstract. The development of lead-free solder alloys suitable for applications that require high- temperatures is essential to fulfilling the growing demand for dependable alternatives to conventional lead-solder alloys. The studies were focused on the effect of indium addition on the morphology alteration and the wettability of commercial Sn10Cu-based solder alloys. The Sn10Cu was added with 0.5, 1, 3, and 5 wt.% of indium and prepared by using the casting method. The microstructure of bulk solder was observed using SEM with EDX, while the wetting angle was measured to evaluate the wettability of solder alloys. Adding indium significantly changes the phase formation of Sn10Cu solder alloys. The Cu_3Sn phase boundary is shifted with indium content up to 5.75 wt.%. The Cu_6Sn_5 IMC was greatly refined and evenly distributed following the addition of 3 wt.% and 5 wt.% indium. The wetting angle of Sn10Cu decreases when indium is added. In comparison to Sn10Cu, the wetting angle of Sn10Cu-5In solder alloy decreased by 30.8%. The lowered melting temperature of indium suggests that it can effectively improve the wettability of Sn10Cu solder. The lowered melting temperature will allow for better fluidity during soldering.

Keywords: Sn-Cu · phase-diagram · high temperature solder · indium and wettability

1 Introduction

The industry of electronic assembly and manufacturing is getting crucial in fulfilling the expanding global need for becoming reliable and compact electronic components. Researchers have shifted from traditional lead-based solders to lead-free alternatives [1]. One of the primary drivers for its popularity is the shift away from traditional lead-based solders, prompted by the implementation of the Restriction of Hazardous Substances Directive (RoHS) in the European Union and similar regulations globally, which aim to reduce the use of hazardous materials in electronic devices. Pb-free solder alloys offer several of interesting families, including Sn–Ag, Sn–Zn, Sn–Ag–Cu and

N. R. Abdul Razak et al. (Eds.): ECOOP 1996, SPPHY 418, pp. 18–24, 2025.
https://doi.org/10.1007/978-981-96-2871-1_3

Sn–Cu. The Sn-Cu solder serves as a lead-free alternative that excellent performance, good mechanical properties and is being widely adopted by various electronic assembly applications [2]. The cost-effectiveness of Sn-Cu solder further enhances its appeal, as it is less expensive than other lead-free alternatives like Ag-containing solders. Therefore, there has been interest in Sn-Cu alloys with increased Cu content as possible options for applications involving high temperatures and soldering. Continuous studies have demonstrated that incorporating small amounts of alloying elements into Sn-Cu solder alloys is a potentially effective and cost-efficient method to enhance their properties [1]. Enhancing the microstructure through transformations in phase is a usual approach employed to improve the performance of solder alloys. Recently, Sn-Cu alloys containing Ni and In showed better performance. Ni and Cu have low solubility in the Sn matrix, thus Ni and Cu forms precipitates as $(Cu,Ni)_6Sn_5$ in the microstructure [3]. Small amounts of Ni added to the Sn-0.7Cu solder alloy have little impact on the solidus and liquidus temperatures, however the introduction of nickel-based IMCs resulting from Ni addition, which dispersed around the Sn matrix contributes to the improvement of its mechanical properties [3]. In the case of the Sn-Cu solder system for high-temperature application, the Cu content must increase to indeed elevate the melting temperature of the alloy. A suitable high temperature solders based on Sn-Cu alloys need to have higher amounts of Cu than 7.6 wt.% [4]. Sn with 10 wt.% of Cu (Sn10Cu) as a base alloy was investigated by researchers as it undergoes equilibrium peritectic solidification. As the Cu content increases, it can be attributed to the formation of intermetallic compounds between Sn and Cu due to the solubility limit of Sn in Cu decreases [5]. The presence of excessive intermetallic compounds (IMCs) can significantly reduce the wettability of solder due to the inherent characteristics of these compounds [6]. Currently, high extensive study on lead-free solders, but only a small percentage of the thousands of papers that have been published recently are specifically about high temperature lead-free solders [7]. Nonetheless, high-temperature solder alloys are a crucial structural element for a variety of industrial assemblies and parts that need to be lead-free and have a high level of reliability and performance. Therefore, it is necessary to understand how adding In affects the microstructure and wettability behavior of high Cu content solder alloys.

2 Experimental Procedure

The Sn10Cu-xIn solder alloys were prepared according to weight percentage (wt.%). The Sn10Cu and Sn-10Cu-xIn ($x = 0.5$, 1, 3, and 5 wt.%) alloys were prepared by melting pure Sn ingots and In bar together at 300 °C using a graphite crucible in an electrical furnace. This temperature was maintained for 15 min to ensure thorough melting of Sn and In. Subsequently, the furnace temperature was raised to 800 °C before introducing Cu. This heating is held for 30 min, and every ten minutes, the mixing materials are stirred to perform homogeneous mixing. The molten solder alloys then be poured into a cylindrical steel mold. Lastly, the molten solder was solidified to room temperature. To prepare the bulk solder sample, the solder alloys were ground using SiC abrasive paper grit 100 to 2000 and then polished using alumina with a particle size of 0.1 and 0.3 μm. The samples were finally polished with an OPS solution to reveal the microstructure before being observed using a scanning electron microscope (SEM) equipped with an energy dispersive X-ray (EDX). The wettability properties of the solder alloys were measured by the contact angle of the solder joint. These samples were prepared by cutting the as-cast solder and rolling it to a thickness of 1 mm. After rolling, the samples were punched into 5 mm-diameter billets by using a puncher. Ethanol was then used to clean the samples, and then distilled water was used to rinse all the samples. Commercial Cu-substrates with dimensions of 15 mm × 15 mm × 0.3 mm were used in this study for the wetting measurement. The Sn10Cu-xIn/Cu solder alloy was reflowed at a reflow temperature of 275 °C using a lead-free reflow oven (Model: Torch T200C +) to form solder joints. For each solder alloy, a total of four samples were reflowed to obtain an average result. After that, the samples were cross-sectioned and mounted in epoxy resin. The cross-sections were subjected to a grinding and polishing process, which is similar for bulk microstructure sample as described before. The wettability of the solder alloys was evaluated by measuring the wetting angle. The angle was analyzed with an optical microscope and measured using ImageJ (version 1.54d) software. The average wetting angle values were then calculated.

3 Results and Discussions

3.1 Phase Diagram of Solder Alloys

The simulate phase transformations and analyze the effects of different alloying elements on material properties was conducted using Thermo-Calc software version 2021b. The equilibrium phase diagram results are depicted in Fig. 1, where (a) represents the Sn-Cu system and (b) represents the binary Sn10Cu-xIn system. Additionally, the Sn-Cu alloy system is seen as a potential option for creating lead-free solder materials. In these systems, increasing the Cu content can indeed elevate the melting temperature of the alloy. This phenomenon can be attributed to the formation of intermetallic compounds between Sn and Cu as the Cu content increases. Noted that, increasing Cu content can cause the alloys to experience peritectic reactions during the solidification process. As shown in Fig. 1 (a), at 10 wt.% of Cu (dashed blue line), the path of equilibrium solidification includes Cu_3Sn nucleation from the liquid. Peritectic solidification includes the formation of a secondary phase (peritectic product) through the direct reaction between the primary phase and the liquid. Moreover, adding In may alter the phase formation in the Sn10Cu solder alloy. The binary phase diagram for the Sn10Cu-xIn alloy system is shown in Fig. 1 (b). Based on the diagram, phase boundaries of Cu_3Sn may change because of the presence of In, which are highlighted by the dashed green line. Furthermore, the equilibrium liquidus line of the solder was seen to be lower with an increase in In concentration, as indicated by the yellow down arrow, which contributed to lowering the melting temperature of the solder alloy and expected to be improving its wettability [2].

3.2 Microstructure of Sn10Cu- xIn

The microstructure of Sn10Cu, Sn10Cu-0.5In, Sn10Cu-1In, Sn10Cu-3In and Sn10Cu-5In bulk alloys is presented in Fig. 2 and Fig. 3. The microstructure reveals that, commercial Sn10Cu solder consisted of darker Cu_3Sn, peritectic Cu_6Sn_5 and Cu_6Sn_5 phase. A

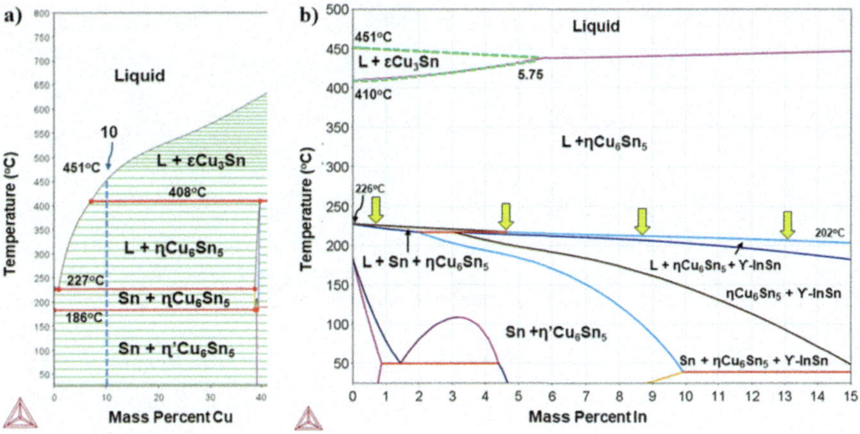

Fig. 1. (a) Binary phase diagram of Sn-Cu alloys with selected area and (b) binary phase diagram of Sn10Cu-xIn alloys.

similar result had been reported by Mehreen et al., 2021 [8]. In addition, the Cu_3Sn IMC phase was found to be randomly distribute around the microstructure. Regarding to the Sn-Cu system, increasing the Cu content up to 40 wt.%, this composition will undergo the peritectic reaction during solidification process. The peritectic product of Cu_6Sn_5 was enclosed with the Cu_3Sn phase, as shown in Fig. 2(b). The EDX result confirms the phase formation as the final microstructure of peritectic Sn10Cu solder alloy, the Cu_3Sn phase at point 1 and the peritectic Cu_6Sn_5 at point 2, Fig. 2 (c). Remarkably, the absence of the Cu_3Sn phase was observed after In addition at all compositions, and the dendritic Cu_6Sn_5 IMC phase was more uniformly distributed, Fig. 3 (a-d). Meanwhile, the Cu_6Sn_5 IMC appear to be refined with the addition of 3 wt.% and 5 wt.%, and it was discovered that the dendritic Cu_6Sn_5 IMC particles were found to be more elongated. According to the Sn-In phase diagram, In has a solubility limit of approximately 7 wt.% in Sn. Consequently, a small quantity of In added to the solder is anticipated to completely dissolve in the Sn-matrix [9]. Therefore, it would suggest that the addition of In up to 5 wt.% significantly alter the microstructure formation of Sn10Cu alloys and finely distribute the Cu_6Sn_5 IMC particle.

3.3 Wetting Angle

Regarding to the results illustrated in Fig. 3 (e), the wetting angle decreased with the addition of In. It is clear that lower wetting resulted with increasing In content. When the In content is 0.5 wt.%, the solder has a wetting angle of 33.5°, which is approximately 7.9% lower than the 36.4° wetting angle of Sn10Cu solder. This indicates that a small addition of In significantly improves the wettability of the solder. However, the wetting angle of Sn10Cu solder dramatically decreased by the addition of In at 1 wt.%, 3 wt.% and 5 wt.%. In details, the wetting angle was measured at 27.4°, 26.4°, and 25.2°, respectively, and it was lower by 24.7%, 27.5%, and 30.8% compared to Sn10Cu solder without adding In. The overall trend is a decrease in wetting angle with the addition of

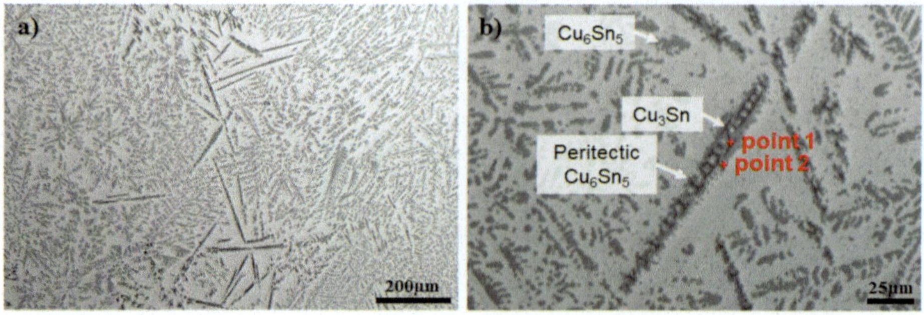

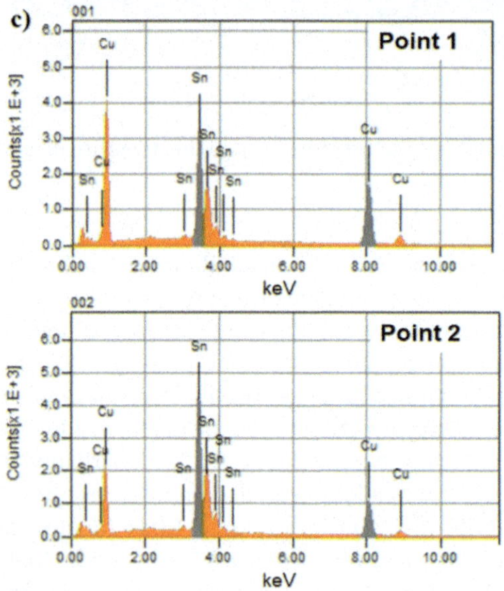

	Cu, at%	Sn, at%	Phase
Point 1	70.1	29.9	Cu_3Sn
Point 2	53.5	46.5	Cu_6Sn_5

Fig. 2. The microstructure of (a) Sn10Cu, (b) Sn10Cu magnified at 2000x, and (c) the analysis of EDX results on primary Cu_3Sn and the peritectic Cu_6Sn_5 IMC.

In, demonstrating that In significantly enhances the wettability of solder joint. This is beneficial for high-temperature soldering applications where good wettability is essential for better reliability. When In is added to a Sn-Cu alloy, it results in a lower melting temperature and primarily reduces the viscosity of the liquid solder at the soldering temperature, enhancing its fluidity under the same conditions [10]. This reduction in viscosity improves the fluidity of the solder, making it easier for the molten solder to wet the substrate.

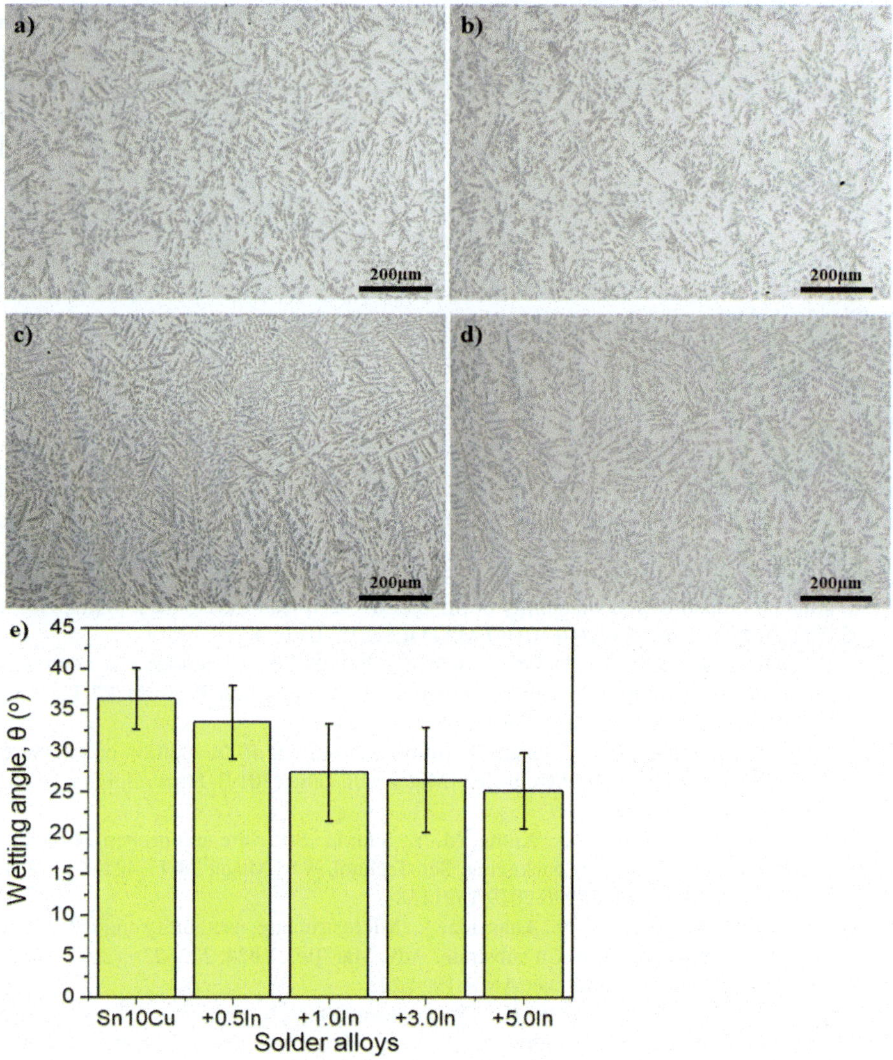

Fig. 3. The microstructure of Sn10Cu-xIn; (a) x = 0.5In, (b) x = 1In, (c) x = 3In, (d) x = 5In and (e) wetting angle of the solder alloys.

4 Conclusion

The effect of adding In on the Sn10Cu-xIn solder alloys was examined, focusing on the phase diagram, microstructure, and wettability. The findings revealed that a small amount of In significantly influences the microstructure and wettability of the Sn10Cu-xIn solder alloy. The conclusions drawn from the study are as follows:

1. As the In content increased, phase boundaries of Cu_3Sn may change. Noted that, the presence of In lowered the equilibrium liquidus line of the solder.

2. The Cu_3Sn phase was not detected after adding In in any wt.% of In composition. The intermetallic compounds seemed to be more refined with the addition of 3 wt.% and 5 wt.% In, and the dendritic Cu_6Sn_5 IMC phase was evenly distributed.
3. The wetting angle of Sn10Cu on a copper pad decreases with the addition of In. Compared to based Sn10Cu solder alloys, the wetting angle of the Sn10Cu-5In solder was 30.8% lower.

Acknowledgments. The authors gratefully acknowledge Nihon Superior Co. Ltd for the materials and lab equipment support. The SEM and EDX analysis were conducted at Nihon Superior (M) Sdn Bhd, Ipoh, Malaysia.

References

1. Zhao, M., Zhang, L., Liu, Z.-Q., Xiong, M., Sun, L.: Structure and properties of Sn-Cu lead-free solders in electronics packaging. Sci. Technol. Adv. Mater. **20**, 421–444 (2019). https://doi.org/10.1080/14686996.2019.1591168
2. Li, L.F., Cheng, Y.K., Xu, G.L., Wang, E.Z., Zhang, Z.H., Wang, H.: Effects of indium addition on properties and wettability of Sn–0.7Cu–0.2Ni lead-free solders. Mater. Des. **64**, 15–20 (2014). https://doi.org/10.1016/J.MATDES.2014.07.035
3. Lai, Y., Hu, X., Jiang, X., Li, Y.: Effect of Ni Addition to Sn0.7Cu Solder Alloy on thermal behavior, microstructure, and mechanical properties. J. Mater. Eng. Perform. **27**(12), 6564–6576 (2018). https://doi.org/10.1007/s11665-018-3734-7
4. Mehreen, S.U., Nogita, K., McDonald, S., Stjohn, D.: Effect of Zn addition on Cu3Sn formation in Sn-10Cu alloys. IOP Conf. Ser. Mater. Sci. Eng. (2019). https://doi.org/10.1088/1757-899X/701/1/012009
5. Zhao, M., Zhang, L., Liu, Z.-Q., Xiong, M.-Y., Sun, L.: Structure and properties of Sn-Cu lead-free solders in electronics packaging. Sci. Technol. Adv. Mater. **20**(1), 421–444 (2019). https://doi.org/10.1080/14686996.2019.1591168
6. Kamarudin, M., Mohd Sharif, N., Anasyida, A.: Microstructure, wettability and IMC Formation of low-Ag SAC solder on Cu substrate. Adv. Mat. Res. **1024**, 223–226 (2014). https://doi.org/10.4028/www.scientific.net/AMR.1024.223
7. Zeng, G., McDonald, S., Nogita, K.: Development of high-temperature solders: review. Microelectron. Reliab. **52**(7), 1306–1322 (2012). https://doi.org/10.1016/J.MICROREL.2012.02.018
8. Mehreen, S.U., Nogita, K., McDonald, S.D., Yasuda, H., StJohn, D.H.: Effect of Ni, Zn, Au, Sb and In on the Suppression of the Cu3Sn Phase in Sn-10 wt.%Cu Alloys. J. Electron. Mater. **50**(3), 881–892 (2021). https://doi.org/10.1007/s11664-020-08709-3
9. Rodrigues, F., Watson, J., Liu, S., Madeni, J.: Study of intermetallic compounds (IMC) that form between indium-enriched SAC solder alloys and copper substrate. Welding in the World **61**, 1–9 (2017). https://doi.org/10.1007/s40194-017-0445-x
10. Hu, X., Lai, Y., Jiang, X., Li, Y.: Effects of In addition on the wettability, interfacial characterization and properties of ternary Sn–Cu–Ni solders. J. Mater. Sci. Mater. Electron. **29**(21), 18840–18851 (2018). https://doi.org/10.1007/s10854-018-0009-x

Effect of High-temperature on SAC305/Sn58Bi Hybrid Solder Alloys on Cu-OSP Surfaces Finish

M. Luqman Mokhtar(✉), Flora Somidin, and Mohd Arif Anuar Mohd Salleh

Faculty of Chemical Engineering Technology, Universiti Malaysia Perlis, 02600 Jejawi, Arau, Perlis, Malaysia
Luqmanmokhtar@studentmail.unimap.edu.my

Abstract. Hybrid solder became the centre of attention after its capabilities to prevent warpage by decreasing the process temperature. Hybrid solder also known as mixed assembly solder, which assemble of high and low melting temperatures of solder in one solder joint. This research focused on the reliability determinisations of the hybrid solder, SAC305/Sn-58Bi, based on the thermal and mechanical behaviour of the solder including intermetallic compound (IMC) layer of the solder were observed. Under morphology observation, the IMC layer of hybrid solder was thinner 11.83 μm than conventional solder. Due to the high content of Bi in Sn-58Bi in hybrid solder, the presence of Bi-rich segregation was found above the interfacial layer and Bi-rich partially warped the SAC305 solder ball. Thermal analysis discovers that the hybrid solder has a lower process temperature, 190.47 °C compared to conventional solder (SAC305), 196.68 °C. Under high and low shear speed tests, the shear strength of hybrid solder was higher than conventional solder after reflowed on Cu-OSP. Hybrid solder shows higher reliability if solder with low process temperature and higher mechanical performance. Thinner IMC layer with segregation of Bi-rich above the layer improve the reliability of the solder.

Keywords: Mixed assembly · Hybrid joining · Warpage · SAC305 · Sn-58Bi

1 Introduction

Electronic packaging has focussed on lead-free solder as the main connection between component and electronic board. Traditional solder, Sn-Pb was banned due to the toxicity [1] of the lead (Pb) that was dangerous for human health and causing pollution [2, 3]. Lead-free solder were category into two group which depend on their application. Low temperature lead-free solder became the main study for researchers due to the capabilities to reduce warpage [4] and cost [5]. Lead-free solder, Sn-58Bi has become the attention due to melting point of 137 °C. In addition, Sn-58Bi has a good wettability and resistance to creep [6–9], thus becoming the best candidates of lead free solder to replace Sn-Pb. For high temperature industries such as aerospace, high melting solder was used. High melting solder, Sn-3.0Ag-0.5Cu (SAC305) has reliability near Sn-Pb with a melting point of 217 °C [10]. SAC305 has better wetting and spreading properties with a higher strength [11, 12].

N. R. Abdul Razak et al. (Eds.): ECOOP 1996, SPPHY 418, pp. 25–31, 2025.
https://doi.org/10.1007/978-981-96-2871-1_4

A new approach of lead-free solder was studied by researchers. A hybrid solder is an assembly of two different solder alloys with different melting points. Hybrid solder SAC305/Sn-58Bi offer a good reliability, which lower process temperature and good mechanical properties [13]. Chantaramanee & Sungkhaphaitoon [14] study hybrid solder of SAC305-xBi/Cu solder joints. From the study, as the Bi content were increase, the thickness of intermetallic compound (IMC) layer decrease and the additional of 2 wt.% of Bi have higher shear stress follow by 1 wt.% and 0 wt.% [14]. Sn–Ag–Cu/Sn–Bi–X hybrid BGA solder joints were studied by. Ren & Huang, 2021, used various reflow process under 200 °C (185–195 °C) were used to ensure the Sn-Ag-Cu remain solid. The Sn-Bi solder partially wrapped the SAC solder ball. As the Bi content increase, the IMC layer will be thinner [15].

In this study, hybrid solder of SAC305/Sn-58Bi were examine on Cu-OSP. To studied the reliability of the solder, compare with the conventional solder, SAC305. Microstructure of solder were observed from the evolution to the morphologies of the IMC layer formed. The thermal and the mechanical of the solder were included to ensure the suitable of hybrid solder as alternative solder for lead-free solder to reduce warpage and reduce cost of fabrication.

2 Experimental Procedure

A solder ball of SAC305 was fabricated from a cylinder with 0.0778 mm of thickness and diameter of 2.5 mm to produce 900 μm solder ball. With the help of flux, the fabrication of the solder ball were free form foreign matter and oxidation. Solder paste Sn-58Bi were used as the low melting point and were standardize the amount of used by using stencil with aperture of 0.76 mm and thickness of 0.24 mm. With the content of flux in the solder paste, prevention of oxidation and joining process were achieved. Organic solderability preservative of copper (Cu-OSP) were used as substrate for hybrid solder in this research. The fabrication of the hybrid solder was completed through reflow process for 1 cycle. The thermal behavior of the hybrid solder was analyzed using differential scanning calorimetry (DSC) to inspect the reaction of the melting and cooling hybrid solder for a single reflow process with the rate of temperature change of 10 °C/min incline to temperature of 300 °C to study the reaction of the hybrid solder joint when both solders were reflowed below their melting point. As the IMC layer grew, the mechanical properties of the solder joint were greatly affected. Shear speed test using DAGE 4000 were used to study the mechanical performance of the hybrid solder with two different speed low and high speed of shear. The mechanical properties of the hybrid solder were support by optical microscope and scanning electron microscope (SEM) to evaluate the evolution of the microstructure after reflow above melting temperature of SAC305 for a single reflow process.

3 Results and Discussions

3.1 Microstructure

Observation of hybrid solder morphology were observed under optical microscope and scanning electron microscopy (SEM). Although optical microscope manages to observed the interfacial solder, SEM used to discover beyond capabilities of optical microscope. Figure 1 show morphology for conventional solder (SAC305) and hybrid solder (SAC305/Sn-58Bi) on Cu-OSP surface finish. Figure 1(a) shows microstructure of SAC305/Cu-OSP after reflowed. With reflow temperature above 200 °C, the solder joint was form after solder were melt and have chemical reaction with Cu-OSP surface finish. IMC layer was shown in the Fig. 1(b). The IMC layer have a scallop likely shape with average thickness of 21.93 μm. In Fig. 1(c) and (d), same observation was shown for SAC305/Cu under SEM observation. Optical observation for hybrid solder were shown in the Fig. 1(e) and (f) after reflow. Segregation of Bi-rich was shown and warp partially Ag-rich were shown in the Fig. 1(e). IMC layer of hybrid in the Fig. 1(f) have uniform likely shape and have lower thickness than SAC305/Cu, 10.10 μm. Further observation was shown under SEM observation in Fig. 1(g) and (h). Distribution of Bi and Sn element was shown above interfacial layer.

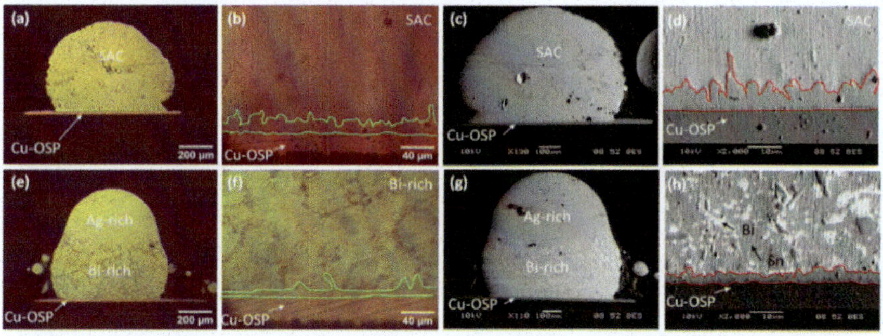

Fig. 1. Morphologies observation of conventional and hybrid solder joint by optical microscope and scanning electron microscope with backscattered electron signal

3.2 Thermal Behaviour

The thermal behavior of hybrid solder resulting a significant reduction in melting point. Same as result of hybrid solder with low temperature of fabrication. In the Fig. 2(a) the analysis of DSC shows a single reflow cycles with a heating and cooling process of the conventional solder joint fabrication. Conventional solder, SAC305/Cu have a melting point of 226.73 °C with onset of 216.53 °C and endset of 233.04 °C resulting pastry range of 16.51 °C. As for the cooling process, the solidification of the conventional solder is start at 197.42 °C with peak of 196.68 °C and end at 192.56 °C. The pastry range for the solidification process are 4.86 °C. overall the undercooling of SAC305/Cu were 23.97 °C. SAC305/Cu have high melting point regarding SAC305 is a high melting lead-free solder. Hybrid solder in a single reflow cycles with higher melting temperature of fabrication resulting several peaks represent solder that involve in the joining as show in the Fig. 2(b). After the hybrid solder were heat for the fabrication of solder joint, a peak of 140.76 °C were appear and started at onset of 139.02 °C and endset of 149.82 °C. This peak shows a reaction of melting process for Sn-58Bi which have a melting temperature approximately at 137 °C. The pastry range of this reaction are 10.8 °C. As the temperature increased along with the time, a peak of melting point for SAC305 was shown with a 217.57 °C peak that slightly appeared with a heat disruption at 190.28 °C, representing the reaction between Sn-58Bi and SAC305 due to different melting point. The melting reaction starts at the onset of 189.08 °C and ends at 222.90 °C. Under heating process for hybrid solder, there was two highest peaks were represented. After reaching 300 °C, DSC analysis were start the cooling process. Due to the single peak of solidification appeared, the solders involved in the hybrid solder were concluded to form a single joint with a new melting and solidification point. The peak started to rise at 189.90 °C with a peak at 190.47 °C and end at 186.10 °C. The pastry range of the solidification peak are 3.8 °C. It appeared that hybrid solder has a lower melting point which led to a lower melting temperature of the fabrication process. The process involves melting point higher that 217 °C also resulting a low melting temperature of fabrication. The solidification of the hybrid solder was lower than conventional solder with lower pastry range.

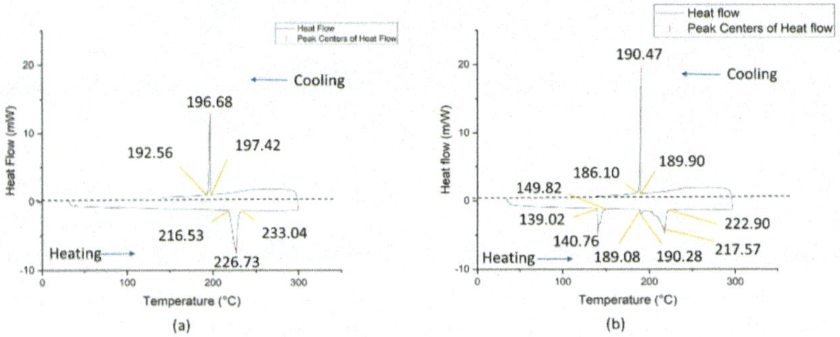

Fig. 2. Thermal analysis using differential scanning calorimetry (DSC) for (a) SAC305/Cu and (b) SAC305/Sn-58Bi/Cu

3.3 Mechanical Behaviour

Reliability of mechanical properties tested with DAGE 4000 shear test with different shear speeds, low and high speed. Figure 3 shows the mechanical properties for both conventional, SAC305/Cu-OSP, and hybrid solder, SAC305/Sn-58Bi/Cu-OSP joints. Hybrid solder shows higher mechanical properties than conventional solder. Low speed shear, 100 mm/s, shows conventional solder has an average shear force of 36.59 N, while hybrid solder has a higher shear force, 52.41N. The same pattern was shown for high-speed shear, 2000 mm/s. Conventional solder have shear strength, 52.24 N, while hybrid solder indicates a higher shear strength, 64.32 N. The use of Sn-58Bi, assembly with SAC305 to create a hybrid solder, produces higher mechanical properties of the solder joint due to the pretense of Bi were known as high mechanical strength with a brittle characteristic.

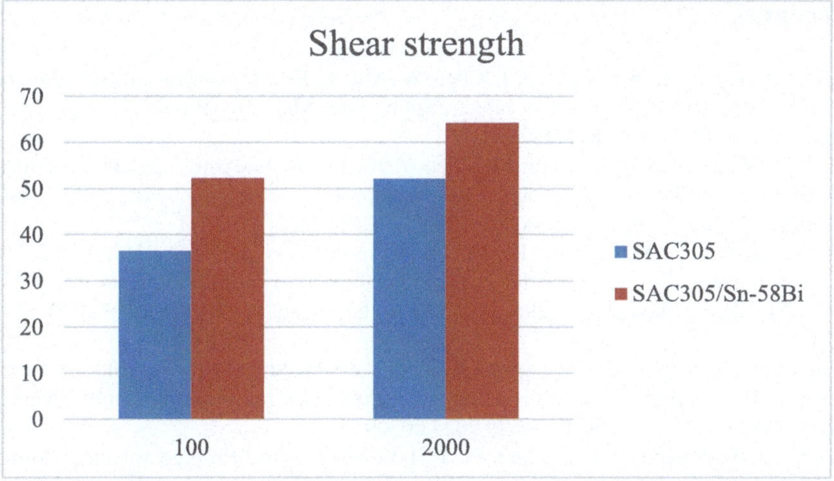

Fig. 3. Shear strength performance of conventional and hybrid solder under low and high-speed shear test

4 Conclusion

Many researchers suggest and study the hybrid solder as the thermal budget concept and the scope of hybrid solder have become trends nowadays. As a result, hybrid solder with a fabrication temperature lower than 200 °C has a joining connection between the solder. In this study, the fabrication temperature was used above melting point SAC305 with one reflow cycle to investigate the properties and the reliability of the hybrid solder. Under thermal analysis, the hybrid solder has a lower melting point for SAC305 and the solder was combined together after the heating process due to a single peak of the solidification process. The fabrication of hybrid solder with temperature above melting point of SAC305 was in the same concept of thermal budget, which consumes low energy for the process where the melting point of SAC305 decreased from 226.73 °C to 217.57 °C. The shear strength performance of the hybrid solder was greatly improved by the presence of Bi content. Conventional solder has a lower shear force, which indicates the mechanical properties of hybrid solder are better. Hybrid solders offer better reliability of solder joints compared to conventional solders.

Acknowledgements. Sincere gratitude for the resources and support provided by CeGeoGTech UniMAP, UniMAP, and the Faculty of Chemical Engineering & Technology. Sincerely thank Nihon Superior for providing the tools and spaces required to carry out and finish this research. The efforts and tremendous help were essential to the completion of this research.

References

1. Zhao, X., Chang, J., Wu, X., Ye, Z.T., Chen, W., Xie, X.: Effect of Sb content on the microstructure and mechanical properties of eutectic SnPb solder. Materials (Basel) **17**(10), 1–13 (2024). https://doi.org/10.3390/ma17102233
2. Wu, M., Wang, S., Sun, W., Hong, M., Chen, Y., Ke, L.: Fracture pattern evolution of SnAgCu–SnPb mixed solder joints at cryogenic temperature. Trans. Nonferrous Met. Soc. China **31**(9), 2762–2772 (2021). https://doi.org/10.1016/S1003-6326(21)65691-1
3. Tunthawiroon, P., Kanlayasiri, K.: Effects of Ag contents in Sn–xAg lead-free solders on microstructure, corrosion behavior and interfacial reaction with Cu substrate. Trans. Nonferrous Met. Soc. China **29**(8), 1696–1704 (2019). https://doi.org/10.1016/S1003-6326(19)65076-4
4. Lu, M., Colgan, E.G., Electromigration kinetics of SAC/SnBiAg hybrid solder. In: Proceedings of the 51st Electronic Components and Technology Conference, pp. 1848–1852 (2024). https://doi.org/10.1109/ECTC51529.2024.00309
5. Hertline, J., Mayberry, R.W., McCoy, J.: Low temperature Pb-Free solder preform technology for molded power module package attach designed for improved thermomechanical performance. In: Conference Proceedings of the APEC 2025 IEEE Applied Power Electronics Conference and Exposition Focusing, pp. 1872–1875 (2024) https://doi.org/10.1109/APEC48139.2024.10509282.
6. Huang, X., Zhang, L., Zhang, J.M., Chen, C., Lu, X., Sun, L.: Intermetallic compound growth behavior and mechanical performance in Sn58Bi/Cu solder joint bearing Mg particles incorporation during thermal aging. Mater. Charact. **208**, 113618 (2024). https://doi.org/10.1016/J.MATCHAR.2023.113618
7. Liu, Y., Ren, B., Zhou, M., Zeng, X., Sun, F.: Microstructure, mechanical, and thermal behaviors of SnBi/Cu solder joint enhanced by porous Cu. J. Mater. Sci. Mater. Electron. **31**(11), 8258–8267 (2020). https://doi.org/10.1007/s10854-020-03361-3
8. Sun, L., Guo, Z., Zhao, X., Liu, Y., Tu, K., Liu, Y.: A New low-temperature solder assembly technique to replace eutectic Sn-Bi solder assembly. Micromachines **13**(6), 1–13 (2022). https://doi.org/10.3390/mi13060867
9. Chen, C., Zhang, L., Wang, X., Lu, X., Guo, Y.: Ultrasonic-assisted connection of Cu/Cu structure using Sn58Bi solder enhanced by B4C nanoparticles. J. Mater. Process. Technol. **319**, 118062 (2023). https://doi.org/10.1016/J.JMATPROTEC.2023.118062
10. Lasky, R., Tevis, I., Lasky, R.: The status of low temperature solders. In: 2024 Pan Pacific Strategic Electronics Symposium, pp. 1–7 (2024). https://doi.org/10.23919/PanPacific60013.2024.10436444
11. Yan, J., Wang, B., Zhao, J., Zhao, L., Guo, R., Sheng, J.: Improvement of heat aging resistance and tensile strength of SAC305/Cu solder joints by multi-element microalloying. Intermetallics **173**, 108428 (2024). https://doi.org/10.1016/J.INTERMET.2024.108428
12. Guo, L., et al.: Dynamic wetting and spreading mechanisms of SAC305 with added Au elements in the laser jet weld ball bonding process. Microelectron. Reliab. **155**, 115362 (2024). https://doi.org/10.1016/J.MICROREL.2024.115362
13. Wang, F., Chen, H., Huang, Y., Liu, L., Zhang, Z.: Recent progress on the development of Sn–Bi based low-temperature Pb-free solders. J. Mater. Sci. Mater. Electron. **30**(4) (2019). https://doi.org/10.1007/s10854-019-00701-w
14. Chantaramanee, S., Sungkhaphaitoon, P.: Influence of bismuth on microstructure, thermal properties, mechanical performance, and interfacial behavior of SAC305-xBi/Cu solder joints. Trans. Nonferrous Met. Soc. China **31**(5), 1397–1410 (2021). https://doi.org/10.1016/S1003-6326(21)65585-1

15. Ren, J., Huang, M.L.: Board-level drop reliability and fracture behavior of low-temperature soldering Sn–Ag–Cu/Sn–Bi–X hybrid BGA solder joints for consumer electronics. J. Mater. Sci. Mater. Electron. **32**(11), 15453–15465 (2021). https://doi.org/10.1007/s10854-021-06094-z

Effects of Zinc Alloy Addition on Sn0.7Cu Solder Properties

Kamrosni Abdul Razak(✉), Athirah Khemar, Aimi Noorliyana Hashim, Mohd Arif Anuar Mohd Salleh, Dewi Suriyani Che Halin, and Nurhidayah Khairul Nizam

Faculty of Chemical Engineering & Technology, Universiti Malaysia Perlis (UniMAP), Taman Muhibah, Jejawi, 02600 Arau, Perlis, Malaysia
kamrosni@unimap.edu.my

Abstract. As lead-free solders gain prominence in electronic packaging, Sn-Cu lead-free solder has emerged as a focal point due to its commendable performance and cost-effectiveness. Relative to Sn-Pb solders, Sn-Cu variants exhibit a higher melting point. In this study, zinc alloys are explored as an alternative alloy. The investigation employs the Micro Vickers Hardness test alongside comprehensive analysis of intermetallic compounds (IMC) and microstructures using an optical microscope. The Micro Vickers Hardness test evaluates the solder's resistance to indentation, offering quantitative insights into its hardness. Examination of IMC aims to discern the bonding nature between the added zinc alloy and the base Sn0.7Cu solder. This scrutiny extends to microstructural observations, leveraging the optical microscope to examine solder details. The study's findings deepen our understanding of how the introduction of zinc alloy impacts the mechanical properties and microstructural attributes of Sn0.7Cu solder. Sn0.7Cu samples, infused with zinc additions of 0.2%, 0.4%, and 0.6%, undergo a reflow process at 200 °C before hardness testing, IMC and microstructure analysis. Results indicate that increased zinc addition correlates with heightened hardness, refined microstructure, and reduced growth layer of IMC. Such insights are pivotal for refining solder formulations, ensuring superior performance in practical applications, and advancing soldering technology.

Keywords: Sn-Cu solder · hardness · zinc addition · intermetallic compound

1 Introduction

Lead-containing solders have long been utilized for attaching components and printed circuit boards (PCBs) in the electronics industry due to their excellent economic efficiency, practicality, and reliability, making them well-suited for various applications. However, the environmental impact of lead, including contamination of land and water supplies and resistance to recycling, necessitates the replacement of lead-containing solder alloys with lead-free alternatives [1]. Among the various alloy formulations within the tin-silver-copper family, it is essential to gather background information to determine

N. R. Abdul Razak et al. (Eds.): ECOOP 1996, SPPHY 418, pp. 32–37, 2025.
https://doi.org/10.1007/978-981-96-2871-1_5

the most suitable alloy for a wide range of applications [2]. The performance and quality of solders are crucial for electronics assembly due to their relatively low melting temperature, good solderability, and high tensile strength. Among the proposed alternatives to traditional lead-based solders, zinc is a notable option.

The development of environmentally friendly, lead-free solder has become a major focus in recent years due to the toxicity of lead and its detrimental effects on human health. Among the many tin-based alloys suggested as replacements for traditional tin-lead solder, the Sn-0.7Cu eutectic alloy stands out as a prominent candidate. This alloy is commonly used in soldering applications because of its advantageous balance of melting temperature, mechanical properties, and cost [3].

However, there are situations where Sn-0.7Cu solder may need to be combined with other alloying elements or additives for specific purposes. For instance, the melting point of the solder alloy may need to be lowered depending on the components being soldered. This can be achieved by introducing lower-melting-point elements such as zinc. Zinc additions can reduce the melting point of solders [4, 5]. Moreover, to improve the flow of molten solder, ensure strong contact with the surface, and enhance the quality of solder joints, it is essential to improve wetting and spreading properties. Zinc is effective in improving wettability, which enhances the ability of molten solder to spread over the copper substrate. Additionally, zinc is an excellent choice due to its antioxidant properties.

2 Experimental Procedure

After obtaining the desired weight of 0.2%, 0.4%, and 0.6% for zinc powder, the cut piece bar of Sn0.7Cu and zinc is then melted using a solder pot at 230–240 °C. After removing the bulk samples, each sample are rolled until reach the desired thickness which is 20–30 μm. After cleaning the copper substrate, the flux is applied evenly on the copper substrate before placing the solder metal sheet on them. The flux is then applied again on top of the metal sheet. Then, the samples are then proceeded with reflowing process using reflow oven at 230 °C.

The hardness behavior of the sample was measured using a Micro Vickers Hardness tester. Meanwhile, the thickness of IMC formation was determined using ImageJ software, and the microstructure properties such as the length between the grain boundaries were studied using optical microscope.

3 Results and Discussions

3.1 Microstructures Analysis

The length between the grain boundaries was obtained using ImageJ application. The result for the length is summarized in Table 1. The average length is obtained by dividing the total length from the five trials. The microstructures of every parameter which is 0% zinc, 0.2% zinc, 0.4% zinc, and 0.6% zinc is shown in Fig. 1. The brighter region in the microstructures is β-Sn and the darker region is the eutectic alloy. The minimum value of length is the sample with the highest amount of addition zinc which is 10.99μm while the maximum value of length is the sample without any addition of zinc which is 87.05μm.

Table 1. The length obtained between grain boundaries of the microstructure observed.

Sn0.7Cu	0wt%Zn (μm)	0.2wt%Zn (μm)	0.4wt%Zn (μm)	0.6wt%Zn (μm)
1	80.88	35.70	41.84	9.89
2	64.32	55.32	42.29	7.98
3	66.79	51.68	28.71	8.19
4	103.88	51.88	33.49	19.18
5	119.40	51.97	41.55	9.72
Average (μm)	87.05	49.31	37.58	10.99

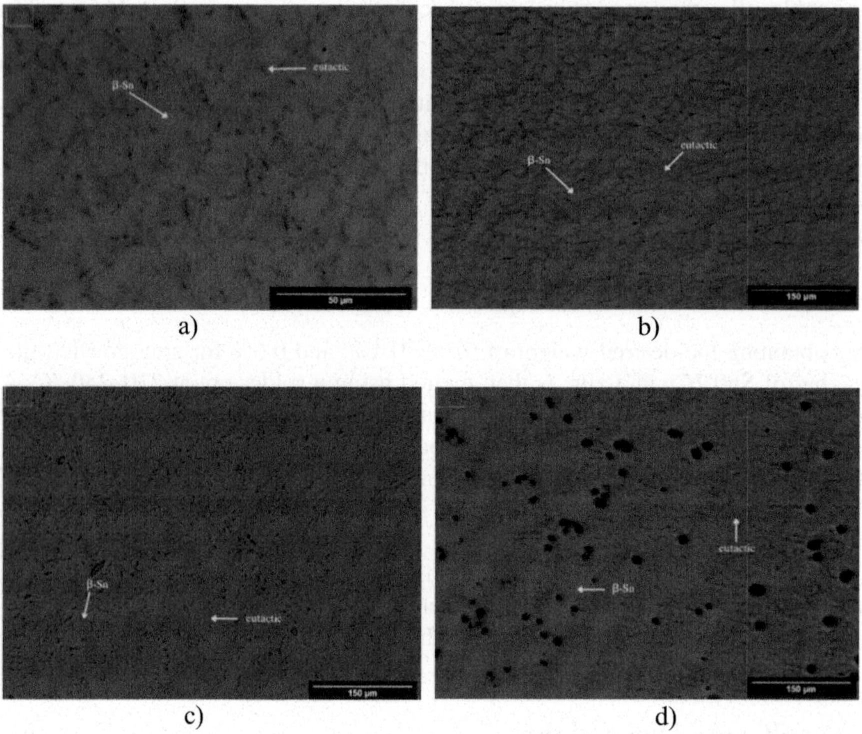

Fig. 1. The microstructures of a) Sn0.7Cu, b) Sn0.7Cu 0.2wt% zinc, c) Sn0.7Cu with 0.4wt% zinc and d) Sn0.7Cu with 0.6wt% zinc

The results indicated that as the weight percentage of zinc increases, the length of the grain boundaries decreases. From the microstructures obtained, Sn0.7Cu exhibits the coarsest structures. In contrast, Sn0.7Cu with a 0.6 wt% addition of zinc has the most refined microstructures compared to other samples. This implies that the microstructures become more refined with increasing zinc addition. Previous research by Jun Shen et al.

(2014) found that the microstructures of the Bi-rich phase in both Sn-40Bi-0.1Cu and Sn-40Bi-2Zn-0.1Cu solder alloys were refined with the addition of Cu and Zn [6].

3.2 Mechanical Properties

For mechanical properties, the test used was Micro Vickers Hardness Test. The average hardness for Sn0.7Cu solder bulk with different weight percentages is summarized in Fig. 2 below. The lowest hardness value gained by the sample with only Sn0.7Cu which is 11.34HV. The addition of zinc into the solder alloy enhanced the hardness value. Sn0.7Cu with 0.2wt% zinc shows the hardness value increase to 11.7 HV, while the maximum value is 12.7 HV shows by the highest amount of addition zinc which is 0.6wt%.

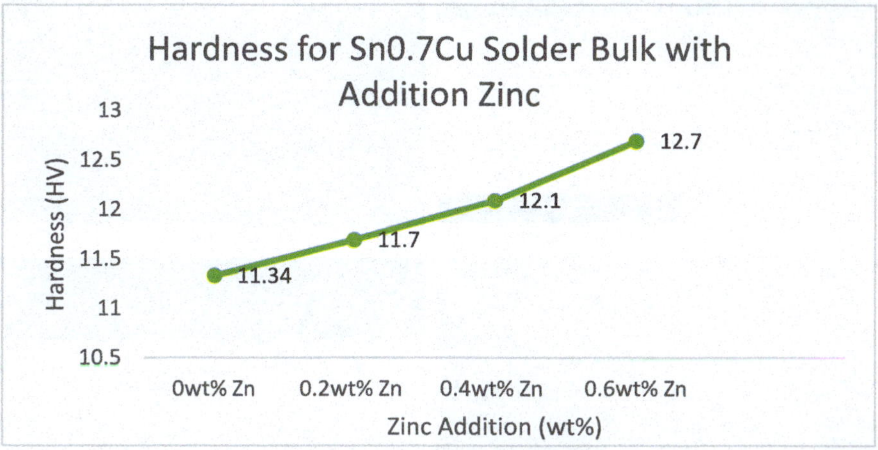

Fig. 2. The average hardness for Sn0.7Cu solder bulk with different weight percentages addition zinc.

The results revealed that an increase in weight percentages of addition zinc results in increased hardness for the bulk samples. From the results obtained, zinc can enhance and improve the mechanical properties of the sample. The achieved strength is caused by the refinement of microstructure and formation of secondary particles in solder matrix [6]. The previous study also supports the result for this finding that zinc addition can promoted the mechanical property of solder by refining the microstructures.

3.3 Intermetallic Compound (IMC) Analysis

From Table 2, the average thickness is decreasing as the weight percentage of addition zinc increases. The minimum value of average thickness for IMC is 0.38 μm, the sample with the highest amount of addition zinc which is 0.6wt% while the maximum value of average thickness is 0.94 μm, the sample without any addition zinc.

The results showed that as the weight percentage addition zinc increases, the average thickness decreases. This is because it is stated that the addition of zinc was found

Table 2. The average thickness for Sn0.7Cu solder bulk with different weight percentages addition zinc.

Sn0.7Cu	0wt%Zn (μm)	0.2wt%Zn (μm)	0.4wt%Zn (μm)	0.6wt%Zn (μm)
1	0.85	0.90	1.14	0.35
2	0.86	0.88	0.54	0.33
3	1.11	1.00	0.63	0.46
Average Thickness	0.94	0.93	0.77	0.38

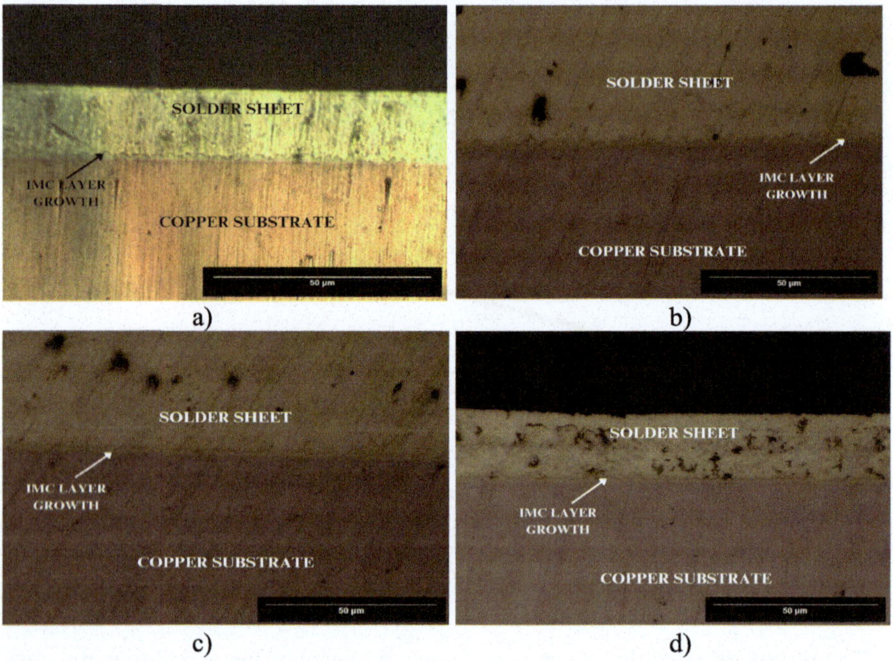

Fig. 3. The IMC thickness of a) Sn0.7Cu, b) Sn0.7Cu 0.2wt% zinc, c) Sn0.7Cu with 0.4wt% zinc and d) Sn0.7Cu with 0.6wt% zinc

to inhibit IMC layer growth. By comparing Fig. 3, the IMC layer thickness of the Sn0.7CuZn solder is found to be thinner than the original Sn0.7CuZn solder, which indicates that the growth of Cu-Sn IMCs in liquid/solid state reaction is inhibited with the addition of 0.8% Zn [7]. The previous study also supports the result for this finding that zinc addition can promoted the inhibit IMC layer growth.

4 Conclusion

This paper discusses the effect of zinc alloy addition on Sn0.7Cu solder properties. The conclusions involve the following points:

1. The microstructures analysis revealed that an increase in the weight percentage of the zinc correlates with a decrease in the length of grain boundaries. The coarsest microstructures were observed in Sn0.7Cu alone, while the addition of 0.6wt% zinc resulted in the most refined microstructures. This suggests that zinc plays a crucial role in refining the microstructures of the solder.
2. The hardness value of the solder was found to be enhanced with the addition of zinc. The hardness of the bulk samples increases as the weight percentage of zinc increases, indicating an improvement in mechanical properties. This improvement in hardness is attributed to the refined microstructure resulting from zinc addition. The correlation between mechanical properties and microstructure highlights the importance of zinc as a beneficial additive in enhancing the overall performance of the Sn0.7Cu solder.
3. The IMC analysis demonstrated that the addition of zinc inhibits the growth of the IMC layer. As the weight percentage of zinc increases, the average thickness of the IMC layer decreases.

Acknowledgments. The author would like to acknowledge the support from the Fundamental Research Grant Scheme (FRGS) under a grant number of FRGS/1/2020/TKO/UNIMAP/03/18 from the Ministry of Higher Education Malaysia also Faculty of Chemical Engineering & Technology for the materials, facilities and finance support.

References

1. Hwang, J.S.: Lead Free Solder: The Sn/Ag/Cu System (pp. 18–21). IPC Publishing Group Inc. (2000)
2. Li, X.-y., Wu, B.-s., Yang, X.-h.: The Formation and Evolution of IMC and Its Effect on the Solder Joint Properties (pp. 273–278) (2005)
3. Zhao, M., Zhang, L., Liu, Z.Q., Xiong, M.Y., Sun, L.: Structure and properties of Sn-Cu lead-free solders in electronics packaging. Sci. Technol. Adv. Mat. **20**(1), 421–444 (2019)
4. Öztürk, E., Aksöz, S., Keşlioğlu, K., Maraşlı, N.: The measurement of thermal conductivity variation with temperature for Sn-20 wt.% in based lead-free ternary solders. Thermochimica Acta **554**, 63–70 (2013)
5. Kiryukhina, K., Le Trong, H., Tailhades, P., et al.: Silver oxalate-based solders: New materials for high thermal conductivity microjoining. Scripta Materialia **68**(8), 623–626 (2013)
6. Shen, J., Yayun, P., Yin, H., Luo, D., Chen, J.: Effects of minor Cu and Zn additions on the thermal, microstructure and tensile properties of Sn–Bi-based solder alloys. J. Alloy. Compd. **614**, 63–70 (2014)
7. Wang, F.J., Gao, F., Ma, X., Qian, Y.-Y.: Depressing effect of 0.2wt.%Zn addition into Sn-3.0Ag-0.5Cu solder alloy on the intermetallic growth with Cu substrate during isothermal aging. J. Elect. Mat. **35**, 1818–1824 (2006)

Epoxy Mold Compound Cracking in BGA Components: Impact of Moisture Content During Assembly

Maria Abu Bakar[1(✉)], Azman Jalar[1,3], Syed Mohamad Mardzukey Syed Mohamed Zain[2,1], Fakhrozi Che Ani[3], and Mohamad Riduwan Ramli[4]

[1] Institute of Microengineering and Nanoelectronics, Universiti Kebangsaan Malaysia, 43600 Bangi, Selangor, Malaysia
{maria,azmn}@ukm.edu.my

[2] Western Digital, SanDisk Storage Malaysia Sdn.Bhd., Plot 301A, Persiaran Cassia Selatan 1, Taman Perindustrian Batu Kawan, MK13, 14100 Batu Kawan, Seberang Perai Selatan, Penang, Malaysia
SyedMohamadMardzukey.SyedMohamedZain@wdc.com

[3] Department of Applied Physics, Faculty of Science and Technology, Universiti Kebangsaan Malaysia, 43600 Bangi, Selangor, Malaysia
Fakhrozi.CheAni@wdc.com

[4] School of Materials and Mineral Resources Engineering, Universiti Sains Malaysia, 14300 Nibong Tebal, Pulau Penang, Malaysia
riduwan@usm.my

Abstract. Electronic devices have been transformed significantly due to technological advancements. Ball grid array (BGA) component must be properly encapsulated by epoxy mold compound (EMC) as a packaging material. However, crack occurrence on the EMC of the BGA component during reflow soldering process has attracted much attention. The moisture content has potential to induce crack occurrence at the EMC encapsulation. Therefore, this study used different BGA component with variation of moisture content from 0.04% to 0.20% as test vehicle to investigate the crack occurrence of the EMC encapsulation during assembly process. All BGA components with and without moisture content were assembled on the printed circuit board (PCB) via reflow soldering process. It is shown that the moisture trapped in the BGA component of amount 0.20% likely play a significant role in inducing the 60% of crack formation on the EMC encapsulation. Therefore, moisture content in the BGA components must be addressed prior to reflow soldering process in addition to establish good BGA metallurgical joining between the component and PCB.

Keywords: Ball grid array component · crack occurrence · epoxy mold compound · moisture content · BGA component assembly

N. R. Abdul Razak et al. (Eds.): ECOOP 1996, SPPHY 418, pp. 38–44, 2025.
https://doi.org/10.1007/978-981-96-2871-1_6

1 Introduction

Semiconductor packaging is one of the important processes in semiconductor manufacturing. Semiconductor packaging roles are to ensure electrical connectivity for device functionality, thermal and heat dissipation management and protect the chip from various physical and environmental impact. This is important to ensure electronic component reliability, performance and integration into electronic devices during assembly process. Driven by the persistent activity of miniaturization, speed, and efficiency in electronic devices, semiconductor packaging technologies have undergone a remarkable evolution. From their humble beginnings with traditional leaded packages, these technologies have advanced to involve complex techniques such as flip-chip, system-in-package, and 3D packaging. These packaging innovations play a vital role in powering a wide range of modern applications, from smartphones, IoT devices to automotive electronics. Selection of a specific packaging method represents a crucial decision in electronics manufacturing especially considering the device complexity, desired performance levels and manufacturing cost.

One of the popular semiconductor packaging is hermetic packaging. Hermetic packaging is used to enclose semiconductor devices and protect them from outside elements like dust, moisture, and corrosives. Hermetic packaging is commonly used in the military, aerospace and medical devices for high performance, high-reliability and long-term stability. However, in the other hand, epoxy mold compound (EMC) is differed from the hermetic packaging. EMC is typically used for non-hermetic packaging especially for consumer electronics where the environmental conditions are less harsh. In semiconductor packaging, EMC has been widely used for the ball grid array (BGA) component. The role of the EMC is to encapsulate and protect the die in the BGA component. There are various types of EMC in the market such as high-temperature EMC, low-temperature EMC, thermally conductive EMC, electrical insulating EMC, moisture-resistant EMC, flame-retardant EMC, conductive EMC and many more. The usage of EMC is dependent on the application, reliability and performance of the electronic components. EMC is made up of epoxy resin, curing agent to facilitate the cross-linking process during curing, filler materials that are used to improve the physical and mechanical properties of EMC and additives such as modifiers and colorants.

Various research has been conducted to study the EMC properties as encapsulation materials for semiconductor packaging. To such, Shao et al. [1] investigate the effects of filler shape, type, and size on the properties of encapsulation molding components. From their work has shown that the filler shape and size affected the thermal conductivity of the EMC. Inamdar et al. [2] studied the high temperature aging of epoxy-based molding compound and its effect on mechanical behavior of molded electronic package. Some researchers also studied the compatibility of the EMC for many types of electronic applications such as large-area power devices, memory device, automotive device [3–5].

The failure of EMC encapsulation in electronic packages is a critical issue that can contribute to shorter product lives, decreased reliability, and malfunctioning devices. Typical failures of the EMC encapsulation such as delamination, cracking, popcorn effects, thermal stress and warpage, chemical degradation and electrical degradation. Among of these failures, the cracks occurrence of the EMC encapsulation during assembly process has gotten much attention. Cracks have the potential to significantly impede the EMC's protection, causing electronic devices to break down before their time. This is due to the cracks that enable outside elements like moisture, air, and pollutants penetrate the package, which can lead to internal component degradation, corrosion, or electrical shorting. Once cracks begin, they typically propagate due to heat or cyclic stress. A minor crack has the potential to propagate quickly and cause larger cracking, package delamination, or total failure. In the other hand, cracks are easily penetrated by moisture, thus may lead to further issues like the popcorn effect during solder reflow. When the package is subjected to high temperatures, the moisture it contains evaporates, producing an internal pressure buildup that might lead to further cracking or delamination.

BGA components are widely used in electronic packaging. However, the BGA components with EMC encapsulation and moisture content are susceptible to cracking during the reflow soldering process. Certain amounts of moisture content can lead to internal stresses during the reflow soldering process and causing the crack occurrence. Consequently, the crack in the EMC encapsulation will significantly affect both the structural integrity and performance of the BGA component. As the electronic packaging manufacturing processes have advanced, the relationship between moisture content and crack formation in BGA components during the reflow soldering process remains inadequately explored. This paper is aimed to investigate the moisture content effect of BGA component on the cracks occurrence during reflow soldering process. This paper will use BGA components with different amounts of moisture content to understand the relationship between moisture content and the EMC encapsulation crack occurrence during assembly process. This study will provide some insight of moisture content effect on the structural integrity of BGA components.

2 Materials and Method

This work used 30 pieces of BGA component as samples. 5 pieces of BGA component are used as controlled sample while another 25 pieces of the BGA were varied with moisture content from 0.04% up to 0.20% as shown in Table 1. The dimension of the BGA is 17.0 mm (length) × 17.0 mm (width) × 0.8 mm (thickness). The BGA component consists of EMC as encapsulation materials, Si die, copper lead frame and Au wire bonding to connect the die and lead frame and lead-free solder as interconnection material. The EMC for the BGA component is made from 30% epoxy and 70% filler.

All the BGA component samples undergo reflow soldering process (BTU Pyramax12ZN) at 255 °C to assemble the BGA components onto the Bismaleimide-Triazine (BT) substrate. The solder paste (SAC305) was printed on the BT substrate via stencil printing process [6] and the SAC305 solder balls were attached on the solder paste prior to the reflow soldering process. After reflow soldering process was completed, the BGA components on the BT substrate were taken out from the reflow oven for observation and characterization. The samples were visually inspected using an optical microscope to observe any abnormalities that occurred on the EMC encapsulation.

Table 1. BGA component with moisture content

Sample group	Number of BGA component (pcs)	Moisture content (%)
A	5	0
B	5	0.04
C	5	0.08
D	5	0.12
E	5	0.16
F	5	0.20

3 Results & Discussion

Figure 1 shows the images of top-view BGA components after the reflow soldering process for all samples group A-F. It is shown that BGA components from group A to E shown no visible abnormalities, indicating that these samples with lower moisture content of less than 0.16% were able to withstand the thermal stresses during reflow soldering process without damage. However, in group F, it is observed that the cracks have occurred at the EMC encapsulation of the BGA component for sample F3, F4 and F5. Table 2 shows the number of BGA component with crack occurrence. It is clearly shown in Table 2 that for sample in group F, 3 out of 5 samples experience cracks. This indicates that 60% of the samples in group F were affected by crack occurrence.

The finding shows that the moisture content of 0.20% was sufficient induced the crack occurrence of the EMC encapsulation. The cracks observed in group F likely resulted from the absorption of moisture by the EMC of the BGA component. During the reflow soldering process, the rapid heating contributed to the vaporizing phenomenon of the absorbed moisture. This has led to internal pressure which known as the popcorn effect. The popcorn effect typically occurs when moistures were absorbed by the EMC encapsulation and leading to electronic package vaporizes during high temperature, in this case, reflow soldering process. Thus, this ultimately caused the crack to form at the EMC encapsulation. This is-line with research work from Chen et al. [7] who reported that the electronic package with moisture content is susceptible to crack propagation under reflow condition. The pressure inside the BGA component can cause cracks or delamination, mainly when the moisture content is high [8–11].

Figure 2 shows the images of the top-view BGA components after the reflow soldering process for sample F3 in group F. Figure 2(b) shows the enlarged image of Fig. 2(a) showing crack line on the EMC encapsulation. The observation in this work is in-line with the Lee et al. [12] who reported the crack occurrence at the EMC encapsulation for SiP package module undergo high temperature reflow soldering at 260 °C. This finding in this work highlights the need of to assess the moisture content and stringent moisture control of the BGA components and devices encapsulated with EMC.

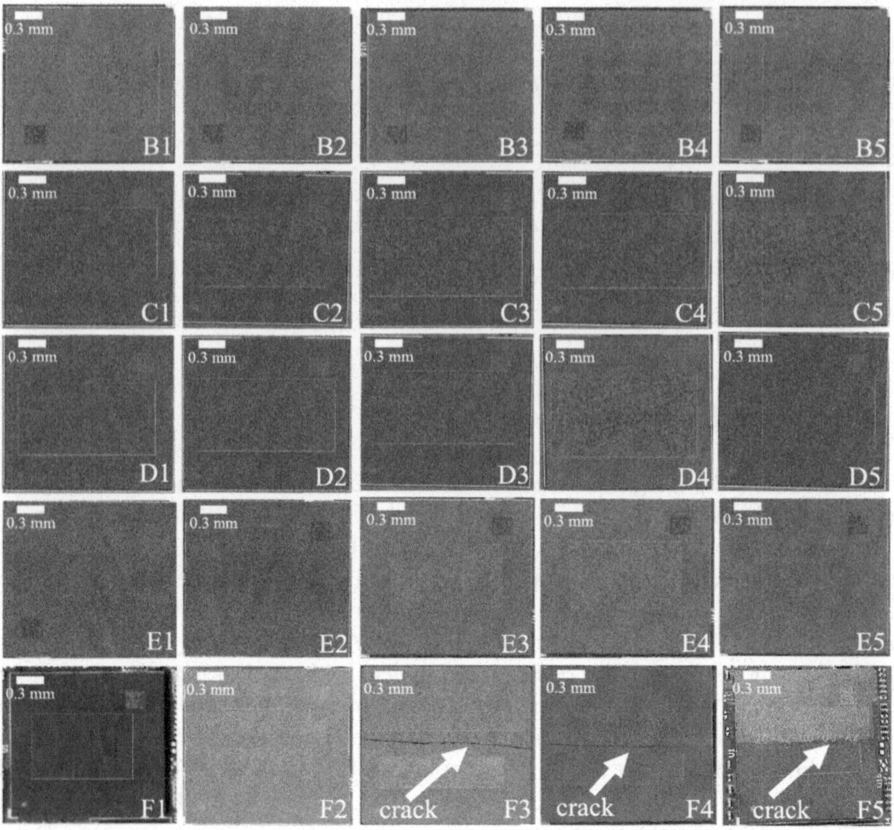

Fig. 1. Images of the top-view BGA components after the reflow soldering process for sample group A (A1-A5), group B (B1-B5), group C (C1-C5), group D (D1-D5), group E (E1-E5) and group F (F1-F5)

Table 2. Crack occurrence at EMC encapsulation of the BGA component

Sample group	Number of BGA component (pcs)	Number of BGA component cracked (pcs)
A	5	0
B	5	0
C	5	0
D	5	0
E	5	0
F	5	3

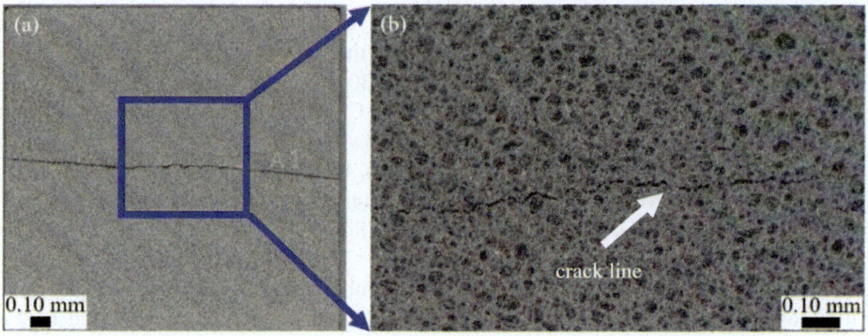

Fig. 2. Images of the top-view BGA components after the reflow soldering process: (a) crack occurrent of sample F3 in group F and (b) enlarged image of (a) showing crack line

4 Conclusion

This study highlights the critical impact of moisture content on crack formation in the EMC encapsulation of BGA components during the reflow soldering process. It was found that a moisture content of 0.20% significantly contributed to a 60% occurrence of cracks in the EMC encapsulation. These findings emphasize on the important of addressing moisture levels in BGA components prior to reflow soldering to avoid the risk of crack occurrence. It is suggested that in order to prevent crack occurrence and propagation, the BGA components should be undergoing baking processes before reflow soldering to ensure the moisture content is below than 0.20%. This is important to achieve good and reliable metallurgical joining between the BGA components and the PCB.

Acknowledgements. The authors would like to acknowledge the financial support and research facilities provided by Western Digital via SanDisk Storage Malaysia Sdn. Bhd (research grant: RR-2023-007) through collaboration activities with Universiti Kebangsaan Malaysia (research grant: TAP-K022551).

References

1. Chao, S., Liaw, Y., Chou, J.H.: The effects of filler shape, type, and size on the properties of encapsulation molding components. Electronics **10**(2), 98 (2021)
2. Inamdar, A., Yang, Y.H., Prisacaru, A., Gromala, P., Han, B.: High temperature aging of epoxy-based molding compound and its effect on mechanical behavior of molded electronic package. Polym. Degrad. Stab. **188**, 109572 (2021)
3. Dobrzynska, J.A., Vobecký, J., Gradinger, T.B., Guillon, D., Corvasce, C.: Epoxy mold compound encapsulation concept for large-area power devices. IEEE Trans. Compon. Packag. Manufact. Technol. **12**(4), 602–609 (2022)
4. Gan, C.L., Chung, M.H., Lin, L.F., Huang, C.Y., Takiar, H.: Evolution of epoxy molding compounds and future carbon materials for thermal and mechanical stress management in memory device packaging: a critical review. J. Mater. Sci. Mater. Electr. **34**, 2011 (2023)
5. Sakib, M.N., Asif Iqba, A.K.M.: Epoxy based nanocomposite material for automotive application- a short review. Int. J. Autom. Mech. Eng. **18**(3), 9127–9240 (2021)

6. Mohamed Sunar, M.S., Abu Bakar, M.A., Jalar, A., Muhamed Mukhtar, M.A.F., Che Ani, F.: Effect of PVD-coated wall aperture roughness on the life span of fine-pitch stencil printing. Soldering Surface Mount Technol. **36**(1), 51–59 (2024)
7. Chen, B.S., Chiu, T.C., Yin, W.J., Kao, C.L., Hung, C.P.: Evaluation of vapor pressure induced debonding failure in fan-out package under reflow condition. 2024 IEEE 74th Electronic Components and Technology Conference (ECTC), 150–156 (2024)
8. Heng E., Abdullah, M.Z.: Influence of moisture concentration and hydrophobic material on induced stress in FCBGA package under reflow. Soldering Surface Mount Technol. **34**(2), 103–115 (2022)
9. Haas, L., Döpper, F., Schmidt, K., Franke, J., Reinhardt, A.: Studies and analyses of moisture conditioned SMT-components. 2021 23rd European Microelectronics and Packaging Conference & Exhibition (EMPC), 1–8 (2021)
10. Jalar, A., Zain, S.M.M.S.M., Ani, F.C., Ramli, M.R., Bakar, M.A.: Effect of potting encapsulation on crack formation and propagation in electronic package. In: Mat Jizat, J.A., et al. Advances in robotics, automation and data analytics. iCITES 2020. Advances in Intelligent Systems and Computing. 1350, pp. 351–357 (2021)
11. Shameem, S., Sony, K., B, K., Babu, K.R., Bollaram, S., Pullela, B.D.: Design of MEMS model to analyse pressure and hygroscopic swelling variables for EMC (Epoxy Mold Compound). 2021 IEEE International Conference on Mobile Networks and Wireless Communications (ICMNWC), 1–6 (2021)
12. Lee, J.C., Chen, C.-C., Lee, D., Chen, J.: Moisture effect on physical failure of plastic molded SiP module. 2020 IEEE 70th Electronic Components and Technology Conference (ECTC), 2124–2132 (2020)

Influence of Epoxy Molding Compounds on Crack Severity in Lead-Free Solder Joints of Mirrored NAND Flash BGA Packages Under Thermal Cycling Conditions

Muhammad Nizam Ilias[1,2], Maria Abu Bakar[2(✉)], Azman Jalar[2,3], Adlil Aizat Ismail[1,2], and Erwan Basiron[1,2]

[1] Western Digital®, SanDisk Storage Malaysia Sdn.Bhd., Plot 301A, Persiaran Cassia Selatan 1, Taman Perindustrian Batu Kawan, MK13, Batu Kawan, 14100 Seberang Perai Selatan, Penang, Malaysia

[2] Institute of Microengineering and Nanoelectronics (IMEN), Selangor, Malaysia
maria@ukm.edu.my

[3] Department of Applied Physics, Faculty of Science and Technology, Universiti Kebangsaan Malaysia, 43600 UKM Bangi, Selangor, Malaysia

Abstract. This paper investigates the influence of different epoxy molding compound (EMC) materials on the reliability of mirrored NOT-AND gate (NAND) component placement (Mirrored NAND) in Flash Ball Grid Array (BGA) packages, addressing a critical issue in electronic packaging. The project aims to close the knowledge gap regarding how variations in EMC materials impact the reliability of lead-free solder joints under thermal cycling conditions. By isolating the effects of EMC material, we aim to provide precise insights into its impact on crack severity within the solder joint array. Our methodology involves subjecting the Mirrored NAND Flash BGA packages to thermal cycling tests (TCT) ranging from 500 to 800 cycles, followed by dye and pull (DnP) testing which refer to observe and analyze crack severity. Despite the occurrence of cracks, both types of EMC materials used in this study enabled the packages to meet the stringent reliability criterion of less than 50% solder joint cracking on functional solder balls. The key findings from this study reveal differences in crack severity behavior based on the EMC material used under thermal stress. These insights not only advance knowledge in the field but also contribute to the development of more sustainable and reliable electronic devices.

Keywords: Thermal cycling · BGA · EMC · Reliability · Lead-free solder · cracks severity · Mirrored NAND

1 Introduction

The reliability of solder joints in Ball Grid Array (BGA) packages is a critical concern in the electronics industry, especially under thermal cycling conditions. Solder joints serve two functions: mechanical support and electrical connectivity, making their integrity

N. R. Abdul Razak et al. (Eds.): ECOOP 1996, SPPHY 418, pp. 45–51, 2025.
https://doi.org/10.1007/978-981-96-2871-1_7

vital to the overall performance and lifetime of electronic systems. However, when subjected to thermomechanical stress, these joints are prone to fatigue failure due to a mismatch in the coefficient of thermal expansion (CTE) between the package and the printed circuit board assembly (PCBA). Thermal cycling causes repeated expansion and contraction of the materials involved, resulting in considerable mechanical stresses and strains in the solder joints. The accumulated plastic strain energy density is usually larger at the solder/component interface than at the solder/PCBA interface, resulting in fatigue failures that often initiate on the component side. This failure mechanism occurs through damage accumulation, crack initiation, and crack propagation, resulting in joint failure [4].

Epoxy molding compounds (EMC) are commonly used to encapsulate electronic devices, providing protection against environmental and mechanical stresses. However, EMCs are hygroscopic and absorb moisture from the ambient environment, which can cause swelling and local stress build-up, similar to thermal stresses. This absorbed moisture can worsen crack propagation inside solder connections, reducing their reliability under thermal cycling conditions [5]. The fatigue loading experienced by the solder joints increases during thermal cycling [7]. Therefore, understanding the influence of EMC on crack propagation in lead-free solder joints is essential for improving the reliability of mirrored NOT-AND gate (NAND) component placement (mirrored NAND) Flash BGA packages.

This paper investigates the impact of epoxy molding compound (EMC) on crack severity in lead-free solder joints of mirrored NAND Flash BGA packages under thermal cycling conditions. By examining the interactions between EMC and solder joints, this study aims to provide insights into the mechanisms driving fatigue failure and suggest strategies for enhancing the durability of these critical components in electronic assemblies.

2 Materials and Method

The primary objective of this study is to examine the influence of EMC materials on the reliability of lead-free solder joints in mirrored NAND Flash Ball Grid Array packages under thermal cycling conditions. Two different types of EMC materials were selected based on their CTE values, which are 9 ppm/°C (CTE9) and 11 ppm/°C (CTE11).

This experiment used Solid State Drive Enterprise and Datacenter Standard Form Factor (SSD EDSFF) 9.5mm E1.L Form Factor with a PCBA that has a CTE value of 17.2 ppm/°C. The main components for drive assembly were enclosures, screws, PCBA, and Thermal Interface Material (TIM). The E1.L PCBA was sourced from a single supplier from the same batch, and the TIM was also from the same batch. The main components of the SSD are the controller, Dynamic Random Access Memory (DRAM), and a few passive components, which are not visible in Fig. 1(a) due to product design layout intellectual property concerns. All of them were identical across all drives, with populated NAND Flash BGA packages having a size of 12 mm width x 18 mm length x 1.4 mm height, 170 balls in total, and a 16-die single stack. The ball diameter was 0.49 mm, and the ball pitch was 1 mm. Table 1 below shows the material properties for CTE9 and CTE11 of the NAND EMC material:

Table 1. Material Properties for CTE9 and CTE11 of NAND EMC material

Material Properties (unit)	CTE9	CTE11
Density (kg/m-3)	1820	1820
Young's Modulus (Pa)	2.6E + 10	2.4E + 10
Poisson's Ratio	0.3	0.3
Bulk Modulus (Pa)	2.2E + 10	2E + 10
Shear Modulus (Pa)	1E + 10	9.2E + 9

Lead-free solder joints using SAC305 Type 5 were used in this experiment, and a total of 128 BGA component samples were prepared for each type of EMC material, resulting in 256 samples in total. This sample size was chosen based on the SSD PCBA E1.L form factor, Flash NAND locations, and the thermal cycling test plan. The thermal cycling tests (TCT) were conducted using a thermal cycling chamber (Make: Espec, Model: EGNX12–7.5NWL). The samples were subjected to a thermal cycling profile at temperatures ranging from -40 °C to 85 °C, with each cycle have a duration of 60 min to ensure uniform thermal exposure. The number of cycles ranged from 500 to 800, in increments of 100, simulating typical operational stress conditions encountered in actual applications.

One unit of drive was equipped with NAND Flash BGA CTE9, and another drive was equipped with NAND Flash BGA CTE11, subject to each TCT cycle ranging from 500 to 800 cycles (T500 to T800), with a total of 8 drives allocated for this experiment. All drives were subjected to component image capturing and 2D X-ray inspection prior to dye and pull (DnP) testing according to IPC-TM-650 standard to ensure no abnormalities. Selected components shown in Fig. 1(a), based on 3D structure simulation through finite element analysis (FEA), indicated high-risk locations determined by strain energy density (SED) values. Hence, these components within this area were selected for physical failure analysis using the DnP test to validate the actual experimental results based on crack severity.

Figure 1 (b) shows the methodology used to quantify the dye penetration percentage and crack mode of the solder joint. Qualitative and quantitative data will be compiled from this methodology for further analysis.

3 Results and Discussions

3.1 Dye and Pull Pattern Analysis

The images from Fig. 2(a) show that each NAND BGA consists of 2 main boundaries: the anchor balls area, colored in white, and the functional balls area, colored in light blue. Those images generated from DnP result which refer to dye penetration at BGA from PCBA side showed in Fig. 2(b) and dye penetration at BGA from component side as showed in Fig. 2(c). Anchor balls in the outer area are used to reduce the impact on functional balls in the inner area, with higher strain expected on the anchor balls area without concern for drive functionality. Functional balls exhibited less severe cracks,

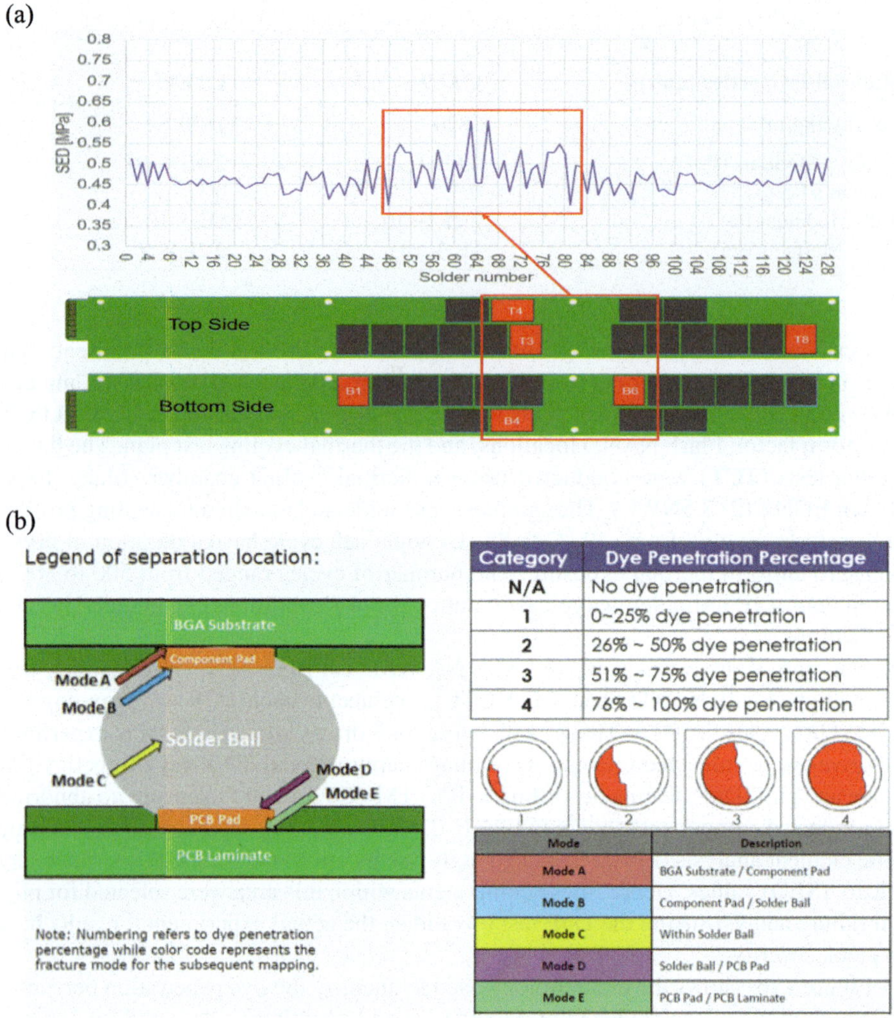

Fig. 1. (a) 3D structure simulation showed high risk location which determined by strain energy density (SED) value. (b) Crack Mode category based on separation location of solder ball and severity of crack.

indicating effective design strategies to protect critical connections. The comparison of CTE materials across TCT (T500, T600, T700, T800) reveals distinct patterns in ball crack severity and crack modes. For CTE9 and CTE11, the severity and complexity of cracks progressively increase from T500 to T800, with higher severity and a notable presence of green crack modes, indicating more severe and complex cracking as the thermal cycles advance. The observed slight difference in crack progression by visual inspection required further analysis using JMP statistical software.

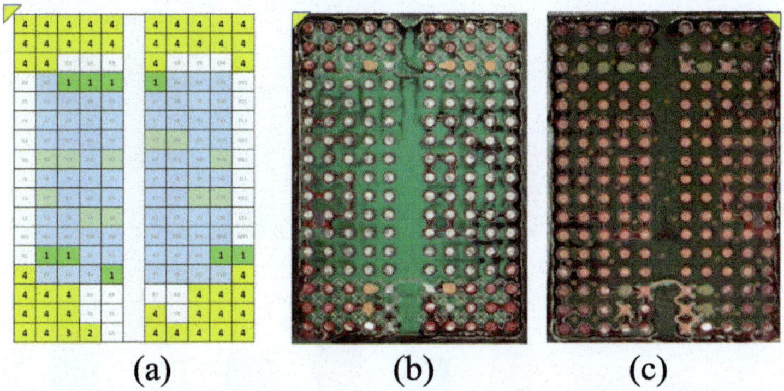

Fig. 2. One of DnP result (T800, B6) (a) is ball crack matrix 170BGA; (b) Dye penetration at BGA (PCBA side) and (c) Dye penetration at BGA (component side).

3.2 Statistical Analysis of Ball Crack Severity

Ball crack severity in Fig. 3(a) revealed a clear progression of ball cracks as the number of thermal cycles increased from T500 to T800 for both CTE materials. Thermal cycling is known to cause repeated expansion and contraction of materials, leading to mechanical stresses and strains that accumulate over time. Our findings support this understanding, showing progressive crack development in solder joints with increasing thermal cycles.

Next, we examined the component side effects, dividing them into two rows. The top row refers to three components on the bottom side (B1, B4, and B6), while the bottom row refers to three components on the top side (T3, T4, and T8), as shown in Fig. 3(a). The top side components, exhibit slight ball crack severity with CTE11 colored in blue compared to CTE9 colored in red. Conversely, the bottom side components show a similar trend only for T500 and T600; however, T700 and T800 reveal higher ball crack severity with CTE11. An additional box plot shown in Fig. 3(b) is supporting observation from all ball crack without segregate by number of thermal cycles. Further investigation on this paper to understand component side placement correlation with solder ball crack severity. It showed that the bottom and top sides have almost similar patterns of crack severity. The median crack severity values showed a reduction for CTE11 compared to CTE9 where this kind of trend clearly visible from Fig. 3(a), suggesting that CTE11 experiences less severe cracks compared to CTE9 due to a lower CTE mismatch with PCBA.

(a)

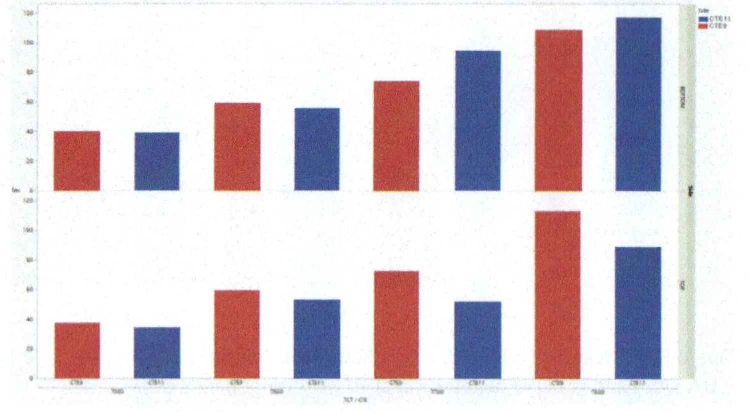

(b)

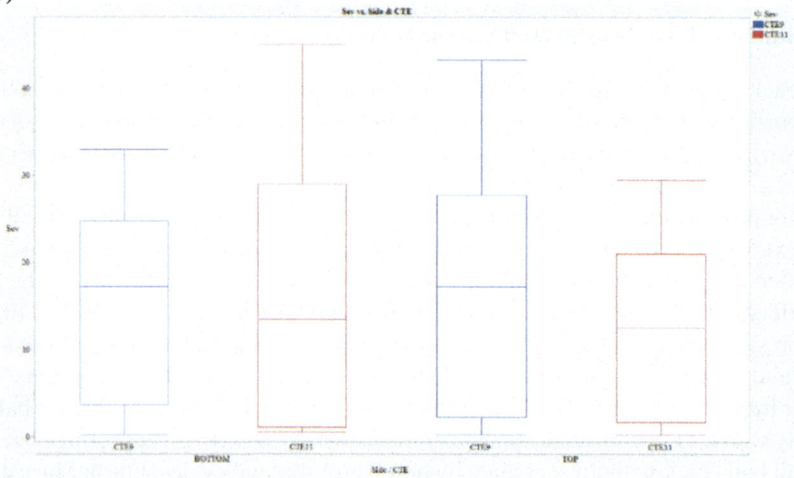

Fig. 3. (a) Ball Crack severity progression by Thermal cycling vs CTE based on component of location as showed in Fig. 1(a), (b): Overall ball crack severity vs CTE based on component side.

4 Conclusion

In conclusion, this study provides valuable insights into the role of EMC materials in influencing the reliability of lead-free solder joints in electronic packaging. The findings suggest that selecting EMC materials with lower CTE mismatch values can improve the durability of solder joints under thermal cycling, contributing to the development of more robust and reliable electronic devices. Future work could expand on this research by exploring additional EMC materials and varying thermal cycling conditions to further optimize electronic packaging reliability. On top of that, this study also highlighting the cumulative damage from repeated thermal stress where thermal cycling led to crack severity increased with the number of cycles.

Acknowledgments. The authors would like to acknowledge the financial support and research facilities provided by Western Digital via SanDisk Storage Malaysia Sdn. Bhd (research grant: RR-2023–007) and Universiti Kebangsaan Malaysia (research grant: TAP-K022551).

References

1. Al Ahsan, M., et al.: Effect of different thermal cycling profiles on the mechanical behavior of sac305 lead free solder. 2020 19th IEEE Intersoc. Conf. Therm. Thermomechanical Phenom. Electron. Syst. (ITherm), 1170–1179 (2020). https://doi.org/10.1109/itherm45881.2020.9190169
2. Roumanille, P., et al.: Evaluation of thermomechanical fatigue lifetime of BGA lead-free solder joints and impact of isothermal aging. Microelectron. Reliab. **126**, 114201 (2021). https://doi.org/10.1016/j.microrel.2021.114201
3. Ramli, M.I., et al.: Formation and growth of intermetallic compounds in lead-free solder joints: A Review. Materials **15**(4), 1451 (2022). https://doi.org/10.3390/ma15041451
4. Depiver, J.A., Mallik, S., Amalu, E.H.: (2021) "Thermal fatigue life of ball grid array (BGA) solder joints made from different alloy compositions." Eng. Fail. Anal. **125**, 105447 (2021). https://doi.org/10.1016/j.engfailanal.2021.105447
5. Jansen, K.M.B., et al.: Effect of temperature and humidity on moisture diffusion in an epoxy moulding compound material. Microelectron. Reliab. **107**, 113596 (2020). https://doi.org/10.1016/j.microrel.2020.113596
6. Jiang, N., et al.: Reliability issues of lead-free solder joints in electronic devices. Sci. Technol. Adv. Mater. **20**(1), 876–901 (2019). https://doi.org/10.1080/14686996.2019.1640072
7. Ismail, A.A., et al.: Cardinal and ordinal directions approach in investigating arrayed solder joints crack propagation of ball grid array semiconductor packages. IEEE Trans. Compon., Packag. Manuf. Technol. **12**(9), 1492–1501 (2022). https://doi.org/10.1109/tcpmt.2022.3198406
8. Ismail, A.A., et al.: Effects of underfill materials on behavior of intermetallic compound thickness of ball grid array solder joints using ANOVA. 2022 IEEE 39th Int. Electron. Manuf. Technol. Conf.e (IEMT) **47**, 1–5 (2022). https://doi.org/10.1109/iemt55343.2022.9969487
9. Ab Rahim, R.A., et al.: Effect of isothermal aging and copper substrate roughness on the SAC305 solder joint intermetallic layer growth of high temperature storage (HTS). Sains Malaysiana **49**(12), 3045–3054 (2020). https://doi.org/10.17576/jsm-2020-4912-16

Influence of Microwave Hybrid Heating on Soderability Performance of SAC305/Cu Solder Joints

Ezzatul Farhain Azmi[1,2,3], Abdullah Aziz Saad[1](✉), and Ahmad Azmin Mohamad[2]

[1] School of Mechanical Engineering, Universiti Sains Malaysia, Nibong Tebal 14300, Pulau Pinang, Malaysia
azizsaad@usm.my

[2] Energy Materials Research Group (EMRG), School of Materials and Mineral Resources Engineering, Universiti Sains Malaysia, Nibong Tebal 14300, Pulau Pinang, Malaysia
aam@usm.my

[3] Faculty of Mechanical Technology and Engineering, Universiti Teknikal Malaysia Melaka, Hang Tuah Jaya, 76100 Durian Tunggal, Melaka, Malaysia

Abstract. Solder joints are important in electronic assemblies as the quality affects the reliability of the electronic devices. Conventional reflow heating (CR) is widely used but presents challenges as suboptimal wettability characteristics. This encourages alternative heating methods like microwave hybrid heating (MHH). This study investigates the influence of MHH on wetting angle, spreading ratio, and microstructural properties of solder joint SAC305 on Cu substrate compared to the CR method. The findings reveal that MHH significantly enhances the wetting properties of SAC305 solder by reducing the wetting angle from 26.38° (CR) to 15.94° and enhancing the spreading ratio from 86.77 to 93.38%, suggesting improved solder wettability. Surface roughness analysis shows a smoother joint surface with an average roughness (Ra) of 0.953 μm for MHH, indicating more uniform and consistent soldering. These findings suggest that MHH provides valuable insights into the optimization of soldering processes in the electronics manufacturing industry, highlighting the advantages of MHH in achieving a more efficient and reliable soldering process.

Keywords: SAC305 · Microwave Hybrid Heating · Wetting Angle · Spreading Ratio

1 Introduction

Solder joints are essential in electronics as they influence the overall durability and reliability connectivity by providing mechanical and electrical continuity in electronic assemblies [1]. Lead (Pb)-free solders have been growing rapidly in the past decades replacing the Tin (Sn)-Pb as it harmful to the environment [2]. Sn-3.0Ag-0.5Cu (SAC305) emerges as a standout choice for many applications among the diverse range of lead-free solder alloys. SAC305 is a lead-free solder that adheres to environmental regulations, making

N. R. Abdul Razak et al. (Eds.): ECOOP 1996, SPPHY 418, pp. 52–58, 2025.
https://doi.org/10.1007/978-981-96-2871-1_8

it a sustainable and responsible choice for modern electronics manufacturing [3]. It has a low melting point, exceptional wetting properties, and mechanical strength, which have solidified its position as a reliable and versatile soldering material in a variety of applications [4].

However, optimizing the reflow method is essential to ensure the reliability and performance of solder joints in electronic assemblies. Microwave hybrid heating (MHH) shows a potential benefit as it offers more uniform and rapid heating [5, 6]. The conversion of electromagnetic energy from the microwave into heat energy happens directly, which offers advantages such as shortening the processing time in comparison to conventional heating (CR) [7]. Further, a comparison study between these two methods needs to be conducted to determine the solderability performance of the solder.

Wettability is a fundamental material property that describes the ability of a liquid to spread and make contact with a solid surface [8]. Wettability can be measured by the contact angle and spreading ratio. Contact angle (θ) in soldering refers to the angle formed at the interface between a molten solder droplet and the solid substrate on which it rests. The spreading ratio in solder refers to the ratio of the wetted area of the molten solder to the total area of the solder joint. It is a measure of how well the solder flows and wets the surface of the joint. The contact angle and spreading ratio are important for determining interconnection reliability between solder and substrate [9].

This study aims to investigate the influence of microwave hybrid heating on the wetting angle, spreading ratio, and surface observation of SAC305/Cu solder joints. By comparing these parameters with those obtained through conventional reflow heating, the study seeks to determine the potential advantages of MHH in enhancing solder joint quality.

2 Experimental Procedure

SAC305 (comprising 96.5 wt.% Sn, 3.0 wt.% Ag, and 0.5 wt.% Cu) is the solder alloy that used in this study. Copper (Cu) with dimension 2 × 3 cm is used as the substrate. The Cu substrates were cleaned with ethanol to remove any contaminant and ensure a clean surface for soldering. The SAC305 solder paste was then applied to the Cu substrates using a stencil printing method with a stencil hole diameter 0.5 cm and a thickness 0.1 cm.

The deposited SAC305 then reflow with two types of heating conditions. The first condition is the CR which was performed using Nitrogen Lead-free Reflow Oven T200N +. The reflow oven heated the deposited SAC305/Cu sample at a temperature of 250 °C for six minutes which allowed the complete reflow process [10].

The second condition is MHH using a Microwave Oven (Panasonic NN-ST34HM) with the assistance of thermal energy from a Silicon (Si) wafer as a susceptor. This is the reason the heating is called hybrid as the use of Si in the microwave. The system operates at a frequency of 2.45 GHz with a microwave power of 800W. The microwave heat is at medium mode settings for 80 s, which is the fastest time to reflow for the medium mode operating power. The MHH setup consists of SAC305/Cu placed on top of a Si wafer, followed by a ceramic pad and microwave glasses used as insulators. The use of a Si wafer susceptor in the MHH setup significantly reduced temperature fluctuations compared to setups without the susceptor [10], resulting in a more stable and uniform temperature profile under medium mode heating conditions.

The reflowed solder was then cross-sectioned, mounted, ground, and polished to capture the image using scanning electron microscope (SEM). The wetting angle and spreading ratio were determined by analyzing images using imageJ processing software. The formula used for the spreading ratio [11] is:

$$S = \frac{d - h}{d} \times 100\%$$

where S is the spreading ratio, d is the diameter and h is the height of the solder sample. Furthermore, the non-cross section sample of MHH analyzed the surface roughness value using a Confocal Laser Scanning Microscopic. The machine scans the surface and provides a detailed surface map and average surface roughness (Ra).

3 Results and Discussions

3.1 Analysis of Wetting Angle

The combined bar chart compares the wetting angle and spreading ratio of SAC305/Cu solder joints reflowed by CR heating for 6 min and MHH medium mode for 80 s (Fig. 1(a)). By CR method, the wetting angle of SAC305/Cu using CR method is 26.38° which is higher compared to MHH with 15.94°. This result suggests that MHH has a more effective wetting process. Lower angles correlate with better adhesion and stronger joints [12]. Furthermore, the result is supported by the spreading ratio. The spreading ratio for the MHH is higher at 93.38% compared to CR at 86.77%. The higher spreading ratio implies that the solder spreads well, which can lead to improve joint formation and reliability [13]. MHH provides more uniform heating, reducing temperature gradients and promoting better solder wetting. In conventional reflow, heat is transferred from the external environment to the solder through conduction and convection, which can lead to temperature gradients within the solder. These gradients can cause uneven melting, leading to potential reliability issues in the solder joints [14].

3.2 Microstructural Analysis

The surface image of the SAC305 reflowed by CR (Fig. 1(b)(i)) shows a smaller rounded solder compared to MHH (Fig. 1(c)(i)). The joint visible convex shape, which is consistent with a higher wetting angle of 10.44° compared to the MHH method. Additionally, the SEM image of the cross-section of the solder joint reflowed by MHH (Fig. 1(c)(ii)) displays a flatter morphology and a more spread-out joint. This result corresponds to the spread of the MHH, which is 6.61% higher than the CR method (Fig. 1(b)(ii)). These findings indicate that the MHH method could enhance the mechanical properties and reliability of the solder joint.

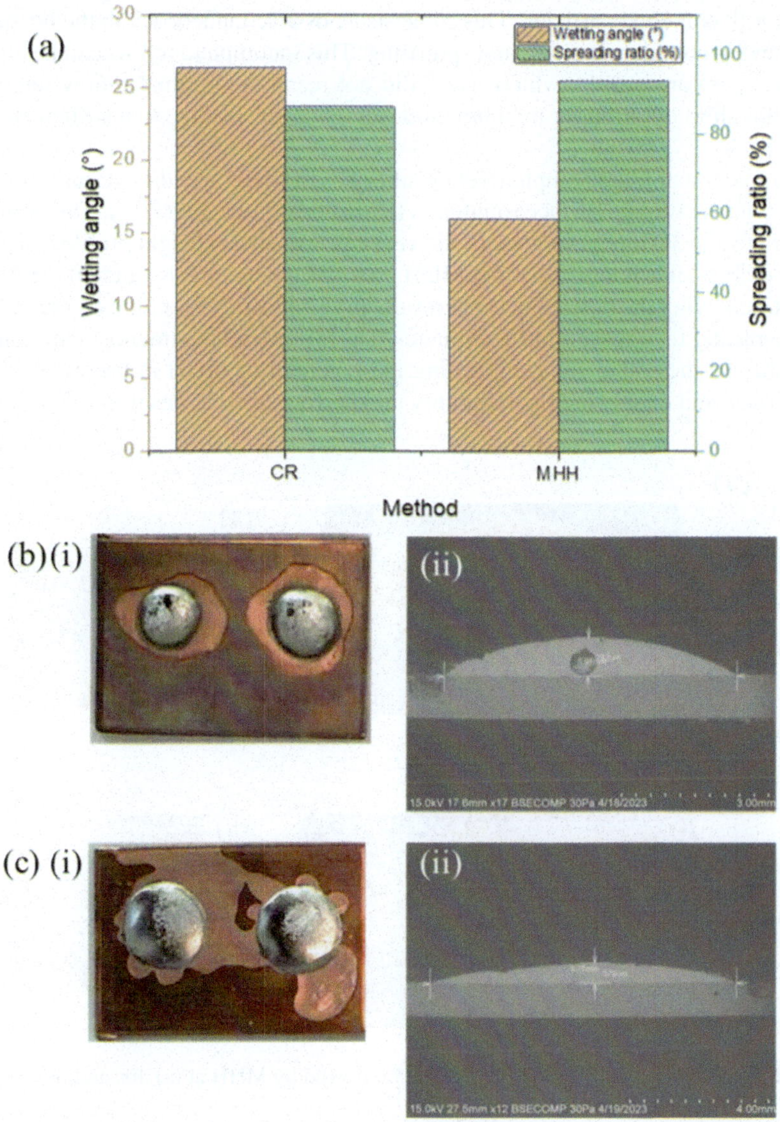

Fig. 1. (a) Wetting angle and spreading ratio graph of SAC305/Cu comparison between CR and MHH method, (b) images of SAC305 reflowed by CR; (i) its surface image and (ii) its microstructural image, and (c) images of SAC305 reflowed by MHH (i) its surface image and (ii) its microstructural image.

3.3 Surface Measurement Analysis

The diagram in Fig. 2 presents the surface morphology and roughness of the SAC305/Cu solder joints after exposure to MHH for periods of 40s and 80s. It is observed that MHH after 40s, the solder did not reflow completely (Fig. 2(a)), resulting in a rough, uneven surface with unmelted particles. This suggests a possible inadequacy in the heating time for achieving complete melting and spreading. This incomplete reflow can be attributed to the temperature profile, which likely did not reach the optimal reflow temperature within this short duration for medium mode temperature, leading to insufficient melting [15].

The solder exhibited complete reflow after 80s of MHH, yielding a smooth, uniform surface with an average surface roughness (Ra) of 0.953 μm (Fig. 2(b)). This was further validated by the 3D topography map, showcasing a consistent height profile and minimal surface defects. It is noteworthy that MHH provides a more precise and efficient heating method, achieving the desired thermal results in significantly less time compared to CR, which typically requires around six minutes. The 80-s MHH treatment notably improved the quality of the solder joint, indicating the potential of MHH in producing superior solder joints and enhancing the reliability and performance of electronic devices.

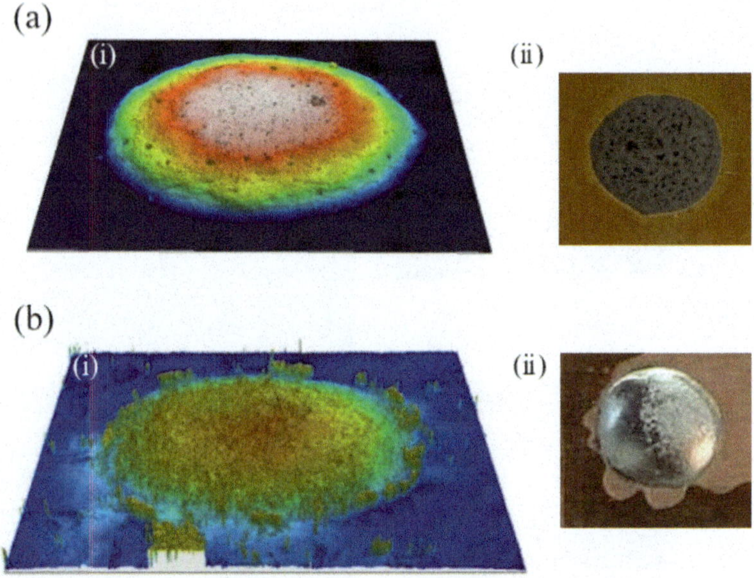

Fig. 2. Surface morphology of SAC305 reflowed by MHH at (a) 40s and (b) 80s.

4 Conclusion

The research highlights the potential benefits of utilizing MHH for soldering in electronic assemblies. According to the study, MHH could potentially create higher quality joints in a shorter time compared to conventional reflow heating methods, thereby potentially leading to more reliable electronic devices. The differences in joint formation between MHH and CR methods suggest that MHH produces flatter, more spread-out joints with better wetting properties, potentially resulting in stronger and more reliable joints. Additionally, the findings indicate that the uniform and rapid internal heating mechanism from

MHH improves the wetting properties and spreading behavior of the solder, resulting in stronger and more reliable solder joints compared to CR. Furthermore, the surface roughness of MHH-processed solder joints reveals low average surface roughness and uniform 3D topography, indicating that MHH effectively enhances the wettability and spreading of the solder, further suggesting its potential superiority over conventional reflow heating for solder joint formation in electronic assemblies.

Acknowledgements. The authors express their sincere gratitude for the support provided by the Ministry of Higher Education Malaysia through the Fundamental Research Grant Scheme (project code: FRGS/1/2022/TK10/USM/02/17) to the School of Mechanical Engineering at Universiti Sains Malaysia (USM) and the Academic Staff Training Scheme Program Scholarship. Additionally, Universiti Sains Malaysia and Universiti Teknikal Malaysia Melaka are gratefully acknowledged.

References

1. Tian, R., et al.: Brittle fracture of Sn-37Pb solder joints induced by enhanced intermetallic compound growth under extreme temperature changes. J. Mater. Process. Technol. **268**, 1–9 (2019)
2. Jiang, N., et al.: Reliability issues of lead-free solder joints in electronic devices. Sci. Technol. Adv. Mater. **20**(1), 876–901 (2019)
3. Pongvittayapanu, K., Srisrual, A., Fakpan, K.: Effect of thermal cycling and vibration on cracking in Sn-3.0 Ag-0.5 Cu solder bump. Mater. Today: Proc. **52**, 2372–2376 (2022)
4. Jang, J.-H., et al.: Microstructural evolution and mechanical properties of SAC305 with the intense pulsed light soldering process under high-temperature storage test. in 2021 IEEE 71st electronic components and technology conference (ECTC). IEEE (2021)
5. Lutfi, M., et al.: Metallurgical and shear strength of brazed bulk pure copper by microwave hybrid heating. in AIP Conference Proceedings. AIP Publishing (2023)
6. Bhatt, S.C., Ghetiya, N.D.: 3D Multiphysics simulation of microwave heating of bulk metals with parametric variations. Chem. Eng. Process.-Process. Intensif. **184**, 109271 (2023)
7. Said, M., et al.: Microwave hybrid heating for lead-free solder: a review. J. Mater. Res. Technol. (2023)
8. Jothi Prakash, C., Prasanth, R.: Approaches to design a surface with tunable wettability: a review on surface properties. J. Mater. Sci. **56**, 108–135 (2021)
9. Jung, D.-H., Jung, J.-P.: Review of the wettability of solder with a wetting balance test for recent advanced microelectronic packaging. Crit. Rev. Solid State Mater. Sci. **44**(4), 324–343 (2019)
10. Said, M., et al.: Effect of microwave operating power and reflow time on the microstructure and tensile properties of Sn–3.0 Ag–0.5 Cu/Cu solder joints. Solder. & Surf. Mt. Technol. **34**(1), 31–44 (2022)
11. Yang, W., et al.: Effect of graphene nanosheet addition on the wettability and mechanical properties of Sn-20Bi-xGNS/Cu solder joints. Materials **13**(18), 3968 (2020)
12. Sarkar, M., Hossain, R., Sahajwalla, V.: Unraveling the structural and morphological properties of thermally transformed microzonal carbon from waste hard rubber by conventional heating and microwave irradiation via single step selective thermal transformation. SSRN 4306039
13. Rajendran, S.H., Kang, H., Jung, J.P.: Ultrasonic-assisted dispersion of ZnO nanoparticles to Sn-Bi solder: a study on microstructure, spreading, and mechanical properties. J. Mater. Eng. Perform. **30**, 3167–3172 (2021)

14. Zhang, P., Xue, S., Wang, J.: New challenges of miniaturization of electronic devices: electromigration and thermomigration in lead-free solder joints. Mater. Des. **192**, 108726 (2020)
15. Jemaa, S.B., Gagnon, P., Sylvestre, J.: In situ measurement method for temperature profile optimization during thermocompression bonding process. IEEE Trans. Compon., Packag. Manuf. Technol. **10**(11), 1929–1937 (2020)

Microstructure Evolution and IMC Growth of SAC Solder with Addition of TiO_2 and Al_2O_3 Nanoparticle

Nur Haslinda Mohamed Muzni[1], Ervina Efzan Mhd Noor[1,2](✉), Mohd Mustafa Al-Bakri Abdullah[3], and Nurliyana Mohamad Arifin[2]

[1] Faculty of Engineering and Technology, Multimedia University, Ayer Keroh, 75450 Malacca, Malaysia
ervina.noor@mmu.edu.my

[2] Centre of Manufacturing, Engineering and Sustainability, Multimedia University, Ayer Keroh, 75450 Malacca, Malaysia

[3] Faculty of Chemical Engineering Technology, Universiti Malaysia Perlis, 02600 Arau, JejawiPerlis, Malaysia

Abstract. SAC305 nanocomposite solders were successfully produced with addition of different weight percentage (0.25, 0.50, 0.75 and 1.0 wt.%) of TiO_2 and Al_2O_3 nanoparticles. This study examined how the addition of TiO_2 and Al_2O_3 nanoparticles affects the microstructural changes during the growth of intermetallic compounds (IMCs) in the SAC305 solder alloy. The results show the addition of nanoparticles alter the morphology of IMC layer and restrict the formation of the IMC layer. The morphology of the IMC layer transitions from an elongated scallop-like shape to a rounded scallop-like shape, and subsequently evolves into a wicker shape with additional additions of TiO_2 and Al_2O_3 nanoparticles. The thickness of the IMC layer significantly decreases by 22.84% with the addition of 0.50 wt.% of TiO_2 and Al_2O_3 nanoparticles. Inhibition of intermetallic compound growth of SAC305 nanocomposite solder effectively resulted at 0.50 wt.% addition of TiO_2 and Al_2O_3 nanoparticles. However, excessive addition of weight percentage of TiO_2 and Al_2O_3 nanoparticles promotes the growth of IMC layer.

Keywords: Nanocomposite · nanoparticles · intermetallic compound

1 Introduction

Lead-free solder has gained widespread use in the electronic packaging industry due to the European Union's Restriction of Hazardous Substances (RoHS), which limited the use of toxic lead starting in 2006. In response to the lead restrictions, numerous studies have been conducted to identify the optimal lead-free solder alternative to traditional Sn-Pb solder, with the Sn-Ag-Cu (SAC) solder alloy emerging as a promising candidate [1]. Among SAC lead-free solders group, SAC305 was seen to be the best candidate that exhibits low melting point (217 °C), less expensive, its availability, good mechanical properties and solderability [2, 3].

N. R. Abdul Razak et al. (Eds.): ECOOP 1996, SPPHY 418, pp. 59–65, 2025.
https://doi.org/10.1007/978-981-96-2871-1_9

The high potential SAC lead-free solders were developed with the high interest and extensive attention towards the miniaturization of electronic device in electronic packaging industries [4]. Nevertheless, these devices have issues to completely meet all the requirements for a high-performance electronic device. High-performance electronic devices should exhibit excellent wettability, strong creep resistance, a low melting temperature, good electrical conductivity, and high strength. In response to this, efforts to develop high performance electronic device have been driven by considering the importance of solder joint reliability. High-performance SAC lead-free solders can be enhanced by incorporating nanoparticles to create nanocomposite lead-free solders.

Ceramic nanoparticles have been identified as promising reinforcement candidates, providing substantial improvements in the properties of lead-free solder alloys include microstructures, thermal behaviours, mechanical properties, and growth of IMCs [1, 5]. SAC lead-free solder reinforced with TiO_2 was found to refine the growth of Cu_6Sn_5 and Ag_3Sn IMC. The refined IMCs phases prove TiO_2 nanoparticles significantly restrict the growth of IMCs by acting as barrier between solder and substrate. Thickness and grain size of IMCs decreases significantly when reinforced with 0.10 wt.% of TiO_2 nanoparticles but further addition of TiO_2 nanoparticles could promotes growth of the IMC [6]. Additionally, the incorporation of Al_2O_3 nanoparticles into SAC lead-free solder effectively enhanced the solder's properties and inhibited the growth of the IMC layer [7].

A limited study on the effects of both TiO_2 and Al_2O_3 nanoparticles reinforced to SAC305 solder has been done. Therefore, the purpose of this paper is to investigate the impact of adding both TiO_2 and Al_2O_3 nanoparticles to SAC305 solder, resulting in SAC305 nanocomposite solder, on the microstructural evolution and intermetallic growth.

2 Experimental Procedure

The SAC305 solder was formulated with a composition of 96.5% Tin (Sn), 3.0% Silver (Ag), and 0.5% Copper (Cu) by weight percentage (wt.%). TiO_2 and Al_2O_3 nanoparticles, in powder form and averaging 25–50 nm in size, were added at varying weight percentages (0.25, 0.50, 0.75, and 1.0 wt.%) to the lead-free SAC305 solder. This mixture was heated in a graphite crucible at 250 °C for 1 h to create SAC305 nanocomposite solder. The resulting nanocomposite solder was then re-melted to ensure a homogeneous composition. Subsequently, the SAC305 nanocomposite solder alloy was shaped into billets with a diameter of 5 mm and a thickness of 1 mm using a punch. Copper plates measuring 10 x 10 x 0.01 mm served as substrates for this study. Both the SAC305 solder and the SAC305 nanocomposite solder alloy were reflowed for 60 s at a temperature of 250 °C to form solder joints. The prepared SAC305 nanocomposite alloy was then mounted in epoxy resin and cross-sectioned. The cross-sectional surfaces of the SAC305 nanocomposite alloy were ground with silicon carbide sandpaper (P2000 grit) to achieve a flat surface. Following this, a polishing pad was employed with 6, 3, and 1 μm diamond paste, along with alumina, to polish the alloy for observation of the solder joint between the solder and the Cu substrate. A Kemet FORCIPOL 2V Grinder and Polisher machine was utilized for grinding and polishing the samples. The microstructural analysis and measurement of the IMC layer thickness in the nanocomposite solders were conducted using an optical microscope, with measurements facilitated by VIS Pro software.

3 Results and Discussions

3.1 Microstructural Evolution of IMC

The microstructure of lead-free solders will be altered by the addition of nanoparticles, subsequently influencing the properties of the solders. The typical microstructures presented in SAC305 lead-free solder alloys are consist of primary β-Sn grains, eutectic network, and Cu_6Sn_5 IMC. The effect of adding TiO_2 and Al_2O_3 nanoparticles into SAC305 lead-free solder on the microstructure are investigated. Figure 1 shows the microstructure of non-reinforced SAC305 lead-free solder alloy by optical microscope. As illustrated in Fig. 1, The white area with a dendritic shape represents the Sn-rich (β-Sn) phase, while the dark area corresponds to the eutectic phase. It was noted that both the Sn-rich and eutectic phases exhibit a random distribution within the solder bulk.

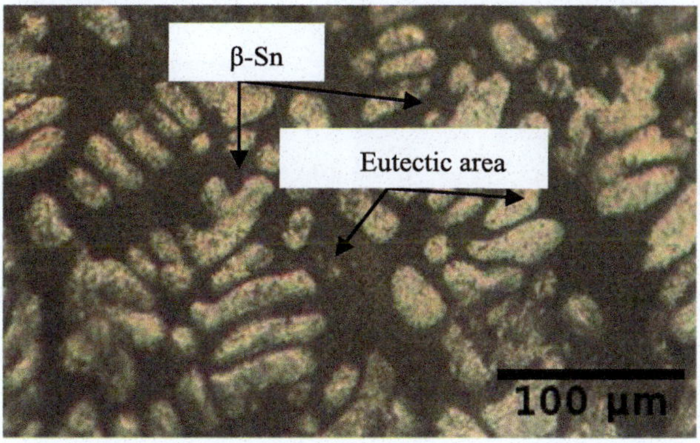

Fig. 1. Microstructure of SAC305 lead-free solder alloy on Cu substrate

Figures 2(a)-(d) illustrate the microstructure of the reinforced SAC305 lead-free solder alloy with the addition of 0.25, 0.50, 0.75, and 1.0 wt.% of TiO_2 and Al_2O_3 nanoparticles, respectively. The morphology of the β-Sn dendrites in the reinforced SAC305 lead-free solder alloy was observed to be smaller than in the non-reinforced SAC305 lead-free solder alloy. As the concentration of TiO_2 and Al_2O_3 nanoparticles increased from 0.25 to 0.50 wt.%, the number of β-Sn dendrites increased while their size decreased, and a reduction in the area of the eutectic phase was noted, as shown in Figs. 2(a) and 2(b). With further increases in TiO_2 and Al_2O_3 nanoparticles, the refinement of the β-Sn phase improved. However, at concentrations exceeding 0.50 wt.%, the size of the β-Sn phase increased slightly, though it remained smaller than that in the non-reinforced SAC305 lead-free solder matrix, as depicted in Figs. 2(c) and 2(d). This finding aligns with the observations of Tang et al. [8], who reported that the growth and size of Ag_3Sn grains in the solder matrix decreased with an increase in TiO_2 nanoparticles up to 0.1 wt.%, but beyond that concentration, both the size and spacing between Ag_3Sn grains slightly increased. This phenomenon was attributed to the agglomeration and segregation of high concentrations of TiO_2 and Al_2O_3 nanoparticles in the solder matrix [9]. The addition of TiO_2 and Al_2O_3 nanoparticles enhances the nucleation density of the second phase in the eutectic area during solidification. Furthermore, the coarsening and growth of the IMC layer can be controlled with the addition of TiO_2 and Al_2O_3 nanoparticles up to 0.50 wt.%, as illustrated in Figs. 2(a) and 2(b).

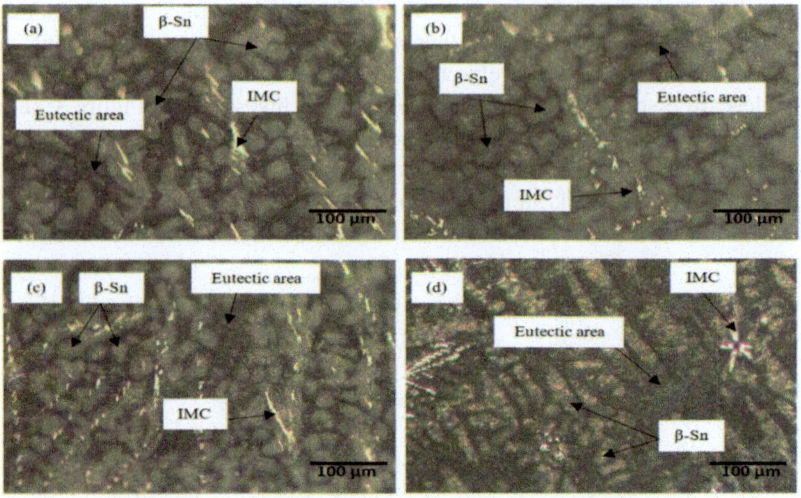

Fig. 2. Microstructure of SAC305 nanocomposite lead-free solder alloy on Cu substrate at different wt% of TiO_2 and Al_2O_3 nanoparticles (a) 0.25 wt%, (b) 0.50 wt%, (c) 0.75 wt%, and (d) 1.0 wt%

3.2 Intermetallic Compound Growth

Figures 3(a)-(e) display the cross-sectional optical micrographs of SAC305 lead-free solder and SAC305 nanocomposite solders with additions of 0.25 to 1.0 wt.% of TiO_2 and Al_2O_3 nanoparticles before the reflow soldering process. The SAC305 lead-free solder alloy shows an elongated scallop-like Cu_6Sn_5 IMC layer forming at the interface between the molten solder and the Cu substrate, as seen in Fig. 3(a). When TiO_2 and Al_2O_3 nanoparticles are added up to 0.50 wt.%, the morphology of the Cu_6Sn_5 IMC layer transitions from an elongated scallop-like shape to a rounded scallop-like shape, as illustrated in Figs. 3(b) and 3(c). However, with the further addition of TiO_2 and Al_2O_3 nanoparticles (1.0 wt.%), the Cu_6Sn_5 IMC layer evolves into a wicker shape, as shown in Fig. 3(e). This wicker-shaped Cu_6Sn_5 IMC layer may compromise the reliability of solder joints by inducing cracks and increasing the brittleness of the solder joints. The morphology of the Cu_6Sn_5 IMC layer in SAC305 nanocomposite lead-free solder joints is more uniform and evenly distributed compared to that in non-reinforced SAC305 lead-free solder joints. A similar phenomenon was reported by Tang et al. [8], noting that the morphology of the Cu_6Sn_5 IMC layer changes from elongated scallop-like to continuous rough scallop-like, eventually becoming wicker-shaped with increasing weight percentages of TiO_2. The Cu_6Sn_5 IMC layer forms at the solder/Cu interfaces due to the dissolution of Cu atoms from the Cu substrate into the molten solder during the soldering process [10].

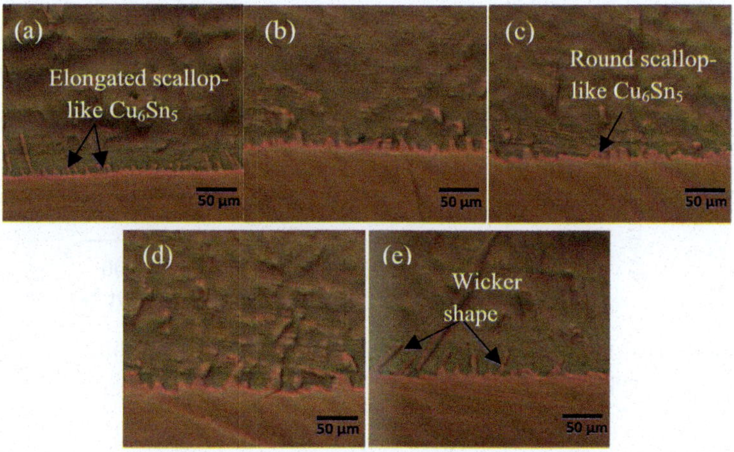

Fig. 3. Optical Micrograph of cross-sectional view of SAC305 nanocomposite solder at different weight percentage of TiO_2 and Al_2O_3 nanoparticles; (a) 0 wt.%, (b) 0.25 wt.%, (c) 0.50 wt.%, (d) 0.75 wt.% and (e) 1.0 wt.%

Figure 4 presents the thickness of the Cu_6Sn_5 IMC layer in SAC305 solder and SAC305 nanocomposite solder with varying weight percentages of TiO_2 and Al_2O_3 nanoparticles. The thickness of the Cu6Sn5 IMC layer decreases with the addition of 0.25 to 0.50 wt.% of these nanoparticles. Notably, the thickness of the IMC layer significantly decreases by 22.84% with the addition of 0.50 wt.% of TiO_2 and Al_2O_3 nanoparticles. However, as the weight percentage of TiO_2 and Al_2O_3 nanoparticles increases from 0.75 to 1.0 wt.%, the thickness of the IMC layer begins to increase. The thickest IMC layer is observed at the highest concentration of 1.0 wt.% of TiO_2 and Al_2O_3 nanoparticles. This suggests that the nanocomposite solder reacts more actively with the Cu substrate when the weight percentage of the nanoparticles exceeds 0.50 wt.%. Despite this increase, the IMC thickness in the SAC305 nanocomposite solder remains thinner than that in the SAC305 solder, even with the addition of excessive nanoparticles. This indicates that TiO_2 and Al_2O_3 nanoparticles effectively inhibit IMC growth, with the most significant effect occurring at weight percentages up to 0.50 wt.%. The inhibition of IMC growth can be explained through the adsorption theory of surface-active TiO_2 and Al_2O_3 materials, as well as the heterogeneous nucleation theory of IMCs. The presence of surface-active nanoparticles increases the number of adsorbing particles at the interface between the molten solder and the Cu substrate, thereby reducing the surface energy of the Cu_6Sn_5 IMC and inhibiting its growth rate [11]. The TiO_2 and Al_2O_3 nanoparticles serve as reinforcing agents and diffusion barriers, effectively hindering the dissolution of Cu atoms into the solder matrix and the formation of the IMC layer. TiO_2 and Al_2O_3

nanoparticles addition provide more nucleation sites, increase nucleation rate and refines Cu_6Sn_5 IMC grain and thus reducing grain ripening effect and retard the growth of IMC [12]. This confirms TiO_2 and Al_2O_3 nanoparticles effectively can inhibit the growth rate of IMC.

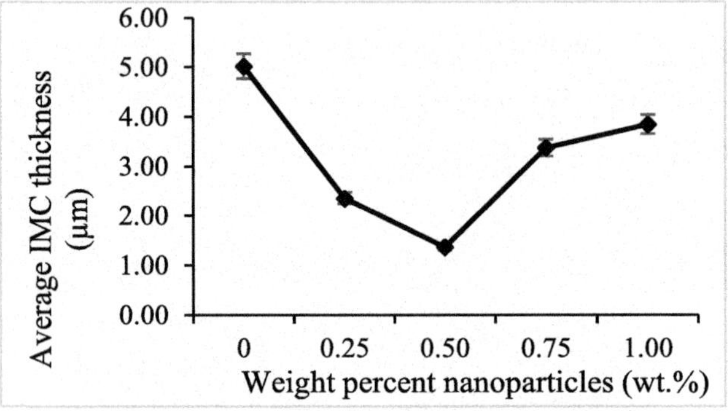

Fig. 4. Average IMC thickness of SAC305 nanocomposite solder with different weight percent TiO_2 and Al_2O_3 nanoparticles

4 Conclusion

This study investigated the impact of adding TiO_2 and Al_2O_3 nanoparticles on the microstructural evolution and IMC growth in SAC305 nanocomposite solder. It can be concluded that the addition of these nanoparticles alters the morphology of the Cu_6Sn_5 IMC layer from an elongated scallop-like shape to a rounded scallop-like shape when the concentration reaches 0.50 wt.%. With further addition to 1.0 wt.%, the Cu_6Sn_5 IMC layer evolves into a wicker shape. The IMC layer becomes thinner with addition of TiO_2 and Al_2O_3 nanoparticles. Significant reduction of IMC layer thickness is 22.84% with addition of 0.50 wt.% TiO_2 and Al_2O_3 nanoparticles. However, a further addition of TiO_2 and Al_2O_3 nanoparticles weight percent will increase the thickness and growth rate of IMC layer. TiO_2 and Al_2O_3 nanoparticles actively act as reinforcement and effectively inhibits the formation of IMC layer by the theory of adsorption and heterogeneous nucleation of IMCs. During reflow soldering process, TiO_2 and Al_2O_3 nanoparticles disperse and dissipate at the interface between solder/Cu. The number of adsorbing particles maximized and reduces the surface energy of Cu_6Sn_5 IMC and also increase nucleation rate with more nucleation sites. Besides, the Cu_6Sn_5 IMC grain refines resulting in grain ripening effect and thus the growth of IMC can be inhibited.

Acknowledgments. The authors would like to acknowledge Ministry of Higher Education Malaysia on the financial support under FRGS grant reference no: FRGS/1/2018/TKMUE180043.

References

1. Xiong, M.Y., Zhang, L.: Interface reaction and intermetallic compound growth behavior of Sn-Ag-Cu lead-free solder joints on different substrates in electronic packaging. J. Mater. Sci. **54**, 1741–1768 (2019)
2. Ervina, M.N.E., Siti, I.N.: Effect of different aging times on Sn-Ag-Cu solder alloy. Transaction on Electrical and Electronic Materials **16**(3), 112–116 (2015)
3. Hongyu, Q., Xiaowu, H., Shuang, L., Yongqiang, W., Qinglin, L.: Shear strength and fracture surface analysis of lead-free solder joints with high fraction of IMCs. Vacuum **180**, 109611 (2020)
4. Jiang, N., Zhang, L., Liu, Z.Q., Sun, L., Long, W.M., He, P., Xiong, M.Y., Zhao, M.: Reliability issues of lead-free solder joints in electronic devices. Sci. Technol. Adv. Mater. **20**, 876–901 (2019)
5. Fathian, Z., Maleki, A., Niroumand, B.: Synthesis and characterization of ceramic nanoparticles reinforced lead-free solder. Ceram. Int. **43**, 5302–5310 (2017)
6. Al-sorory, H., Gumaan, M.S., Shalaby, R.M.: Effect of TiO_2 nanoparticles on the microstructure, mechanical and thermal properties of rapid quenching SAC355 lead-free solder alloy. Soldering & Surface Mount Technology **35**, 18–27 (2023)
7. Tikale, S., Prabhu, K.N.: Development of low-silver content SAC0307 solder alloy with Al_2O_3 nanoparticles. Mater. Sci. Eng., A **787**, 139439 (2020)
8. Tang, Y., Li, G.Y., Pan, Y.C.: Influence of TiO_2 nanoparticles on IMC growth in Sn–3.0Ag–0.5Cu–xTiO_2 solder joints in reflow process. Journal of Alloys and Compounds, **554**, 195–203 (2013)
9. Tan, A.T., Tan, A.W., Farazila, Y.: Influence of nanoparticle addition on the formation and growth of intermetallic compounds (IMCs) in Cu/Sn–Ag–Cu/Cu solder joint during different thermal conditions. Sci. Technol. Adv. Mater. **16**, 033505 (2015)
10. Lee, L.M., Habsah, H., Ahmad, M. A.: Interfacial reaction of a Sn-3.0Ag-0.5Cu thin film during solder reflow. Soldering & Surface Mount Technology **25**(1), 15–23 (2013)
11. Nur, M.M.H., Ervina, M.N.E., Mohd, A.A.M.: Influence of TiO_2 and Al_2O_3 nanoparticle addition on thermal, wettability and IMC growth of Sn-305Ag-0.5Cu lead-free solder at different reflow temperature. Journal of Advanced Research in Fluid Mechanics and Thermal Sciences **109**(1), 188–195 (2023)
12. Tikale, S., Prabhu, K.N.: Effect of multiple reflow cycles and Al_2O_3 nanoparticles reinforcement on performance of SAC305 lead-free solder alloy. J. Mater. Eng. Perform. **27**, 3102–3111 (2023)

Solution Hardening Effect of Bi Additions to the β-In_3Sn Phase in In-Sn-Bi Low Temperature Solder Alloys via Alloy Design and Heat Treatment

Jiye Zhou, Xin Fu Tan, Stuart D. McDonald, and Kazuhiro Nogita(✉)

Nihon Superior Centre for the Manufacture of Electronic Materials, School of Mechanical and Mining Engineering, University of Queensland, St. Lucia, QLD 4072, Australia
k.nogita@uq.edu.au

Abstract. The binary In-Sn low temperature solder alloys with a majority of β In-rich phase in the microstructure exhibit exceptional electrical conductivity and ductility. However, at room temperature or above, they require higher strength to match the performance of common solder alloys. This study investigates the potential of Bi to be used as an alloying addition to optimise the strength of a single phase β-In_3Sn alloy through solution hardening. The alloy selected, based on microstructural design using CALPHAD methods, was In-13wt%Sn-13wt%Bi which is characterised by a 100% β-In_3Sn(Bi) phase being present in the microstructure above 55 °C. The microstructure was examined using scanning electron microscopy (SEM) and energy dispersive X-ray (EDX) spectrum analysis before and after annealing. The mechanical properties of the alloy before and after annealing were characterised using Vicker's hardness and compared with the hardness of In-30wt%Sn and In-25wt%Sn binary alloys.

Keywords: Low temperature solder · Solution hardening · Heat treatment · In-Sn-Bi alloy · Microhardness

1 Introduction

Solder alloys play a crucial role in ensuring the durability and reliability of electronic device connections by acting as the buffer between the Cu substrate and the components. Compared to conventional lead-free solder alloys like Sn-Ag-Cu and Sn-Cu alloys, indium-based solder alloys stand out for their low melting points, good ductility and excellent wetting properties, making them ideal for delicate soldering processes that require minimizing the thermal stress between substrates and components, as well as those involving thermally sensitive electronic devices [1–3]. However, In-Sn binary alloys with higher indium content often contain a low strength β phase (tetragonal crystal structure space group I_4/mmm) in the microstructure which contributes to their ductility and good wetting behavior [4, 5]. It has been reported that In-Sn binary alloys with a 100% β phase in the microstructure have lower Vickers hardness of around 2.6HV and

N. R. Abdul Razak et al. (Eds.): ECOOP 1996, SPPHY 418, pp. 66–72, 2025.
https://doi.org/10.1007/978-981-96-2871-1_10

ultimate tensile strength around 5MPa [6] compared to 30MPa to 80MPa reported in conventional lead-free solders [6–8]. The relatively lower strength of these In-Sn binary solder alloys makes them mechanically weaker in providing support for electronic components compared to other solders and less suitable for applications where mechanical loads or stresses are significant. Recently, the application of indium-based solder alloys has been typically limited to flexible devices and some low-temperature applications [9, 10], rather than applications subjected to mechanical stress.

Solid solution hardening represents a reliable strategy for enhancing the mechanical strength of certain phases through the addition of one or more alloying elements without significantly compromising other properties, such as ductility and electrical conductivity [11, 12]. Solid solution hardening can provide a barrier to dislocation motion by introducing lattice distortions which are caused by the dissolved atoms. Furthermore, solid solution hardening can be more stable compared to precipitation or particle hardening after thermal cycling or at higher temperatures [13], as it has been proven that the coarsening of Ag_3Sn particles in SAC solder alloys with thermal aging can cause the loss of hardening effect [14, 15]. The solid solution hardening of β-Sn in lead-free solder alloys has been achieved by adding solute atoms such as Bi or Sb which have extended solubility in Sn with increasing temperature [12, 16, 17]. These solute atoms are effective in impeding the movement of dislocations and thus improving the mechanical properties of a solder alloy [11]. Studies found that a commercial Sn-0.7Cu-0.05Ni (all compositions are in wt% unless specified otherwise) alloy was strengthened after adding 1.5 wt% Bi and that amount of Bi remained in solution in the Sn phase and increased the lattice parameters of Sn [16]. Another study showing that an addition of 2 wt% Bi to eutectic Sn-Ag solder increased the tensile strength by approximately 40% [18]. However, Bi and Sb have a temperature dependent solubility limit in Sn and alloying beyond the solubility limit can result in the formation of a second phase which result in an unstable strengthening effect due to the dissolution or precipitation associated with temperature changes [13, 15]. Thus, understanding the solubility of trace elements over a temperature range of interest is important when considering the solid solution strengthening effect.

In this paper, the expected microstructure of the selected composition was manipulated based on the solubility changes at variable temperatures. Through comparing the Vickers hardness between samples with different microstructures, the authors investigate the solid solution hardening effect of the β phase in indium-tin solder alloys associated with Bi solute atoms. This study of solid solution strengthening in indium-based solders offers a new strategy to overcome the limitations imposed by the soft β phase, enabling these alloys to achieve higher strength at room temperature and above, thereby widening their applicability in electronics manufacturing.

2 Experimental Procedure

The In-13wt%Sn-13wt%Bi alloy composition was selected using a CALPHAD method. The composition exhibits a microstructure of $BiIn_2(Sn)$ and β-phase at room temperature, which transforms to a pure β-phase at elevated temperatures. Therefore, this alloy provides a model to help investigate both precipitation hardening by $BiIn_2$ and solid solution hardening by Bi atoms within the β-phase in the In-Sn-Bi alloy system. To make the

In-13wt%Sn-13wt%Bi sample, appropriate quantities of pure indium (99.995%, supplied by Nihon Superior), pure tin and pure bismuth (99.99%, supplied by Northern Smelters Pty Ltd., Australia) ingots were alloyed and melted using graphite crucibles in an electric resistance furnace at temperatures approximately 220 °C above the equilibrium melting point for one hour. Bulk samples were remelted at 200 °C and solidified between two glasses with a spacer to make a sheet sample with a thickness of approximately 0.8mm. To achieve a near 100% β phase in the microstructure, a piece of sheet sample was cut and aged for 1 h at 55 °C in the oven.

As the dissolution and reprecipitation of $BiIn_2$(Sn) is time-temperature dependent, the timeframe of the whole experimental procedure is summarised in Fig. 1(a). The microstructural analysis used an in-situ heating scanning electron microscope (SEM) (Hitachi SU3500-A) fitted with a Deben enhanced cooling stage (−25 °C to + 160 °C temperature range with a calibrated ± 1 °C error). An ion milling machine (Hitachi IM 4000 Plus Ion mill) with a 2kV acceleration voltage, was used to obtain a clean surface finish on the samples. Each sample underwent ion milling with a cycle of 10 s beam on and 20 s beam off, repeated for a total of 3 min. The microstructure images, EDX mappings and spectrum analysis of the prepared samples were obtained under variable pressure (VP) backscatter electron (BSE) mode at a 15kV accelerating voltage with a 50 nm spot size. The aged sample was further cooled to −20 °C to observe the reprecipitation of $BiIn_2$. The Vickers hardness (HV) test was performed approximately one hour after solidification and immediately after heat treatment using a Leco Vickers MicroHardness Machine with a 10gf (0.098N) test load and 5s dwell time. The reported results are the average of ten individual indents on each sample.

3 Results and Discussions

3.1 Solidification and Microstructural Characterisation

Figure 1(b) shows the predicted volume fraction change of each phase within the microstructure of an In13Sn13Bi alloy during solidification. Based on the CALPHAD calculation, a 100% β phase microstructure can be obtained in the temperature range of 55 °C to 85 °C, while $BiIn_2$(Sn) starts to precipitate from the β phase at temperatures below 50 °C with nearly 20% $BiIn_2$(Sn) exhibited in the microstructure at room temperature.

As shown in Fig. 2, the microstructures of the as-cast and heat treated In13Sn13Bi samples were analyzed by SEM and EDX. The as-cast sample (Fig. 2(a) and (b)) have needle-like $BiIn_2$(Sn) precipitates (bright) either along the grain boundaries of the primary β phase (dark) or within the β phase which was confirmed by EDX analysis. EDX point analyses of three $BiIn_2$(Sn) sites in this as-cast In13Sn13Bi sample show a Sn concentration ranging from 6.3wt% to 8.9wt%. The overall volume fraction of $BiIn_2$(Sn) and β phase is consistent with the Thermo-Calc prediction shown in Fig. 2. The expected microstructure of almost 100% β phase was achieved through heat treatment, and a small amount of $BiIn_2$(Sn) precipitates can be found in the microstructure (see Fig. 2(d)) after cooling. Furthermore, stronger reprecipitation and larger needle like $BiIn_2$(Sn) precipitates were observed at grain boundaries or within the β phase when the temperature decreased to −20 °C (see Fig. 2(e) and (f)). By comparing Fig. 2(c) and (e), it is evident that the reprecipitation of $BiIn_2(Sn)$ from the β phase occurs more rapidly at lower temperatures than at room temperature. As such, we conclude that the β phase with non-equilibrium Bi compositions starts approaching thermodynamic equilibrium immediately after heat treatment or after solidification in In13Sn13Bi alloy by precipitation of the Bi-rich $BiIn_2$ phase, and lower temperatures can accelerate this reprecipitation process.

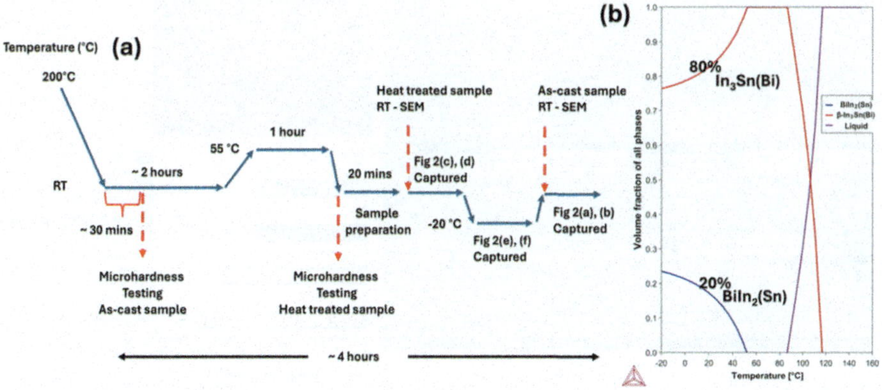

Fig. 1. (a) Timeline of the experimental procedure. (b) Predicted equilibrium volume fraction of phases during solidification in In13Sn13Bi alloy based on Thermo-Calc 2024a [19] database (TCSLD4: Solder Alloy v4.1).

3.2 Vickers Hardness

The Vickers hardness testing of In13Sn13Bi samples was done before and directly after heat treatment and the results are presented in Fig. 2(g). Compared to the In-25Sn and In-30Sn binary alloys which consist of 100% β phase within their microstructure, both In13Sn13Bi samples exhibit higher hardness. This is attributed to either the second phase strengthening effect by the harder hexagonal $BiIn_2(Sn)$ phase or solid solution hardening due to the presence of Bi solute atoms within the β phase. After heat treatment, the sample exhibits an increase in hardness compared to the as-cast sample. As the hardness testing was performed immediately after heat treatment, coupled with the microstructural analysis shown in Fig. 2(c) and (d), the hardening effect of heat-treated sample is expected to be dominated by solid solution strengthening, though a small amount of $BiIn_2(Sn)$ may begin to reprecipitate in the microstructure after heat treatment despite not being observed in SEM. As such, a microstructure that contains mostly β phase enriched with dissolved $BiIn_2(Sn)$ exhibits a higher hardness compared to a microstructure that consists of both normal $BiIn_2(Sn)$ + β phase or a microstructure comprising 100% β-In_3Sn. In the future, further characterization, such as tensile testing of bulk-cast samples with varying Bi concentrations, may provide a better understanding of the strength and ductility of In-Sn-Bi alloys.

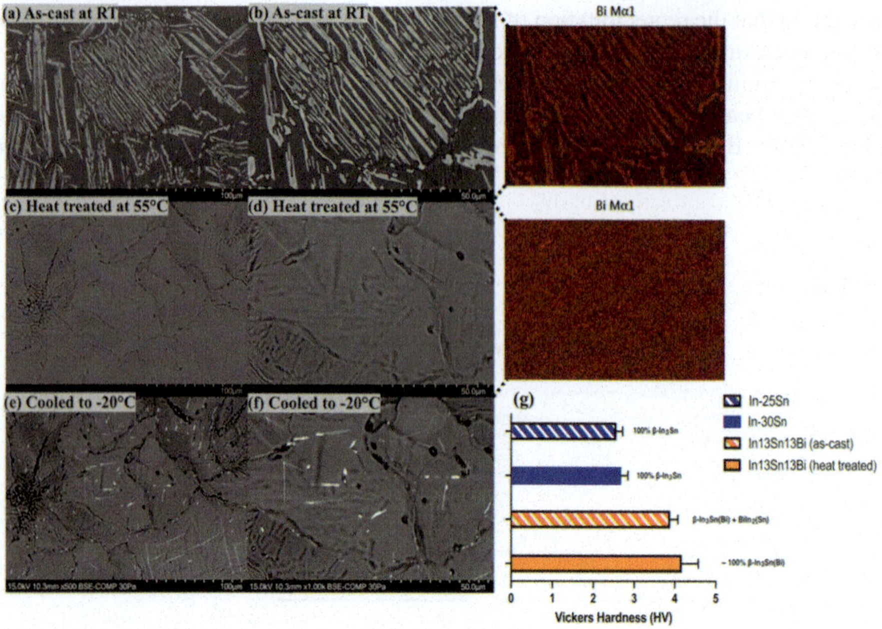

Fig. 2. SEM images of In13Sn13Bi (a) and (b) before heat treatment, captured at room temperature under different magnifications with EDX mapping of Bi; (c) and (d) after heat treatment at 55 °C, captured at room temperature under different magnifications with EDX mapping of Bi element; (e) and (f) heat treated and cooled in-situ to −20 °C, captured at −20 °C under different magnifications. (g) Vickers hardness of as-cast and heat-treated In13Sn13Bi samples compared with In-Sn binary alloys with 100% β phase in the microstructure [5].

4 Conclusion

Binary In-Sn alloys that contain 100% β phase have limited strength and an experimental alloy that consists of 100% β phase after solidification but has substantial $BiIn_2(Sn)$ precipitate during cooling to room temperature was examined as a pathway to improve the mechanical performance. The $BiIn_2(Sn)$ precipitates have a needle like morphology from SEM images or may be plate-like in 3 dimensions and the reprecipitation of $BiIn_2(Sn)$ from the β phase in In13Sn13Bi is faster at lower temperatures compared to room temperature. In addition to the $BiIn_2(Sn)$ phase, Bi was also present in solid solution within the β phase. In In13Sn13Bi, the $BiIn_2(Sn)$ phase is fully dissolved during heat treatment at 55 °C. The Vickers hardness of the resultant near-100% β alloy that is supersaturated with Bi solute, significantly increases to nearly double that of In-Sn binary alloys which are entirely composed of Bi-free β phase. This study suggests that with the addition of Bi in In-Sn binary alloys, the microhardness has been improved compared to In-Sn binary alloys with similar indium concentration. By increasing the concentration of Bi in the alloy to beyond 13% and enhancing the dissolution of Bi in the β phase, the strength of the indium-based solder alloys may be further improved. However, the introduction of excessive Bi could lead to the formation of brittle intermetallics, such as

BiIn, leading to phase segregation or embrittlement, limiting the overall performance of the solder alloy.

Acknowledgments. We wish to acknowledge 1. The Centre for Microscopy and Microanalysis at The University of Queensland for providing instruments for experiment and sample preparation; and Dr Jordan Poitras for assisting with the SEM cooling stage installation. 2. Nihon Superior Company Ltd for their support of this research and the supply of the indium ingots. 3. Funding: This research was funded by the Australian Research Council (ARC), Australia (LP210301248) and The University of Queensland Research Training Program Stipend.

References

1. Dalton, E., Ren, G., Punch, J., Collins, M.N.: Accelerated temperature cycling induced strain and failure behaviour for BGA assemblies of third generation high Ag content Pb-free solder alloys. Mater. Des. **154**, 184–191 (2018)
2. Jiye, Z., Xin, F.T, Qinfen, Gu., Stuart, D., McDonald and Kazuhiro Nogita.: The temperature-dependent phase transformation and microstructural characterisation in In-Sn solder alloys JOM (2023)
3. Chaowasakoo, T., Ng, T.H., Songninluck, J., Stern, M.B., Ankireddi, S.: Indium solder as a thermal interface material using fluxless bonding technology, 2009 25th Annual IEEE Semiconductor Thermal Measurement and Management Symposium, pp. 180–185 (2009)
4. Chang, M.S., Salleh, M.A.A.M., Halin, D.S.C., Ramli, M.I.I., Yasuda, H., Nogita, K.: The effect of Ni on the growth morphology of primary β-phase in an In-35 wt%Sn alloy. J. Alloy. Compd. **897**, 163172 (2022)
5. Zhou, J., Tan, X.F., McDonald, S.D., Nogita, K.: Mechanical properties and microstructure of binary In-Sn Alloys for flexible low temperature electronic joints, materials (2022)
6. Hao, Q., Tan, X.F., Gu, Q., Sweatman, K., McDonald, S.D., Nogita, K.: The effects of temperature and solute diffusion on volume change in Sn-Bi solder alloys. JOM **74**(4), 1739–1750 (2022)
7. Pang, J.H.L., Xiong, B.S.: Mechanical Properties for 95.5Sn-3.8Ag-0.7Cu Lead-Free Solder Alloy. IEEE Trans. Compon. Packag. Technol. **28**(4), 830–840 (2005)
8. Darveaux, R., Reichman, C.: Proceedings 57th Electronic Components and technology conference, IEEE, NEW YORK, pp. 695–706 (2007)
9. Reed, R.P., McCowan, C.N., Walsh, R.P., Delgado, L.A., McColskey, J.D.: Tensile strength and ductility of indium, Materials science & engineering. A. Struct. Mater.: Prop., Microstruct. Process. **102**(2) (1988) 227–236
10. Nah, J.W., Ren, F., Tu, K.N., Venk, S., Camara, G.: Electromigration in Pb-free flip chip solder joints on flexible substrates. J. Appl. Phys. **99**(2), 023520 (2006)
11. Callister, W.D., Rethwisch, D.G.: Fundamentals of materials science and engineering: an integrated approach: an integrated approach, Wiley Global Education (2011)
12. Li, G.Y., Chen, B.L., Shi, X.Q., Wong, S.C.K., Wang, Z.F.: Effects of Sb addition on tensile strength of Sn–3.5Ag–0.7Cu solder alloy and joint. Thin Solid Films **504**(1) (2006) 421–425
13. Zhou, J., Tan, X.F., McDonald, S.D., Sweatman, K., Akaiwa, T., Nogita, K.: Thermal stability of Bi in Sn-Cu based high strength Pb-Free solder alloys. Int. Conf. Electron. Packag. (ICEP) **2024**, 81–82 (2024)
14. Kumar, P., et al.: Microstructurally adaptive model for primary and secondary creep of Sn-Ag-based solders. IEEE Trans. Compon., Packag. Manuf. Technol. **2**(2), 256–265 (2012)

15. Liu, S., McDonald, S., Sweatman, K., Nogita, K.: The effects of precipitation strengthening and solid solution strengthening on strain rate sensitivity of lead-free solders: Review. Microelectron. Reliab. **84**, 170–180 (2018)
16. Nogita, K., et al.: Effects of Bi in Sn-Cu based lead-free solder alloys and interconnects. Int. Conf. Electron. Packag. (ICEP) **2017**, 377–380 (2017)
17. Ye, X., Tao, L., McDonald, S.D., Tan, X.F., Sweatman, K., Nogita, K.: The effects of solution treatment and room temperature ageing on mechanical properties of Sn-37wt%Bi and Sn-57wt%Bi. International Conference on Electronics Packaging (ICEP) **2023**, 215–216 (2023)
18. Kariya, Y., Otsuka, M.: Effect of bismuth on the isothermal fatigue properties of Sn-3.5mass%Ag solder alloy. J. Electron. Mater. **27**(7) (1998) 866–870
19. Andersson, J.O., Helander, T., Höglund, L., Shi, P., Sundman, B.: Thermo-Calc & DICTRA, computational tools for materials science. Calphad **26**(2), 273–312 (2002)

Temperature Dependent Crystallographic Changes in Hypoeutectic Sn-Bi Alloys

Qichao Hao, Xin F. Tan, Stuart D. McDonald(✉), and Kazuhiro Nogita

Nihon Superior Centre for the Manufacture of Electronic Materials, School of Mechanical and Mining Engineering, The University of Queensland, St. Lucia, QLD 4072, Australia
s.mcdonald1@uq.edu.au

Abstract. Hypoeutectic Sn-Bi solder alloys have melting points in the range of 139–232 °C, dependent on the concentration of Bi in the alloy. Typical solidification structures of these alloys are characterised by Sn-rich dendrites surrounded by the lower melting point eutectic Sn-Bi phases. Furthermore, as the solubility of Bi in Sn varies significantly with temperature, depending on the composition, Bi can also be present in the form of precipitates within the Sn-rich phases. The present work uses in-situ synchrotron powder X-ray diffraction to examine how the lattice parameters of Sn and Bi, present in a multicomponent solder alloy based on the hypoeutectic Sn-Bi system, vary in the temperature range of −100 °C to 180 °C. For comparative purposes, the lattice parameters of pure Sn and pure Bi were characterised in the same temperature range. The results confirm the significant influence of Bi dissolution/precipitation that occurs with temperature variations on the lattice parameters of Sn and highlights complexities associated with the multi-length scale microstructure of Sn-Bi alloys.

Keywords: Sn-Bi alloys · Low Temperature Solder · Synchrotron powder X-ray diffraction · Electronics Packaging

1 Introduction

In recent years there has been an increased focus on Sn-Bi alloys for implementation as solders in electronic assemblies. The low melting point of the Sn-Bi eutectic, at 139 °C [1], can minimise soldering temperatures, with associated benefits including reduced package warping and energy consumption. Bi is also considered non-toxic and is relatively low cost compared to other elements often present in solder alloys. However, despite Sn-Bi alloys having a relatively simple binary eutectic system, they display a complex multi-scale microstructure, dependent on composition, temperature and processing history. In this research we characterize the solidification sequence and mechanical properties of a multicomponent Sn-Bi alloy and compare them to those of pure Bi and Sn. In-situ synchrotron powder X-ray diffraction (PXRD) was performed over a large temperature range to investigate the behaviour of the Sn and Bi phases in the alloy, compared to those of the pure metals.

N. R. Abdul Razak et al. (Eds.): ECOOP 1996, SPPHY 418, pp. 73–79, 2025.
https://doi.org/10.1007/978-981-96-2871-1_11

2 Methodology

Ingots of a solder alloy[1] of composition Sn-37Bi-0.5Sb-0.5Cu-0.03Ni (compositions in wt%) were supplied by Nihon Superior Co., Ltd., while pure Sn and Bi ingots were supplied by Hayes Metals Pty Ltd. These ingots were remelted in a clay graphite crucible in an electric resistance furnace at a temperature approximately 215 degrees above the liquidus and cast into an aluminium mould at 120 °C, (details elsewhere [2]). The solidification sequence was characterized by sampling a melt with a tapered (9°) stainless steel crucible of 40 mm height with a base diameter of 24 mm and monitoring the temperature with stainless steel K-type thermocouple while cooling, and then during rapid reheating. Additional tensile samples were made by casting and sectioning (<1 mm) a block with the cross section of a flat tensile bar with reduced length dimensions of 4.34 * 2 mm. Tensile testing was performed on an Instron 4464 at 0.2 mm/min. The tensile bars were polished to a final finish of 0.5 μm colloidal silica for metallographic examination, prior to testing. Micrographs were captured in backscatter electron mode in a Hitachi TM3030 at 15 kV.

In situ synchrotron powder X-ray diffraction (PXRD) was performed at the Australian Synchrotron. The cast samples were first comminuted into a powder using an agate mortar before mixing with ground quartz powder to decrease X-ray attenuation (required due to the high mass absorption coefficient of the Bi phase) and being placed in a quartz capillary of 500 μm internal diameter. The capillary was mounted on a rotary holder and exposed to temperatures ranging from −100 °C to 180 °C in an otherwise standard atmosphere, while PXRD patterns were recorded at 20 °C intervals. For pure Bi and Sn, the total scan time was one minute at each temperature, while for the hypoetuectic alloy, ten scans were performed at each temperature at one minute for each scan. A temperature ramp rate of 6 °C/min was used for all samples. In addition, room temperature PXRD patterns were collected as the first step in each experiment prior to cooling of the sample. A 21 keV monochromatic incident beam, calibrated using a standard LaB_6 sample (NIST660b) for precise wavelength determination at room temperature, resulted in a wavelength of 0.5904 Å. The diffracted patterns were captured in transmission mode using a Mythen-II strip detector, maintaining a consistent comparison of peak positions.

Phase identification and the lattice parameters of each phase were derived from the Rietveld refinement to the 2theta 8° to 45° of each sample using TOPAS Academic V6 (Bruker-AXS). Tetragonal βSn (PDF number: 00-004-0673, space group I41/amd, a = 5.831 Å, c = 3.182 Å), and trigonal Bi (PDF number: 00-044-1246, space group R¯3 m, a = 4.547 Å, c = 11.863 Å) structures were used as the starting structure for Rietveld refinement. The RWP of all refinements was controlled below 10 to ensure a good fit between the experimental and refined data.

[1] For brevity, this alloy will be referred to as the 'hypoeutectic Sn-Bi alloy' throughout this paper.

3 Results and Discussions

3.1 Solidification Characterisation and Mechanical Properties

Figure 1(a) shows the cooling curves for the pure Bi, pure Sn and the hypoeutectic Sn-Bi alloy used in this research. Both the pure Bi and Sn are characterized by a single thermal arrest on cooling with around 10 and 3.5 degrees of recalescence respectively, with plateau temperatures (inset Fig. 1(a)) consistent with literature values (within the limits of thermocouple accuracy) [1]. The hypoeutectic alloy is characterized by two major reactions on cooling, one high temperature reaction associated with the formation of primary Sn, with recalescence of around 3.5 degrees and a plateau temperature of 174.2 °C and a second longer thermal arrest associated with the formation of the Sn-Bi eutectic with recalescence of 3 degrees and a plateau temperature of 136.4 °C. The heating curves confirm that there is little undercooling associated with the growth of pure Sn or Bi, however, growth of the Sn-Bi eutectic in the hypoeutectic alloy occurs with a few degrees of undercooling. The thermal arrest on heating associated with the melting of the Sn-rich phase was spread over a temperature interval of some 15 degrees and as such no defined plateau temperature could be established.

The microstructure of the hypoeutectic alloy is shown in Fig. 1(c) and consists of Sn dendrites with small Bi precipitates and interdendritic Sn-Bi eutectic. While some of the eutectic regions have well defined alternating lamellae of Sn and Bi, others have a 'divorced' appearance with Bi particles that do not appear to have grown cooperatively with the adjacent Sn. The hypoeutectic alloy also contains Cu, Ni and Sb and there are numerous crystals of Cu_6Sn_5 present in the microstructure. The location of the Sb and Ni was not determined, although they are likely to be present in the Sn and Cu_6Sn_5 phases respectively.

The tensile testing results in Fig. 1(b) show the Sn and Bi phases are easily deformed and have similar and low tensile strengths at around 12–13 MPa. The hypoeutectic alloy displayed a significantly higher tensile strength at around 72 MPa. Similar observations have been made using nanoindentation with Sn and Bi shown to be relatively soft compared to Sn-Bi alloys [3]. At the relatively low strain rate used ($\approx 0.0008^{-1}$), pure Sn had the most elongation to failure, in excess of 40% strain, while pure Bi and hypoeutectic samples were similar at around 17% strain. Some discontinuities were observed during deformation of pure Bi with sudden small decreases in load as seen at 3 and 5% strain in Fig. 1(b) (green curve). In single crystal Bi samples, similar observations at levels of strain $<10\%$ have been attributed to twinning [4]. The microstructure of the polished surface of the hypoeutectic tensile bar after testing, is shown in Fig. 1(c) and shows deformation has been accommodated by decohesion of the interface between the eutectic Bi and dendrite arms as well as recrystallisation and deformation of the Sn phase at a scale comparable to that of the Bi precipitates.

3.2 Powder X-Ray Diffraction: Dependence of Crystallography on Time and Temperature in Sn-Bi Alloys

The synchrotron PXRD pattern for the hypoeutectic Sn-Bi alloy at 20 °C, prior to cooling and a significant time after the original casting of the alloy (minimum several weeks) is

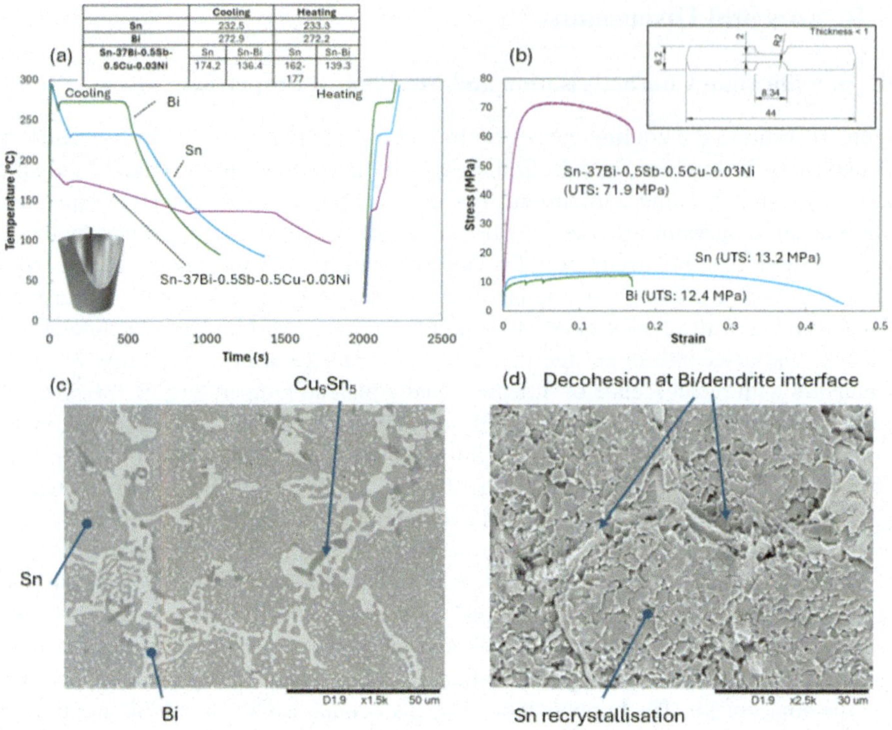

Fig. 1. (a) Cooling and heating curves for pure Bi and Sn compared to that of the hypoeutectic alloy. The insets show the geometry of the crucible used and the plateau temperatures. (b) The tensile curves for pure Bi and Sn compared to that of the hypoeutectic alloy, the tensile strength is shown in parenthesis and geometry of the tensile bars in the inset. (c) The microstructure of the hypoeutectic alloy tensile bar prior to testing, showing the Sn (light grey), Bi (bright white) and Cu_6Sn_5 (dark grey) phases. (d) The surface of the tensile bar near the region of fracture, after tensile testing. Arrows indicate Sn recrystallisation and decohesion at the dendrite/Bi interface.

shown in Fig. 2(a). The results confirm the sample contains tetragonal β-Sn and trigonal Bi, consistent with the PDF numbers detailed in the Methodology as well as small $(Cu,Ni)_6Sn_5$ peaks. The Rietveld refined lattice parameters were calculated for pure Sn and pure Bi over the temperature range of −100 to 180°C and all parameters varied in a relatively linear fashion over the entire temperature range, as shown in Table 1.

For the hypoeutectic Sn-Bi alloy, both Sn_a and Sn_c parameters were larger than those of the pure Sn for the entire range of temperatures tested and had a complex non-linear relationship with temperature. The behaviour is similar for both Sn_a and Sn_c and is discussed with respect to unit cell volume, as shown in Fig. 2(b). In this figure, the temperature is divided into 3 regimes, the first (light blue) is the range −100 to 60 °C where there is a relatively linear increase in unit cell volume as temperature increases. In the second region between 80 and 140 °C (yellow region) there is a departure from this linear behaviour and a sharper increase in cell volume with temperature. Finally, as the temperature increases above 140 °C (red region) the unit cell volume rapidly decreases.

Table 1. Linear fit of the lattice parameters for the pure Sn and pure Bi in the temperature range −100 to 180 °C ($R^2 > 0.99$ for all equations)

Linear Fit of Lattice Parameter (Å) with Temperature, T (°C)
$Sn\_a = 1 \times 10^{-4}\ T + 5.8297$
$Sn\_c = 1 \times 10^{-4}\ T + 3.1792$
$Bi\_a = 5 \times 10^{-5}\ T + 4.5453$
$Bi\_c = 2 \times 10^{-4}\ T + 11.859$

The unit cell volume is represented by two separate series of data in the region from 80 to 140 °C. This is because the PXRD patterns in this temperature range display a broadening and separation of the Sn peaks as shown in the inset of Fig. 2 (b) for the Sn(200) peak and observed elsewhere [5]. The splitting of peaks is due to differences in Bi concentration and the small red circles in Fig. 2 (b) indicate Sn with a higher concentration of Bi, while the small light blue circles indicate Sn with a lower concentration of Bi.

The difference between the unit cell volume of pure Sn, and the Sn phase of the hypoeutectic alloy occurs due to the temperature dependent solubility of Bi in the Sn phase. This is readily apparent when comparing the unit cell volumes in Fig. 2(b) with a plot of the maximum solubility of Bi in Sn as a function of temperature. While the solvus and solidus lines are digitized from available phase diagram data [1] and overlaid as the green solid lines in Fig. 2(b), they need to be considered as a schematic only as we cannot quantify the concentration of Bi at each of the analysis temperatures. That said, from this data it is seen that as the solubility of Bi in Sn increases from around 2 wt% at room temperature to 20.6 wt% at the eutectic temperature, there is a corresponding increase in the unit cell volume that closely follows the shape of the solvus line and further temperature increases mirror the decreasing Bi concentration in the shape of the solidus line.

The increase in Bi concentration in the Sn phase relies on the dissolution of eutectic Bi and Bi precipitates. While it is unclear from the data the extent to which the Sn has reached the maximum solubility limits for Bi, the broadening and splitting of the Sn peaks indicates there is significant inhomogeneity in this process throughout the PXRD sample. This could arise due to different compositions of individual powder particles but may also be expected in bulk samples, due to the large distribution in the length scale of individual microstructural features.

The change in the lattice parameters of the Bi phase in the hypoeutectic Sn-Bi alloy are similar to those reported elsewhere [6] although the temperature range investigated here is broader. The Bi_a lattice parameters in the hypoeutectic alloy are mostly lower than those of pure Bi for temperatures of 140 °C and below while those of Bi_c are slightly higher, however in both cases the values converge with those of pure Bi at higher temperatures. The combined effect on the unit cell volume is shown in Fig. 2(c), where at low temperatures the cell volume is higher for pure Bi but the values converge at higher temperatures. The differences between the pure metal and the alloy are likely to arise from the solubility of Sn in Bi. Lee et al. [7] discuss the maximum solubility of Sn in

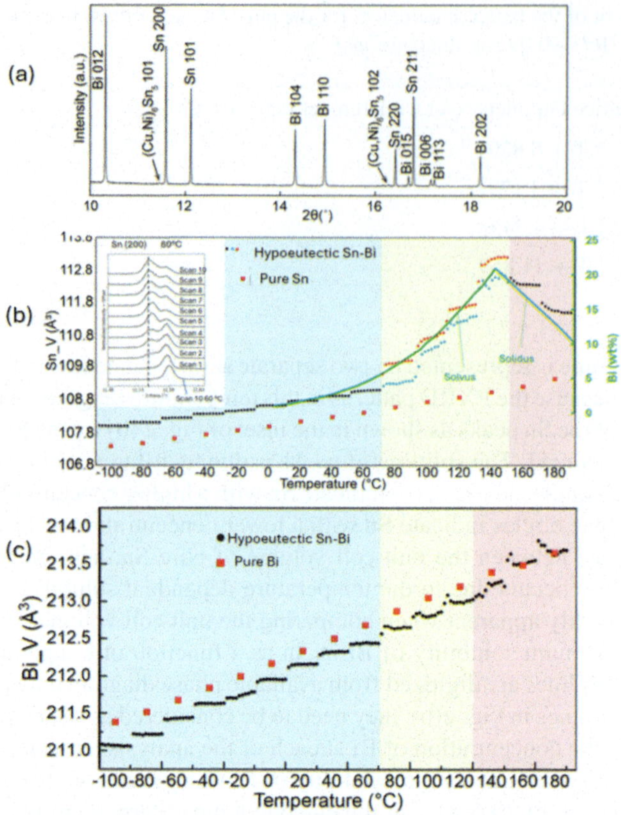

Fig. 2. (a) The synchrotron PXRD pattern of the hypoeutectic alloy at 20 °C and (b) the unit cell volume change of Sn vs temperature for the pure Sn and the Sn in the hypoeutectic alloy. The inset shows an example of the broadening and splitting of Sn (200) peak, in this case during heating at 80 °C, the *schematic* solvus and solidus lines (green) showing the maximum solubility of Bi in Sn are shown on a secondary axis. (c) The unit cell volume change of Bi vs temperature for the pure Bi and the Bi in the hypoeutectic alloy.

Bi is difficult to ascertain but very limited and used a value of 0.2 at% (0.11wt%) in their analysis. Slightly higher values have been found using electron probe microanalysis, where it has also been shown that the presence of Sb can influence the concentration of Sn in the Bi phase [8]. The Bi peaks were present in the PXRD, even above the eutectic temperature, likely due to some of the isolated powder particles being rich in Bi concentration. These particles may be more likely to come from eutectic regions (due to the larger size of the Bi) and the convergence of the cell volume at high temperatures may indicate a difference in the solute concentration of the Bi precipitates vs eutectic Bi.

The homogeneity and times required for diffusion within hypoeutectic Sn-Bi alloys will be influenced by numerous length scales and will influence the degree to which the crystallographic changes in this research are applicable to commercial solders. These include the Sn-grain size and dendrite arm spacing, the eutectic cell size and interlamellar spacing of the eutectic, as well as the Bi precipitate size and the intermetallic size and distribution. In turn, these will depend strongly on the exact composition, cooling rates and interactions with substrates as well as any recrystallisation, grain growth or particle ripening that occurs on events such as deformation, heating or cooling. Many of these phenomena remain the subject of current research endeavours.

4 Conclusion

In this research, a multicomponent hypoeutectic Sn-Bi alloy was compared to pure Sn and Bi. The combination of elements in the alloy results in a large reduction in melting temperature and a significant increase in the tensile strength. The crystal structure of both the Sn and Bi phases was examined over an extended temperature range from −100 to 180 °C. The relatively simple relationship between structure and temperature in the pure metals was contrasted with that of the alloy. The complexity of the alloy behaviour was discussed in terms of the temperature dependent solubility of Bi in Sn and the multitude of microstructural length scales present in the alloy.

Acknowledgements. This work was supported by Nihon Superior Co., Ltd, Japan, the Australian Research Council, Australia (LP210301248) and ANSTO, Australia (Grant number AS241/PD/21311). The authors would also like to acknowledge the assistance of Dr Wen Liang Tan from the Australian Synchrotron Powder Diffraction beamline in conducting these experiments.

References

1. NIST Phase Diagrams & Computational Thermodynamics, Bi-Sn System (2024). https://www.metallurgy.nist.gov/phase/solder/bisn.html
2. Zhou, J., Tan, X.F., McDonald, S.D., Nogita, K.: Mechanical properties and microstructure of binary In-Sn alloys for flexible low temperature electronic joints. Materials (Basel) **15**, 8321 (2022)
3. Shen, L., Septiwerdani, P., Chen, Z.: Elastic modulus, hardness and creep performance of SnBi alloys using nanoindentation. Mater. Sci. Eng. A Struct. Mater. Prop. Microstruct. Process. **558**, 253–258(2012)
4. Yanaka, Y., Kariya, Y., Watanabe, H., Hokazono, H.: Plastic deformation behavior and mechanism of bismuth single crystals in principal axes. Mater. Trans. **57**, 819–823 (2016)
5. Tan, X.F., Hao, Q., Zhou, J., Gu, Q., McDonald, S.D., Sweatman, K., Ikeda, M., Yasuda, K., Bermingham, M.J., Nogita, K.: In-situ investigation of the time-temperature dependent lattice and microstructure of Sn-Bi alloys. Materialia **33**, 101974 (2024)
6. Hao, Q., Tan, X.F., Gu, Q., Sweatman, K., McDonald, S.D., Nogita, K.: The effects of temperature and solute diffusion on volume change in Sn-Bi solder alloys. JOM **74**, 1739–1750 (2022)
7. Lee, B.-J., Oh, C.-S., Shim, J.-H.: Thermodynamic assessments of the Sn-In and Sn-Bi binary systems. J. Electron. Mater. **25**, 983–991 (1996)
8. Tan, X.F., Hao, Q., Gu, Q., McDonald, S.D., Sweatman, K., Bermingham, M., Nogita, K.: The effects of Sb on the lattice and microstructure characteristics of hypo-eutectic Sn-Bi alloys. Mater. Charact. **201**, 112934 (2023)

The Influence of Indium to the Microstructure and Solidification of Sn-58Bi Low Temp Lead-Free Solder

Muhammad Hafiz Zan Hazizi, Mohd Arif Anuar Mohd Salleh(✉), Flora Somidin, and Nur Syahirah Mohamad Zaimi

Faculty of Chemical Engineering Technology, Universiti Malaysia Perlis, 02600 Arau, Perlis, Jejawi, Malaysia
arifanuar@unimap.edu.my

Abstract. The demand for low temperature soldering in electronic manufacturing has surged, driven by the necessity for a hierarchy of solder alloys in packaging technology with the need for enhanced reliability. Among the potential solutions, the Sn-Bi series solder emerges as a promising candidate due to its low melting point, cost-effectiveness, and eco-friendliness. However, its brittleness has hindered widespread adoption. Consequently, extensive efforts have been directed towards enhancing the properties of Sn-Bi based solders. Third element additions have offered partial improvements but have not provided complete solutions. The Sn-58Bi based solder were prepared via the casting method. Indium (In) were then added to the based solder at different weight percentages which were 6.5, 10.5 and 12. Microstructure of the bulk solder were then examined by using Scanning Electron Microscope (SEM). Solidification graphs were simulated by using Thermo-Calc Software to further discuss the results of the microstructure. It is expected that by adding In to Sn-58Bi, the microstructure will consist primarily of β-Sn and Bi phases with the existence of InBi phases, where it will be consistent with the phase diagram simulated by Thermo-Calc Software.

Keywords: Lead-free solder · Sn-58Bi

1 Introduction

Solder materials are essential for the reliability of solder joints. In the past, lead-based solders were commonly employed in the electronics and electrical industries, especially for assembling circuits. However, as technology progresses rapidly, most products become disposable when new versions are introduced, causing obsolete, lead-containing products to end up in landfills.

Lead leaching from landfills into water sources poses a serious health hazard, given its toxicity. Recognizing the risks, many leading nations have prohibited the use of lead in electrical and electronic goods. The European Union (EU) has enacted two key regulations. The first is the Restriction of Hazardous Substances (RoHS), and the second is the Waste Electrical and Electronic Equipment (WEEE) directive. Under the WEEE

N. R. Abdul Razak et al. (Eds.): ECOOP 1996, SPPHY 418, pp. 80–85, 2025.
https://doi.org/10.1007/978-981-96-2871-1_12

directive, lead must be removed from electronic and electrical waste, while the RoHS directive forbids the use of lead in any products made after July 1, 2006 [1].

As a result, various research on lead-free solders have been done by researchers all over the world. Several lead-free solders have been developed and introduced to the industry such as Sn-Cu, Sn-Ag, Sn-Ag-Cu, Sn-Bi, Sn-Zn and Sn-Sb. Of all these lead-free solders, Sn-Bi is considered as one of the excellent substitutes, because of its cost-effectiveness, lower melting point and outstanding resistance to creep [2]. Lately, there has been an increasing need for solder materials that have a melting point lower than 180 °C. Electronic packages are often been reflowed several times and thus the warpage issue becomes serious. So far, there are not many options for a good lead-free solder that feature a low melting point and one of the available options is the eutectic Sn-Bi, with a melting point of 138 °C [3]. Nonetheless, Sn-Bi solders are characterized by brittleness. Additionally, the mechanical properties might deteriorate. This deterioration could be due to the segregation of Bi at the interface. As a result, efforts have been made to enhance its mechanical properties by introducing a third element. Numerous past studies have concentrated on investigating the properties of Sn-Bi-In solders and the impact of adding small amounts of Indium into it [4].

In this paper, the effects of adding In to Sn-58Bi solders to its phase diagram and solidification curves were simulated by using Thermo-Calc software and further discussed.

2 Experimental Procedure

In this study, Tin (Sn) was sourced from Nihon Superior Co., Japan, while Bismuth (Bi) and Indium (In) were acquired from Changsha Santech Materials Co., Ltd. The alloys were produced through a casting process by melting the Sn, Bi, and In in a boron nitride-coated graphite crucible, heated in an electric furnace set at 350 °C. After 15 min of heating, the alloy mixture was stirred to ensure homogeneity. After melting, the alloys were poures into a steel mold and left to solidify naturally at room temperature. The reference alloy underwent the same process but without the inclusion of In. The chemical compositions of the solder alloys are shown in Table 1.

Table 1. Example of table

Samples	In (wt.%)	Sn (wt.%)	Bi (wt.%)
0In	0	42	58
6.5In	6.5	35.5	58
10.5In	10.5	31.5	58
12In	12.0	30	58

Samples for metallography observations were prepared according to standard metallographic methods, which involved cold mounting the samples in epoxy resin, followed by grinding with silicon carbide (SiC) papers of varying grit sizes, and polishing was performed using a 0.3 μm alumina suspension. Prior to any grinding or polishing, all samples were ultrasonically cleaned with deionized water. A scanning electron microscope (SEM), the model is JEOL JSM-6010LA, with energy dispersive x-ray (EDX) was utilized for conducting observations and performing elemental analysis.

Thermo-Calc software was utilized to simulate the phase diagram for each alloy composition, identifying the phases present and their solidus and liquidus temperatures. Using the Scheil method, the solidification curve was also simulated within the same software. The phase diagram and the solidification paths for the alloys were calculated with Thermo-Calc® 2021a software with the TCSLD v3.3 database.

3 Results and Discussions

3.1 Phase Diagram

From Thermo-Calc software, the simulated phase diagram is as shown in Fig. 1(a). The eutectic Sn-58Bi solder features a eutectic melting temperature of 138 °C, which is quite low making it a suitable candidate for low temperature soldering [5]. Figure 1(a) depicts the ternary phase diagram of Sn-Bi-In, indicating the melting points of ternary systems at varying combinations of In elements ranging from 0% to 15% In (in wt.%). Ternary alloys can be thermodynamically complex because they involve a large number of phases. Moreover, these phases coexist within equilibrium fields that are generally limited to extremely narrow ranges of temperature and composition [3].

In this study, the weight percentages of Indium selected were 6.5%, 10.5%, and 12%. These particular ranges were chosen because they represent the area of interest where changes in the solder phases occur during heating and cooling, involving solidification, as depicted in Fig. 1(a). According to the simulated phase diagram, the melting points of the Sn-Bi-In solder alloys are 133.2 °C, 122.8 °C, and 119.0 °C for 6.5 wt.%, 10.5 wt.%, and 12 wt.% of In addition, respectively.

3.2 Solidification Curve

It has been commonly observed that non-equilibrium solidification takes place in solder alloy systems while solidifying. Hence, for precise alloy design, it is essential to consider this non-equilibrium effect to accurately simulate the actual solidification process in real-world applications. The Scheil model can simulate this solidification process. It is assumed that local equilibrium takes place at the liquid/solid interface and there is no diffusion within the solid phase [6].

Figure 1(b), 1(c), and 1(d) shows the calculated mole fraction of the solid phases versus the temperature variation of Sn-Bi-xIn alloy under equilibrium (dashed curve) and Scheil-model (solid curve) solidification conditions. Under equilibrium conditions, the solidification of the 6.5In and 10.5In samples initiates at 133.21 °C and 122.8 °C, respectively, and completes at 105.35 °C and 84.2 °C. According to the Scheil simulation, a considerable portion of the liquid phase still exists at 76.4 °C. Having a liquid phase at a temperature this low is regarded as a disadvantage since it may lead to the lift-off defect at the connection point between the solder, Cu substrate and the connecting wire [7].

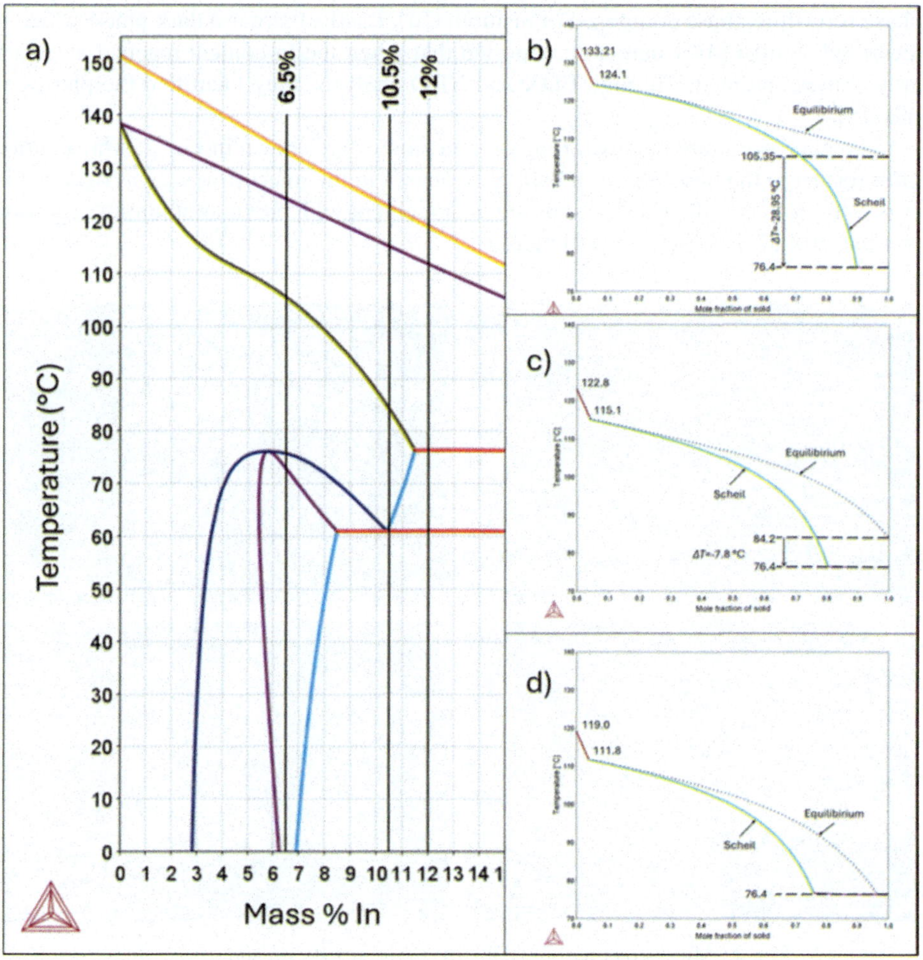

Fig. 1. Thermo-Calc simulation results of (a) Sn-Bi-In phase diagram, (b) 6.5In solidification curve, (c) 10.5In solidification curve and (d) 12In solidification curve.

For sample 12In, the solidification initiates at 119.0 °C and completed when the temperature reaches 76.4 °C for both Scheil and equilibrium curves. The disparity between these two seems to be minimal. All 6.5In, 10.5In and 12In curves showed that during the solidification process, the reaction begins with Liquid (L) transforming into L + Bi, then continue to transform into L + βSn + Bi and finally into βSn + InBi. It was observed that InBi phases were present in all of the samples with In addition.

3.3 Microstructural Examination

Research has demonstrated that three distinct phases are present in the microstructure of the Sn-Bi-In solder alloy. These phases are evenly distributed across the entire solder matrix [3]. One of these phases is a black intermetallic phase known as β-Sn. Another phase is a white phase composed of bismuth (Bi). Lastly, there is a dark phase which is identified as InBi [4]. Chen et al., reported that when the In content reached 5%, dark grey phases emerged. Through EDX and XRD analysis, they identified the phases as InBi [8].

The microstructure results for the current studies are shown in Fig. 2. The addition of In results in the appearance of dark grey phases in the microstructure, as depicted in Figs. 2(b), (c), and (d). These results are in agreement with previous studies, suggesting that the dark grey phases are InBi phases.

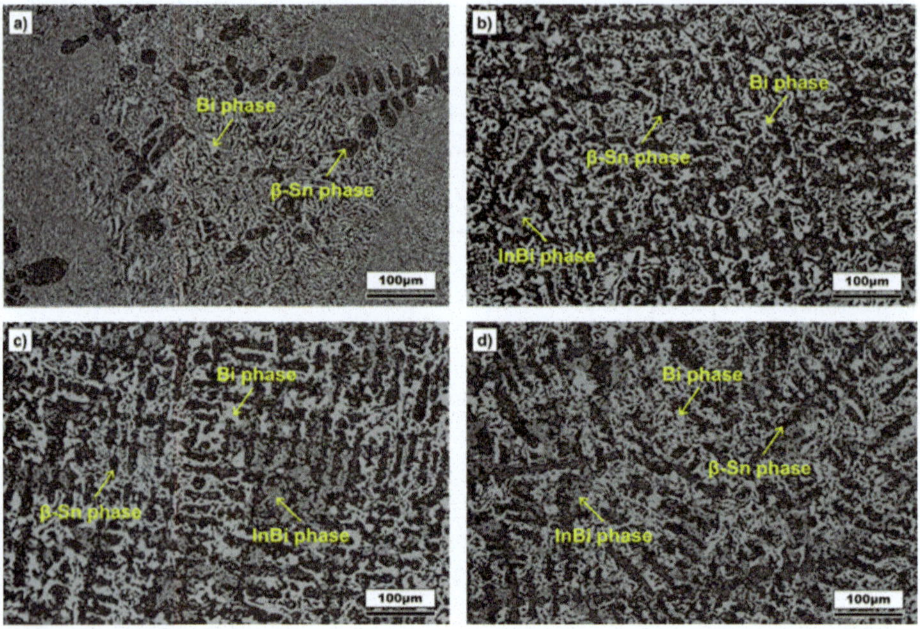

Fig. 2. Microstructure of (a) 0In, (b) 6.5In, (c) 10.5In and (d) 12.0In.

4 Conclusion

In conclusion, these preliminary studies have proven that the simulation data from the Thermo-Calc software is highly beneficial to the research input in getting the initial expectation for the experimental results. According to the phase diagram and solidification curves that were simulated by Thermo-Calc, it has shown that by adding In to the Sn-58Bi solder, InBi phases will start to exist. These hypotheses were then further confirmed from the microstructural examination of the solder alloys and supported by several others previous research. It is noted that with more In addition, the Bi phase will become coarsened and the interlamellar spacing became thicker. Solidification curves also showed that with more In added, the reactions and phases present remains the same. However, it is worth noting that with 12 wt.% of In added, the differences between Scheil and equilibrium curves are the most lowest.

Acknowledgements. This work was supported by Ministry of Higher Education, Malaysia under the fundamental research grant scheme (FRGS) FRGS/1/2024/TK09/UNIMAP/02/8.

References

1. Ma, H., Suhling, J.C.: A review of mechanical properties of lead-free solders for electronic packaging. J. Mater. Sci. **44**(5), 1141–1158 (2009). https://doi.org/10.1007/s10853-008-3125-9
2. Jiang, N., Zhang, L., Gao, L.L., Song, X.G., He, P.: Recent advances on SnBi low-temperature solder for electronic interconnections, Springer. Sep. 01 (2021). https://doi.org/10.1007/s10854-021-06820-7
3. Liu, Y., Tu, K.N.: Low melting point solders based on Sn, Bi, and In elements. Mater Today Adv. **8** (Dec. 2020). https://doi.org/10.1016/j.mtadv.2020.100115
4. Li, Q., Ma, N., Lei, Y. P., Lin, J., Fu, H.G., Gu, J.: Characterization of low-melting-point Sn-Bi-In lead-free solders. J. Electron. Mater. **45** (11) (2016). https://doi.org/10.1007/s11664-016-4366-z
5. Xu, K.K., Zhang, L., Gao, L.L., Jiang, N., Zhang, L., Zhong, S.J., Review of microstructure and properties of low temperature lead-free solder in electronic packaging, Taylor and Francis Ltd (Jan. 31, 2020). https://doi.org/10.1080/14686996.2020.1824255
6. Ohnuma, I., Miyashita, M., Liu, X.J., Ohtani, H., Ishida, K.: Phase equilibria and thermodynamic properties of Sn-Ag based Pb-free solder alloys. IEEE Trans. Electron. Packag. Manuf. **26**(1), 84–89 (2003). https://doi.org/10.1109/TEPM.2003.813003
7. Suganuma, K.: Advances in lead-free electronics soldering. Curr. Opin. Solid State Mater. Sci. **5**(1), 55–64 (2001). https://doi.org/10.1016/S1359-0286(00)00036-X
8. Chen, X., Xue, F., Zhou, J., Yao, Y.: Effect of in on microstructure, thermodynamic characteristic and mechanical properties of Sn-Bi based lead-free solder. J. Alloys Compd. **633**, 377–383 (Jun.2015). https://doi.org/10.1016/j.jallcom.2015.01.219

The Effect of Thermal-Aging on the Microstructure of Sn0.7Cu and Sn0.7Cu0.05Ni Solder Joint

Aimi Noorliyana Hashim[1,2(✉)], Mohd Arif Anuar Mohd Salleh[1,2], and Muhammad Faqih Ikhwan Norhalwi[1]

[1] Faculty of Chemical Engineering and Technology, Universiti Malaysia Perlis (UNIMAP), Taman Muhibah, 02600 Jejawi, Arau, Perlis, Malaysia
aimiliyana@unimap.edu.my

[2] Centre of Excellence Geopolymer and Green Technology, Universiti Malaysia Perlis (UNIMAP), Taman Muhibah, 02600 Jejawi, Arau, Perlis, Malaysia

Abstract. This paper studied the effects of thermal aging conditions on interfacial IMC layer of Sn0.7Cu and Sn0.7Cu0.05Ni solders joint. Solder joint was formed using the reflow soldering method and were thermal aged for 24 and 120 h at ambient temperature, 85 ^{0}C and 150 ^{0}C respectively. The characterization emphases on the microstructure evaluation of interfacial intermetallic compound (IMC) layer during thermal aging using optical microscopy (OM) and Image-J software. The study found that adding 0.05 wt.% Ni to Sn0.7Cu solder changes the properties of the interfacial IMC layer. Sn0.7Cu0.05Ni reduced the interfacial IMC thickness with thinner and more refined IMC layer compared to Sn0.7Cu at the solder/substrate interface. Additionally, the presence of Ni significantly improves wettability by decreasing the wetting angle and stabilizes the IMC structure, preventing the formation of brittle Cu_3Sn layers and more suitable for high temperature applications.

Keywords: Thermal Aging · Interfacial IMC · High temperature · Reflow Soldering

1 Introduction

Due to the toxic nature of lead (Pb) poses significant health and environmental risks, the restriction of hazardous substances (RoHS) directive prohibits the use of Pb containing solder in consumer electronics, prompting the industry to seek lead-free alternatives that maintain desired properties [1–3]. One promising lead-free solder option is Sn0.7Cu, noted for its cost-effectiveness and robust electrical conductivity, making it suitable for wave and dip soldering applications. However, it has limitations in wetting properties and the brittle failure mechanism of solder joints. To improve these characteristics, solder alloying techniques involving elements like nickel (Ni), silver (Ag), bismuth (Bi) and rare earths have been employed.

N. R. Abdul Razak et al. (Eds.): ECOOP 1996, SPPHY 418, pp. 86–92, 2025.
https://doi.org/10.1007/978-981-96-2871-1_13

The intermetallic compounds (IMC) growth at interface of solder/substrate is essential for ensuring the reliability of solder joints [4]. Excessive intermetallic compounds (IMC) growth can attain brittle failure by reducing the strength of the solder joint, ultimately impacting its long-term reliability [1, 5]. Ni is one of the alloying elements that can improve the reliability of solder joints. It has been reported that Sn0.7Cu0.05Ni can reduce the development of thick intermetallic compounds (IMC) and help diminish the crack propagation at the IMC layer [6–8]. Additionally, Ni improves mechanical properties, refines the grain structure, and takes part in interfacial reactions during the soldering process [9].

This study investigates the effect of thermal-aging on the IMC interfacial layer of Sn0.7Cu and Sn0.7Cu0.05Ni solder joint on copper (Cu) substrates using reflow soldering techniques. The emphasis of this study was discussed on the growth of interfacial IMC layer during thermal aging at of 85 °C and 150 °C for durations of 24 h and 120 h.

2 Experimental Procedures

Sn0.7Cu and Sn0.7Cu0.05Ni solders were preferred as solder materials. The experimental route was start with samples preparation of Sn0.7Cu and Sn0.7Cu0.05Ni solders joint. Both solders are initially in sheet form of 0.3mm thickness and punched to 0.3mm diameter. Then positioned with flux and on a Cu substrate with dimensions of 1.5mm × 1.5mm and a 0.8mm thickness. Following this, both types of solders undergo a reflow process at 250 °C for 10 min. After the reflow process, the samples were dipped into a beaker containing acetone and cleaned using a digital ultrasonic cleaner to remove any residual of flux. Subsequently, these two samples were placed in a thermal aging oven at temperatures; 85 °C and 150 °C for durations of 24 h and 120 h. After the thermal aging process, a standard metallographic process was performed on the samples. This involves using an epoxy hardener and epoxy resin in a 1:1 ratio, allowing the mixture to cure for 24 h. This curing process helps to make the samples easier to handle and prevents surface tension on the substrate during the next stages of preparation. Finally, the samples were ground and polished to prepare them for subsequent microstructural analysis. This detailed preparation ensures that the samples are suitable for an in-depth examination of their microstructural properties.

The wettability analysis of solder joints was conducted to evaluate the fluidity properties of the samples. The punched sample was reflowed on Cu substrate in an F4N reflow oven at 250 °C for 127 s to form a solder joint. After reflowing, the samples were mounted using epoxy resin to facilitate handling during the subsequent grinding and polishing steps. Micrographic analysis was conducted using an optical microscope (OM) to observe the morphology IMC interfacial layer. The Image J software used to calculate the IMC thickness formed at the interface as illustrated in Fig. 1. The Image-J software allows for accurate tracing and measurement of the IMC areas. The IMC thickness can be calculated using the Eq. 1 [10]:

$$Thickness\,of\;IMC\;(T) = \frac{Area\,of\;IMC\;(A)}{Length\,of\;IMC\;(L)} \tag{1}$$

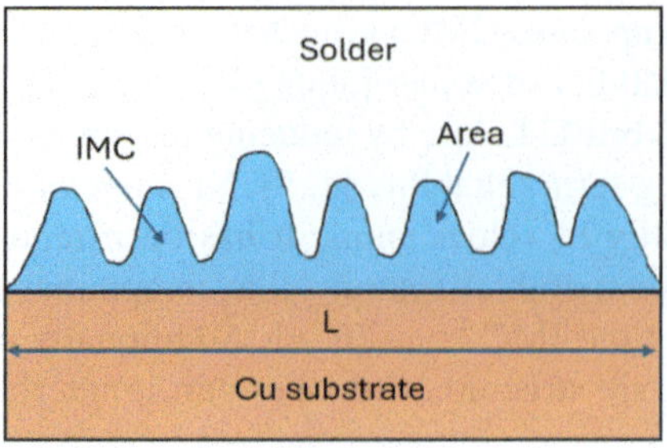

Fig. 1. Schematic diagram of measuring of interfacial IMC thickness.

3 Results and Discussions

3.1 Wettability Analysis

Wettability, is the ability of molten solder to spread on a substrate during the soldering process, is usually evaluated by quantifying the contact angle [3]. Wetting behavior is vital for solder alloys, as it ensures strong metallic connection in electronic packaging. In the soldering process, a smaller contact angle of the solder indicates better wettability of the interconnections [2, 3]. Figure 2 represents the cross-sectional samples of Sn-0.7Cu and Sn-0.7Cu-0.05Ni solder joint on Cu substrate. Sn-0.7Cu-0.05Ni solder joint demonstrated superior wettability with a smaller contact angle of 26.52° compared to the Sn-0.7Cu solder joint, which had a contact angle of 29.59°. The smaller contact angle of Sn-0.7Cu-0.05Ni indicates a greater affinity for the substrate surface, resulting in better wetting. The improved wettability allows the solder to spread more easily and form a thin, uniform layer between surfaces, leading to stronger bonding and enhanced electrical and thermal conductivity [11]. Conversely, a larger contact angle suggests poor wetting, which can cause incomplete or weak solder joints, potentially prominent to the formation of voids, cracks, or other defects that undermine the integrity of solder joint [12].

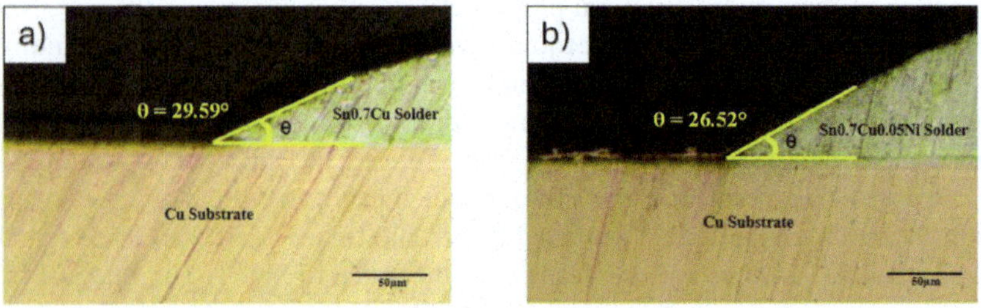

Fig. 2. The contact angle of (a) Sn-0.7Cu (b) Sn-0.7Cu-0.05Ni on Cu substrate.

3.2 The Interfacial IMC Layer

Thermal aging is typically employed to accelerate the growth of the interfacial IMC layer by exposing it to an elevated temperature environment [11]. Figure 3 shows the formation and growth of interfacial IMC of Sn-0.7Cu and Sn-0.7Cu-0.05Ni solder joints during thermal aging at 85 °C and 150 °C for durations of 24 h and 120 h. It was observed at ambient temperature interfacial Cu_6Sn_5 IMC (Fig. 3(a)) and $(Cu,Ni)_6Sn_5$ IMC (Fig. 3(f)) form simultaneously as a result of the interfacial reaction that occurs at the interface during the reflow process. The interfacial Cu_6Sn_5 IMC forms a thin, scallop-shaped due to slow diffusion. Meanwhile $(Cu,Ni)_6Sn_5$ IMC interfacial layer creates a more refined and thinner structures compared to Cu_6Sn_5 IMC layer. The brittle nature of IMC poses a significant risk to the reliability of solder joints, particularly when excessive IMC thickness forms at the solder/substrate interface, leading to degradation of impact resistance [5]. The ideal interface bonding process occurs when a thin layer of IMC is formed, whereas a thick IMC layer can lead to defects and compromise the integrity of the solder joint.

As well, the thickness of interfacial Cu_6Sn_5 IMC and $(Cu,Ni)_6Sn_5$ IMC layer of Sn-0.7Cu and Sn-0.7Cu-0.05Ni solder joints become thicker with the aging temperature is upsurge to 85 °C and 150 °C. It was observed that the Cu_6Sn_5 IMC interfacial layers in the round scallop-like (Fig. 3 (c-e)) morphology and interfacial $(Cu,Ni)_6Sn_5$ IMC layers in the uniform planar-like (Fig. 3 (g-j)) morphology. Typically, interfacial IMC layer composed by Cu_6Sn_5 and Cu_3Sn phases principally once the thermal aging. However, Cu_3Sn phase is not identified for both solder joint as the Cu_3Sn layers are emaciated to be observed with an optical microscope in an objective manner.

Figure 4 shows the average thickness of Sn-0.7Cu and Sn-0.7Cu 0.05Ni interfacial IMC during thermal aging at 85 ^{0}C and 150 ^{0}C for 24 h and 120 h. Noticeably, the interfacial IMC of Sn-0.7Cu indicates thicker IMC layer compared interfacial IMC of Sn-0.7Cu-0.05Ni for both conditions of temperature and aging time. At 85 ^{0}C for aging time 24 h, the thickness Cu_6Sn_5 IMC layers is 4.57μm and $(Cu,Ni)_6Sn_5$ IMC layers is 4.01μm. Meanwhile with the increasing of thermal aging up to 150 ^{0}C, the thickness Cu_6Sn_5 IMC layers is 4.64 μm and $(Cu,Ni)_6Sn_5$ IMC layers is 4.28 μm. The similar increasing trends of the thickness interfacial IMC layer can be seen while aging time up to 120 h. Predominantly in case of Sn0.7Cu0.05Ni solder joint, it is believed that addition of Ni greatly reduced the growth and production of the Cu_3Sn phase and stabilized the crystal structure of the $(Cu,Ni)_6Sn_5$ grains, maintaining the low misfit of Cu atoms between $(Cu,Ni)_6Sn_5$ [4]. The $(Cu,Ni)_6S_5$ phase has better mechanical properties and thermal stability, improving the solder joint's reliability by avoiding brittle IMC layers [11].

4 Conclusion

The properties of thermal-aging on Sn0.7Cu and Sn0.7Cu0.05Ni interfacial IMC layer have been studied in this work. The following conclusion can be reached:

i. The interfacial IMC thickness of Sn0.7Cu0.05Ni reduced with thinner and more refined IMC layer compared to Sn0.7Cu solder joint.

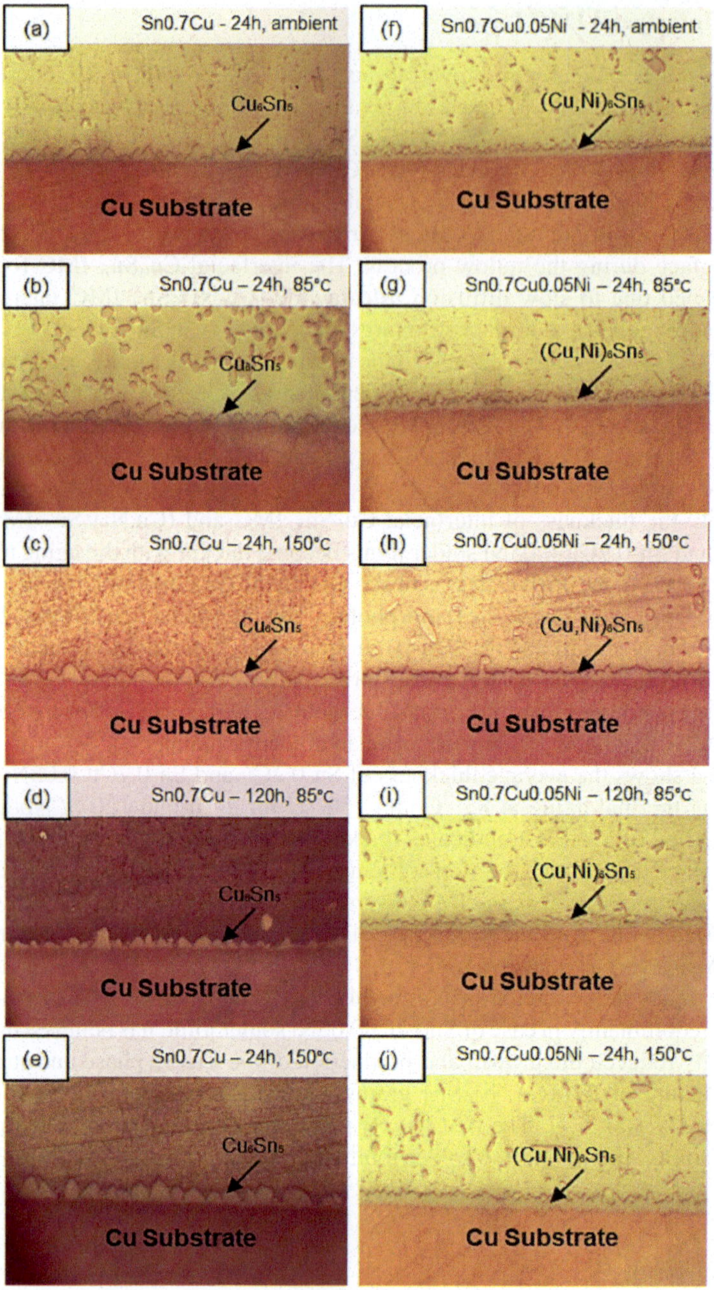

Fig. 3. The interfacial IMC growth of Sn-0.7Cu (a-e) and Sn-0.7Cu-0.05Ni (f-j) solder joints during thermal aging at 85 °C and 150 °C for 24 h and 120 h.

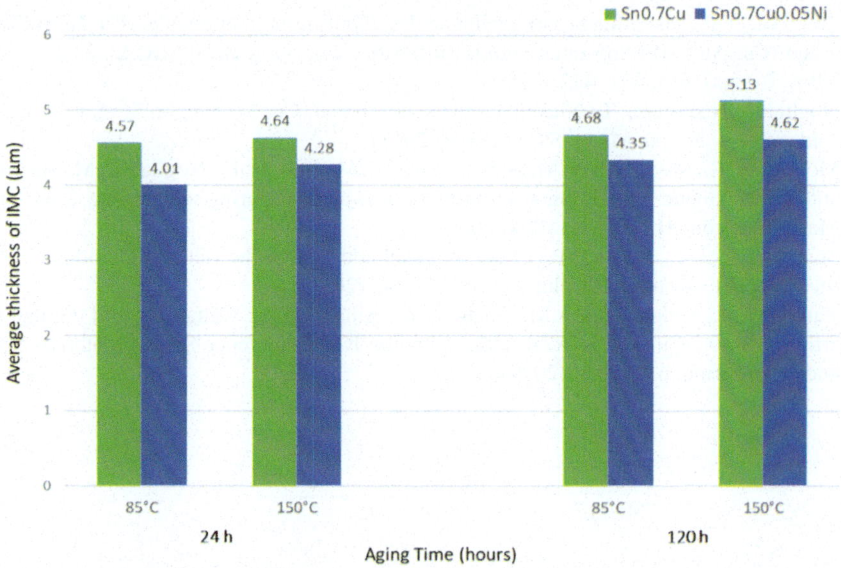

Fig. 4. Average thickness of Cu_6Sn_5 IMC and $(Cu,Ni)_6Sn_5$ IMC interfacial layer during thermal aging at 85 ^{0}C and 150 ^{0}C for 24 h and 120 h.

ii. The total IMC thickness for both Sn0.7Cu and Sn0.7Cu0.05Ni solders joint increased linearly during thermal aging time and temperature.
iii. The presence of 0.05 wt.% nickel significantly improves wettability by reducing the wetting angle compared to Sn0.7Cu solder joint.

References

1. Yang, C., Song, F., Lee, S.R.: Impact of Ni concentration on the intermetallic compound formation and brittle fracture strength of Sn–Cu–Ni (SCN) lead-free solder joints. Microelectron. Reliab. **54**, 435–446 (2014)
2. Hashim, A.N., et al.: Effect of isothermal annealing on Sn whisker growth behavior of Sn0. 7Cu0. 05Ni solder joint. Materials **16**, 1852 (2023)
3. Amli, S.F.M., et al.: Effect of surface finish on the wettability and electrical resistivity of Sn-3.0 Ag-0.5 Cu solder. In: IOP Conference Series: Materials Science and Engineering, p. 012029 (2019)
4. Lai, Y., Chen, S., Ren, X., Qiao, Y., Zhao, N.: Growth behavior and morphology evolution of interfacial (Cu, Ni) 6Sn5 in (001) Cu/Sn/Ni micro solder joints. Mater. Charact. **186**, 111803 (2022)
5. Chen, W.-L., Yu, C.-Y., Ho, C.-Y., Duh, J.-G.: Effects of thermal annealing in the post-reflow process on microstructure, tin crystallography, and impact reliability of Sn–Ag–Cu solder joints. Mater. Sci. Eng., A **613**, 193–200 (2014)
6. Nogita, K., Gourlay, C., Nishimura, T.: Cracking and phase stability in reaction layers between Sn-Cu-Ni solders and Cu substrates. JOM **61**, 45–51 (2009)
7. Salleh, M.M., McDonald, S., Gourlay, C., Belyakov, S., Yasuda, H., Nogita, K.: Effect of Ni on the formation and growth of primary Cu 6 Sn 5 intermetallics in Sn-0.7 wt.% Cu solder pastes on Cu substrates during the soldering process. J. Electron. Mater. **45**, 154–163 (2016)

8. Zeng, G., et al.: The influence of ageing on the stabilisation of interfacial (Cu, Ni) 6 (Sn, Zn) 5 and (Cu, Au, Ni) 6Sn5 intermetallics in Pb-free Ball Grid Array (BGA) solder joints. J. Alloy. Compd. **685**, 471–482 (2016)
9. Nogita, K., Nishimura, T.: Nickel-stabilized hexagonal (Cu, Ni) 6Sn5 in Sn–Cu–Ni lead-free solder alloys. Scripta Mater. **59**, 191–194 (2008)
10. Mohd Said, R., Mohd Salleh, M., Saud, N., Ramli, M., Yasuda, H., Nogita, K.: Microstructure and growth kinetic study in Sn–Cu transient liquid phase sintering solder paste. J. Mater. Sci. Mater. Electron. **31**, 11077–11094 (2020)
11. Somlyai-Sipos, L., Baumli, P.: Effect of nickel addition on the wettability and reactivity of tin on copper substrate. Resolut. Discov. **2**, 9–12 (2017)
12. Ramli, M.I.I., Salleh, M.A.A.M., Saleh, N.A., Amli, S.F.M.: Wettability and thermal properties of Sn-0.7 Cu-0.05 Ni-xZn solder alloy. In: IOP Conference Series: Materials Science and Engineering, p. 012024 (2019)

Wetting Ability of Lead-Free Sn-Based Low-Temperature Solder: A Short Review

Y. P. Tan[1,2](✉), Mohd Arif Anuar Mohd Salleh[1,2], and Flora Somidin[1,2]

[1] Centre of Excellence Geopolymer and Green Technology, Universiti Malaysia Perlis, 01000 Kangar, Perlis, Malaysia
yipeitan97@gmail.com

[2] Faculty of Chemical Engineering Technology, Universiti Malaysia Perlis, Taman Muhibbah, 02600 Jejawi, Arau, Perlis, Malaysia

Abstract. The wetting ability of low-melting-point lead-free Sn-based solder alloys on copper substrates is a critical factor influencing the reliability and performance of electronic assemblies. This review examines the fundamental principles of wetting, the characteristics of Sn-based solder alloys, particularly those with low melting points, and their interactions with copper substrates. The surface energy of both the solder and the intermetallic compounds plays an important role in the wetting characteristics. The effects of solder composition, substrate surface preparation and environmental conditions on wetting behaviour are discussed, along with recent advancements in solder technology aimed at enhancing wetting performance.

Keywords: Wetting ability · low-temperature solder · Sn-based solder · Cu substrate · contact angle

1 Introduction

The rapid advancements in modern electronic devices have driven the need for solders with better overall performance for electronic packaging. The phase-out of lead-based solders, due to growing environmental concerns and legislation, has catalyzed the development of alternative solder materials that provide excellent reliability for electronic devices [1–3]. The miniaturization of advanced electronic devices initiated the development of low-melting point solders which were suitable for multi-level packaging and soldering of heat-sensitive components [4, 5]. Among the promising lead-free solder systems, tin-bismuth (Sn-Bi) and tin-indium (Sn-In) alloys have gained significant attention for their relatively low melting points (<150°C). These attributes make them particularly attractive for applications requiring compatibility with temperature-sensitive components and substrates, such as heat-stabilized polyethylene terephthalate (PET) and polyethylene naphthalate (PEN) [6–8].

The wetting behavior of solder alloys on substrate surfaces is a critical factor in determining the quality and reliability of the solder joints [9]. The wetting process is characterized by the contact angle (θ) formed between the liquid solder and the solid

N. R. Abdul Razak et al. (Eds.): ECOOP 1996, SPPHY 418, pp. 93–100, 2025.
https://doi.org/10.1007/978-981-96-2871-1_14

substrate, which can be quantitatively described by Young's equation and illustrated in Fig. 1:

$$\gamma_{SV} = \gamma_{SL} + \gamma_{LV}\cos\theta$$

where γ_{SV} represents the solid-vapor interfacial energy, γ_{SL} denotes the solid-liquid interfacial energy and γ_{LV} indicates the liquid-vapor interfacial energy. A smaller contact angle is more favorable as it signifies better spreading of the solder on the substrate. Effective wetting ensures intimate contact between the solder and the substrates, enabling the formation of robust metallurgical bonds and facilitating the transfer of electrical, thermal and mechanical loads across the interconnect.

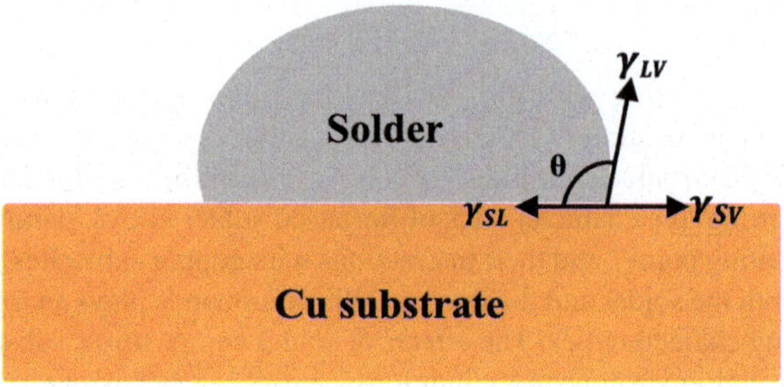

Fig. 1. Illustration of contact angle measurement between solder with Cu substrate.

Achieving reliable wetting on commonly used substrates, Copper (Cu) with lead-free, low-temperature solders presents unique challenges that must be addressed to ensure long-term performance and integrity of solder joints [10]. This short review paper aims to provide an overview of the fundamental principles of wetting of Sn-Bi and Sn-In solder on Cu substrate, the factors influencing their wettability, the challenges and, strategies associated with achieving consistent and robust solder-Cu wetting.

2 Wetting Characteristics of Low-Temperature Sn-Bi and Sn-In Solder Alloys

The development of lead-free low-temperature solder alloys has been a significant focus in the electronics industry. The interaction between the lead-free low-temperature solder on Cu substrates is illustrated in Fig. 2. The intermetallic compounds (IMCs) formed between the solder and substrate play an important role in solder joint performance. Among the commonly studied systems, Sn-Bi and Sn-In have emerged as potential alternatives to traditional lead-based solders.

Sn-Bi solder alloys typically have a melting point of approximately 139°C at eutectic composition Sn-58Bi, making them attractive for applications that require reduced thermal budgets [11, 12]. The eutectic Sn-58Bi solder paste achieved a wetting angle of 35° on the Cu substrate with the use of flux which consisting of a dilute solution of zinc chloride and hydrochloric acid [12–14]. The addition of Bi to Sn helps lower the

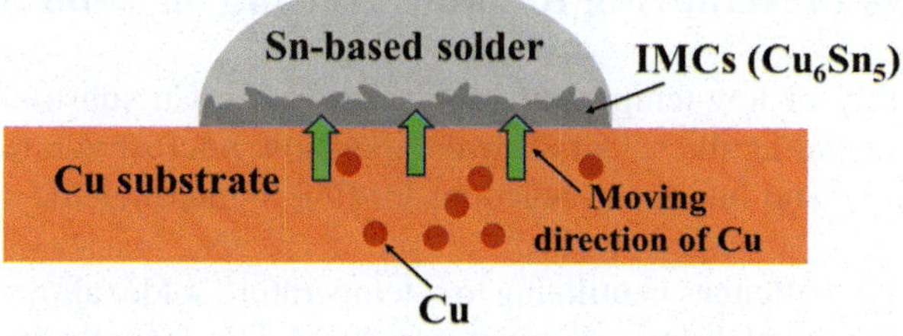

Fig. 2. Illustration of IMCs formation between Sn-based solder and Cu substrate.

overall melting point by reducing the surface tension of the liquid solder and enhanced the fluidity of the solder during the reflow process. As reported by Ramli et al. [15], the addition of 1.5wt% Bi into Sn-0.7Cu-0.05Ni decreased the contact angle by 5.8% and comparable discovery was reported by Liu et al. [16], where the addition of Bi into Sn-0.3Ag-0.7Cu showed significant improvement in wettability as well.

Sn-In solder alloys, on the other hand, exhibit even lower melting points, typically in the range of approximately 118 ~ 180°C, depending on the Sn-In composition [17]. The wetting angle of Sn-50In is reported at 35°on Cu substrate [18]. The addition of In helps further reduce the melting point and enhance the wetting characteristics of the solder on various substrates, including Cu. Fima et al. [19] reported that the wettability of Sn-8.8Zn improved by introducing up to 1.5wt% In. The similar effects of indium on improving solder wettability also found in Sn-0.7Cu-0.2Ni solder reported by Li et al. [20]. Nabihah and Nurulakmal [21] also reported the improvement in wetting ability of SN100C on Cu substrate by minor In addition. The effect of Bi and In elements on wetting behavior of some Sn-based solder on Cu substrate was shown in Table 1.

The observed wetting characteristic of the discussed solder demonstrates their ability to form reliable joints when optimized through appropriate flux formulations and alloying strategies. However, the formation of brittle Bi-rich and the potential of In for oxidation remain challenges in the widespread adoption of Sn-In solders.

Table 1. Contact angle of Sn-based solder on Cu with addition of Bi and In element.

Solder composition	Contact angle (°)
Sn	Approximately 39.0 to 57.0 [22]
Sn-58Bi	35.0 [12–14]
Sn-50In	35.0 [18]
Sn-0.7Cu-0.05Ni	18.2 [15]
Sn-0.7Cu-0.05Ni-1.5Bi	17.2 [15]
Sn8.8Zn	46.5 [19]
Sn-8.8Zn-1.5In	34.0 [19]
SN100C	28.58 [21]
SN100C-2In	22.45 [21]

3 Challenges in Achieving Reliable Wetting on Solid Substrates

The wetting ability of low-temperature solder alloys on Cu substrates is affected by several factors, including the solder alloy composition, microstructure, surface properties of the substrate and interfacial reactions leading to intermetallic compound (IMC) formation.

One of the key challenges in utilizing low-temperature solder alloys on Cu substrates is their inherent poor wettability compared to traditional tin-lead (Sn-Pb) solders [9]. The composition and the microstructure of the solder alloy significantly impact its wettability on Cu substrates. The surface energies of the phases present in the solder are directly related to its crystal orientation [23, 24]. Thus, the brittle Bi-rich layer formation in Sn-Bi/Cu joint may act as a barrier to inhibit the wettability of the solder and influence the quality of the interconnect formed [25, 26]. A lead-free low-temperature solder with lower surface energy is favorable for enhancing the wetting ability of the solder on Cu substrate.

The poor wettability of low-temperature solders on Cu surfaces is primarily attributable to the formation of surface oxides, the presence of contaminants and the development of brittle IMCs at the solder-substrate interface. These factors can impede the solder's ability to spread and disturb the intimate contact of solder with Cu surface, leading to incomplete wetting and the defects formation such as voids or discontinuities in the solder joints. The surface properties of the Cu substrate, such as roughness, oxidation and contamination, can significantly influence the wetting behavior of low-temperature solder alloys [27]. Bhat et al. [28] and Satyanarayan et al. [29] reported that the higher the roughness, the better the wettability. The presence of native oxides or surface impurities on the Cu surface can hinder the solder-substrate interactions by acting as barrier at the interfacial between the solder and substrate. Thus, it is important to prepare a contaminants-free with optimum surface roughness substrate for better solder-substrate bonding.

The interfacial reactions between solder alloys and copper substrates, as well as the formation and growth of IMCs, significantly impact the wetting behavior and the long-term reliability of solder joints. The growth of brittle IMCs, such as Cu_6Sn_5 and Cu_3Sn, at the solder-Cu interface can adversely affect the mechanical properties and reliability of the solder joints. According to Eustathopoulos' theory [30], the wettability is primarily influenced by the properties of the interfacial reaction products, rather than the intensity of the interfacial reaction. Consequently, when the surface energy of the IMC layer is higher, the wetting behavior of the soldering system improves, assuming the surface energy of the liquid solder remains constant [31]. The IMCs formed which exhibit higher surface energy compared to that of solder is preferred to further improve the spreading of solder on Cu substrate. To regulate interfacial reactions and the growth of IMCs at the solder-Cu interface remains a significant challenge in the context of low-temperature solder applications.

4 Strategies for Improving Solder Wettability on Solid Substrates

One of the effective strategies to enhance the wettability of low-temperature solder alloys on substrates is the implementation of surface pretreatment and modification of solder composition techniques. These methods aim to prepare the substrate surface by removing contaminants, reducing surface oxides and providing a more favourable environment for solder wetting and spreading.

Mechanical cleaning methods, such as abrasive polishing or brushing, can help remove surface oxidation and expose the underlying clean Cu surface, improving the solder's ability to wet and bond to the substrate. Chemical cleaning techniques, involving the use of acidic or alkaline solutions, can also be employed to etch and remove surface contaminants and oxides, also increase the surface roughness of the substrates, promoting better wettability [32–34]. Thermal pretreatment such as annealing or reflow in controlled atmospheres, can also be utilized to modify the Cu surface and promote the formation of a thin, stable oxide layer that is more receptive to solder wetting. Zhang et al. [35] reported that the wetting behaviour of SAC305 and Sn on Cu substrate are improved by coating the substrate with a layer of preferred-orientation Cu_5Sn_6.

Modifying the composition of low-temperature solder alloys can be an effective strategy to enhance their wettability on Cu substrates. The addition of minor alloying elements and IMC formation, leading to improved wetting characteristics. The addition of Sb and rare earth elements such as tellurium (Te) to Sn-Bi solder alloy has been shown to enhance the wetting behaviour by influencing the formation and growth of the Bi-rich layer at the solder-Cu interface [36–38]. The presence of these alloying additions can help mitigate the adverse effects of the Bi-rich layer and promote better solder spreading and bonding on the Cu surface. These alloying additions can modify the solder's fluidity, surface tension and interfacial reactions, leading to enhance wettability and improved solder joint formed. The strategies and limitations in enhancement of the wettability of discussed Sn-based solder on Cu substrate was summarized in Table 2.

In summary, the wettability of low-temperature solder alloys on Cu can be significantly enhanced through a combination of surface treatment and compositional modifications. Surface texturing and the incorporation of minor alloying elements have demonstrated their efficacy in enhancing solder wettability. Nonetheless, challenges persist regarding the long-term stability and reliability of these improvements, highlighting the necessity for further investigation into the influence of additional alloying elements on solder joint performance. Ongoing research in this field is critical for the development of more robust and reliable soldering solutions suitable for advanced electronic applications.

5 Conclusion

This review highlights an overview of the fundamental principles governing the wetting behavior of Sn-Bi and Sn-In solder on Cu substrates, the factors affecting their wettability, the associated challenges, and strategies for achieving consistent and reliable solder-Cu wetting. The inherent poor wettability of these solders, influenced by factors such as alloy composition, surface preparation and interfacial reactions, poses significant

Table 2. Summaries of the research reviewed for solder wettability enhancement.

Strategies reported	Limitations
Surface treatment by laser surface texturing to produce dimple-microtextured copper substrate. The wettability of Sn-based on the treated Cu substrate increased with increasing roughness [34]	Insufficient data on long-term stability of surface modifications and their effects on potential low-temperature solder
Surface treatment by magnetron sputtering Cu substrate with Cu_6Sn_5 preferred orientation coating such as $Cu_6Sn_{5(40\text{-}2)}$ and $Cu_6Sn_{5(132)}$ improved the wettability of Sn and SAC305 on Cu substrate [35]	Limited studies of their effects on solder joint performance in terms of mechanical properties and reliability of the solder joints
The addition of RE and Te enhance the wettability of Sn-Bi-based solders on Cu substrates by reducing interfacial tension and modifying IMC formation [37, 38]	Limited understanding of the long-term stability of wettability improvements. Insufficient data on the effects of other potential alloying elements

challenges for reliable solder joint formation. However, through effective surface pre-treatment methods and compositional optimization, it is possible to enhance the wetting characteristics of these solder alloys, thereby improving their performance in electronic packaging applications. Future research should prioritize refining these strategies and investigating additional alloying elements in various low-temperature solders to enhance the long-term reliability and stability of solder joints. This approach will contribute to the development of more robust and efficient lead-free soldering solutions for advanced electronic applications.

Acknowledgments. The authors would like to acknowledge the financial support from the collaboration between Nihon Superior Co Ltd. and Universiti Malaysia Perlis, UniMAP. The authors are truly appreciated the support from the Centre of Excellence Geopolymer & Green Technology (CEGeoGTech), Universiti Malaysia Perlis, UniMAP.

References

1. Doorbar, P.J., Kyle-Henney, S.: 4.19 development of continuously-reinforced metal matrix composites for aerospace applications. Compr. Compos. Mater. **II**, 439–463 (Jan. 2018)
2. Zhao, M., Zhang, L., Liu, Z.-Q., Xiong, M.-Y., Sun, L.: Structure and properties of Sn-Cu lead-free solders in electronics packaging. Sci. Technol. Adv. Mater. **20**(1), 421–444 (Dec. 2019)
3. Felberbaum, M., Ventura, T., Rappaz, M., Dahle, A.K.: Microstructure formation in Sn-Cu-Ni solder alloys. JOM **63**(10), 52–55 (2011)
4. Negm, S.E., Mady, H., Bahgat, A.A.: Influence of the addition of indium on the mechanical creep of Sn–3.5%Ag alloy. J. Alloys Compd. **503**(1), 65–70 (2010)
5. Khan, H.N., Hounshell, D.A., Fuchs, E.R.H.: Science and research policy at the end of Moore's law. Nat. Electron. **1**(1), 14–21 (2018)

6. Luo, X., Peng, J., Zhang, W., Wang, S., Cai, S., Wang, X.: CALPHAD-guided alloy design of Sn–In based solder joints with multiphase structure and their mechanical properties. Mater. Sci. Eng. A **860** (Dec. 2022)
7. Le Han, D., Shen, Y.A., Huo, F., Nishikawa, H.: Microstructure evolution and shear strength of tin-indium-xCu/Cu joints. Metals (Basel) **12**(1) (Jan. 2022)
8. Davila-Frias, A., Yadav, O.P., Marinov, V.: A review of methods for the reliability testing of flexible hybrid electronics. IEEE Trans. Components Packag. Manuf. Technol. **10**(11), 1902–1912 (Nov. 2020)
9. Abtew, M., Selvaduray, G.: Lead-free solders in microelectronics. Mater. Sci. Eng. R Reports **27**(5–6), 95–141 (Jun. 2000)
10. Králová, I., et al.: Wettability in lead-free soldering: effect of plasma treatment in dependence on flux type. Appl. Surf. Sci. **668**, 160447 (2024)
11. Tan, X.F., et al.: In-situ investigation of the time-temperature dependent lattice and microstructure of Sn-Bi alloys. Materialia **33**, 101974 (Mar. 2024)
12. Liu, Y., Tu, K.N.: Low melting point solders based on Sn, Bi, and In elements. Mater. Today Adv. **8** (Dec. 2020)
13. Shen, J., Pu, Y., Yin, H., Luo, D., Chen, J.: Effects of minor Cu and Zn additions on the thermal, microstructure and tensile properties of Sn–Bi-based solder alloys. J. Alloys Compd. **614**, 63–70 (2014)
14. Shalaby, R.M.: Effect of silver and indium addition on mechanical properties and indentation creep behavior of rapidly solidified Bi–Sn based lead-free solder alloys. Mater. Sci. Eng. A **560**, 86–95 (2013)
15. Ramli, M.I.I., Mohd Salleh, M.A.A., Yasuda, H., Chaiprapa, J., Nogita, K.: The effect of Bi on the microstructure, electrical, wettability and mechanical properties of Sn-0.7Cu-0.05Ni alloys for high strength soldering. Mater. Des. **186**, 108281 (2020)
16. Liu, Y., Sun, F.L., Yan, T.L., Hu, W.G.: Effects of Bi and Ni addition on wettability and melting point of Sn-0.3Ag-0.7Cu Low Ag Pb-free solder. In: Proceedings, 2008 Int. Conf. Electron. Packag. Technol. High Density Packag. ICEPT-HDP 2008, pp. 8–11 (2008)
17. Zhou, J., Tan, X.F., McDonald, S.D., Nogita, K.: Mechanical properties and microstructure of binary In-Sn alloys for flexible low temperature electronic joints. Materials (Basel), **15**(23) (2022)
18. El-Bediwi, A.B., El-Bahay, M.M.: Influence of silver on structural, electrical, mechanical and soldering properties of tin-indium based alloys. Radiat. Eff. Defects Solids **159**(2), 133–140 (2010)
19. Fima, P., Gancarz, T., Pstruś, J., Sypień, A.: Wetting of Sn-Zn-xIn ($x = 0.5, 1.0, 1.5$ wt%) alloys on cu and Ni substrates. J. Mater. Eng. Perform. **21**(5), 595–598 (May 2012)
20. Li, L.F., Cheng, Y.K., Xu, G.L., Wang, E.Z., Zhang, Z.H., Wang, H.: Effects of indium addition on properties and wettability of Sn-0.7Cu-0.2Ni lead-free solders. Mater. Des. **64**, 15–20 (Dec. 2014)
21. Nabihah, A., Nurulakmal, M.S.: Effect of In addition on microstructure, wettability and strength of SnCu Solder. Mater. Today Proc. **17**, 803–809 (2019)
22. Sheng, J., et al.: Investigation on wetting behavior of Cu/Sn systems by molecular dynamics simulation. Results Surf. Interfaces, 100219 (2024)
23. Wang, Y., et al.: First-principles calculations on physical properties of Ni3Snx binary system intermetallic compounds and Ni/Ni3Sn interfaces in Nickel-Tin TLPS bonding layer. Intermetallics **101**, 27–38 (May 2018)
24. Mayahi, R.: An investigation concerning generalized stacking fault behavior of AlCoxCrFeNi ($0.25 \leq x \leq 2$) high entropy alloys: Insights from first-principles study. J. Alloys Compd. **818**, 152928 (2020)

25. Liu, Y., Fu, H., Sun, F., Zhang, H., Kong, X., Xin, T.: Microstructure and mechanical properties of as-reflowed Sn58Bi composite solder pastes. J. Mater. Process. Technol. **238**, 290–296 (2016)
26. Yang, C.H., Zhou, S., Lin, S.K., Nishikawa, H.: A computational thermodynamics-assisted development of Sn-Bi-In-Ga quaternary alloys as low-temperature Pb-free solders. Mater. **12**(4), 631 (Feb. 2019)
27. Ismail, N., et al.: A systematic literature review: the effects of surface roughness on the wettability and formation of intermetallic compound layers in lead-free solder joints. J. Manuf. Process. **83**, 68–85 (2022)
28. Bhat, K.N., Prabhu, K.N., Satyanarayan: Effect of reflow temperature and substrate roughness on wettability, IMC growth and shear strength of SAC387/Cu bonds. J. Mater. Sci. Mater. Electron. **25**(2), 864–872 (2014)
29. Satyanarayan, S., Prabhu, K.N.: Solder joint reliability of Sn-Cu and Sn-Ag-Cu lead-free solder alloys solidified on copper substrates with different surface roughnesses. Mater. Sci. Forum **830–831**, 265–269 (2015)
30. Eustathopoulos, N.: Dynamics of wetting in reactive metal/ceramic systems. Acta Mater. **46**(7), 2319–2327 (1998). https://doi.org/10.1016/S1359-6454(98)80013-X
31. Young, T.: An essay on the cohesion of fluids. Philos. Trans. R. Soc. London **95**, 65–87 (Aug. 1805)
32. Zou, J., et al.: Exploration of pretreatment process and wettability of high bonding Ni-P coatings on W-Cu composites for electronic packaging. Appl. Surf. Sci. **673**, 160911 (2024)
33. Lei, P., et al.: Research status of laser surface texturing on tribological and wetting properties of materials: a review. Optik (Stuttg) **298**, 171581 (2024)
34. Roduan, S.F., et al.: Effectiveness of dimple microtextured copper substrate on performance of Sn-0.7Cu solder alloy. Materials (Basel) **16**(1), 1–15 (2023)
35. Zhang, Z., et al.: Magnetron sputtering preparation of Cu6Sn5 preferred-orientation coating and its influence on wettability of Sn-based lead-free solder. Surf. Coatings Technol. **450**, 129014 (2022)
36. Hua, F., Mei, Z., Glazer, J.: Eutectic Sn-Bi as an alternative to Pb-free solders. In: 1998 Proceedings. 48th Electronic Components and Technology Conference (Cat. No.98CH36206), pp. 277–283 (1998)
37. Dong, W., Shi, Y., Xia, Z., Lei, Y., Guo, F.: Effects of trace amounts of rare earth additions on microstructure and properties of Sn-Bi-based solder alloy. J. Electron. Mater. **37**(7), 982–991 (2008)
38. Hou, Z., et al.: Enhancement mechanism of Te doping on microstructure, wettability and mechanical properties of Sn–Bi-based solder. Mater. Sci. Eng. A **848**, 143445 (2022)

The Effect of Modified Calcium Carbonate on Hardness and Microstructure of Tin-Zinc Composite Solder

Wan Hafizah Mohd Saufee[1], Juyana A. Wahab[1], Farah Farhana Zainal[1], Khairel Rafezi Ahmad[1], Ahmad Azmin Mohamad[2], and Muhammad Firdaus Mohd Nazeri[1](✉)

[1] Surface Technology Special Interest Group, Faculty of Chemical Engineering Technology, Universiti Malaysia Perlis, 02600 Jejawi, Arau, Perlis, Malaysia
firdausnazeri@unimap.edu.my

[2] Energy Materials Research Group (EMRG), School of Materials and Mineral Resources Engineering, Universiti Sains Malaysia, Penang, Malaysia

Abstract. This work investigates the effect of adding nickel (Ni)-coated micro-sized precipitated calcium carbonate (PCC) on the hardness performance of near-eutectic tin-zinc (Sn-9Zn) solder and its corresponding microstructure. The surface modification of PCC was acquired by implementing electroless Ni plating, resulting in the deposition of a Ni layer onto the PCC as filler in the Sn-9Zn composite solder. The sample is subsequently subjected to analysis by employing Zeta potential and microstructure analyses to evaluate the effect of the Ni coating to the PCC. The addition of 0.5 wt.% micro-PCC and Ni-coated micro-PCC revealed a significant increase in hardness, particularly with Ni-PCC (SZNMP), reaching 18.5 HV. The enhancement is attributed to the effective dispersion of Ni-PCC within the solder matrix, which mitigates agglomeration issues which associated with uncoated PCC. The Ni coating promotes better filler-matrix interaction, enabling a more uniform distribution of reinforcement particles that helps in producing finer Zn-rich phase. This uniformity is crucial for the Zener pinning effect, where Ni-PCC particles effectively pin grain boundaries, refining the microstructure and increasing hardness.

Keywords: Precipitated calcium carbonate · Ceramic filler · Electroless plating · Nickel · Composite solder

1 Introduction

The eutectic tin-zinc (Sn-9Zn) alloy is a promising lead-free solder (LFS) with a melting point around 198 °C and benefits from its single melting temperature [1]. However, Zn presence contributes to low corrosion resistance and poor wettability [2]. To improve performance, metals and non-metal additives are introduced, including non-metal fillers in composite solders. Ceramic particles like alumina [3], zinc oxide [4], and kaolin [5] have been used to enhance properties such as thermal conductivity, wettability, and mechanical strength.

N. R. Abdul Razak et al. (Eds.): ECOOP 1996, SPPHY 418, pp. 101–107, 2025.
https://doi.org/10.1007/978-981-96-2871-1_15

Recently, precipitated calcium carbonate (PCC) has gained attention as a cheap, accessible composite solder filler [6]. However, ceramic fillers like PCC tend to agglomerate due to density differences, leading to poor homogeneity in the solder [7]. This aggregation weakens the interfacial interaction between the matrix and filler and causes uneven filler distribution during the reflow process. To address this, dispersion techniques are essential, such as ball milling [8], high-energy mixing [9], or using dispersants [10] and surfactants [11]. Additionally, coating ceramic fillers with metal is a viable method to improve wettability and distribution [12].

Nickel (Ni) is widely used as a coating material for its durability and oxidation resistance, making it an effective electrocatalyst [13]. Its applications span various industries, including enhancing corrosion resistance on copper surfaces [14]. Additionally, Ni coating on ceramic fillers is effective in reducing agglomeration, thanks to its strong affinity for Sn. Recent studies indicate that a 5% Ni coating provides optimal coverage on micro-sized PCC fillers [15], with successful interaction between Ni and the Sn matrix observed during the soldering process [16].

To our knowledge, there have been very limited comprehensive studies on the effectiveness of adding a Ni-coated layer to ceramics such as micrometer-sized PCC (micro-PCC) in relation to critical reliability aspects of LFS, including the hardness and its relationship with the microstructure. In this work, we investigate the impact of Ni-coating on micro-PCC ceramic filler on the hardness of eutectic Sn-9Zn composite solders, supported by microstructural and elemental analyses.

2 Experimental Procedure

To produce Ni-PCC, hydrated nickel chloride ($NiCl_2{\cdot}6H_2O$), 20 g of micro-PCC (average size is 2.8 μm) and hypophosphite (NaH_2PO_2) were mixed in a 1:1:1 ratio in heated water (80 °C) and stirred at 400 rpm. Sodium hydroxide drops were added to maintain a pH of 8 while keeping the temperature constant. Subsequently, NaH_2PO_2 was added in a 1:1 ratio with $NiCl_2{\cdot}6H_2O$ at a concentration of 5%. After cooling, the filtered residues were pulverized into a fine powder using a pestle and mortar. To produce the composite solders, 0.5 wt.% of either PCC or 5% Ni-coated PCC powders were added to the Sn-9Zn master alloy. The mixture was heated in an inert crucible at 550 °C, then allowed to cool at ambient temperature. Various samples were prepared, including Sn-9Zn (SZ), Sn-9Zn-micro-PCC (SZMP), and Sn-9Zn-5 wt.% Ni-PCC (SZNMP). The composite solders were then manually rolled to a thickness of 0.5 mm and mechanically punched to create 5 mm-diameter solder pellets. The resulting composite solders were analyzed for microstructural and elemental properties using a scanning electron microscope (SEM, JEOL JSM-6460-LA equipped with energy dispersive spectroscopy (EDS)). Vickers microhardness testing (Sinowon) was conducted at 0.1 kgf with a 10-s dwell time to assess the hardness. The Zeta potential was measured by using Zeta potential analyzer (Malvern Panalytical).

3 Results and Discussions

3.1 Hardness

The Vickers microhardness shows increasing trend with the additions of 0.5 wt.% PCC or Ni-PCC (Fig. 1a). Pristine SZ produced 15.8 ± 1.7 HV and significantly increased to 18.1 ± 0.3 HV for the SZMP. The highest reading was recorded for SZNMP (18.5 ± 0.2 HV). In general, improvement in the hardness and mechanical properties were contributed by the increased ability to provide additional support and resistance against deformation [15]. The presence of these particles within the matrix creates multiple barriers to dislocation motion, which is a primary mechanism of deformation in alloys [17].

The ceramic particles bear a significant portion of the applied load. This reduces the stress on the softer matrix material, leading to increased hardness [18]. The Zeta potential of −3.39 mV for micro-PCC particles suggests that the particles or surfaces in the system carry a net negative charge, with relatively strong attraction between particles (Fig. 1b). This highlights that agglomeration is seen for this reinforcement. In contrast, the Ni-coated micro-PCC particles have a Zeta potential of −11.9 mV, indicating a considerably stronger repulsive force between the particles [19, 20]. This difference highlights that the Ni coating on micro-PCC successfully achieves better particle separation. Moreover, the Ni coating on PCC particles further enhance the strength of the composite. The effect of metallic coating such as Ni effectively allows for better reinforcement distribution [21]. The more efficient reinforcement distribution permits for the best hardness seen for the SZNMP through the Zener pinning effect.

The well dispersed particles allow for better interaction with the grain boundaries, and can physically block the movement of the grain boundaries. Hence, higher energy is needed to move around or through these particles [22], and gives the highest hardness value for the SZNMP.

3.2 Microstructure

The eutectic SZ solder is represented with the distribution of a Zn-rich phase in the form of a black needle-like structure, randomly distributed in the light-contrast matrix of β-Sn (Fig. 2a). The black needle-like structure was produced due to the limited solubility of both elements in each other, as previously encountered [6]. For SZ solder, the average Zn-rich phase lengths are 36.9 ± 5.9 μm. Corresponding EDS shows that the weight proportion of Sn (90.49 wt.%) and Zn (9.51 wt.%) in pristine SZ solder. This composition matched the raw materials and was properly produced since there were no impurities detected [12].

The composite solder containing micro-PCC without Ni coating exhibits an increased in the Zn-rich phase length to 157.1 ± 20.2 μm (Fig. 2b). Interestingly, additions of non-coated PCC slightly increased the weight of Zn to 9.93 wt. %. This is strikingly concurrent with the enlarged Zn-rich phase formation. In the case of the presence of Ni coating, the amount of Zn detected is lower (5.82 wt.%) in SZNMP, of which is consistent with the smaller Zn-rich phase seen in SZNMP (Fig. 2c).

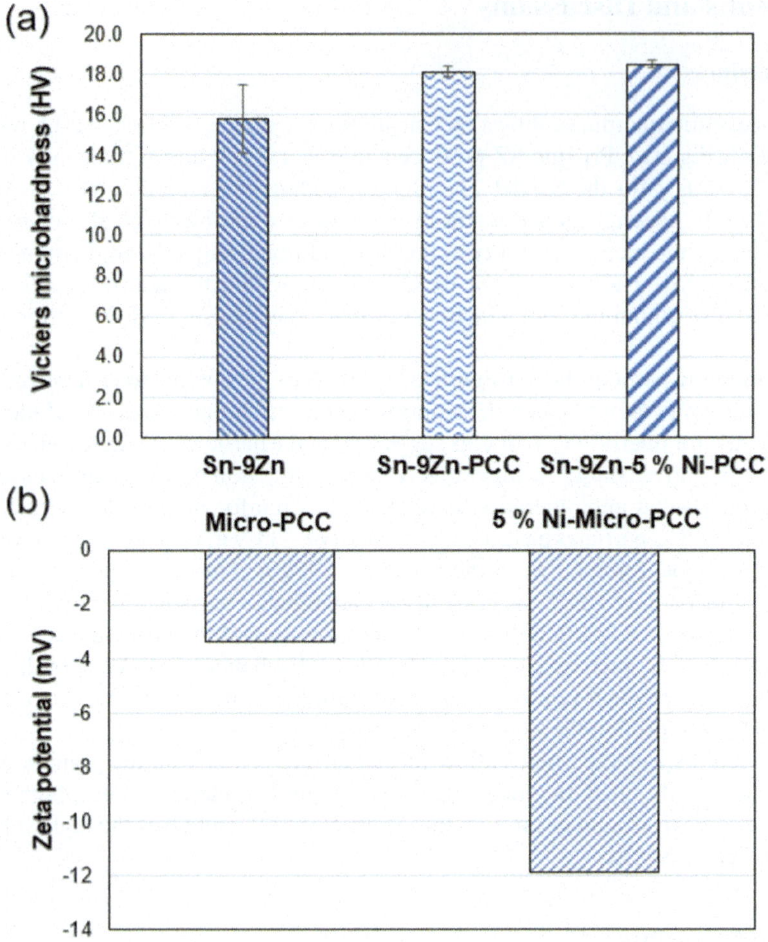

Fig. 1. (a) Average Vickers microhardness value of composite solders and (b) effect of Ni coating on Zeta potential of PCC

The increase in the size of the Zn-rich phase in SZMP can be attributed to the less efficient distribution of PCC within the composite solders and the presence of agglomeration. The addition of ceramics typically promotes heterogeneous nucleation during the solidification process. Agglomerated reinforcements, especially with high surface energy may serve as nucleation sites for solidification [23]. As more Zn atoms cluster around these agglomerated sites, the Zn-rich regions can grow larger compared to those in a PCC-free matrix.

On the other hand, the successful deposition of Ni on micro-PCC forms a metallic surface that exhibits enhanced compatibility with Sn in the solder [24], through the formation of core-shell structure of the reinforcement [25] and improve the filler-matrix interaction. This facilitates the mitigation of agglomeration concerns in the ceramic filler. In this case, the individual particles can act as effective nucleation sites during the

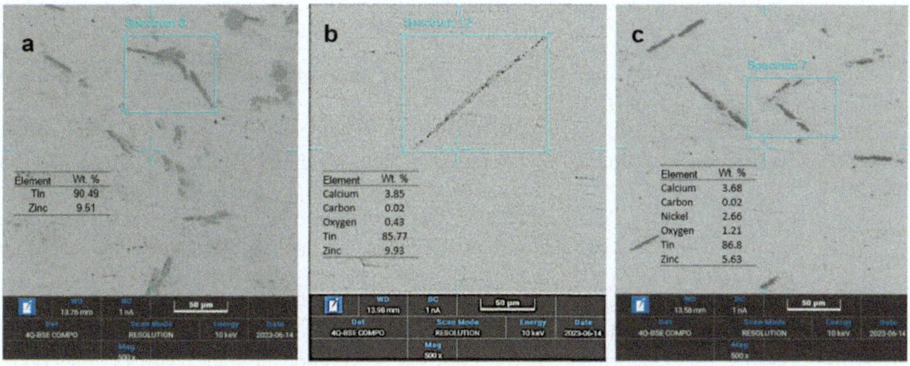

Fig. 2. The effect of sample variations on microstructure (a) Sn-9Zn, (b) Sn-9Zn + 0.5 wt. % PCC and (c) Sn-9Zn + 0.5 wt. % Ni-PCC

solidification process. Uniform dispersion of these particles within the molten solder lowers the energy barrier for the nucleation [26]. This promotes the formation of more refined microstructure of Zn-rich regions for SZNMP. Comparable finding on grain refinement was also previously reported [15].

4 Conclusion

The investigation of Ni coating on micro- and nano-PCC, and its effect on the hardness and microstructure of Sn-9Zn composite solder have been investigated. It was discovered that:

i. The addition of 0.5 wt.% PCC contributes to the increased hardness by providing additional support and resistance against deformation. The presence of PCC particles within the solder matrix introduces multiple barriers to dislocation motion, which is the primary mechanism of deformation in alloys. These barriers force the dislocations to bypass or overcome the ceramic particles, requiring more energy and resulting in an overall increase in hardness.
ii. However, the distribution of PCC in SZMP is less efficient, due to agglomeration, proven by the Zeta potential value.
iii. Better filler-matrix interaction allows for finer microstructure and higher Zeta potential. This is evident from the microstructure analysis, where the SZNMP shows smaller Zn-rich phase compared to SZMP.
iv. The addition of nickel-coated PCC enhances the hardness of the Sn-Zn solder. Better PCC distribution is seen responsible for this positive trait.

Overall performance shows that SZNMP solder provides the most advantageous outcomes, as Ni coating on PCC notably augments the hardness of pure SZ and SZMP solders.

Acknowledgments. The authors would like to thank the Ministry of Higher Education Malaysia for financial support through the Fundamental Research Grant Scheme FRGS/1/2019/TK05/UNIMAP/02/5.

References

1. Jiang, N., et al.: Reliability issues of lead-free solder joints in electronic devices. Sci. Technol. Adv. Mater. **20**(1), 876–901 (2019)
2. Jung, D.-H., Jung, J.-P.: Review of the wettability of solder with a wetting balance test for recent advanced microelectronic packaging. Crit. Rev. Solid State Mater. Sci. **44**(4), 324–343 (2019)
3. Wang, B., Wu, Y., Wu, W., Wang, H., Zhang, K.: Fabrication and properties of NiO-modified Al2O3 reinforced Sn1. 0Ag0. 5Cu composite solder and soldering performance, J. Mater. Sci. **57**(36), 17491–17502 (2022)
4. Skwarek, A., et al.: Application of ZnO nanoparticles in Sn99Ag0. 3Cu0. 7-based composite solder alloys. Nanomaterials **11**(6), 1545 (2021)
5. Zaimi, N.S.M., et al.: Performance of Sn-3.0 Ag-0.5 Cu composite solder with kaolin geopolymer ceramic reinforcement on microstructure and mechanical properties under isothermal ageing. Materials **14**(4), 776 (2021)
6. Leong, W.K., Mohamad, A.A., Nazeri, M.F.M.: Effect of the nickel coated precipitated calcium carbonate addition on microstructure, phase and wettability of Sn-9Zn solder. J. Phys. Conf. Ser. IOP Publishing, p. 012031 (2022)
7. Wan, Y.-J., Li, G., Yao, Y.-M., Zeng, X.-L., Zhu, P.-L., Sun, R.: Recent advances in polymer-based electronic packaging materials. Compos. Commun. **19**, 154–167 (2020)
8. Duo, L., et al.: Interfacial characteristics in CNTs-AgCuTi systems. Chin. J. Aeronaut. **35**(4), 450–460 (2022)
9. Pal, M.K., Gergely, G., Koncz-Horváth, D., Gácsi, Z.: Characterization of the interface between ceramics reinforcement and lead-free solder matrix. Surf. Interfaces **20**, 100576 (2020)
10. Wang, S., Zhao, Q., Xing, P., Zhuang, Y., Wang, L.: Influence of dispersant on the microstructure and performance of the hot-pressed B4C-YB4 ceramics. J. Aust. Ceramic Soc., 1–13 (2023)
11. Morales-Florez, V., Dominguez-Rodriguez, A.: Mechanical properties of ceramics reinforced with allotropic forms of carbon. Prog. Mater. Sci. **128**, 100966 (2022)
12. Leong, W.K., Mohamad, A.A., Nazeri, M.F.M.: Surface Modifications on Ceramic Reinforcement for Tin-Based Composite Solders, Recent Progress in Lead-Free Solder Technology: Materials Development, pp. 53–75. Springer, Processing and Performances (2022)
13. Luczak, J., Lieder, M.: Nickel-based catalysts for electrolytic decomposition of ammonia towards hydrogen production. Adv. Colloid Interface Sci., 102963 (2023)
14. Yang, Z., Liu, X., Tian, Y.: Fabrication of super-hydrophobic nickel film on copper substrate with improved corrosion inhibition by electrodeposition process. Colloids Surf. A **560**, 205–212 (2019)
15. Leong, K., et al.: Wettability and hardness investigation of nickel-coated precipitated calcium carbonate Sn-9Zn composite solder. Arch. Metallurgy and Materials (2023) 1035–1040
16. Mokhtari, O., et al.: Disabling of Nanoparticle effects at increased temperature in nanocomposite solders. J. Electron. Mater. **41**(7), 1907–1914 (2012)
17. Alhussein, W.A.: Effect of Nano-Aluminium Oxide Percentage of A-TIG on mechanical properties of welded joint of 7075 Aluminum Alloy. Int. J. Mater. Technol. **3**(2), 74–79 (2023)
18. Yan, Y.-F., et al.: Ceramic particles reinforced copper matrix composites manufactured by advanced powder metallurgy: preparation, performance, and mechanisms. Int. J. Extreme Manuf. **5**(3), 032006 (2023)
19. Farcas, C.G., et al.: Thermosensitive betulinic acid-loaded magnetoliposomes: A promising antitumor potential for highly aggressive human breast adenocarcinoma cells under hyperthermic conditions. Int. J. Nanomed., 8175–8200 (2020)

20. Gaba, B., et al.: Vitamin E loaded naringenin nanoemulsion via intranasal delivery for the management of oxidative stress in a 6-OHDA Parkinson's disease model. J. BioMed. Res. Int. (2019)
21. Ge, Y.X., et al.: Towards high performance in Ti-based composite through manipulating nickel coatings on graphene reinforcement. J. Alloy. Compd. **893**, 162240 (2022)
22. Singh, T., Tiwari, S., Shukla, D.: Effect of nano-sized particles on grain structure and mechanical behavior of friction stir welded Al-nanocomposites, **234**(2), 274–290 (2020).
23. Greer, A.L., Bunn, A.M., Tronche, A., Evans, P.V., Bristow, D.J.: Modelling of inoculation of metallic melts: application to grain refinement of aluminium by Al–Ti–B. Acta Mater. **48**(11), 2823–2835 (2000)
24. Ma, Y., Wang, X., Dong, Z., Ni, P., Li, Z., Liu, H.: Microstructure and mechanical performance of nickel coating on copper sheet via laser shock processing. Appl. Surf. Sci. **592**, 153366 (2022)
25. Gyökér, Z., Gergely, G., Horváth, D.K., Bodnár, E., Gácsi, Z.: Role of reinforcement surface treatment on the SnAg3Cu0. 5 microelectronic joints. J. Appl. Surf. Sci. **475**, 982–985 (2019)
26. Li, M.-l., Zhang, L., Jiang, N., Zhang, L., Zhong, S.-j.J.M.: Design, materials modification of the lead-free solders incorporated with micro/nano-sized particles: a review. Mater. Design **197**, 109224 (2021)

The Effect of Silver (Ag) Addition on Sn0.7Cu Solder Alloy Properties

Athirah Khemar[1], Kamrosni Abdul Razak[2(✉)], Aimi Noorliyana Hashim[2], Norsuria Mahmed[2], and Suwetha R. Naidu[2]

[1] Centre of Excellence Geopolymer and Green Technology, Faculty of Chemical Engineering Technology, Universiti Malaysia Perlis (UniMAP), 02600 Taman Muhibah, Arau, Perlis, Malaysia

[2] Faculty of Chemical Engineering Technology, Universiti Malaysia Perlis, 02600 Jejawi, Arau, Perlis, Malaysia

kamrosni@unimap.edu.my

Abstract. The study explores the effects of adding silver (Ag) to a Sn0.7Cu solder alloy on its mechanical properties and microstructures. Ag was introduced at concentrations of 3 wt. %, 4 wt. %, and 5 wt. %. The solder alloy was melted at 330 °C for bulk samples and 217 °C for solder sheets on aluminium plates. After cleaning with acetone and ultrasonic treatment, the solder sheets were rolled to 30 μm thickness and cut into 12 mm^2 pieces. Copper substrates were prepared by etching, rinsing, and drying, and then solder flux paste was applied. The solder sheets were placed on the substrates, covered with another layer of flux paste, and reflowed in an oven to complete the process. After reflow, samples were mounted for observation of intermetallic compounds (IMC) and analysed for microstructure using an optical microscope. The hardness of bulk samples was evaluated using a micro-Vickers tester. Results showed that silver addition increased hardness progressively with higher concentrations in the solder alloy. Microstructural analysis revealed that while the original Sn0.7Cu alloy had mainly eutectic networks, the addition of silver introduced a combination of eutectic networks and β-Sn grains. The spacing between these structures decreased with 3 wt. % silver and increased with higher concentrations (4 wt. %, 5 wt. %). Overall, the study highlights how silver modification alters the microstructure and mechanical properties of Sn0.7Cu solder alloys, enhancing solder joint performance through a dispersion-strengthening mechanism.

Keywords: Solder alloy · hardness test · intermetallic compound · reflow

N. R. Abdul Razak et al. (Eds.): ECOOP 1996, SPPHY 418, pp. 108–114, 2025.
https://doi.org/10.1007/978-981-96-2871-1_16

1 Introduction

Soldering is a critical process in both electronics and metalworking industries, where metal components arejoined using molten filler metal called solder [1]. Traditionally, Sn-Pb solder was widely used but has been phased out due to its toxicity and environmental impact, leading to a shift to lead-free alternatives mandated since 2006 in Europe, such as tin-silver-copper (Sn-Ag-Cu), tin-silver (Sn-Ag), tin-copper (Sn-Cu), tin-zinc (Sn-Zn), and tin-bismuth (Sn-Bi) alloys [2]. Among these, Sn0.7Cu stands out for its simpler production and recycling processes, making it cost-effective and environmentally friendly, although it suffers from drawbacks like low tensile strength, poor wetting properties, and a relatively high melting point [3].

To improve the performance of lead-free solders, micro-alloying techniques have been employed effectively, with silver (Ag) being one of the key additives used. Silver enhances wettability, thermal conductivity, and mechanical strength in solder joints, particularly in popular alloys like SnAgCu (SAC) [4]. However, SAC alloys still fall short of the reliability provided by lead-based solders, especially as the trend towards miniaturization in microelectronic devices demands superior thermal management, wetting characteristics, and mechanical durability [5]. The formation and stability of intermetallic compounds (IMCs), which naturally form at the interface between solder and substrate, play a critical role in determining the long-term reliability of solder joints, as IMCs are inherently brittle and can compromise mechanical properties [5].

Despite ongoing research efforts into the properties of Sn-xAg0.7Cu (where x represents various percentages of silver) solder alloys, there remains a significant gap in understanding their interfacial interactions and the kinetics of IMC layer development as-reflowed. This study investigation proceeds by using optical microscopy and Vickers hardness of ASTM B32-20 testing methods. By elucidating how silver (Ag) content affects microstructure, hardness, and mechanical behavior, the research seeks to advance the understanding and application of Sn-xAg0.7Cu solder alloys in modern electronics manufacturing, where reliability and performance are paramount.

2 Experimental Procedure

2.1 Raw Materials Preparation

The raw materials for the Sn0.7Cu solder alloy were sourced from Nihon Superior (M) Sdn. Bhd, located in Ipoh, Perak, Malaysia. The solder alloy preparation involved melting the ingots in a solder pot at a temperature of 250 °C. The substrate used was copper with a purity of 99.9%, measuring 15 mm × 15 mm in dimension and 0.3 mm in thickness. By using the casting method, where the Sn0.7Cu solder alloy was melted at 330 °C, and Ag granular was filled into the molten Sn0.7Cu alloy. Three different weight percentages (wt%) of Ag granules content, 3 wt%, 4 wt%, and 5 wt%, were added to the molten Sn0.7Cu alloy. Subsequently, the molten mixture was poured onto a ceramic plate and button shape mould for bulk sample to enable the cooling process. As a result of the casting process, Sn-xAg 0.7Cu solder alloys were formed and utilized as the solder material (x = 0 wt%, 3 wt%, 4 wt%, 5 wt%). Then, a microhardness test was conducted with 10 s of holding time with a 10kgs load for the bulk solder sample.

2.2 Sample Preparation

The bulk solders were then proceeded for a hardness test to see the effect of different Ag percent contained in Sn0.7Cu solder alloy. Then, thin sheet soldering was applied to prepare the Sn0.7Cu-xAg solder alloy. All the alloys were rolled to form a thin sheet layer with an approximate thickness of ~30 μm. Then, the sheets were cut into dimensions of 12 mm × 12 mm as they were to be placed onto the cleaned substrate. For substrate preparation, copper (Cu) substrate was used and had to be cleaned first. The substrate was manually cleaned by dipping into a 5% hydrochloric acid (HCl) solution and washed with distilled water at an ambient temperature to remove the oxidized layer on the substrate surface. Then, the substrate was cleaned using acetone (C_3H_6O) and washed with distilled water. The substrates were then left to air-dry, and after drying, flux was spread on the substrate to aid in solder joining. The flux was applied in the form of paste. Then, the thin sheets of Sn0.7Cu, Sn3Ag 0.7Cu, Sn4Ag 0.7Cu, and Sn5Ag 0.7Cu were placed on the substrates spread with flux, and flux paste was also applied to the top of the solder alloys to enhance the solder joint during reflow. The samples were then reflowed using a reflow oven at 230 °C for 300 s. After that, the samples were cleaned in acetone using an ultrasonic bath for 5 min. The samples then were prepared for the microstructure and IMC layers study of the solder joint. Lastly, the samples were mounted and polished to proceed with the initial study on the IMC layer with and without the addition of Ag.

3 Results and Discussions

3.1 Hardness Test Properties on the Sn0.7Cu-XAg Solder

The results of Ag addition in Sn0.7Cu solder alloy have changed the mechanical properties of the solder, especially the hardness properties. The addition of silver (Ag) to the Sn0.7Cu alloy affected not only its hardness but also its melting temperature. The microhardness of the solder is often connected with how the metallic material resists wearing or abrasion. It also determines the applicability under various circumstances [6].

Figure 1 shows that the hardness data of the alloy increased with the addition of Ag in Sn0.7Cu solder. The data revealed that Sn0.7Cu without the addition of Ag has a hardness of 11.7 HV, while adding 3% Ag reduces the hardness to 9.5 HV. However, the hardness of solder that contained 4 wt% and 5 wt% of Ag increased to 12.6 HV and 14.8 HV respectively. This indicates that while a 3 wt% Ag addition slightly softens the alloy, increasing the Ag content beyond this amount leads to greater hardness. Another study supports these findings, showing that adding Ag initially decreases the hardness of Sn0.7Cu solder, but with further increases in silver content, the hardness gradually increases [1]. In addition, Fig. 1 shows the average reading of the Vickers hardness test of all samples, as the increased of Ag contained caused changes in the mechanical properties of the solder alloy. As in previous studies concluded that addition Ag element into Sn0.7Cu solder influenced against the hardness of the Sn0.7Cu alloy [1]. Consequently, alloys with higher Ag content became harder and required higher temperatures to melt. The more Ag was added, the melting temperature of the alloy increased because silver had a higher melting point than Sn0.7Cu solder alloy.

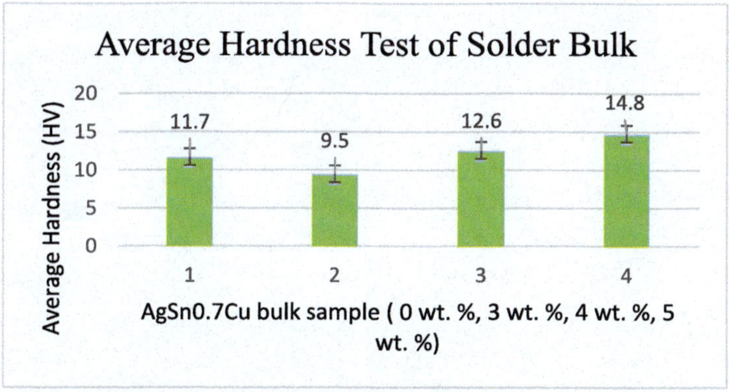

Fig. 1. Average Vickers microhardness of bulk solder

3.2 Microstructures and Intermetallic Compound (IMC)

The microstructure and IMC of the Sn0.7Cu and SnxAg0.7Cu were examined using an optical microscope (OM). OM images of microstructure in AgSn0.7Cu solder alloys containing 3 wt. %, 4 wt. % and 5 wt. % Ag respectively after being ground and polished, are shown the Fig. 2 (a) illustrates the microstructure of bulk solder Sn0.7Cu without the addition of Ag element and showing the formation of eutectic region of primary β-Sn. Figure 2 (b) shows that after the addition 3 wt% Ag to Sn0.7Cu solder alloys, the microstructure of the solder were changed, forming a new region containing Ag element and primary Sn. Meanwhile, Fig. 2 (c) shows the formation of Sn-Ag microstructure have change the shape of the primary Sn. This happened due to the good alloying properties of Ag have changed the microstructure of the Sn0.7Cu solder alloy which varied after the addition of the element. For Fig. 2 (d) the apparent region of Sn-Ag (5 wt%) microstructure and it believes that the changes also happened on the intermetallic properties of the solder. As increasing contained of Ag, the microstructure formed a stronger solder alloy that have the advantages that used mostly in electronic devices [8]. The amount of Ag contained in Sn0.7Cu solder effect the microstructure that occurs in the solder alloy.

The material properties of solder Sn0.7Cu-xAg depend on the change in the microstructure. Since the addition of Ag element into Sn0.7Cu solder has found differential results in the microhardness, thus also changed in its microstructure. The microstructures of SnAg0.7Cu lead-free solder alloys are made up of primary β-Sn grains as in Fig. 2. According to a dispersion strengthening mechanism, uniform microstructure positively affects the solder joints by enhanced mechanical properties [7].

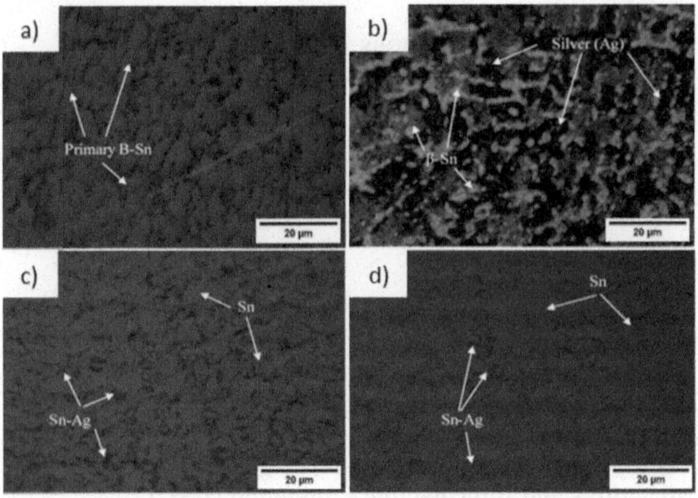

Fig. 2. Microstructure of (a) Sn0.7Cu solder alloy, (b) Sn3Ag0.7Cu solder alloy, (c) Sn4Ag0.7Cu solder alloy and (d) Sn5Ag0.7Cu solder alloy

The cross-sectional views of the solder alloy were also observed to study the effect of Ag to the layers for without and after the addition of Ag element into Sn0.7Cu solder alloy. Hence, the Fig. 3 show the IMC layers of the solder alloys as-reflowed. Figure 3 (a) show the cross-sectional of IMC formation for Sn0.7Cu solder that have a scallop-shape of Cu_6Sn_5 IMC morphology whereas the solder molten during the reflow. However, it also happens for solder with addition of Ag with increasing in percent of Ag as in Fig. 3 (b) Fig. 3 (c) and Fig. 3 (d) where the IMC layer found also formed scallop shape of Cu/Sn-Ag. This results also correspond to the previous study done by Lee et al. [9] on the effect of IMC on SAC solder alloy as-reflow at 260 °C and formed an interfacial reaction of Cu/Sn-Ag solder joints with typical scallop-types of Cu_6Sn_5 IMC layers were formed for the sample SAC interfaces.

The average thickness of the IMC layer in AgSn0.7Cu were observed. The 3 wt. % Ag in Sn0.7Cu have the thinnest IMC layer which is 0.508 μm compared to the others and 0 wt. % Ag in Sn0.7Cu shows that it has the thickest IMC layer which is 0.744 μm. Ag has a lowering thickness of IMC properties. As according to (Xu et al., 2017), silver has softening properties, thus it causes the layer of IMC to be thinner compared to 0 wt. % Ag in Sn0.7Cu.

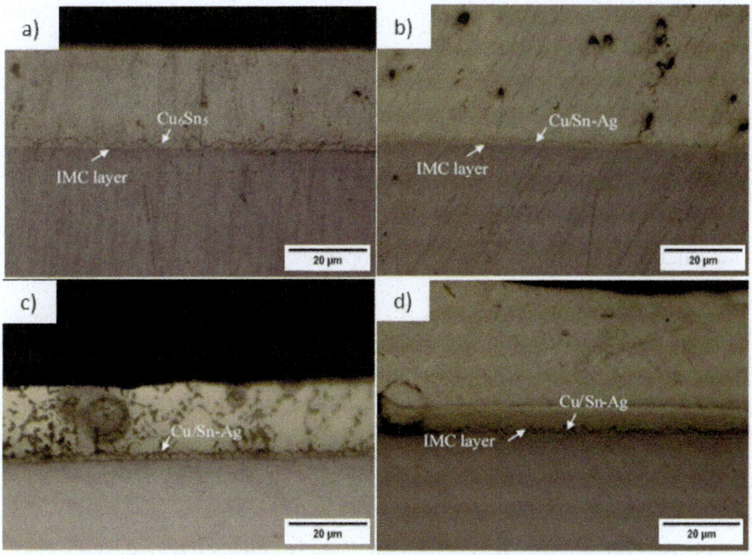

Fig. 3. IMC layer of (a) Sn0.7Cu solder alloy, (b) Sn3Ag0.7Cu solder alloy, (c) Sn4Ag0.7Cu solder alloy and (d) Sn5Ag0.7Cu solder alloy.

4 Conclusion

In conclusion, the study on the addition of Ag into Sn0.7Cu solder alloy affects ~~to~~ the soldering and process of solder joint reliability in several ways. A wide range of Ag-containing alloys have been studied by electronic manufacturers, materials suppliers, universities and other research groups [10]. With the addition of a small amount of Ag to Sn0.7Cu eutectic alloy, mechanical properties which are hardness properties and the thickness of the IMC layer of SnxAg0.7Cu solder were enhanced obviously. However, when adding Ag to Sn0.7Cu the hardness slightly drops as Ag carries softening properties by altering the bulk solder microstructure. In addition, the microstructure of SnAg0.7Cu solders and retarding effect of elements on the IMC in the solder is also demonstrated. The highest hardness value is shown by 5 wt % of Ag in Sn0.7Cu which was 14.8 HV. The IMC thickness shows that by adding the Ag, it drops and increases slowly based on the amount of silver in Sn0.7Cu.

Acknowledgments. The author would like to acknowledge the support from the Fundamental Research Grant Scheme (FRGS) under a grant number of FRGS/1/2020/TKO/UNIMAP/03/18 from the Ministry of Higher Education Malaysia also Special Interest Group (MetaPrime Research Group) Faculty of chemical engineering technology for the materials, facilities and finance support.

References

1. Siahaan, E., Riza, A.: The influences of silver content in lead-free solder Sn-0.7CuxAg (SAC) in physical and mechanical properties. In IOP Conference Series: Materials Science and Engineering, IOP Publishing Ltd (2020)
2. Li, Y., Chen, C., Yi, R., Ouyang, Y.: Review: Special brazing and soldering. J. Manuf. Process. **60**, 608–635 (2020)
3. Lai, Y., Hu, X., Jiang, X., Li, Y.: Effect of Ni addition to Sn0.7Cu solder alloy on thermal behavior, microstructure, and mechanical properties. J. Mater. Eng. Perform. **27**(12), 6564–6576 (Dec.2018)
4. Mao, X., Zhang, R., Yi, X., Hu, X., Li, Y., Jiang, X.: Study on the performance of Cu foam with different porosity on SAC305 solder joints under ultrasonic-assisted soldering. J. Mater. Sci. Mater. Electron. **32**(24), 28108–28118 (Dec.2021)
5. Zhao, M., Zhang, L., Liu, Z.Q., Xiong, M.Y., Sun, L.: Structure and properties of Sn-Cu lead-free solders in electronics packaging. Taylor and Francis Ltd (2019)
6. Ramli, M.I.I., Salleh, M.A.A.M., Amli, S.F.M., Razak, N.R.A.: Effect of bismuth additions on wettability, intermetallic compound, and microhardness properties of sn-0.7cu on different surface finish substrates. Sains Malays **49**(12), 3255–3259 (2020)
7. Adetunji, O.R., Ashimolowo, R.A., Aiyedun, P.O., Adesusi, O.M., Adeyemi, H.O., Oloyede, O.R.: Tensile, hardness and microstructural properties of Sn-Pb solder alloys. In Materials Today: Proceedings, pp. 321–325, Elsevier Ltd (2021)
8. Yang, W., Felton, L.E., Messler, R.W.: The effect of soldering process variables on the microstructure and mechanical properties of Eutectic Sn-Ag/Cu Solder Joints. J. Electron. Mater. **24**(10), 1465–1472 (1995)
9. Lee, L.M., Mohamad, A.A.: Interfacial reaction of Sn-Ag-Cu lead-free solder alloy on Cu: a review (2013)
10. Juan Zhong, S., Zhang, L., Lan Li, M., Min Long, W., Jiang Wang, F.: Development of lead-free interconnection materials in electronic industry during the past decades: Structure and properties. Mater. Des. **215**, 110439 (2022)

Advanced Materials

A Computational Study on the Impact of AgNP Nanocoating Thickness on Heat Sink Dissipation Capacity Inside Electronic Packaging

Xing Qi Lim[1], Mohd Sharizal Abdul Aziz[1(✉)], and C. Y. Khor[2]

[1] School of Mechanical Engineering, Engineering Campus, Universiti Sains Malaysia, 14300 Nibong Tebal, Seberang Perai Selatan, Penang, Malaysia
msharizal@usm.my

[2] Faculty of Mechanical Engineering & Technology, Universiti Malaysia Perlis (UniMAP), 02600 Arau, Perlis, Malaysia

Abstract. Nanoparticles (NP) with sizes ranging from 1 to 100 nm give high surface area properties and have a high potential to improve the thermal performance of the heat sink. Good heat-conducting metals, such as silver (Ag) and gold (Au), are rarely used in heat sinks due to their high cost. With nanocoating techniques, only a small amount of Ag or Au is needed to coat a layer on the heat sink, and the cost can be reduced. In this paper, the application of AgNP coating onto a heat sink is modelled to predict the heat dissipation performance of the heat sink. The performance of the heat sink with AgNP coating under natural convection is modeled using the ANSYS FLUENT software. The simulation result is verified by other researchers with the experimental findings. A parametric study on the effect of AgNP coating thickness is conducted by using a nanocoating thickness of 10–500 μm. The presence of the AgNP coating promotes heat dissipation of the heat sink but reduces when the thickness increases.

Keywords: CFD · coated heat sink · thermal performance · thermal resistance · emissivity

1 Introduction

In the present era, electronic devices have become indispensable to the community. The development of these devices can be attributed to the trend toward portability, size reduction and weight reduction. The miniaturization of electronic devices such as circuit boards has led to excessive thermal energy generation [1]. Heat accumulation that leads to circuit's temperature rise is the main cause of chip failures [2]. A suitable cooling system is necessary to guarantee the performance of the appliances over a prolonged duration while preserving their efficiency.

To enhance heat dissipation, the heat sink is commonly used to increase the available surface area for heat transfer [3]. Researchers have attempted many methods to enhance the performance of the heat sink, including the heat sink design [4–18], forced cooling [1, 18, 19], nanofluids [20, 21], and phase change material (PCM) [22, 23].

N. R. Abdul Razak et al. (Eds.): ECOOP 1996, SPPHY 418, pp. 117–125, 2025.
https://doi.org/10.1007/978-981-96-2871-1_17

Lately, researchers have sought to incorporate nanotechnology into cooling systems. It is hypothesised that the nanoparticles (NP) can be used as a coating on a heat sink to improve its thermal efficiency. The thermal conductivity of silver nanowires was enhanced when applied zinc oxide coating [24] and silver coating [25]. Silver NP [26, 27], copper NP [28, 29], and aluminum NP [30, 31] were applied onto chosen substrates with improvement in thermal performance spotted.

The simulation model is initially validated using the experimental results obtained by Zu et al. [32]. In the following study, the AgNP coating thickness will be varied to investigate its influences. This numerical simulation study aims to evaluate the impact of the AgNP coating (low emissivity [33] and high thermal conductivity) on the thermal dissipation performance of a heat sink using ANSYS Fluent 2021 R1. The findings of this research have practical implications for engineers and designers, as they can use this knowledge to improve the efficiency and reliability of electronic systems by implementing better heat dissipation strategies.

2 Numerical Methods (Model, Boundary Conditions & Material)

The solid and fluid domains were modelled using the actual size to simulate the conjugate heat transfer in real life. Figure 1a shows the plate-fin heat sink modeled with dimensions labelled. Figure 1b shows the thin layer created to represent the AgNP nanocoating coated onto the heat sink. Figure 1c depicts the full model of the coated heat sink with the air domain (height 72 mm, width 120 mm & length 100 mm) surrounding. Figure 1d depicts the boundary conditions employed, and detailed settings are shown in Table 1. The surface-to-surface (S2S) radiation model is adapted to consider the radiation effect of the heat sink.

For validation, since the result is compared with Zu et al. [32], material properties were retrieved from their paper except for thermal conductivity and specific heat capacity of the coating, which were not specified. Since the composite coating used acrylic as a matrix, the property of acrylic is taken where thermal conductivity is 0.1870 W m^{-1} K^{-1} and specific heat capacity 1460 J kg^{-1} K^{-1}. The properties of air, heat sink material (aluminum) and the coatings used are summarized in Table 2.

3 Results and Discussions

3.1 Grid Independence Test

To ensure the accuracy and reliability of the simulation model, grid independence test and validation must be conducted. The computational setup of the heat sink with 20 μm silver coating was meshed with different numbers of nodes and elements, named 'A' to 'E'. The maximum and minimum temperature of the coated heat sink after 100 min for each mesh was recorded in Table 3. The grid independence test revealed that mesh 'D' only shows a maximum deviation of 0.0026% from mesh 'E', thus, the model with 404k nodes and 347k elements is sufficiently accurate.

The simulation model was then validated with the experimental result for the uncoated and coated heat sink [32]. The temperature measurements were taken at three

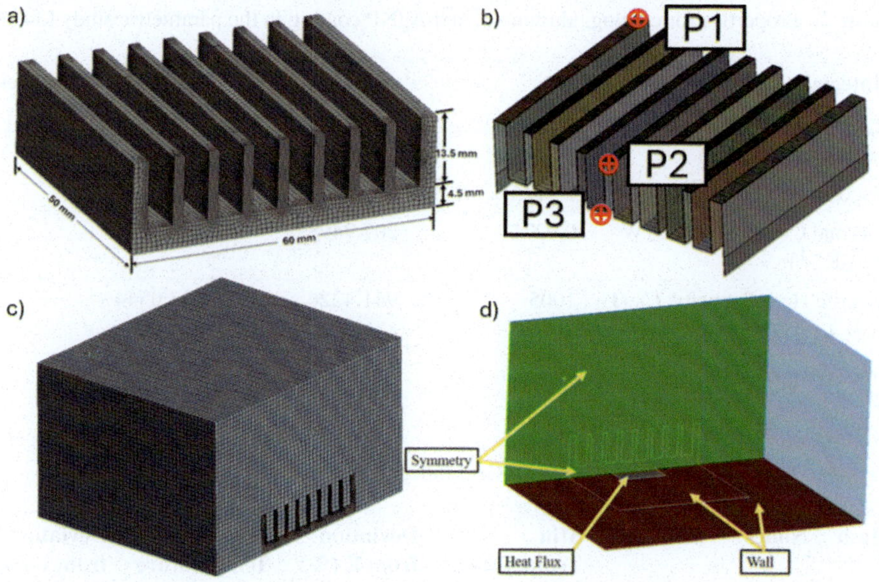

Fig. 1. a) Plate-fin heat sink model with dimensions, b) Coating with measurement points for validation; c) Coated heat sink with air domain, and d) Boundary condition assigned.

Table 1. Summary of the boundary conditions of the setup

Boundary surface	**Hydrodynamic Boundary Condition**	**Thermal Boundary Condition**
Air (Side, Back & Top)	Pressure inlet (101325 Pa)	Temperature (Validation - 24.3 °C (non-coated) & 23.8 °C (coated); Parametric study - 26.85 °C)
Air (Bottom), Heat Sink (Bottom) & Coat (Bottom)	No slip wall	Adiabatic (zero heat flux)
Coat-Air interface & Coat-Heat Sink Interface	No slip wall	Coupled
Heat flux	-	Constant heat flux 45000 W/m^2

points as shown in Fig. 1b. The data was recorded, and the percentage of error for each point was calculated in Table 4. The highest error percentage is merely 3.972%, with the majority of errors hovering around 3%.

Table 2. Properties for coating, aluminum, and AgNP coating in the parametric study [34, 35]

Material	Air	Aluminum (Al)	Silver Nanoparticles (AgNP) Coating
Density, ρ (kg m^{-3})	1.127	2570	10450
Viscosity, μ (kg m^{-1} s^{-1})	1.9125×10^{-5}	-	-
Thermal Conductivity, k (W m^{-1} K^{-1})	0.0271	207.9498	420.9205
Specific Heat Capacity, C_p (J kg^{-1} K^{-1})	1005	941.4226	233.0544
Emissivity, ε	-	0.030	0.020

Table 3. Grid Independence Test

Mesh	Nodes	Elements	Min Temperature (°C)	Deviation from E (%)	Max Temperature (°C)	Deviation from E (%)
A	222901	177707	122.469	4.6824	127.287	4.0930
B	243513	197263	117.181	0.1624	122.357	0.0613
C	323619	268249	117.180	0.1616	122.349	0.0548
D	404461	346718	116.988	0.0026	122.282	0.0000
E	833211	731512	116.991	-	122.282	-

Table 4. Validation result using the experimental data from Zu et al. [32]

Coating	Ambient Temperature (°C)	Position	Experiment, T (°C)	Simulation, T (°C)	% Error
Uncoated	24.3	P1	110.6	114.72	3.725
		P2	114.0	118.32	3.789
		P3	116.2	119.35	2.711
Coated	23.8	P1	84.6	87.96	3.972
		P2	92.6	94.04	1.555
		P3	94.6	95.07	0.497

3.2 Effect of AgNP Coating Thickness on Heat Sink Maximum, Minimum Temperature and Thermal Resistance

The steady state temperature of the heat sink is altered when it is coated with a thin layer of AgNP coating. From Table 5, the maximum and minimum temperatures of the uncoated heat sink recorded are 122.385 °C and 116.941 °C respectively, and its thermal resistance is 5.3075 K/W. Compared with coated heat sink of different thickness, it is observed that only 10 μm and 20 μm coated heat sinks perform slightly better than uncoated heat sink while the rest lower down the heat transfer efficiency of the heat sink. AgNP coating with high thermal conductivity facilitates the heat conduction and heat distribution from the heat source rapidly as the temperature difference of uncoated heat sink is the largest, 5.444 °C among all heat sinks. Nevertheless, the low emissivity of the AgNP coating layer, on the other hand limits the heat lost through radiation compared to the uncoated aluminum heat sink with a higher emissivity value. This finding aligns with the outcome reported by Gao et al. [33], which AgNP film with low emissivity would suppress thermal radiation.

The AgNP coating thickness of the heat sink affects the thermal performance of the coated heat sink. The plots of maximum and minimum temperature of the coated heat sink against coating thickness are depicted in Fig. 2 (left). The trendlines show insignificant increments at thin coating thickness, but the trends climb more rapidly when the coating gets thicker. The trendline of the minimum temperature described a steeper growth at thin coating thickness compared to the maximum temperature trendline. The coated heat sinks are sorted following their AgNP coating thickness from the thinnest to the thickest in Fig. 2 (right). The heat sinks with coating thickness ranging from 10 μm to 100 μm have shown similar temperature contours because the temperature rise is insignificant. When the coating thickness increases from 150 μm to 300 μm, the cyan and green colored region expand towards the ends of the heat sink and the centre of the heat sink being heated up starts turning yellow. Further thickening the AgNP coating to 500 μm only brings the heat sink to a higher temperature. These unveil a visible elevation of the heat sink temperature when the AgNP coating gets thicker as depicted by the plots in Fig. 2 (left).

The actual readings of the maximum and minimum temperature of the AgNP coated heat sink are listed in Table 5, along with the temperature difference and their thermal resistance calculated. The heat sink with 10 μm AgNP coating recorded the lowest maximum temperature of 122.181 °C and minimum temperature of 116.843 °C, while 500 μm AgNP coated heat sink showed the poorest performance, where it reached the highest maximum temperature of 128.197 °C and minimum temperature of 124.243 °C. The 10 μm coated heat sink has the lowest thermal resistance of 5.2962 K/W, while the 500 μm coated heat sink possesses the highest thermal resistance of 5.6304 K/W. Due to the high thermal conductivity value of AgNP coating, the maximum and minimum temperature difference of coated heat sinks reduces marginally when the coating thickness increases. However, this marginal improvement brought in another issue where the smaller fin gap resists the passage of cooler air along the channels. This finding is in accordance with Ghandouri et al. [36], where more space between two adjacent fins will allow more air in the main flow stream.

In short, the findings from these results are important for coating design for heat sinks in electronic devices. The observed variations highlight the significance of considering coating thickness in heat sink design. A small adjustment of the coating thickness might improve the heat dissipation performance of the heat sink.

Table 5. Coating thickness of heat sink, its maximum and minimum temperature, temperature difference, ΔT, and thermal resistance

Coating Thickness (μm)	Maximum Temperature (°C)	Minimum Temperature (°C)	Temperature Difference, ΔT (°C)	Thermal Resistance (K W^{-1})
Uncoated	122.385	116.941	5.444	5.3075
10	122.181	116.843	5.338	5.2962
20	122.282	116.988	5.294	5.3018
50	122.422	117.288	5.154	5.3107
100	122.468	117.515	4.953	5.3121
150	123.091	118.313	4.778	5.3467
200	123.846	119.222	4.624	5.3887
250	124.490	119.994	4.496	5.4244
300	125.215	120.849	4.366	5.4647
350	125.682	121.467	4.215	5.4907
400	126.595	122.474	4.121	5.5414
450	127.225	123.191	4.034	5.5764
500	128.197	124.243	3.954	5.6304

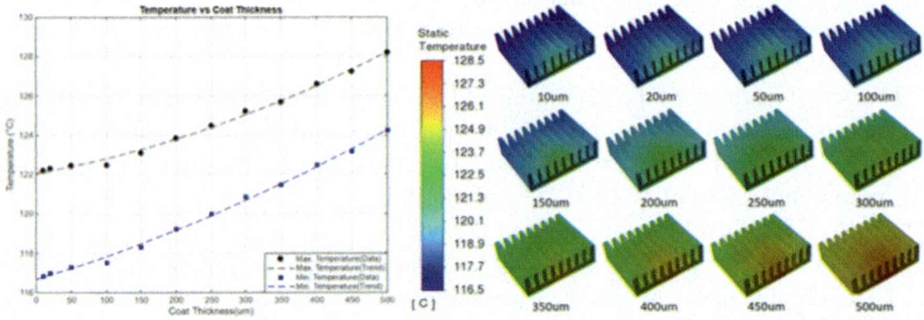

Fig. 2. Plots of maximum temperature and minimum temperature of heat sink against coating thickness (left), and Temperature contours of the heat sink (steady state) with different coating thickness (right).

4 Conclusion

This paper utilized ANSYS FLUENT to numerically simulate the heat dissipation performance of a heat sink coated with AgNP. The simulation model was verified using the experimental data, and the percentage error is a maximum of 3.972%, far below the threshold of 10%. Subsequently, several thicknesses of AgNP coating were put into the identical heat sink. Their heat dissipation performances are assessed based on the recorded maximum temperature, minimum temperature, and thermal resistance. As the coating thickness grows, the maximum and minimum temperatures exhibit a divergent trend by rising. Increasing the thickness of the coating results in a decrease in the temperature differential between the highest and minimum temperatures. This is due to the greater amount of AgNP, which has superior heat conductivity, applied to the heat sink. Including an AgNP coating does not adequately enhance the thermal dissipation of the heat sink. The best-performing heat sink, coated with a 10 μm AgNP coating, only enhances the performance of an uncoated heat sink by 0.214%. The occurrence resulted from the AgNP coating's low emissivity, which decreases the heat lost by radiation. The present study provides valuable information on evaluating the heat dissipation efficiency of a heat sink coated with AgNP, which has high thermal conductivity but low emissivity. The study validated the findings through experimental results and offers insights into how different coating thicknesses affect the heat sink's thermal resistance, thereby achieving optimal heat dissipation properties.

Acknowledgments. The authors would like to acknowledge that the study covered in this manuscript was funded by Ministry of Higher Education (MoHE) Malaysia on the financial support under the Fundamental Research Grant Scheme, FRGS (Grant number FRGS/1/2023/TK10/USM/02/5). The authors would also like to thank Universiti Sains Malaysia for providing technical support.

References

1. Naufal bin Samsudin, A., Salami Tijani, A., Thottathil Abdulrahman, S., Kubenthiran, J., Kolawole Muritala, I.: Thermal-hydraulic modeling of heat sink under force convection: investigating the effect of wings on new designs. Alexandria Eng. J. (2022)
2. Băjenescu, T.M.I.: Miniaturisation of electronic components and the problem of devices overheating. EEA—Electrotehnica, Electron. Autom. **69**(2), 53–58 (2021)
3. Ahmad, E.Z., Fazlizan, A., Jarimi, H., Sopian, K., Ibrahim, A.: Enhanced heat dissipation of truncated multi-level fin heat sink (MLFHS) in case of natural convection for photovoltaic cooling. Case Stud. Thermal Eng. **28**, 101578 (2021)
4. Mousavi, H., Rabienataj Darzi, A.A., Farhadi, M., Omidi, M.: A novel heat sink design with interrupted, staggered and capped fins. Int. J. Therm. Sci. **127**, 312–320 (2018)
5. Feng, S., Shi, M., Yan, H., Sun, S., Li, F., Lu, T.J.: Natural convection in a cross-fin heat sink. Appl. Therm. Eng. **132**, 30–37 (2018)
6. Liu, Y.B.: On thermal structure optimization of a power LED lighting. Procedia Eng **29**, 2765–2769 (2012)
7. Jeon, D., Byon, C.: Thermal performance of plate fin heat sinks with dual-height fins subject to natural convection. Int. J. Heat Mass Transf. **113**, 1086–1092 (2017)
8. Selvan, M., et al.: Influence of fin thickness on the thermal performance and selection of coating method for a bus duct conductor. Coatings **13**(1), 1–25 (2023)
9. Yuan, C.S., Mohd, F.A.: Modeling of natural convection heat transfer of inline pin fin and plate heat sink by computational fluid dynamics, pp. 131–132 (2022)
10. Huang, R.T., Sheu, W.J., Wang, C.C.: Orientation effect on natural convective performance of square pin fin heat sinks. Int. J. Heat Mass Transf. **51**(9–10), 2368–2376 (2008)
11. Huang, Y., Shen, S., Li, H., Gu, Y.: Numerical analysis on the thermal performances of different types of fin heat sink for high-power led lamp cooling. Therm. Sci. **23**, 233 (2017)

12. Joo, Y., Kim, S.J.: Comparison of thermal performance between plate-fin and pin-fin heat sinks in natural convection. Int. J. Heat Mass Transf. **83**, 345–356 (2015)
13. Habib, N., Siddiqi, M.U.R., Tahir, M.: Thermal analysis and optimization of L-shape fin heat sink under natural convection using Anova and Taguchi. Therm. Sci. **26**(2), 1519–1530 (2022)
14. Shadlaghani, A., Tavakoli, M.R., Farzaneh, M., Salimpour, M.R.: Optimization of triangular fins with/without longitudinal perforate for thermal performance enhancement. J. Mech. Sci. Technol. **30**(4), 1903–1910 (2016)
15. Ray, R., Mohanty, A., Patro, P., Senapati, S.K.: Performance improvement of a heat sink with triangular slotted and interrupted fins: a computational study. Appl. Therm. Eng. **230**, 120783 (2023)
16. Ahmadi, M., Mostafavi, G., Bahrami, M.: Natural convection from rectangular interrupted fins. Int. J. Therm. Sci. **82**(1), 62–71 (2014)
17. Kalyan Charan, A., Uday Kumar, R., Balunaik, B.: Evaluation of thermal properties by incorporating perforations of various shapes in fins made with aluminium. Mater. Today Proc. **46**, 609–614 (2021)
18. Maji, A., Bhanja, D., Patowari, P.K.: Numerical investigation on heat transfer enhancement of heat sink using perforated pin fins with inline and staggered arrangement. Appl. Therm. Eng. **125**, 596–616 (2017)
19. Pua, S.W., Ong, K.S., Lai, K.C., Naghavi, M.S.: Natural and forced convection heat transfer coefficients of various finned heat sinks for miniature electronic systems. Proc. Inst. Mech. Eng. Part A: J. Power Energy **233**(2), 249–261 (2019)
20. Choong, J.Y., Yu, K.H., Abdullah, M.Z.: Numerical assessment on heat transfer performance of double-layered oblique fins micro-channel heat sink with Al_2O_3 nanofluid. Therm. Sci. **26**(1), 477–488 (2022)
21. Baby, T.T., Sundara, R.: Synthesis of silver nanoparticle decorated multiwalled carbon nanotubes-graphene mixture and its heat transfer studies in nanofluid. AIP Adv. **3**(1) (2013)
22. Farzanehnia, A., Khatibi, M., Sardarabadi, M., Passandideh-Fard, M.: Experimental investigation of multiwall carbon nanotube/paraffin-based heat sink for electronic device thermal management. Energy Convers. Manag. **179**, 314–325 (2019)
23. Coetzee, D., Venkataraman, M., Militky, J., Petru, M.: Influence of nanoparticles on thermal and electrical conductivity of composites. Polymers (Basel) **12**(4) (2020)
24. Zhang, L., Zhu, W., Qi, G., Li, H., Qi, D., Qi, S.: Highly thermal conductive and electrically insulating epoxy composites based on zinc-oxide-coated silver nanowires. Polymers (Basel) **14**(17), 1–14 (2022)
25. Tran, Q.H., et al.: Corrigendum to enhancement of electrical and thermal properties of silver nanowire transparent conductive electrode by Ag coating. Mater. Sci. Eng. B **279**, 115671 (2022)
26. Armstrong, M., Sivasubramanian, M., Selvapalam, N.: Experimental investigation on the heat transfer performance analysis in silver nano-coated double pipe heat exchanger using displacement reaction. Mater. Today Proc. **45**, 2482–2490 (2021)
27. Kwak, W.G., Oh, M.H., Gong, M.S.: Preparation of silver-coated cotton fabrics using silver carbamate via thermal reduction and their properties. Carbohydr. Polym. **115**, 317–324 (2015)
28. Pungaiah, S.S., Kailasanathan, C.K.: Thermal analysis and optimization of nano coated radiator tubes using computational fluid dynamics and Taguchi method. Coatings **10**(9) (2020)
29. Khabari, A., Zenouzi, M., O'Connor, T., Rodas, A.: Natural and forced convective heat transfer analysis of nanostructured surface. Lect. Notes Eng. Comput. Sci. **1**, 317–319 (2014)
30. Pongiannan, S.K., Ramalingam, V., Nagendran, L.: Natural convection heat transfer enhancement of aluminum heat sink using Nano coating by electron beam method. Therm. Sci. **2018**(5), 3129–3141 (2018)

31. Lotfizadeh, H., Mehrizi, A.A., Motlagh, M.S., Rezazadeh, S.: Thermal performance of an innovative heat sink using metallic foams and aluminum nanoparticles-experimental study. Int. Commun. Heat Mass Transfer **66**, 226–232 (2015)
32. Zu, H., Dai, W., Li, Y., Li, K., Li, J.: Analysis of enhanced heat transfer on a passive heat sink with high-emissivity coating. Int. J. Therm. Sci. **166** (2021)
33. Gao, T., Jelle, B.P.: Silver nanoparticles as low-emissivity coating materials for window glazing applications. NSTI: Adv. Mater. TechConnect Briefs 2015 **2**, 242–245 (2015)
34. Wang, E.: Thermal Emittance (2013)
35. Forsythe, W.E.: Smithsonian Physical Tables (9th Revised Edition). Knovel. [Online] (2003)
36. El Ghandouri, I., El Maakoul, A., Saadeddine, S., Meziane, M.: Thermal performance of a corrugated heat dissipation fin design: a natural convection numerical analysis. Int. J. Heat Mass Transf. **180**, 121763 (2021)

Effect of Al Addition in CoCrFeMnNi High Entropy Alloy

Nurlyana Izyan Mohd Ali[1], Nur Izzati Muhammad Nadzri[1(✉)], Mohd Arif Anuar Mohd Salleh[1], Arnita Surieya Sangar[1], Tea-Sung Jun[2], Vinh-Dat Vuong[3], and Nurul Razliana Abdul Razak[1]

[1] Centre of Excellence Geopolymer and Green Technology (CEGeoGTech), Faculty of Chemical Engineering Technology, Universiti Malaysia Perlis (UniMAP), Taman Muhibbah, 02600 Arau, Perlis, Malaysia
izzatinadzri@unimap.edu.my

[2] Department of Mechanical Engineering, Incheon Univeristy, (Songdo-Dong) 119 Academy-Ro, Teonsu-Gu, Incheon 22012, Republic of Korea

[3] VNU-HCM Key Laboratory for Material Technologies (MTLab), Ho Chi Minh City University of Technology (HCMUT), 268 Ly Thuong Kiet Street, District 10, Ho Chi Minh City, Vietnam

Abstract. The Cantor alloy has emerged as one of the most promising alloy systems in recent years, distinguishing itself as a high-entropy alloy (HEA) with exceptional ductility and fracture toughness. This study focuses on a series of CoCrFeMnNi and the addition of 9.43 wt.% of Al in CoCrFeMnNi HEAs, aiming to explore the impact of aluminium (Al) as an alloying element on their structure. These HEAs were fabricated using casting techniques, and their phase transformation was analyzed through X-ray diffraction (XRD) and field emission scanning electron microscopy (FESEM). Hardness values were determined via Vickers hardness tests. The findings reveal that Al significantly influences the alloy's solidification process and grain growth kinetics. These alterations in microstructure contribute to a notable enhancement in hardness, increasing from 1.798 GPa to 5.007 GPa (178.52% increase). This improvement primarily stems transformation from FCC to BCC crystal structure that leads to strong lattice distortion.

Keywords: High Entropy Alloys · Microstructure · Hardness

N. R. Abdul Razak et al. (Eds.): ECOOP 1996, SPPHY 418, pp. 126–132, 2025.
https://doi.org/10.1007/978-981-96-2871-1_18

1 Introduction

High entropy alloys (HEAs) have been receiving significant interest not only because of the large variety of alloy bases for exploration that they provide, but also because of their remarkable properties [1]. Traditional alloys usually consist of one or two principal elements, with other elements added in minor quantities. In contrast, HEAs comprised of five or more multiple principal elements in roughly equiatomic proportion with elemental concentration of each in the range of 5–35 at. % [2]. CoCrFeMnNi HEA is one the most frequently used model systems in HEA research due to its well-studied properties and relatively simple phase structure. They have been known to break the limits of traditional alloys, including having a combination of strength, ductility and corrosion resistance that makes them attractive for a variety of applications such as aerospace components, structural materials, and parts exposed to harsh environments. CoCrFeMnNi HEA predominantly exhibits simple face-centered cubic (FCC) solid solutions rather than complex phases and intermetallic compounds [3]. CoCrFeMnNi HEAs exhibit excellent ductility, which is the ability of the material to undergo significant plastic deformation before fractures. This high ductility is primarily attributed to their FCC crystal structure, which allows for extensive slip systems, facilitating plastic deformation [4]. The HEA possess high fracture toughness, demonstrating their capacity to withstand crack propagation and absorb energy before fracturing [5].

Despite being highly malleable and durable, this alloy's usefulness is restricted by insufficiently low yield and ultimate tensile strength. This has been attributed mostly to a single-phase microstructure and insufficient lattice distortion, which causes solute solution hardening [4]. The Cantor alloy, a HEA with a base face-centered cubic (FCC) structure, can be strengthened by adding various elements to modify its microstructure. Adding elements like titanium (Ti) can induce a phase transition from FCC to body-centered cubic (BCC), further enhancing the mechanical properties of the alloy. Incorporating tungsten (W) improves corrosion resistance; specifically, CoCrFeMnNi with 2.5 wt.% WC shows reduced corrosion current density and increased corrosion potential, indicating better stability in corrosive environments. This improvement is attributed to the formation of a protective passive film enriched with chromium oxides (Cr2O3 and Cr(OH)3). Additionally, the introduction of carbon (C) increases the alloy's strength through solid solution strengthening due to the interstitial nature of carbon atoms. [5–7].

Even though CoCrFeMnNi is the most researched HEA and Al has a bigger atomic size than the elements in CoCrFeMnNi which may result in larger lattice distortion and strengthening, this study aims to examine the impact of adding Al on the microstructures and mechanical properties of CoCrFeMnNiAl HEA. The mechanism of phase structure, mechanical properties, and their correlations were analyzed, and the comparison between CoCrFeMnNi and CoCrFeMnNiAl HEAs was also made.

2 Experimental Procedure

The five component CoCrFeMnNi and six component CoCrFeMnNiAl HEAs were synthesized by vacuum melting. Table 1 shows the composition of all elements in the HEAs. The samples were poured into a mould to create rectangular ingots measuring

250 mm x 20 mm x 25 mm. XRD Shimadzu 6000 diffractometer was utilized for X-ray diffraction tests with Cu-Kα radiation (λ = 1.5406 Å). Tests were conducted on polished specimens for diffraction angles 2θ ranging from 20° to 80°. The obtained XRD patterns were analyzed using the Panaltical HighScore Plus software to identify the phases present in the alloy. Field emission scanning electron microscopy (FESEM) (Model FESEM FEI Nova Nanosem 450) was utilized to characterize the microstructure and surface morphology of the alloy. Before imaging, the samples were mechanically polished. The FESEM images were obtained at an accelerating voltage of 20 kV. Energy-dispersive X-ray spectroscopy (EDS) analysis was performed simultaneously. Hardness measurements were performed using Sinowon WH series Digital Auto-Turret Vickers Hardness Tester with a Vickers indenter with a load of 0.1 kgf.

Table 1. Composition of both CoCrFeMnNi and CoCrFeMnNiAl HEAs in weight percent (Wt.%)

Samples	Weight percentage (Wt.%)					
	Co	Cr	Fe	Mn	Ni	Al
CoCrFeMnNi	20.00	19.27	18.39	21.4	20.49	0.00
CoCrFeMnNiAl	18.34	17.66	16.65	18.55	19.30	9.43

3 Results and Discussions

3.1 Phase Analysis

Figure 1 illustrates the X-ray diffraction (XRD) patterns of the CoCrFeMnNi and CoCrFeMnNi HEAs. The diffraction peaks located at around 44.55°, 51.59°, and 75.48° in 2θ were identified as the (111), (200), and (220) planes of the FCC structure, and the diffraction peaks at around 45.66° were assigned to the (110) plane of the BCC crystal. As observed, all the diffraction peaks for CoCrFeMnNi HEA were identified to be a single FCC structure. The (110)BCCdiffraction peak appeared at around 45.66° in Fig. 1 with the addition of 9.43 wt.% Al, suggesting the formation of a new crystalline structure of BCC. All FCC diffraction peaks disappeared and only the BCC peaks could be identified in CoCrFeMnNiAl HEA. The effect of Al addition in CoCrFeMnNi HEA also shows a 2.5% peak shift to a higher 2θ value from 44.55° to 45.66°. The shift in peaks and transformation of FCC to BCC crystal structure caused by lattice distortion due to Al has a significant atomic size difference compared to the other elements in the CoCrFeMnNi alloy. Al has a larger atomic radius (0.142 nm) compared to other constituents (Co, Cr, Fe, Mn, and Ni, which have atomic radii around 0.124 to 0.128 nm). Increasing atomic size difference resulted in high lattice distortion that caused the peak to shift to a higher 2θ value as the crystal structure adjusted to accommodate larger Al atoms. The effect on peak shift and phase transformation has been reported previously by research in CoCrFeMnNiAlx HEAFs where higher Al content resulted in 1.6% peak shift from 43.6° to 44.3° and transformation of FCC to BCC crystal structure at Al0.3 [8]. The observation is aligned with the findings that changes in peak position are consistent with the impact of adding Al which indicates that Al influences the lattice parameter, which in turn affects the diffraction angle. Therefore, it could be concluded that the phase transformation from FCC to BCC occurred with the addition of Al. This transformation is evidenced by the appearance of BCC peaks and the shift of diffraction peaks to higher 2θ values, reflecting the adjustment of the crystal structure to accommodate the larger Al atoms. This observation highlights the strong influence of alloying elements on the phase stability and lattice parameters of high-entropy alloys.

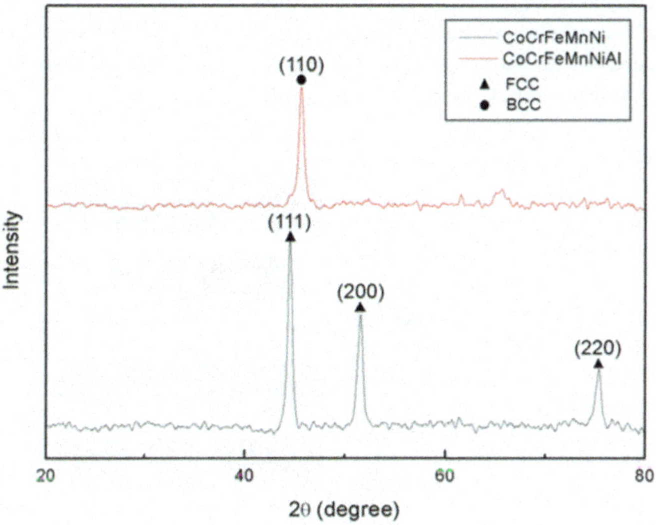

Fig. 1. X-ray diffraction (XRD) patterns of the CoCrFeMnNi and CoCrFeMnNiAl samples

3.2 Microstructure Analysis

Figure 2 shows the microstructures of CoCrFeMnNi and CoCrFeMnNiAl HEAs. In these micrographs, the brighter areas indicate Co-Fe rich regions while the darker areas indicate Mn-Ni rich regions. Both CoCrFeMnNi and CoCrFeMnNiAl exhibit uniform surfaces with equiaxed grains. The addition of Al to the CoCrFeMnNi alloy introduces significant changes to the microstructure due to its distinct atomic radius compared to the other constituent elements. Al has a larger atomic radius, contributing to increased lattice distortion when incorporated into the alloy's crystal structure. This lattice distortion is a critical factor in altering alloy's the phase stability and microstructural evolution. As mentioned in the XRD results, lattice distortion promotes the formation of a BCC structure alongside the FCC structure. The presence of a BCC structure in the alloy matrix reduces the number of interphase boundaries which allows for the formation of larger grains as there are fewer obstacles to impede grain growth during solidification. The addition of Al in the alloy affects the dislocation dynamics within the alloy in which it enhances the solid solution strengthening. The addition of Al reduces the dislocation mobility due to transformation from FCC to BCC which affects the slip system as BCC the slip planes in BCC slip systems is more closely packed compared to FCC structure. The BCC structure is also less restrictive where higher diffusion mobility occurs. Higher diffusion mobility accelerates grain boundary migration which lead to faster grain growth as atoms move more freely from smaller to larger grains, allowing them to migrate and consume smaller grains which lead to formation of larger grains. This phenomenon has been observed in similar studies, where the addition of Al to CoCrFeMnNi, specifically in Al0.5CoCrFeMnNi, resulted in reduced dislocation mobility and consequently larger grain sizes [9]. The observation is aligned with this study in which the addition of Al affects the microstructure, particularly how it influences dislocation dynamics and grain growth. It can be concluded that the addition of 0.93 wt.% Al to CoCrFeMnNi HEA significantly affects the microstructure by introducing lattice distortion, which alters phase stability and promotes the formation of a BCC structure.

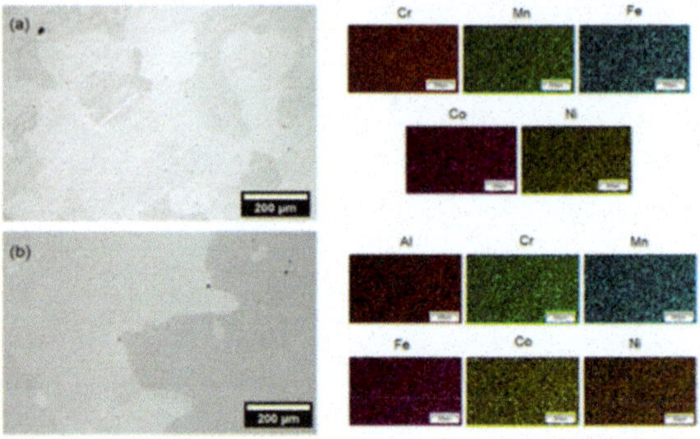

Fig. 2. FESEM images and elemental mapping of (a) CoCrFeMnNi and (b) CoCrFeMnNiAl HEAs.

3.3 Microhardness Analysis

Table 2 shows the Vickers hardness value of both CoCrFeMnNi and CoCrFeMnNiAl HEAs. The hardness of the CoCrFeMnNiAl HEA is measured to be 5.007 GPa, which is almost triple the value compared to the hardness of CoCrFeMnNi HEA (1.798 GPa) with 178.52% increases. The significant increase in hardness of the CoCrFeMnNiAl HEA can be attributed to the transformation from an FCC to a BCC structure. The changes in crystal structure and reduction in dislocation mobility associated with this transformation directly impact the alloy's mechanical properties. The BCC structure generally has fewer slip systems than the FCC structure, which restricts the movement of dislocations, making it more difficult for the material to deform and resulting in increased hardness. The addition of Al, which has a larger atomic radius compared to the other elements in the CoCrFeMnNi HEA, causes significant lattice distortion. These lattice distortions generate local stress fields around the Al atoms. These stress fields interact with moving dislocations, making it more difficult for dislocations to propagate through the crystal lattice, thereby increasing the alloy's strength. The value of hardness in FeCoCrNiMn10Al10 was also reported to be 5.124 GPa. Additionally, the addition of 0.6 at.% Al results in 5.71 GPa, which is around the same value as the sample with the addition of Al [10, 11]. Similar hardness increases are observed which supports that lattice distortion caused by Al is a key mechanism contributing to the increased hardness. This aligns with the concept that Al-induced lattice distortions interfere with dislocation movement, thereby increasing the alloy's strength. It can be concluded that the increase in hardness is primarily attributed to the transformation from an FCC to a BCC structure. This transformation leads to changes in dislocation mobility, which directly enhance the mechanical properties of the alloy. Additionally, incorporating Al, with its larger atomic radius, induces significant lattice distortions that create local stress fields. These stress fields interact with dislocations, hindering their movement through the crystal lattice and further contributing to the alloy's increased strength.

Table 2. Vickers hardness value of both CoCrFeMnNi and CoCrFeMnNiAl HEAs

Samples	Hardness Value (GPa)
CoCrFeMnNi	1.798
CoCrFeMnNiAl	5.007

4 Conclusion

The addition of Al in CoCrFeMnNi HEA is proposed here to explore the impact of Al addition in CoCrFeMnNi HEA. The main conclusions are listed as follows.

1. XRD analysis shows that Al addition transforms the crystal structure from FCC to BCC in CoCrFeMnNiAl HEA, changes lattice parameters due to its larger atomic radius, and dissolves into the matrix to form a solid solution, altering the local atomic arrangements and XRD pattern.
2. The addition of Al to CoCrFeMnNi high-entropy alloys introduces lattice distortion due to its larger atomic radius, promoting the formation of a BCC structure alongside FCC and leading to larger grain sizes and enhanced solid solution strengthening.
3. The hardness of CoCrFeMnNiAl HEA (5.007 GPa) is nearly triple that of CoCrFeMnNi HEA (1.798) due to Al creating a more complex, distorted lattice that enhances resistance to deformation, and inducing a phase transformation from FCC to BCC, which has higher yield strength from different slip systems and more complex dislocation interactions.

The main focus of this research is to investigate the effect of Al addition in CoCrFeMnNi HEA. Al has the potential to significantly alter the mechanical properties of HEAs. However, it's essential to acknowledge that this study does not take into account other properties. This work offers a pathway to expand the research by experimenting with other alloying elements or exploring the different fabrication techniques to optimize the properties of HEAs.

Acknowledgments. The authors would like to acknowledge the support from the Fundamental Research Grant Scheme under grant number FRGS/1/2022/STG05/UNIMAP/02/3 from the Ministry of Education Malaysia, Center of Excellence Geopolymer & Green Technology (CEGeoGTech), Faculty of Chemical Engineering and Technology, Universiti Malaysia Perlis (UniMAP) for their partial support.

References

1. Zhang, Y., Li, R.: New advances in high-entropy alloys. Entropy **22**(10), 1158 (2020)
2. Ibrahim, P.A., Özkul, İ, Canbay, C.A.: An overview of high-entropy alloys. Emerg. Mat. **5**(6), 1779–1796 (2022)
3. Hsieh, K.T., Lin, Y.Y., Lu, C.H., Yang, J.R., Liaw, P.K., Kuo, C.L.: Atomistic simulations of the face-centered-cubic-to-hexagonal-close-packed phase transformation in the equiatomic CoCrFeMnNi high entropy alloy under high compression. Comput. Mater. Sci. **184**, 109864 (2020)
4. Kumar, J., Linda, A., Sadhasivam, M., Pradeep, K.G., Gurao, N.P., Biswas, K.: The effect of Al addition on solid solution strengthening in CoCrFeMnNi: Experiment and modelling. Acta Mater. **238**, 118208 (2022)
5. Shi, P., Li, R., Li, Y., et al.: Hierarchical crack buffering triples ductility in eutectic herringbone high-entropy alloys. Science **373**(6557), 912–918 (2021)
6. Wang, Y., Zhang, J., Li, X., et al.: Microstructural evolution and mechanical properties of CoCrFeMnNi high-entropy alloy during high-temperature deformation. Materials **16**(13), 4701 (2023)
7. Akhavan, J., Moradi, A., Khosroshahi, E.S.: Effect of titanium addition on the microstructure and mechanical properties of CoCrFeMnNi high-entropy alloy. SN Appl. Sci. **3**(12), 1041 (2021)
8. Grzybowski, A., Kolwas, M.: Entropy production in nonequilibrium processes. Entropy **22**(1), 2 (2020)
9. Chen, X., Zhou, H., Fu, Y., Liu, T., Liu, C.T.: Revealing the deformation mechanism of a dual-phase high-entropy alloy through in situ neutron diffraction and acoustic emission. J. Mater. Sci. Technol. **137**, 59–68 (2023)
10. Alleg, S., Bekhouche, A., Hachache, H., Sunol, J.J.: Effect of aluminum addition on the microstructure, magnetic, and mechanical properties of FECRCONIMN high-entropy alloy. Crystals **13**(10), 1483 (2023)
11. Liu, Y., Zhang, Y.: The effects of grain size on the mechanical properties of high-entropy alloys. Entropy **22**(1), 2 (2020)

Effects of Acid Functionalization on Mechanical and Electrical Properties of Melt Compounded Acrylonitrile Styrene Butadiene (ABS)/Recycled Carbon Black (RCB) Composites

Azlin Fazlina Osman[1(✉)], Norshah Aizat Shuaib[2], Shahrul Izwan Neessa Sapuan[1], and Mohd Hanif Mohd Pisal[1]

[1] Faculty of Chemical Engineering and Technology, Universiti Malaysia Perlis, 02600 Jejawi, Arau, Perlis, Malaysia
azlin@unimap.edu.my

[2] Faculty of Mechanical Engineering and Technology, Universiti Malaysia Perlis, Kampus Tetap Pauh Putra, 02600 Arau, Perlis, Malaysia

Abstract. Acrylonitrile butadiene styrene (ABS) terpolymer is a low-cost engineering thermoplastic that can be easily processed. However, its insulation property restricts its use in electronic applications. The addition of conductive filler can transform the ABS into conductive polymer composite. The recycled carbon black (rCB) powder that is extracted from the waste tyres can be a potent conductive filler for polymer composite. This research focused on producing more effective rCB filler as a reinforcing material in an ABS terpolymer. Acid functionalization was done to improve the compatibility of the rCB with the ABS and facilitate its dispersion throughout the ABS matrix. The modified rCB filler; the so-called "acid functionalized recycled carbon black (f-rCB)" was employed to produce ABS/f-rCB with improved mechanical and electrical properties. The effects of rCB and f-rCB loadings (0, 5, 10, 15, 20 and 25 wt. %) on the tensile and electrical properties of the ABS composite were studied. Results indicated that the 10 and 15 wt. % of f-rCB filler has greater capability to reinforce and enhance the electrical conductivity of the ABS as compared to the rCB filler due to improved compatibility between filler and matrix, thus forming homogeneous composite when used at optimum loading.

Keywords: Acrylonitrile butadiene styrene (ABS) · recycled carbon black (rCB) · composites · electrical conductivity · mechanical properties

1 Introduction

Acrylonitrile butadiene styrene (ABS) is a type of thermoplastic terpolymer that is resistant to impact. The three monomers that made up ABS are acrylonitrile, butadiene, and styrene. ABS have a wide processing window and can be processed on a variety of equipment. Injection moulding, blow moulding, and extrusion are some of the methods

N. R. Abdul Razak et al. (Eds.): ECOOP 1996, SPPHY 418, pp. 133–141, 2025.
https://doi.org/10.1007/978-981-96-2871-1_19

that can be used to process the ABS into products [1]. The ABS possesses important properties needed for common applications such as good strength, toughness and impact resistant [1, 2]. Helmets, pipes, and fittings, as well as the bodies of appliances such as vacuum cleaners, sports equipment, safety helmets, luggage, furniture, medical equipment, tubes, caps, telephone sets, and cameras, are all examples of the common applications of the ABS [1, 2]. Recently, ABS has been incorporated with various types of filler to improve its properties for various applications. For instance, ABS composites containing bismuth oxides and tungsten as fillers were fabricated for lead free x-ray radiation shielding application [3]. ABS composites containing permalloy nanoparticles were developed for their application in magnetoactive systems, such as magnetic sensors and actuators [4]. In this work, the ABS was used as the matrix of ABS composite for future development as organic solar cell panel on the roof of vehicles. The good impact strength of the ABS is beneficial for this application since the solar cell panel will be installed on the roof part of the vehicle that is moving and is likely to experience vibrations and collisions [5].

Carbon black (CB) has long been used as filler in polymeric materials due to its ability to improve the mechanical and electrical properties of the resultant composites. As a consequence of the short inter-aggregate distances, CB is extremely electrically conductive. Electrical percolation and a conductive carbon network are accomplished at lower filler concentrations because to the high surface area of CB particles [6]. Thus, the polymer/CB composite has great potential to be commercialized as the conductive polymer composite (CPC). One of the CPC applications is for solar cell. Despite the great potential of the CB as a conductive filler, current market shows that the costs of CB are raising continuously, necessary to address the substantially increasing costs to provide reliability and supply assurance to customers caused by the Covid-19 crisis [7]. Furthermore, there is increased capacity of the CB utilization by various industries, in which the global demand for carbon black grows by about 4.5% per year [6, 7]. With regard to these matters, the use of recycled carbon black (rCB) will reduce the dependency of communities and industries on the CB. The rCB can be obtained from the scrap or waste of various types of plastics. However, the most important rCB source is the tire waste, where around four billion waste tires are deposited in landfills worldwide due to the growth of the automotive industry [7, 8]. Researchers have invented an innovative process such as pyrolysis and cavitational vortex milling to enhance the quality of the rCB [8]. Turning those rCB into various value-added products will contribute to the world's sustainable development goal (SDG). The electrical conducting ability of the rCB when embedded in the matrix of the ABS terpolymer has never been revealed so far. Furthermore, the ability to being mixed well with polymeric materials may depend on the level of the rCB dispersion in the matrix phase. Theoretically, it is not easy to disperse the rCB in polymeric matrices, especially multiphase polymer like the ABS [9]. Therefore, chemical modification on the rCB should be done to improve the dispersion of the rCB in the ABS matrix in order to obtain the ABS/rCB composite with improved mechanical and electrical performance [10]. In this study, acid functionalization of the rCB using nitric acid was applied. The effect of rCB loading with acid-functionalization on the mechanical and electrical properties of the resulting ABS composites were investigated towards optimization of this ABS/rCB composite system for possible use in solar cell panel application.

2 Experimental Procedure

Materials. Acrylonitrile butadiene styrene (ABS) was manufactured by Toray Plastics (Malaysia) Sdn. Berhad. situated in the Prai Free Industrial Zone in Penang, Malaysia. ABS was utilized as the matrix, the main component of the polymer composites. Recycled carbon black (rCB) was supplied by ECO Power Synergy Sdn. Bhd, situated at Puchong, Selangor, Malaysia. In this study, the rCB was used as a filler in the production of the ABS composites. Nitric acid was manufactured by Fisher Scientific (M) Sdn Bhd and supplied by A.R Alatan Sains Sdn. Bhd., Alor Setar, Kedah, Malaysia. Nitric acid was employed in the chemical treatment of the rCB, specifically known as acid functionalization process. This treatment was performed to reduce the inorganic impurities and improve the compatibility between the rCB and the ABS matrix.

Acid functionalization of the recycled carbon black. The rCB in powder form with particle size of 250 μm was sieved using a 63 μm Gratech Sieve Shaker for 3 h to obtain more uniform and purer rCB particles (by removal of impurities/dust). The following procedures to obtain the acid-functionalized rCB were according to the method reported by Debnath et al.,2020. First, the 1 g of rCB powder was added into 10 mL concentrated nitric acid and stirred vigorously. Then, the solution was refluxed for approximately 8 h at 70–80 °C with slow stirring. The refluxed solution was then diluted with water, filtered, and rinsed with distilled water until the sample's pH were 5. At this stage, the acid functionalized rCB (f-rCB) suspension obtained. Finally, the f-rCB was poured into a petri dish and filtered by using filter paper then dried for 24 h at 80 °C in a vacuum oven.

Preparation of ABS/rCB and ABS/f-rCB composites. Different loading percentages were used to produce the ABS/f-rCB composites (0, 5, 10, 15, 20 and 25 wt. %). The control sample of the ABS composites with rCB were also prepared in order to compare and realize the effectiveness of the acid functionalization process in enhancing both mechanical and electrical performance of the composite. Twin screw extruder machine (capacity: 200gm, heating temperature: up to 210 °C, RPM: 50) were used to manufacture the composites. At the end of the process, the compounded ABS/f-rCB composite was collected. The same procedures were performed for the sample preparation of the ABS composite with the rCB. Table 1 shows the formulation of the ABS/rCB composites prepared for this study.

The electrically heated hydraulic press model GT 7014 A was used to produce composite sheet samples with a thickness of 1 mm. The composites sheet was cut into desired shape and size for testing and characterization.

Tensile test. The ABS/rCB and ABS/f-rCB composites were tensile tested using the Universal Testing Machine Instron 5569 at a cross-head speed of 50 mm/min to assess their tensile properties according to ASTM D638. Five replicates of each sample were tested. At the end of each test, the average values of tensile strength, elongation at break (Eb), and Young's modulus were calculated for comparison between samples.

Electrical test. Electrical test is performed to determine a material's capacity to conduct electricity. In this study, a Keithley's source measuring unit, Model 4200, was

Table 1. Formulation and acronyms of ABS/rCB and ABS/f-rCB composites

ABS (wt.%)	rCB (wt.%)	f-rCB (wt.%)	Acronym
100	-	-	ABS
95	5	-	ABS/rCB5
95	-	5	ABS/f-rCB5
90	10	-	ABS/rCB10
90	-	10	ABS/f-rCB10
85	15	-	ABS/rCB15
85	-	15	ABS/f-rCB15
80	20	-	ABS/rCB20
80	-	20	ABS/f-rCB20
75	25	-	ABS/rCB25
75	-	25	ABS/f-rCB25

used to measure the electrical conductivity of the ABS/rCB and ABS/f-rCB composites. The samples were chopped into 15 $\times$ 20 $\times$ 1 mm specimens. Using a two-point probe, the analysis was performed at voltages ranging from 0 to 10 V. Electrical conductivity, (S/cm), was calculated using Ohm's Law, where the specimen thickness (cm), R is the resistance, and A (cm2) is the cross-sectional area.

Scanning electron microscopy (SEM). SEM analysis was performed to analyse the tensile fractured surface of the ABS/rCB and ABS/f-rCB composites. Furthermore, the performance of surface modification was confirmed by comparing the surface morphology of the treated acid functionalization recycled carbon black to the original recycled carbon black.

3 Results and Discussions

3.1 Mechanical Properties by Tensile Test

The effect of the rCB filler loading on the tensile strength, elongation at break and Young's modulus of the ABS, ABS/rCB composites, and ABS/f-rCB composites are shown in Fig. 1. The addition of the rCB and f-rCB to the ABS has successfully enhanced the tensile strength, elongation at break and Young's modulus.

The ABS composites with 15 wt.% rCB demonstrated an increase in tensile strength and elongation at break when compared with the neat ABS. The tensile strength and elongation at break of ABS composites have clearly shown a decreasing pattern as the amount of rCB filler contained in the composites increased from 15 wt.% and higher. During the melt compounding process, agglomeration of ABS results in the creation of big particles for melt mixing with the ABS. Instead of filler-matrix interactions, this encouraged filler-filler interactions. Overall, the f-rCB has caused more significant improvement in the tensile strength, Young's modulus and elongation at break of the ABS terpolymer when

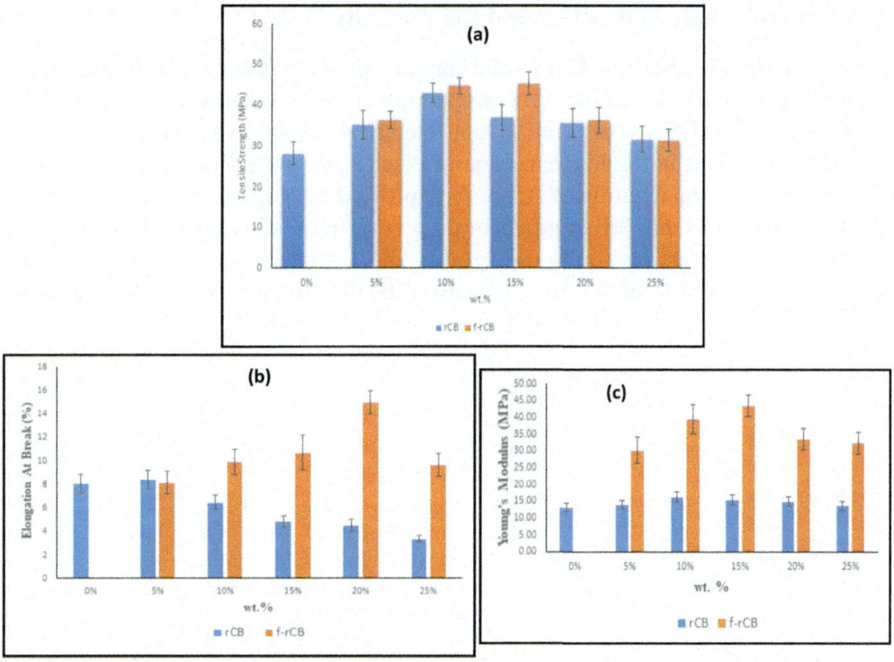

Fig. 1. The effect of rCB loading on the a) tensile strength b) elongation at break and c) Young's modulus of ABS/rCB and ABS/f-rCB composites

added as filler. This shows that acid functionalization has improved the compatibility between the rCB and the ABS, producing a more homogeneous composite system. Thus, regardless how much the filler loading used, the ABS/f-rCB composites demonstrated a higher tensile strength and modulus when compared to the ABS/rCB composite. The highest tensile strength and Young's modulus were obtained when 15wt% f-rCB was added as filler, where the values were remarkably increased by 60.5% and 233.5%, respectively.

The mechanical performance of polymer composites is theoretically dependent on the interfacial adhesion between the matrix and the filler [10, 11]. The addition of up to 15 wt.% f-rCB to the ABS matrix has considerably enhanced the tensile properties of the terpolymer before decreasing off at 25 wt.% f-rCB. ABS had a tensile strength of 28.23 MPa and an elongation at break of 8.04%, respectively. The former and latter values increased to 43.3 MPa and 10.69%, respectively, by adding only 20 wt.% f-rCB, which is equivalent to a 65% and 13% increment over the neat ABS. The reinforcement provided by greater interfacial adhesion and dispersion of f-rCB across the matrix led to the composite's improvement in tensile strength, stiffness, and elongation at break. Acid functionalization process has improved the compatibility between the carbon black with the ABS, producing more homogeneous composite system. This has been proved through SEM analysis (see Sect. 3.3).

3.2 Electrical Analysis by Electrical Conductivity Test.

As shown in Fig. 2, when the rCB is acid functionalized (f-rCB), it can further improve the electrical conductivity of the ABS. Apparently, the f-rCB shows a greater capability in reinforcing the ABS and also improve its electrical conductivity. The best loading of f-rCB that can result in the highest value of electrical conductivity is 10wt.%. This is because, acid functionalization of rCB will improve the compatibility between the filler and ABS matrix, thus this will improve the dispersion of the filler inside the matrix phase. Consequently, the electrical properties of the ABS composite can be further improved due to greater capability of the filler to transfer the electron throughout the matrix [12].

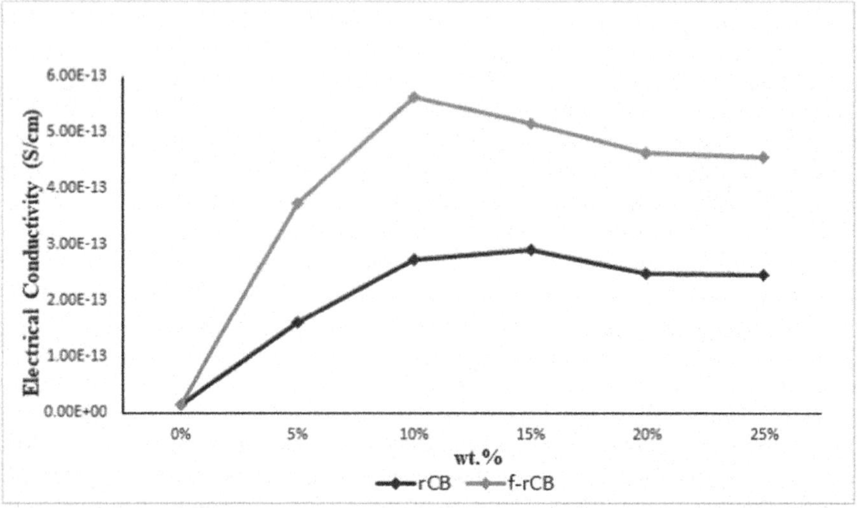

Fig. 2. The effect of different wt. % of rCB and f-rCB on electrical conductivity of ABS/rCB and ABS/f-rCB composites

The composite film prepared with rCB has lower electrical conductivity compared to f-rCB. It is known that the hydroxyl and carboxyl groups were introduced in the carbon black upon acid functionalization. This will cause de-agglomeration of the rCB particles, producing more individual particles of the rCB. This disaggregation of the rCB will produce filler particles with high surface area. According to Gouda et al., high surface area particles will result in much better electron transfer between filler and polymeric matrix [13]. Based on the mechanical and electrical tests data, we can certify that the best loading of the f-rCB filler in the ABS is 10%. This is because; the ABS/f-rCB shows almost similar tensile properties when added in 10 and 15 wt.%. However, in terms of electrical property, the composite shows much higher electrical conductivity value when added in 10wt.% as compared to 15 wt.%. So, 10wt.% f-rCB is enough to allow good improvement in both mechanical and electrical properties of the ABS.

Morphological analysis of neat ABS, ABS/ rCB and ABS/ f-rCB composites. The fractured surface morphology of the ABS, ABS/rCB composites and ABS/f-rCB composites were studied using scanning electron microscopy (SEM). The magnifications of 500× and 1500× were used to analyse the fracture behaviour of the samples and understand the role of the f-rCB in improving the fracture surface morphology of the ABS matrix after subjected to tensile test. SEM micrographs of the neat ABS, ABS containing rCB and f-rCB (at 10wt% loading) are shown in Fig. 3(a), (b) and (c), respectively.

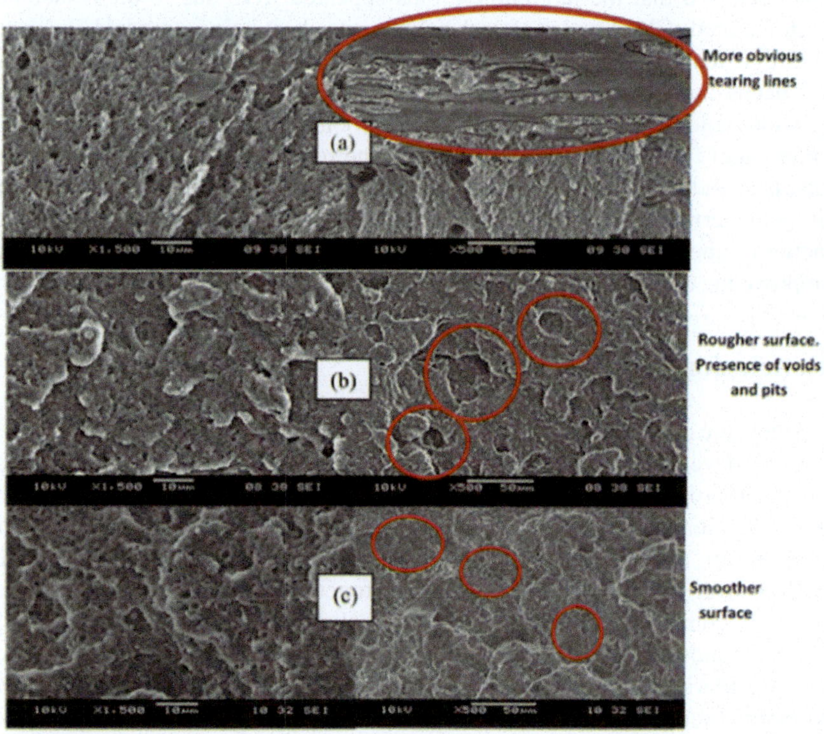

Fig. 3. SEM images of (a) neat ABS, (b) ABS/ rCB and (c) ABS/ f-rCB composites at 1500× magnification (Left) and 500× magnification (Right)

Figure 3(a) indicates that there are clear tearing lines on the surface morphology of neat ABS. This shows that, without the presence of filler, the ABS experienced a more pronounced tearing effect, due to its lower tensile strength. When the ABS stretched, there was no filler particles to enhance the force of the matrix to resist large deformation before fracture, therefore, a deeper and wider tearing area observed. This is in good agreement with the tensile test data obtained. However, when the r-CB was added as filler, no clear tearing lines observed. This suggested that the inclusion of the filler assists the matrix to achieve greater force before it breaks, so that less matrix deformation and tearing effect occur. Improved matrix-filler interfacial adhesion will reduce the matrix deformation and tearing line occurrence [14]. Even though, the rCB filler may help the ABS matrix to achieve a greater strength to resist permanent deformation, the poor dispersion of the rCB may lead to the appearance of voids and pits. These features are clearly seen in Fig. 3(b). Voids and pits occurred due to poor filler-matrix interactions, that produced a non-homogeneous composite. According to Kimura et al., voids and cracks that appear in the matrix of a polymer composite do not simply caused by local stresses, but instead occur due to these two mechanisms: i) filler/matrix interface debonding and ii) matrix's crack initiation. Therefore, good matrix-filler interactions will avoid these heterogeneities that lead to poor composite morphology upon fracture [15]. Furthermore, it is also believed that the large holes/voids appeared were due to the detachment of the agglomerated rCB particles after the composite sheet being stretched and ruptured. Figure 3(c) reveals that the presence of f-rCB as filler leads to the appearance of a more homogeneous composite. In the surface morphology of the ABS/f-rCB composite, there is no obvious voids/pits/cracks, proving the good bonding and mixing between the matrix and filler. Acid functionalization process has improved the compatibility between the carbon black and the ABS through the production of hydroxyl and carboxyl functional groups at the terminals of defective sites of the rCB. This will enhance the filler-matrix interactions rather than filler-filler interactions, thus preventing the rCB agglomeration and improve the dispersibility of the filler in the ABS matrix.

4 Conclusion

In this study, the recycled carbon black (rCB) was employed as an economic filler in the ABS for the development of the conductive polymer composite (CPC). The mechanical and electrical properties of the melt-compound ABS/rCB composites were examined in relation to rCB loading and acid functionalization. Acid functionalization was performed using the reflux method in order to improve the compatibility between the rCB and the ABS matrix. The performance of the acid functionalized rCB (f-rCB) as filler was examined by comparing the tensile and electrical properties data of the neat ABS, the ABS/rCB composite and ABS/f-Rcb composites, and then supported by morphological analysis. Results indicated that the acid functionalization of the rCB filler has transformed it into more efficient filler that can enhance the tensile and electrical properties of the ABS terpolymer.

Acknowledgements. The authors would like to acknowledge Eco Power Sdn. Bhd. for funding research materials used in this research study.

References

1. Margolis, J.M.: Engineering thermoplastics: properties and applications, pp. 135–140. CRC Press, United States (2020)
2. Ziąbka, M., Dziadek, M., Pielichowska, K.: Surface and structural properties of medical acrylonitrile butadiene styrene modified with silver nanoparticles. Polymers **12**(1), 197 (2020)
3. Olivieri, F., et al.: High filler content acrylonitrile-butadiene-styrene composites containing tungsten and bismuth oxides for effective lead-free x-ray radiation shielding. Polymer Composites **45**(3), 2101–2113 (2024)
4. Merazzo, K.J., et al.: Acrylonitrile butadiene styrene-based composites with permalloy with tailored magnetic response. Polymers **15**(3), 626 (2023)

5. Hou, W., Xiao, Y., Han, G., Lin, J.Y.: The applications of polymers in solar cells: a review. Polymers **11**(1), 143 (2019)
6. Donnet, J.-B.: Carbon black: science and technology, 2nd edn. CRC Press, United Kingdom (2018)
7. Amirkhanian, S., Letcher, T., Shulman, V.: Tire waste and recycling, pp. 236–240. Elsevier Science, Netherlands (2021)
8. Rikmann, E., Mäeorg, U., Liiv, J.: Recycling of low-quality carbon black produced by tire pyrolysis. Appl. Sci. **14**(5), 2192 (2024)
9. Chandran, A.J., Rangappa, S.M., Suyambulingam, I., Siengchin, S.: Micro/nano fillers for value-added polymer composites: A comprehensive review. J. Vinyl. Addit. Technol. **30**(5), 1083–1123 (2024)
10. Debnath, N., Panwar, V., Roy, T., Saha, M., Pal, K.: Improved dispersion of carbon black in ABS/PANI blend through its acid functionalization and addition of nanoclay, thereby enhancing mechanical and thermal properties. Polymer Bull. **77**(9), 4759–4772 (2020)
11. Soheilmoghaddam, M., et al.: Development of ethylene-vinyl acetate composites reinforced with graphene platelets. Macromolecular Mater. Eng. **302**(2), 1600260 (2016)
12. Alauddin, S.M., Ismail, I., Zaili, F.S., Ilias, N.F., Kamalul Aripin, N.F.: Electrical and mechanical properties of acrylonitrile butadiene styrene/graphene platelet nanocomposite. Mater. Today: Proc. **5**(2), S125–S129 (2018)
13. Gouda, K., Bhowmik, S., Das, B.: A review on allotropes of carbon and natural filler-reinforced thermomechanical properties of upgraded epoxy hybrid composite. Rev. Adv. Mater. Sci. **60**(1), 237–275 (2021)
14. Mohammed Fitri, T.F., Osman, A.F., Alosime, E.M., Othman, R., Hashim, F., Abdullah, M.A.A.: Biomedical PEVA nanocomposite with dual clay nanofiller: cytotoxicity, mechanical properties, and biostability. Polymers **13**(24), 4345 (2021).
15. Kimura, M., Watanabe, T., Takeichi, Y., Niwa, Y.: Nanoscopic origin of cracks in carbon fibre-reinforced plastic composites. Sci. Report **9**, 19300 (2019)

Fabrication of an Interpenetrating Polymer Network (IPN) Using PANI, PVDF, PEO, and LiTFSI for Solid Polymer Electrolyte (SPE) Applications

Nur Najiha Abd Maliaman[1], Awatif Hassim[1], Mohamad Arif Kasri[1,2], Muhammad Zharfan Mohd Halizan[3], Saiful Arifin Shafiee[2](✉), and Mohd Muzamir Mahat[1](✉)

[1] Faculty of Applied Sciences, Universiti Teknologi MARA, 40450 Shah Alam, Selangor, Malaysia
mmuzamir@uitm.edu.my

[2] Department of Chemistry, Kuliyyah of Sciences, International Islamic University Malaysia, Jalan Sultan Ahmad Shah, 25200 Kuantan, Pahang, Malaysia
sabs@iium.edu.my

[3] Faculty of Applied Science, Universiti Teknologi MARA, 02600 Arau, Perlis, Malaysia

Abstract. Lithium-ion battery (LIB) is essential in electric and hybrid vehicles and high-tech devices, incorporating solid polymer electrolytes (SPE) that conduct lithium ions across the system. Previous study has reported on a Polyethylene oxide (PEO):Polyvinylidene fluoride (PVDF):Lithium bis (trifluoromethanesulfonyl) imide (LiTFSI)-based electrolyte. The objective of that research is to enhance the efficiency of lithium-ion batteries that are all-solid-state by maximizing the formation of electrolytes composed of polyethylene oxide (PEO) and polyvinylidene fluoride (PVDF). The research found that the ionic conductivity significantly improves when PEO/PVDF has a ratio of 1:5 to lithium bis (trifluoromethane sulphonyl) imide (LiTFSI). At 30 ^{0}C, the conductivity reaches 2.98 X 10$^-$5 S/cm, and at 60 ^{0}C, it increases to 5.56 X 0$^-$4 S/cm. Introducing an interpenetrating polymer network (IPN) onto SPE by compositing it with conducting polymer (CP) enhances their ionic conductivity by providing more pathways for ion transport in SPE. In this study, IPN films of polyaniline (PANI):PEO:PVDF:LiTFSI were synthesized with varying PANI concentrations (1 wt.%, 3 wt.%, and 5 wt.%) and composited with 7 wt.% of PEO and 3 wt.% of PVDF. The phase and structure of these polymer composite films were characterized using X-ray diffraction (XRD), which showed that increasing the concentration of PANI will increase the SPE's ionic conductivity by reducing its crystallinity. Electrochemical Impedance Spectroscopy (EIS) showed that the film with 5 wt.% PANI exhibited the highest ionic conductivity, measuring 1.8 x 10^{-5} S/cm. Morphological analysis conducted using a Field Emission Scanning Electron Microscope (FESEM) revealed that the PANI:PEO:PVDF:LiTFSI SPE with 1 wt.% and 3 wt.% of PANI displays a homogeneous surface with an IPN structure, potentially enhancing Li-ion migration rates. These findings indicate a promising direction for the future fabrication of solid polymer electrolytes.

N. R. Abdul Razak et al. (Eds.): ECOOP 1996, SPPHY 418, pp. 142–148, 2025.
https://doi.org/10.1007/978-981-96-2871-1_20

Keywords: Lithium bis (trifluoromethanesulfonyl) imide · polyaniline · polyethylene oxide · polyvinylidene fluoride · conducting polymer · solid polymer electrolyte · conductivity · porosity · interpenetrating network · lithium ions

1 Introduction

In batteries, conventional electrolytes known as liquid electrolytes are made up of chemicals like KOH and $LiPF_6$ mixed with dielectric solvents like water and ethylene carbonate. However, this type of electrolyte presents significant safety risks [1]. SPEs mitigate several safety hazards associated with liquid electrolytes, including internal short circuits, electrolyte leakage, and the formation of reaction products that are flammable at the electrode surface [2]. In 1973, an ion-conducting polymer that acts as a polymer electrolyte, polyethylene oxide (PEO) combined with alkali metal was discovered [3]. Solid polymer electrolytes (SPEs) are used in various applications, including lithium-ion batteries (LIBs), fuel cells, and water desalination. SPE is a crucial part of LIB as it serves as a medium for ion transport in LIBs and as a separator to separate both electrodes in LIBs. For it to serve as a good electrolyte for LIB, SPE should possess characteristics of having high ionic conductivity under both room temperature and colder conditions, strong mechanical resilience, a notable transference number, thermal and electrochemical stability, and compatibility with electrodes [2].

In SPEs, the polymer host facilitates the delivery of Li-ions through the motion of segments within the amorphous sections [4]. SPE dissolves ionic salts like Li-ion salts into a coordinating polar polymer host. The film is then cast using the traditional solution cast method or a recently developed hot-press technique. PEO is a favorable polymer host for SPE as it creates a stable dry compound that demonstrates a comparatively higher ionic conductivity compared to other solvating polymers because of the way the polar groups (-O-, -H-, and -C-H-) and oxyethylene groups (-CH2-CH2-O-) are arranged inside the polymer chains. These groups can make the ionic salts dissolve [3]. One example of this type of electrolyte is a blend of lithium bis(trifluoromethanesulfonyl)imide (LiTFSI), a lithium salt with cross-linked polyethylene oxide (PEO) [1]. Unfortunately, the ionic conductivity of the PEO-based electrolytes is limited by a high concentration of crystalline phase. The PEO-salt systems have relatively low ionic conductivity, between 10^{-7} and 10^{-8} S/cm [5]. To counter this issue, the blending of two polymers is done as it is an easy and efficient way to increase ionic conductivity [6].

Fan et al. reported that it is possible to attain a good compromise between mechanical stability and excellent ionic conductivity by adding a specific amount of PEO to the Poly (vinylidene fluoride-hexafluoropropylene) (PVDF-HFP) matrix. Because of its exceptional dielectric constant, many functional groups in its backbone that have a potent electron-withdrawing action, and exceptional thermal stability, PVDF is widely employed in membrane fabrication [7]. According to Li et al., combining PEO with PVDF and LiTFSI enhances the SPE's ionic conductivity [7].

Furthermore, it is possible to improve SPE's ionic conductivity by inhibiting the crystallization of polymer chains which can be done by introducing an interpenetrating polymer network (IPN) onto SPE [8]. IPN is a type of polymer in which two or more polymer networks are interlaced on a molecular scale and it is used in supercapacitors and solar cell applications. Fong et al. (2017) invented a supercapacitor electrode made up of poly (3,4-ethylene dioxythiophene) (PEDOT) which is an electronically conductive polymer within an ionically conducting polymer matrix [9]. They focused on developing PEDOT's semi-IPNs (sIPNs) with polyethylene oxide (PEO), an ionically conducting polymer, and comparing it to the bare PEDOT. This study found that this sIPN electrode exhibits a greater specific capacitance value than bare PEDOT, due to the morphology of the sIPN. Siddhanta and Gangopadhyay synthesized a new sIPN gel that combines PANI and poly (2-acrylamido-2-methyl propanesulphonicacid) (PAMPS) [10]. The conductivity of the pure PAMPS gel was found to be only $\leq 10^{-5}$ S/cm, an order of 10^{-4} lower than that of the PANI-PAMPS composite. The resulting gel has strong mechanical properties, swellability, and electrical conductivity. All these studies used conductive polymers to form IPN films. One of the most promising ICPs is polyaniline (PANI) given its ability to create a thin layer possessing electrical, magnetic, redox, antioxidant, and sensing properties [11]. Hence PANI is a good choice to be used in fabricating the Solid Polymer Electrolyte-Conducting Polymer (SPE-CP) IPN film. While a great deal of research has been done on different conducting polymers in the context of solid polymer electrolytes (SPEs), relatively there is limited research on PANI as a SPE material.

2 Experimental Procedure

2.1 Materials

The following materials were bought from Sigma-Aldrich: polyaniline (emeraldine salt) (PANI, Mw $\approx$ 15, 000 g/mol), polyethylene oxide (PEO, Mw $\approx$ 400, 0000 g/mol), polyvinylidene fluoride (PVDF, Mw = 534, 000 g/mol), lithium bis(trifluoromethanesulfonyl)imide (LiTFSI, $\geq$ 99%), and N, N-dimethylformamide (DMF, > 99%).

2.2 Compositing PANI-SPE IPN

The solid electrolytes PANI:PEO:PVDF:LiTFSI were prepared by solution casting. PANI's weight percentages varied, with P1, P3, and P5 denoting the 1 wt.%, 3 wt.%, and 5 wt.% [12]. PEO, PVDF, and LiTFSI have a weight ratio of 7:3:2 [7]. First, PEO powder was mixed and dissolved in DMF solvent. Next, the solution was mixed with powdered PVDF. LiTFSI powder was added to the mixture and stirred for two hours. PANI powder was added to the solution after it had been swirled for two hours. The homogeneous solution was then transferred into a petri dish and dried for 24 h at 60 °C in a vacuum oven. This produced a consistent film that was about 200 μm thick.

2.3 Characterization Methods

The conductivity of the PANI-SPE film was determined using electrochemical impedance spectroscopy (EIS) and the HIOKI 3532–50 LCR Hi-TESTER PGZ100.

An electron microscope with field emission scanning capability was used to examine the materials' surface morphology (FESEM). The FESEM was operated at 10.0 kV Electron High Tension (EHT) and pictures were captured in high vacuum mode. The phase analysis study was done using Malvern Panalytical's X'Pert3. The XRD patterns were collected at room temperature with a rotating copper anode, using X-rays with a wavelength of 1.542 Å. The scanning rate was set to 0.5°/min, and the minimum angle step was 0.001°. XRD analysis can show the degree of crystallinity of the polymer based on the peak intensity. Crystalline polymers have sharp peaks in their XRD patterns, while amorphous polymers have broad, diffuse scattering patterns. The polymer's degree of crystallinity is correlated with how sharp the peaks are. The peaks are sharper as the degree of crystallinity increases. It looked as though the peaks were narrowing for the amorphous sample.

3 Results and Discussions

3.1 Structural Properties

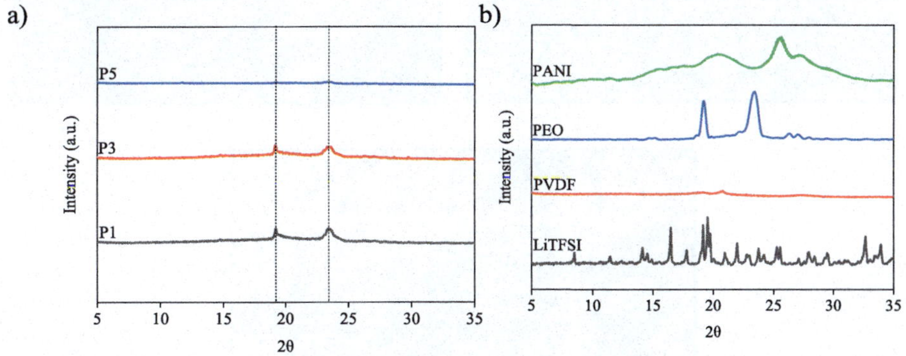

Fig. 1. XRD spectra of (a) P1 film, P3 film, and P5 film (b) pure PANI, PEO, PVDF, LiTFSI.

X-ray diffraction (XRD) analysis was conducted on pure LiTFSI, PVDF, PEO, PANI, and SPE films with different concentrations of PANI to examine the effects of films' crystallinity on the LiTFSI, PVDF, PEO, and PANI interactions. Pure LiTFSI shows characteristic peaks at 11.4°, 14.1°, 16.4°, 19.6°, 20.9°, 21.9°, 23.7°, 25.4°, 27.9°, 29.4°, 32.7° and 33.9° [13]. PVDF shows two broad diffraction peaks at 18.2° and 19.8° due to PVDF's semi-crystalline structure characteristic while PEO exhibits two sharp peaks at 19.3° and 23.4° because of its high crystallinity structure. Due to its semi-crystalline structure, PANI displays two broad diffraction peaks at 20.6° and 25.6°. PANI:PVDF:PEO:LiTFSI films show two broad peaks at 19.3° and 23.4°. The peaks are broader compared to pure PEO, showing that compositing polymers decrease the crystallinity of the composite film. As PANI concentration increases, PEO peaks at 19.3° and 23.4° become broader, showing that PANI disrupted PEO's crystallinity, making the polymer composite less crystalline which may lead to high ionic conductivity (Fig. 1).

3.2 Morphological Properties

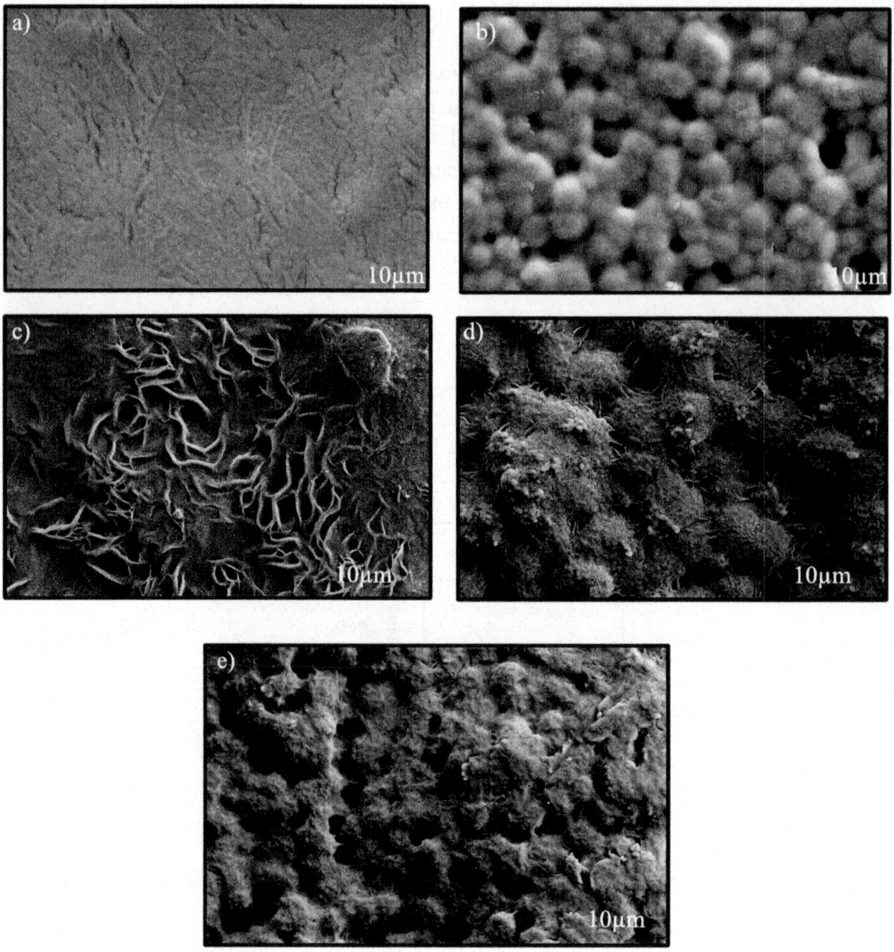

Fig. 2. Morphology of FESEM image of (a) PEO film (b) PVDF film (c) P1 film (d) P3 film and (e) P5 film captured by FESEM

We also examined the films' morphology to understand better how the PANI concentration affected the composite film. Using field emission scanning electron microscopy (FESEM), Fig. 2 shows images of PEO, PVDF, and films made of PANI, PEO, PVDF, and LiTFSI at varying PANI concentrations (1 wt.%, 3 wt.%, and 5 wt.%). Figure 2a shows the micrograph of the pure PEO. The hard and rough structure of the PEO demonstrates its crystalline structure. At a 10,000 × magnification, Fig. 2b's FESEM image of PVDF film demonstrates the noticeable presence of spherical formations strewn throughout its surface. They are densely packed but exhibit no specific pattern in their arrangement. IPN is present in PANI:PEO:PVDF:LiTFSI films with 1 wt.% and 3 wt.% of PANI, which can improve the mechanical characteristics and ionic conductivity of SPEs [14]. The P1, P3, and P5 films all contained a porous network, as seen by FESEM images.

3.3 Electrochemical Properties

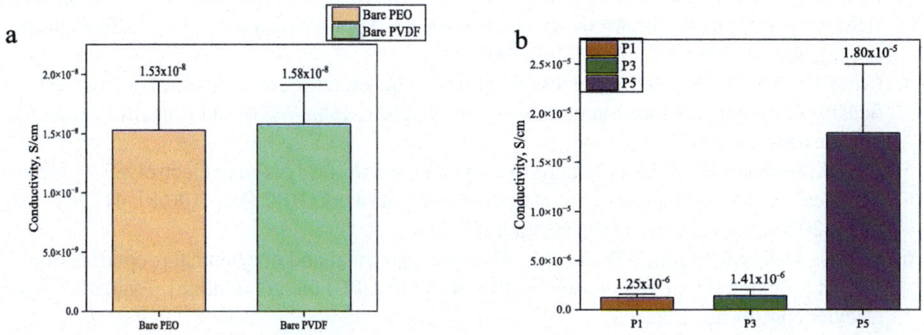

Fig. 3. Conductivity values of (a) PEO and PVDF film (b) P1, P3, and P5 films EIS was used to test the films and measure their conductivities.

PVDF films and bare PEO had conductivities of 1.58×10^{-8} S/cm and 1.53×10^{-8} S/cm, respectively. When compared to PEO and PVDF films alone, the composite's conductivity is improved by the addition of a secondary polymer. The average conductivity values of the P1, P3, and P5 films are 1.80×10^{-5} S/cm, 1.41×10^{-6} S/cm, and 1.25×10^{-6} S/cm, respectively. The composite containing 5 wt.% PANI has the highest conductivity value, 1.80×10^{-5} S/cm. XRD study further confirms this conclusion, which demonstrates that the composite polymer with 5 wt.% content has the least crystallinity which consequently boosts the ionic conductivity. According to the findings, conductivity tends to increase when PANI concentration rises within the studied range. Since there are more pathways for charge carriers to take, an increase in PANI content raises total conductivity (Fig. 3).

4 Conclusion

This study explored the fabrication of interpenetrating polymer networks (IPN) within a PEO:PVDF:LiTFSI solid polymer electrolyte (SPE) through the incorporation of varying PANI concentrations. The results indicate that the inclusion of 5 wt.% PANI enhanced conductivity to 1.80 x 10^{-5} S/cm. The improvement in conductivity is likely due to the formation of an IPN structure, supported by XRD and FESEM characterizations. These findings demonstrate the potential of utilizing PANI-based IPNs to improve SPE performance within lithium-ion batteries.

Acknowledgments. The authors express gratitude to PETRONAS Research Sdn. Bhd for financing their research. Many thanks to UiTM Shah Alam's Faculty of Applied Science for providing the facilities.

References

1. Hallinan, D.T., Balsara, N.P.: Polymer electrolytes. Jul. (2013). https://doi.org/10.1146/annurev-matsci-071312-121705
2. Stephan, A.M.: Review on gel polymer electrolytes for lithium batteries. Jan. (2006). https://doi.org/10.1016/j.eurpolymj.2005.09.017
3. Agrawal, R.C., Pandey, G.P.: Solid polymer electrolytes: Materials designing and all-solid-state battery applications: An overview. J. Phys. D. Appl. Phys. **41**(22), Nov. (2008). https://doi.org/10.1088/0022-3727/41/22/223001
4. Chen, H., et al.: Functional additives for solid polymer electrolytes in flexible and high-energy-density solid-state lithium-ion batteries. Nov. 1, 2021, John Wiley and Sons Inc. https://doi.org/10.1002/cey2.146
5. Leš, K., Jordan, C.S.: Ionic conductivity enhancement in solid polymer electrolytes by electrochemical: In situ formation of an interpenetrating network. RSC Adv. **10**(68), 41296–41304 (Nov. 2020). https://doi.org/10.1039/d0ra07966a
6. Fan, L., Dang, Z., Nan, C.-W., Li, M.: Thermal, electrical and mechanical properties of plasticized polymer electrolytes based on PEO/P(VDF-HFP) blends [Online]. Available: www.elsevier.com/locate/electacta
7. Li, J., et al.: Optimisation of conductivity of PEO/PVDF-based solid polymer electrolytes in all-solid-state Li-ion batteries. Mater. Technol. **37**(4), 240–247 (2022). https://doi.org/10.1080/10667857.2020.1827873
8. Murata, K., Izuchi, S., Yoshihisa, Y.: An overview of the research and development of solid polymer electrolyte batteries [Online] (2000). Available: www.elsevier.nl/locate/electacta
9. Fong, K.D., Wang, T., Kim, H.K., Kumar, R.V., Smoukov, S.K.: Semi-Interpenetrating Polymer Networks for Enhanced Supercapacitor Electrodes. ACS Energy Lett. **2**(9), 2014–2020 (Sep. 2017). https://doi.org/10.1021/acsenergylett.7b00466
10. Siddhanta, S.K., Gangopadhyay, R.: Conducting polymer gel: Formation of a novel semi-IPN from polyaniline and crosslinked poly(2-acrylamido-2-methyl propanesulphonicacid). Polymer (Guildf) **46**(9), 2993–3000 (Apr. 2005). https://doi.org/10.1016/j.polymer.2005.01.084
11. Beygisangchin, M., Rashid, S.A., Shafie, S., Sadrolhosseini, A.R., Lim, H.N.: Preparations, properties, and applications of polyaniline and polyaniline thin films—a review. Polymers (Basel) **13**(12), Jun. (2021). https://doi.org/10.3390/polym13122003
12. Khapre, S.A., Yawale, S.P., Yawale, S.S.: Structure and properties of processible conductive polyaniline and polyethelene oxide composite
13. Li, Y., et al.: A composite solid polymer electrolyte incorporating MnO2 nanosheets with reinforced mechanical properties and electrochemical stability for lithium metal batteries. J. Mater. Chem. A Mater. **8**(4), 2021–2032 (2020). https://doi.org/10.1039/c9ta11542k
14. Tan, S.J., Zeng, X.X., Ma, Q., Wu, X.W., Guo, Y.G.: Recent advancements in polymer-based composite electrolytes for rechargeable lithium batteries. Jun. 01, 2018, Springer Science and Business Media B.V. https://doi.org/10.1007/s41918-018-0011-2

Investigating the Primary Factors Influencing Conductivity Behavior in Gelatine/Polypyrrole Hydrogels

Nur Hidayah Shahemi[1], Nur Syafiqah Mohd Shabri[1], Dania Adila Mohd Ruzaidi[2], Mohamad Arif Kasri[2], and Mohd Muzamir Mahat[1](✉)

[1] Faculty of Applied Science, Universiti Teknologi MARA, 40450 Shah Alam, Selangor, Malaysia
mmuzamir@uitm.edu.my

[2] Department of Chemistry, Kulliyyah of Science, International Islamic University Malaysia, Jalan Sultan Ahmad Shah, 25200 Kuantan, Pahang, Malaysia

Abstract. Conductive hydrogels are promising for neuroregenerative therapy due to their flexibility and ability to support electrical signal transmission. This study develops a PPy/gelatine conductive hydrogel with optimized electrical properties. Gelatine, known for its biocompatibility, was combined with the conductive polymer PPy and crosslinked using glutaraldehyde. Hydrogels with varying PPy concentrations (0 wt%, 2 wt%, 4 wt%, and 6 wt%) were analyzed. FTIR confirmed chemical bonding between gelatine and PPy, while XRD showed an increased crystallinity index compared to the Gel/0PPy sample. Cyclic voltammetry revealed that a 6% PPy concentration markedly enhanced electrical reactivity compared to lower concentrations. Results indicate that the crosslinking network between PPy and gelatine enhances conductivity by facilitating π-electron mobilization in the PPy backbones through the increased order of crystallinity. Additionally, higher PPy concentrations better facilitate more efficient electron and ionic transport, significantly boosting the hydrogel's electrical activity. Notably, our findings demonstrate that even with similar crystallinity percentages, a higher PPy concentration (6%) markedly enhances electrical activity compared to a lower concentration (4%). These results highlight the crucial importance of optimizing both the crosslinking network and PPy concentration to significantly improve the hydrogel's electrical properties, which is essential for future use in electroactive materials.

Keywords: Crystallinity Structure · Crosslinking · Electroconductive behavior

1 Introduction

Electricity is essential in modern society, driving innovations such as the internet and medical devices like pacemakers. Similarly, our bodies rely on electrical signals for communication, particularly in nerve cells. Disruptions in these signals can lead to issues like cell demyelination and axon degeneration, impairing synaptic transmission [1, 2].

N. R. Abdul Razak et al. (Eds.): ECOOP 1996, SPPHY 418, pp. 149–155, 2025.
https://doi.org/10.1007/978-981-96-2871-1_21

Conductive hydrogels have emerged as promising materials for neuroregenerative therapy due to their ability to provide a pro-regenerative microenvironment and their flexible structure, which can adapt to complex lesions and reconnect neuronal pathways [3, 4]. However, most conductive hydrogels face challenges in achieving optimal conductivity and stability, especially under physiological conditions [5–7]. While conductive polymers such as polypyrrole (PPy) exhibit promising biocompatibility and electrical stability in aqueous environments, their conductivity can be hindered over time by structural defects or disordered regions that disrupt efficient electron transport along the polymer backbone [8, 9]. Hence, PPy must be integrated into a supporting medium to provide structural stability and biocompatibility while maintaining its conductive properties. Gelatine, a natural polymer derived from collagen, is recognized for its biocompatibility, biodegradability, and ability to mimic the extracellular matrix (ECM), promoting cell adhesion, proliferation, and differentiation. While gelatine itself is not conductive, it provides an ideal scaffold for integrating conductive polymers like PPy, supporting their functionality in biomedical applications [10]. In this study, we aim to optimize the electrical properties of PPy by improving its integration with gelatine in a hydrogel matrix using glutaraldehyde as a crosslinker. Although both PPy and gelatine typically exhibit amorphous characteristics, the crosslinking process is designed to enhance the structural cohesion between the polymers, facilitating more efficient electron movement across the hydrogel [11, 12]. By optimizing the interactions between PPy and gelatine, we aim to enhance the electrical conductivity of the hydrogel, making it more suitable for neuroregenerative therapies. The novelty of this study lies in exploring how PPy's intrinsic conductive properties and the molecular interactions with gelatine influence the hydrogel's overall electrical behavior. By keeping the glutaraldehyde concentration constant, we aim to isolate the key factors affecting the hydrogel's electrical performance. Understanding these factors could provide valuable insights into the design of next-generation conductive hydrogels for biomedical applications.

2 Experimental Procedure

Gelatine (95% purity, MW 260.87 g/mol, Sigma Aldrich, USA), 98% pyrrole monomer (Sigma Aldrich, USA), Glutaraldehyde (ThermoScientific, USA), Iron (III) chloride (ThermoScientific, USA), 99% acetic acid (ThermoScientific, USA), and distilled water were used.

In a total volume of 25 mL, 7 wt% of gelatine powder was prepared in 1.25M acetic acid and mixed with a magnetic stirrer for 2 h to allow the gelatine to be dissolved. Following that, pyrrole monomer was added at varied concentrations (0 wt%, 2 wt%, 4 wt%, and 6 wt%) to the mixture in the presence of $FeCl_3$ as oxidant and dopant at room temperature. The mixture was stirred for another 2 h to allow in-situ oxidative polymerization of PPy. 0.5 wt% glutaraldehyde was added to each sample and the mixture was further stirred for another 1 h to allow crosslinking, before being poured into a 24-well container as mold. The resulting Gel/PPy hydrogel was rinsed with distilled water and freeze-dried for 24 h at -20 °C after 24 h left overnight. The sample was labeled as Gel/0PPy, Gel/2PPy, Gel/4PPy and Gel/6PPy. Gel/0PPy as control.

Fourier-transform infrared (FTIR) analysis was discussed on the chemical bonding formation between gelatine and PPy. Subsequently, X-ray diffraction (XRD) analysis was investigated by calculating the crystallinity index (CI) using the equation below:

$$CrystallinityIndex(CI)percentage = \frac{Areaunderofallthecrystallinepeaks}{Areaunderofallthecrystallineandamorphouspeaks} \times 100$$

Further, the electrical behavior of the hydrogel was studied using CV curves. The possible structural formation and their conductivity data were further discussed results and discussions section.

3 Results and Discussions

3.1 FTIR and XRD Analysis

Figure 1 displayed FTIR spectra for each sample. 3300 cm^{-1} indicates N-H stretching, 1628 cm^{-1} C = O stretching, 1540 cm^{-1} C = N stretching, 1380 cm^{-1} C = C stretching, 1260 cm^{-1} C-H bending, 1000 cm^{-1} C-N stretching in amines, and 616 cm^{-1} C-H bending [10, 13]. These peaks suggest the formation of hydrogen bonds between N-H and C = O groups, dipole-dipole interactions between C = O and C = N groups, and π-π stacking interactions between the aromatic rings of PPy and gelatin [10]. Additionally, van der Waals forces between C-H groups contribute to the overall stability of the composite. These interactions collectively suggest a successful crosslinking formation between the PPy and gelatine through a glutaraldehyde crosslinker. Moreover, this would potentially lead to an enhancement of the crystallinity and electrochemical reactivity of the gelatine/PPy composite.

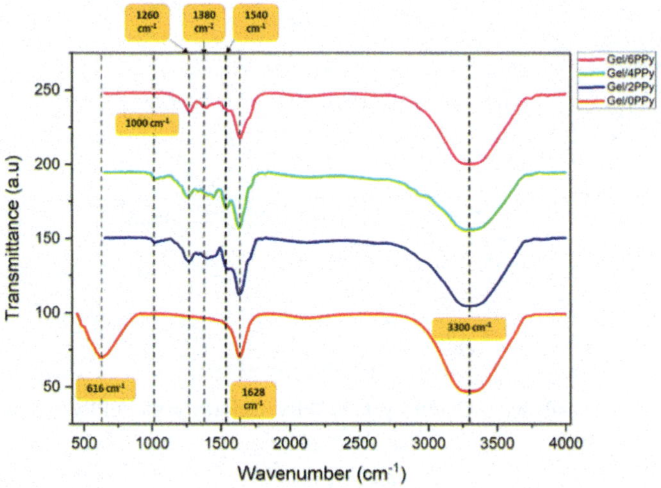

Fig. 1. FTIR analysis of Gelatine/PPy hydrogel at varied PPy concentrations, Gel/0PPy, Gel/2PPy, Gel/4PPy, and Gel/6PPy.

Figure 2 shows the XRD peaks for Gel/0PPy, Gel/2PPy, Gel/4PPy, Gel/6PPy, and pure PPy. The successful crosslinking between gelatine and PPy at varied concentrations is evident from the formation of crystalline peaks. Additionally, the peak intensity increased with higher PPy concentrations. Further analysis using Origin software was conducted to calculate the crystallinity index (CI). The calculated CIs were 12.5, 44.5, 49.8, and 49.7% for sample Gel/0PPy, Gel/2PPy, Gel/4PPy, and Gel/6PPy, respectively. The increasing CI indicates a higher degree of structural order within the material. Higher crystallinity typically signifies more ordered, periodic atomic or molecular arrangements, which can influence material properties. However, the CI values for the samples with 4% and 6% PPy did show almost similar values. This suggests that beyond a certain concentration, additional PPy does not substantially enhance crystallinity. The similar crystallinity indices observed at PPy concentrations above 4% may be due to the exhaustion of available crosslinking sites on glutaraldehyde. As a result, additional PPy does not undergo further crosslinking, leading to an increase in amorphous regions and phase separation, rather than contributing to the enhancement of the ordered crystalline structure. In this work, we aim to correlate the degree of crystallinity with the conductivity behavior. The relationship between structural order and electrical conductivity is crucial for optimizing the performance of conductive hydrogels in biomedical applications.

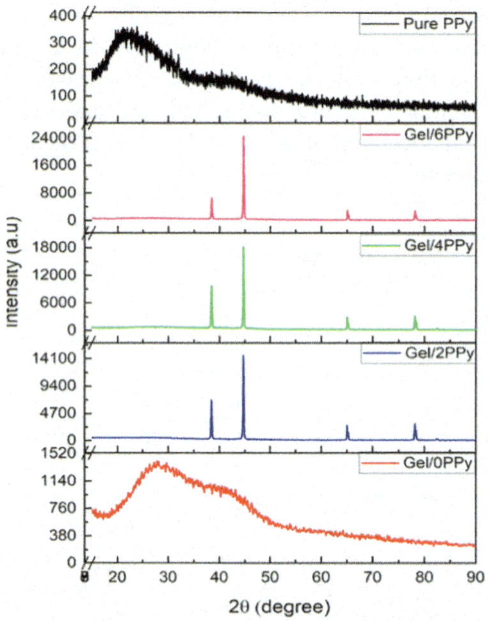

Fig. 2. XRD analysis of Gel/0PPy, Gel/2PPy, Gel/4PPy, Gel/6PPy and pure PPy

3.2 Electrical Behavior by Cyclic Voltammetry

The voltammogram responses for Gel/PPy hydrogel samples with varying concentrations of PPy are shown in Fig. 3. A typical voltammogram displays the current response of a material against the potential applied. Often, the voltammogram would reveal anodic (oxidation) and cathodic (reduction) peaks, arising from the sample's electrical activity due to the potential application [14]. The sample with 0% PPy shows a voltammogram corresponding to material with poor electrical behavior. When PPy is added at 2% and 4%, the sample exhibits electrical reactivity, as indicated by the presence of a small anodic peak at 0.5 V. The sample formulated with 6% of PPy results in a voltammogram with a pronounced anodic peak, initially visible at 0.50 V and then shifted to a smaller value at 0.25 V. Meanwhile, the cathodic peak is initially recorded at around −0.70 V, then lowered to a value of −0.50 V.

The shift of the anodic and cathodic peaks towards smaller values implies the boost in the electrical behavior of the samples endowed by the addition of PPy at higher concentrations [15]. As PPy majorly contributes to the sample's electrical behavior, this paved the way for faster reactions on the sample at a lower energy. The ability of PPy to boost the electrical behavior can be attributed to its effect in enhancing the crystallinity of the samples, providing a rapid pathway of electron transfer from the ordered atomic arrangement in the sample. Although Gel/4PPy and Gel/6PPy exhibit similar crystallinity index (CI) values, indicating comparable structural order, their electrical behavior differs significantly. This difference is attributed to the increased mobility of ionic species in the Gel/6PPy sample, which facilitates more effective ionic reactions on the sample's surface during successive scans. Consequently, this results in shifts in the potential readings of both anodic and cathodic peaks.

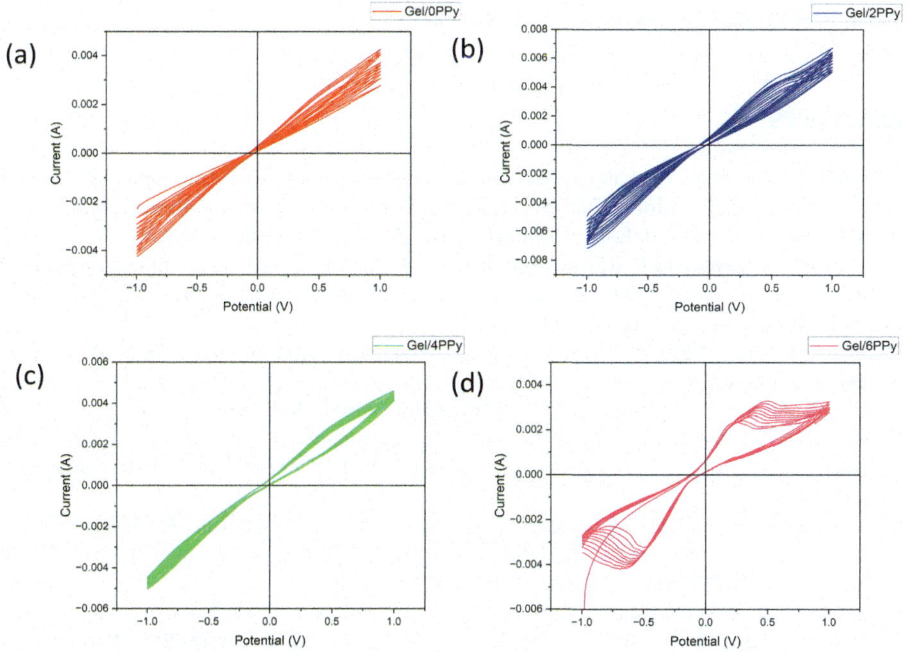

Fig. 3. Cyclic voltammetry scans of Gelatine/PPy hydrogels (a) Gel/0PPy, (b) Gel/2PPy, (c) Gel/4PPy, and (d) Gel/6PP

4 Conclusion

In conclusion, this study reveals that the electrical activity of the hydrogel is critically influenced by both its crystallinity structure and the concentration of PPy. Notably, our findings demonstrate that even with similar crystallinity percentages, a higher PPy concentration (6%) markedly enhances electrical activity compared to a lower concentration (4%). These results highlight the crucial importance of optimizing both the crosslinking network and PPy concentration to significantly improve the hydrogel's electrical properties, which is essential for future use in electroactive materials. Future work should focus on further elucidating the factors influencing conductivity behavior by employing Electrochemical Impedance Spectroscopy (EIS) to provide a more detailed analysis of the hydrogel's electrical properties and to identify key impedance components. Additionally, performing Thermogravimetric Analysis (TGA) and Differential Scanning Calorimetry (DSC) would offer insights into the hydrogel's thermal stability and phase transitions, further elucidating the relationship between PPy concentration, crystallinity, and conductivity. High-resolution Transmission Electron Microscopy (TEM) could also be used to observe the nanoscale structure and dispersion of PPy within the gelatine matrix, offering a clearer understanding of the microstructural contributions to conductivity. These additional analyses will help to refine further and enhance the performance of PPy-gelatine hydrogels for advanced electroactive applications.

Acknowledgements. The authors would like to acknowledge that the study covered in this manuscript was funded by Universiti Teknologi MARA on financial support under grant reference no: 600-RMC/YTR/5/3 (004/2022).

References

1. Safdarian, M., et al.: Global, regional, and national burden of spinal cord injury, 1990–2019: a systematic analysis for the Global Burden of Disease Study 2019. Lancet Neurol. **22**(11), 1026–1047 (Nov. 2023). https://doi.org/10.1016/S1474-4422(23)00287-9
2. Alizadeh, A., Dyck, S.M., Karimi-Abdolrezaee, S.: Traumatic spinal cord injury: an overview of pathophysiology, models and acute injury mechanisms. Front Neurol. **10**, (Mar. 2019). https://doi.org/10.3389/fneur.2019.00282
3. Shahemi, N.H., Mahat, M.M., Asri, N.A.N., Amir, M.A., Ab Rahim, S., Kasri, M.A.: Application of conductive hydrogels on spinal cord injury repair: a review (Jul. 10, 2023). American Chemical Society. https://doi.org/10.1021/acsbiomaterials.3c00194
4. Min, J.H., Patel, M., Koh, W.G.: Incorporation of conductive materials into hydrogels for tissue engineering applications (Sep. 28, 2018). MDPI AG. https://doi.org/10.3390/polym10101078
5. Kougkolos, G., Golzio, M., Laudebat, L., Valdez-Nava, Z., Flahaut, E.: Hydrogels with electrically conductive nanomaterials for biomedical applications. J Mater Chem B **11**(10), 2036–2062 (2023). https://doi.org/10.1039/D2TB02019J
6. Liu, D., et al.: Conductive polymer based hydrogels and their application in wearable sensors: a review. Mater. Horiz. **10**(8), 2800–2823 (2023). https://doi.org/10.1039/D3MH00056G
7. Burnstine-Townley, A., Eshel, Y., Amdursky, N.: Conductive scaffolds for cardiac and neuronal tissue engineering: governing factors and mechanisms. Adv. Funct. Mater. **30**(18), 1901369 (May 2020). https://doi.org/10.1002/adfm.201901369

8. Liang, Y., Goh, J.C.-H.: Polypyrrole-incorporated conducting constructs for tissue engineering applications: a review. Bioelectricity **2**(2), 101–119 (Jun.2020). https://doi.org/10.1089/bioe.2020.0010
9. Ahmad Ruzaidi, D.A., et al.: Advocating electrically conductive scaffolds with low immunogenicity for biomedical applications: a review. Polymers (Basel) **13**(19), 3395 (Oct. 2021). https://doi.org/10.3390/polym13193395
10. Wang, S., et al.: Fabrication of polypyrrole-grafted gelatin-based hydrogel with conductive, self-healing, and injectable properties. ACS Appl. Polym. Mater. **2**(7), 3016–3023 (Jul. 2020). https://doi.org/10.1021/acsapm.0c00468
11. Hu, W., Wang, Z., Xiao, Y., Zhang, S., Wang, J.: Advances in crosslinking strategies of biomedical hydrogels (Mar. 01, 2019). Royal Society of Chemistry. https://doi.org/10.1039/c8bm01246f
12. Augustine, R., Alhussain, H., Zahid, A.A., Raza Ur Rehman, S., Ahmed, R., Hasan, A.: Crosslinking strategies to develop hydrogels for biomedical applications. Springer, pp. 21–57 (2021). https://doi.org/10.1007/978-981-15-7138-1_2
13. Zhao, Y., Liang, Y., Ding, S., Zhang, K., Mao, H., Yang, Y.: Application of conductive PPy/SF composite scaffold and electrical stimulation for neural tissue engineering. Biomaterials **255**, 120164 (Oct. 2020). https://doi.org/10.1016/j.biomaterials.2020.120164
14. Elgrishi, N., Rountree, K.J., McCarthy, B.D., Rountree, E.S., Eisenhart, T.T., Dempsey, J.L.: A practical beginner's guide to cyclic voltammetry. J. Chem. Educ. **95**(2), 197–206 (Feb. 2018). https://doi.org/10.1021/acs.jchemed.7b00361
15. Lee, K.J., McCarthy, B.D., Dempsey, J.L.: On decomposition, degradation, and voltammetric deviation: the electrochemist's field guide to identifying precatalyst transformation. Chem. Soc. Rev. **48**(11), 2927–2945 (2019). https://doi.org/10.1039/C8CS00851E

Impact of Titanium Addition on the Microstructural and Mechanical Properties of CrCoFeMnNi High Entropy Alloy

Arnita Surieya Sangar[1], Nur Izzati Muhammad Nadzri[1(✉)], Muhammad Firdaus Mohd Nazeri[1], Nurlyana Izyan Mohd Ali[1], Nur Athirah Binti Sidek[1], Puvaneswaran Chelvanathan[2], Yoganash Putthisigamany[2], Tea-Sung Jun[3], and Sudha Joseph[4]

[1] Centre of Excellence Geopolymer and Green Technology (CEGeoGTech), Faculty of Chemical Engineering Technology, Universiti Malaysia Perlis (UniMAP), Taman Muhibbah, 02600 Arau, Perlis, Malaysia
izzatinadzri@unimap.edu.my

[2] Solar Energy Research Institute (SERI), Universiti Kebangsaan Malaysia, 43600 Bangi, Selangor, Malaysia

[3] Department of Mechanical Engineering, Incheon National University, Incheon 22012, Republic of Korea

[4] Department of Materials Engineering, Cambridge Institute of Technology, Bengaluru, Karnataka 560036, India

Abstract. The impact of titanium (Ti) addition on cobalt, chromium, iron, manganese, and nickel (CrCoFeMnNi) high entropy alloy (HEA) was studied in the aspect of microstructural and mechanical properties. The CrCoFeMnNiTi HEA was cast via a vacuum melting process and remelted five times to achieve homogeneity. X-ray diffraction (XRD) was used to identify the phases presence where addition Ti introduced new phase (BCC). Field emission scanning electron microscopy (FESEM) analysis discovers a dendritic structure present in the titanium-added CrCoFeMnNi HEA. In addition, the CrCoFeMnNiTi HEA gives a higher hardness value of 949.24 $\pm$ 5 HV compared to the CrCoFeMnNi alloy, which has a hardness value of 184.42 $\pm$ 5 HV. The findings suggest that the addition of Ti has a positive impact such as higher hardness, which has potential applications in a wide range of materials science and engineering fields.

Keywords: High Entropy Alloy · CrCoFeMnNi · Titanium · Microstructure · Mechanical properties

1 Introduction

High-entropy alloys (HEAs) have garnered significant attention in materials science having superior mechanical properties and structural versatility. Among these, the (CrCoFeMnNi HEA), colloquially known as the Cantor alloy, has emerged as a flagship material due to its remarkable ductility, high strength, and superior fracture toughness

N. R. Abdul Razak et al. (Eds.): ECOOP 1996, SPPHY 418, pp. 156–161, 2025.
https://doi.org/10.1007/978-981-96-2871-1_22

even at cryogenic temperatures [1]. However, the alloy's full potential remains unrealized, as it grapples with limitations in yield strength, hardness, and wear resistance, posing obstacles to its widespread utilization in demanding engineering applications.

The microstructure and properties of HEAs are intricately linked with their elemental composition, offering a pathway for customized optimization. Previous studies demonstrated the efficacy of alloy addition in modulating material characteristics, with notable enhancements observed through the addition of elements such as aluminum (Al), molybdenum (Mo), and carbon (C) [2–5].

Within this framework, the current investigation aims to investigate the transformative potential of adding titanium (Ti) to enhance the microstructure and characteristics of CrCoFeMnNi HEA. Titanium is known for its varied impacts on the behavior of alloys and has the potential as a strategic dopant to improve corrosion resistance and mechanical performance. There are still some gabs in studying on the impact of Ti addition in CrCoFeMnNi HEA. This study intends to clarify the complex relationship between alloy composition, microstructural evolution, and mechanical properties by adding Ti. The findings will provide important new information for the design and optimization of advanced HEA systems.

2 Experimental Procedure

A casting CrCoFeMnNiTi HEA ingot was prepared by vacuum melting, re-melting, and repeated 3 times to ensure the final uniform microstructure by removing trapped air. Then, the sample was poured into a mold size 5cm x 5cm and was processed by cutting it into a smaller size 1cm x 1cm, cleaned by ultrasonic cleaning in acetone, and then ground and polished. Microstructural analysis was conducted using a FEI Nova Nanosem 450 Field emission scanning electron microscope (FESEM) equipped with energy-dispersive X-ray spectroscopy (EDX) for composition analysis. Phase analysis was performed using X-ray diffractometer (XRD, Bruker, Phaser D2). Mechanical properties were evaluated through Vickers microhardness (MicroVicky VH1010) tester with a load of 1000 gf and dwell times of 10 s ensuring enough time for the indenter to fully penetrate the sample surface. This comprehensive experimental approach enabled a thorough understanding of the microstructure, mechanical behavior, and corrosion resistance of the CrCoFeMnNiTi HEA, facilitating its potential for diverse engineering applications.

3 Results and Discussions

3.1 Microstructural Analysis

The microstructural analysis of the CrCoFeMnNi and CrCoFeMnNiTi HEA from Fig. 1, conducted through FESEM and EDS mapping, shows the microstructure and compositions. Variations in shades across the microstructure of CrCoFeMnNi HEA indicate different elements and phase distributions, where the brighter regions exhibit higher concentrations of Cr and Fe (Fig. 1.c). Conversely, darker regions were homogenized with all elements, and greyish grains rich in Mn and Ni. This also confirms the findings of related scholars [6]. The larger grain has an average grain area $\sim 18000\ \mu \pm 4$, while darker regions with an average grain area $\sim 84000\ \mu \pm 4$ (Fig. 1.a).

Moreover, the addition of Ti to the CrCoFeMnNi HEA shows a new phase rich in Ti, distributed uniformly in irregular island-like formations with an average grain area $\sim 1800\ \mu \pm 4$ throughout the structure (Fig. 1.b). The matrix structure remains rich in Cr and Fe but depleted in Ti (Fig. 1.d), indicating preferential partitioning of Ti into the precipitated phase. The microstructure of the CrCoFeMnNiTi HEA exhibits dendritic structures mixed with island grains. The structure is characterized by dark and light lines that are rich in Mn and Ni, and greyish island grains rich in Cr and Fe. This confirms the findings from other related studies where they found regions that are rich with Mn and Ni are depleted with Cr and Fe [6]. The presence of Ti in CrCoFeMnNi HEA which has varying atomic sizes and chemical affinities can lead to phase separation or segregation during solidification. This segregation results in the formation of different regions within the microstructure, such as lamellae or laves faces, consisting of different compositions. The observed difference in grain size between the CrCoFeMnNi HEA and CrCoFeMnNiTi HEA is due to the addition of Ti since Ti precipitated and hindered grain growth. The irregular island-like structures were formed due to the uniform distribution Ti precipitations. CrCoFeMnNi HEA shows a single phase while the addition of Ti changes it to BCC structure since there is precipitation occurs as shown in XRD results [7].

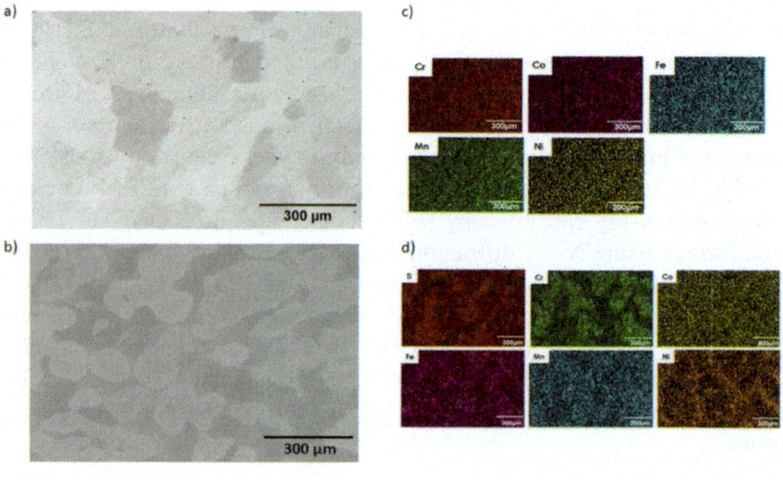

Fig. 1. FESEM pictures of a) CrCoFeMnNi, b) CrCoFeMnNi with Ti addition HEA and EDS-mapping of c) CrCoFeMnNi, d) CrCoFeMnNi with Ti addition

3.2 Phase Analysis

The XRD pattern of CrCoFeMnNi and CrCoFeMnNi with Ti addition HEAs result findings are shown in Fig. 2.

Figure 2 shows the XRD pattern of CrCoFeMnNi and CrCoFeMnNi with Ti addition, there are three obvious diffraction peaks at 43.68°, 50.68° and 74.63° which correspond to (111), (200), and (220) planes indicate to single FCC crystal structure of CrCoFeMnNi,

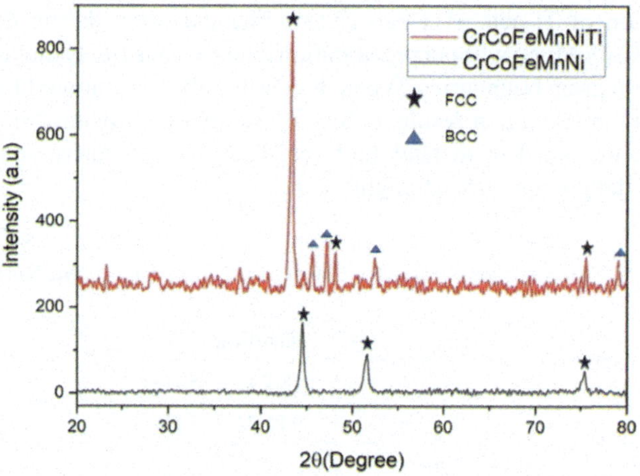

Fig. 2. XRD results of the CrCoFeMnNi and CrCoFeMnNi high entropy alloys.

this phase formation of these distinct peaks indicates that all the peaks were equimolar and confirms the crystalline nature of high entropy alloys. While, other elements were substituted at the positions of diffraction peaks at 43.44°, 45.65°, 48.18°, 52.46°, and 79.12° in XRD pattern corresponded to crystal orientations of (332), (422), (432), (440) and (741). The newly formed phase was a BCC crystal structure rich in Ti, which was called the Ti-rich phase. With the addition of Ti, a new BCC structure was formed and shifted due to the lattice distortion since Ti atoms are larger than other elements and have bigger lattice parameter compared to CrCoFeMnNi [8]. However, with the addition of Ti, the crystal structure of the alloy changes from FCC to FCC + BCC. The result was supported by [6] Chen et al. (2022) stated that Ti addition not only affects grain but also the phase distribution of alloys.

3.3 Mechanical Property

The Vickers microhardness results of the CrCoFeMnNi and CrCoFeMnNiTi HEA is shown in Table 1. The CrCoFeMnNiTi HEA gives a higher hardness value of 949.24 ± 5 HV compared to the CrCoFeMnNi HEA, which has a hardness value of 184.42 ± 5 HV. The addition of Ti atoms to the CrCoFeMnNi HEA, leads to an increase in lattice parameters and lattice distortion within the face-centered cubic (FCC) structure and hinders dislocation movement, thereby increasing the CrCoFeMnNiTi HEA resistance to deformation [6]. Besides, the addition of Ti changes the FCC structure to the BCC structure which naturally has higher strength than the FCC structure. The BCC structures help to increase the hardness since the structure is consist of precipitated of Ti. Moreover, the presence of the dendritic structure in the microstructure of CrCoFeMnNiTi HEA (as discussed in subtopic 3.1) helps to improve the hardness of HEA by giving a brittle property to the material. The addition of Ti affects the strengthening of HEA via several mechanisms such as precipitations, solid solution, and dislocation strengthening [7]. The Microstructure itself reveals the precipitation of Ti which gives the dendritic structure.

The precipitation of Ti acts as a barrier to withstand force during deformation and dislocation. CrCoFeMnNi without precipitation tends to have freely moving grains which produces straight grain boundaries. When there is highly concentrated force applied, the HEA will easily crack which results in lower hardness [8]. Meanwhile, CrCoFeMnNi with Ti precipitations will withstand the force. Therefore, the increase in hardness was contributed by the precipitation strengthening.

Table 1. Microhardness value for CrCoFeMnNi and CrCoFeMnNiTi HEA.

Samples	**Reading (HV)**
CrCoFeMnNi	184.42 ± 5 HV
CrCoFeMnNiTi	949.24 ± 5 HV

4 Conclusion

In conclusion, Ti addition has a positive effect on the properties of the CrCoFeMnNi HEA. The casting method used to fabricate the CrCoFeMnNiTi HEA was successful. The XRD results indicate that CrCoFeMnNiTi HEA has higher peak intensity and exhibits BCC structure by forming new peaks compared to CrCoFeMnNi. The surface morphology of CrCoFeMnNiTi HEA exhibits lamella and dendritic structure which shows the precipitation of Ti in the HEA matrix. Meanwhile, the microhardness test showed that there is a tremendous increase in the hardness from 184.42 ± 5 HV to 949.24 ± 5 HV for CrCoFeMnNiTi HEA. Precipitation is the main strengthening mechanism for CrCoFeMnNiTi HEA.

Acknowledgments. The authors would like to acknowledge that the study covered in this manuscript was funded by Ministry of Higher Education Malaysia on the financial support under grant reference no: FRGS/1/2022/STG05/UNIMAP/02/3.

References

1. Xu, Z., et al.: Corrosion resistance enhancement of CoCrFeMnNi high-entropy alloy fabricated by additive manufacturing (2020)
2. Fang, Y., et al.: Selective laser melting of AlCoCrFeMnNi high entropy alloy: Effect of heat treatment (2023)
3. Wang, W. et al.: Effect of Mo and aging temperature on corrosion behaviour of (CoCrFeNi) 100-Mo high entropy alloys (2020)
4. Zhang, J. et al.: Manipulation of microstructure and mechanical properties in N-doped CoCrFeMnNi High Entropy alloys (2021)
5. limova, M. et al.: Effect of carbon on recrystallised microstructure and properties of CoCrFeMnNi-type high entropy alloys (2021)

6. Chen, Y. et al.: Effect of Ti content on the microstructure and properties of CoCrFeMnNiTix high entropy alloy (2022)
7. Brito-Garcia, S.J., et al.: Impact of Ti doping on the microstructure and mechanical properties of CoCrFeMoNi High Entropy Alloys (2023)
8. Karimzadeh, M., et al.: Effects of titanium addition on the microstructure and mechanical properties of quaternary CoCrFeNi high entropy alloy (2022)

Investigation of Oxidation Kinetics and Oxide Formation on Heat-Treated Fe-Ni-Cr Alloys at High-Temperature Isothermal Oxidation

Danish Mikael Mohd Zaidi[1] and Noraziana Parimin[1,2](✉)

[1] Faculty of Chemical Engineering and Technology, Universiti Malaysia Perlis, 02600 Arau, Perlis, Malaysia
noraziana@unimap.edu.my
[2] Surface Technology Special Interest Group, Faculty of Chemical Engineering and Technology, Universiti Malaysia Perlis, 02600 Arau, Perlis, Malaysia

Abstract. The investigation of oxidation kinetics and oxide scale formation on heat-treated Fe-Ni-Cr alloys at high temperature isothermal oxidation was studied. Altering the grain size of the Fe-Ni-Cr alloy through heat treatment was performed to investigate its impact on oxidation characteristics. Heat treatment was carried out at 1000 °C and 1200 °C. Subsequently, the heat-treated alloy was subjected to a 150-h isothermal oxidation test at 700 °C. The investigation focused on characterizing the grain structure of heat-treated samples using the linear intercept method, and then utilized oxidation kinetics and X-ray diffraction techniques to analyze the behavior of heat-treated alloys post-oxidation. As the temperature of the heat treatment increases, the grain structure also grows. The oxidation rate of heat-treated Fe-Ni-Cr alloys shows both parabolic and linear rate laws. Heat treatment at 1000°C is preferred as these samples exhibit a diffusion-controlled mechanism and a parabolic rate law, preventing further oxidation. The sample treated at 1200°C displays a linear rate law, resulting in continuous formation and growth of the oxide layer. Furthermore, the oxidation test led to the formation of several oxide phases, including Cr-rich, Cr-Mn-rich and Ti-rich oxide layers.

Keywords: Oxidation kinetic · Heat treatment · Isothermal oxidation · Fe-Ni-Cr alloy · Ni-Based alloy

1 Introduction

Ni-based alloys have the ability to retain excellent combinations of mechanical properties and resistance to oxidation, making them widely used in various applications as materials for structures at high temperatures [1, 2]. Ni-based alloys that can withstand high temperatures form a consistent layer of an oxide that gradually forms and is thermodynamically stable. This layer is crucial for effective oxidation resistance as it allows further oxidation, protecting the metallic alloys.

The application commonly involves the use of Fe-Ni-Cr alloy. At elevated temperatures, the oxidation of Fe-Ni-Cr alloy is impacted by the composition of its oxide phase

N. R. Abdul Razak et al. (Eds.): ECOOP 1996, SPPHY 418, pp. 162–169, 2025.
https://doi.org/10.1007/978-981-96-2871-1_23

due to the presence of various alloying elements, resulting in the development of diverse oxide phases. The addition of alloying elements offers the advantage of enhancing the oxidation resistance of Fe-Ni-Cr alloy through the creation of a protective and firmly adherent oxide scale comprising of Cr_2O_3, $MnCr_2O_4$, and/or TiO_2 oxides [3, 4]. The formation of protective oxide scales is essential for attaining resistance to oxidation at high temperatures [5, 6]. Oxides containing Cr, Ti, and/or Mn are essential for improving surface stability, thereby leading to exceptional oxidation resistance [7, 8].

One method to enhance resistance to oxidation is to create a rapid initial oxide layer through grain boundary diffusion. Modifying the size of the grains can affect the rate of oxidation, leading to greater resistance to the flaking and adherence of oxide scales, and promoting grain boundary diffusion. Studies show that the boundaries between grains serve as a route for the outward movement of cations towards the surface of the alloy, leading to the rapid development of a protective oxide layer [7]. A smaller grain size provides enhanced resistance to the peeling off of scales because smaller grains aid in stress relaxation within the oxide layer and improve the bonding of the oxide layer to the matrix. The refinement of the grain structure can be accomplished through heat treatment [9–11]. This study focused on examining the isothermal oxidation characteristics of different grain sizes produced through a heat treatment process. The ability of Ni-based alloys to form a firmly adhering oxide layer on the metal surface significantly affects their resistance to oxidation at high temperatures.

2 Experimental Procedure

The study used Ni-based Fe-Ni-Cr alloys. The alloy's dimensions are 10 mm x 10 mm x 3 mm. Optical emission spectroscopy was used to analyze the chemical composition, which revealed the following weight percentages: Fe 46.07%, Ni 32.62%, Cr 18.97%, C 0.08%, Mn 0.56%, Ti 0.49%, Si 0.32%, Al 0.54%, Co 0.14%, Mo 0.04%, Cu 0.08%, Nb 0.03%, V 0.03%, and W 0.03%. The alloy is denoted by the materials composition Fe-33Ni-19Cr. Two specific heat treatment temperatures were used: 1000 ºC and 1200 ºC, each for a duration of 3 h with a heating rate of 10 °C, and then the alloy was quenched in water, known as H10 and H12, respectively. The main aim of the heat treatment process is to modify the grain size of the alloy. The grain size of the heat-treated alloy was determined using ASTM's linear intercept methods E112. H10 and H12 specimens yielded an average grain size of 54.05 μm and 135.13 μm, respectively. The fine grain size was observed in the H10 samples, while the coarse grain size was observed in the H12 sample.

The heat-treated alloy was subjected to an isothermal oxidation test in a laboratory air environment. Using a muffle furnace, the oxidation test was conducted at a temperature of 700 °C for a total of 150 h for discontinuous testing exposure. Before the oxidation test, the surface area and weight of all samples were recorded. The samples were subjected to the oxidation conditions at 25-h intervals. The weight of the samples was measured after each oxidation exposure. The oxidation process was then continued for an additional 25 h exposure, resulting in cumulative exposure durations of 50 h, 75 h, 100 h, 125 h, and 150 h. This study utilized discontinuous testing to plot the oxidation kinetics of the sample.

The alloy that had undergone oxidation at constant temperature was analyzed to study the oxidation rate and phase present using X-ray diffraction (XRD). The weight variation of the samples during 25-h intervals was utilized to create the graph for oxidation kinetics. Additionally, the samples oxidized for 25 h and 150 h were selected to identify the phases present at the beginning and end of the oxidation process. The oxidation kinetics were graphed by plotting the weight change per surface area, denoted as x, against time (t) in hours. Equation (1) was used to determine the oxidation rate law, where c is a constant. The rate data represented by the m value in curves with values of 1 and 2 corresponded to the linear and parabolic rate laws, respectively. If the gradient's m value falls between 0.50 and 1.49, the graph will reflect a linear rate law (m equals 1). When the gradient value ranges from 1.50 to 2.49, it indicates a parabolic rate law (m equals 2). The parabolic rate constant, K_P, was determined using Eq. (2). The linear rate constant, K_L, was determined using Eq. (3).

$$\log x = 1/m \log t + c \tag{1}$$

$$X^2 = K_p t + c \tag{2}$$

$$X = K_L t + c \tag{3}$$

3 Results and Discussions

3.1 Oxidation Kinetic

In Fig. 1, the oxidation kinetics of the heat-treated Fe-Ni-Cr alloy at 700 °C are depicted. The increasing trend of the sample weights is evident as the oxidation time progresses. The oxidation kinetic curve for H10 indicates that the average highest weight change recorded after 150 h is 0.00648 mg/cm^2, with the lowest reading observed at 25 h of exposure time being 0.00271 mg/cm^2. In addition, sample H12 experienced the highest weight change of 0.00856 mg/cm^2 after 150 h of heating, while the lowest value of weight change was 0.00177 mg/cm^2 after 25 h. The data indicates that weight changes rise as the exposure time increases. This oxidation kinetic trend mirrors the pattern observed by other researchers, where weight change increases with longer exposure time [12, 13].

The diffusion paths for atoms or molecules are significantly increased by the high grain boundary area in materials with fine grain sizes in the case of the H10 sample. The extended diffusion distance aids in the development and growth of oxide layers, leading to faster diffusion and the enhanced formation of a protective initial oxide layer on the alloy surface. As a result, the oxidation process becomes more effective and is more likely to follow a diffusion-controlled mechanism, resulting in a parabolic rate curve [7, 13].

In the case of the H12 sample, with larger grain sizes, the diffusion path is shortened, resulting in a less efficient oxidation process due to fewer diffusion routes available. This leads to a slower formation of the initial oxide layer compared to a higher diffusion route in fine grain size [7, 13]. The weight gain data at 25 h exposure in Fig. 1 clearly shows a lower weight gain for H12 compared to H10 sample. As the exposure duration increases, the diffusion process will continuously take place over time, resulting in the production of a protective oxide layer on the alloy surface and thereby increasing the weight change data, which enhances the linear rate law. The oxide layer's development and expansion occur at a relatively constant rate, leading to a more uniform spread of the oxidation process across the material's surface. This characteristic is reflected in a linear relationship between the rate of oxidation and time.

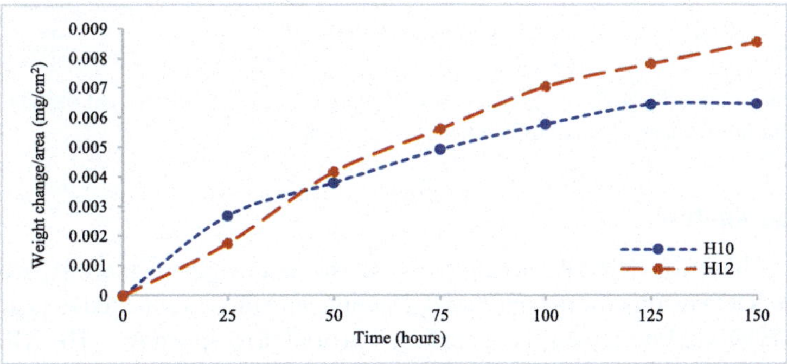

Fig. 1. Oxidation kinetic of heat-treated Fe-Ni-Cr alloy after isothermal oxidation at 700 °C.

The graph in Fig. 2 (a) depicts the log weight changes per surface area plotted against log time exposure to identify the oxidation rate law, which can be linear, or parabolic. Equation (1) is utilized to determine the m value from this graph theoretically. Upon calculating the m value, it was concluded that the oxidation rate law for sample H10 is parabolic, with an m value of 1.93, and for sample H12 is linear, with an m value of 1.14.

The H10 sample was analyzed using Eq. (2), which describes a parabolic rate law, and the parabolic rate constant, K_P value, was obtained from Fig. 2(b). Similarly, Eq. (3) was utilized for the H12 sample, which exhibited a linear rate law, and the linear rate constant, K_L value, was determined from Fig. 2(c). The K_P and K_L values were derived from the gradient of each oxidation kinetic curve. Sample H10 has a K_P value of 8.3787×10^{-11} $mg^2cm^{-4}s^{-1}$, while sample H12 has a linear rate constant, K_L, of 1.4713×10^{-9} $mgcm^{-2}s^{-1}$.

Based on the comparison, sample H10 emerges as the superior candidate among the two samples due to its parabolic rate constant, indicating a diffusion-controlled mechanism. This behavior suggests that the formation of an initial oxide layer acts as a barrier to diffusion, hindering further oxidation of the material. As the oxide layer forms and hinders reactant diffusion, the initial phase of rapid oxidation is succeeded by a reduction in the oxidation rate.

The oxidation rate of sample H12 increases due to diffusion, with the rate of oxidation being directly related to time. According to the findings, sample H10 is the most effective, making sample H12 the least effective among these samples.

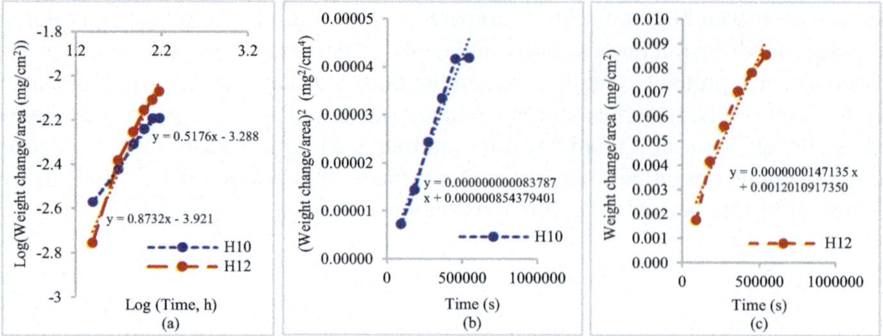

Fig. 2. Heat-treated Fe-Ni-Cr alloy after isothermal oxidation at 700 °C; (a) oxidation rate law, (b) parabolic rate constant, K_P and (c) linear rate constant, K_L.

3.2 Phase Analysis

The X-ray diffraction (XRD) technique was used to conduct phase analysis, as depicted in Fig. 3. The XRD results for the unoxidized sample and those subjected to heat treatment processes H10 and H12 are displayed in Fig. 3 (a) and (b) respectively. The XRD patterns of the two samples are nearly indistinguishable, with the identification of the three highest peaks of the base metal, Fe-Ni-Cr alloy, at 2θ 44.32°, 51.44°, and 75.28°. These three peaks in each sample represent the base metal Fe-33Ni-19Cr alloy.

The XRD pattern of sample H10 after 25 h of exposure is depicted in Fig. 3 (c). In this sample, three distinct peaks were observed for the base metal along with the presence of the oxide phase, specifically Cr_2O_3, $MnCr_2O_4$ and TiO_2. The most prominent peak for the oxide phase is evident at 2θ 36.23°, corresponding to TiO_2. In the XRD pattern of sample H12 after 25 h of exposure, as shown in Fig. 3 (d), a similar peak to that of sample H10 was detected. The primary distinction is the decreased intensity of the TiO_2 peak at 2Θ 36.23°. The XRD pattern in Fig. 3 (e) displays the results for sample H10 after being exposed for 150 h, showing similar peaks to those of the base metal. The analysis detected the presence of Cr_2O_3, $MnCr_2O_4$ and TiO_2 oxide phases in this sample. Specifically, Cr_2O_3 oxide was observed at 2θ angles of 35.81°, 33.32°, and 24.28°, $MnCr_2O_4$ oxide at 2θ angles of 35.1°, 29.80°, and 56.45°, and TiO_2 oxide at 2θ angles of 54.35° and 27.29°. A noticeable decrease in the intensity of the base metal itself was observed between the 25-h and 150-h exposures. The influence on the intensity of the base metal can be inferred from the isothermal oxidation at 700°C after 150 h of exposure, as the thickness of the oxide phases starts to grow over time. Thin oxide scales were indicated by the low intensity of oxide phase after 25 h of exposure on the sample surfaces, while thick oxide scales were indicated by the high intensity of oxide phase after 150 h of exposure. Figure 3 (f) displays the XRD pattern of sample H12 after being exposed for 150 h. In general, this sample exhibited a comparable phase to the H10 sample. However, the detection of the TiO_2 oxide phase was lower in this sample.

Furthermore, the introduction of Mn alloying element into the alloy system enhances the development of the oxide scale by forming an external Cr/Mn-spinel layer, which effectively decreases the evaporation of volatile Cr-species [14]. Moreover, the existence of a Cr/Mn-spinel layer, such as $MnCr_2O_4$, indicates the swift diffusion of Mn within

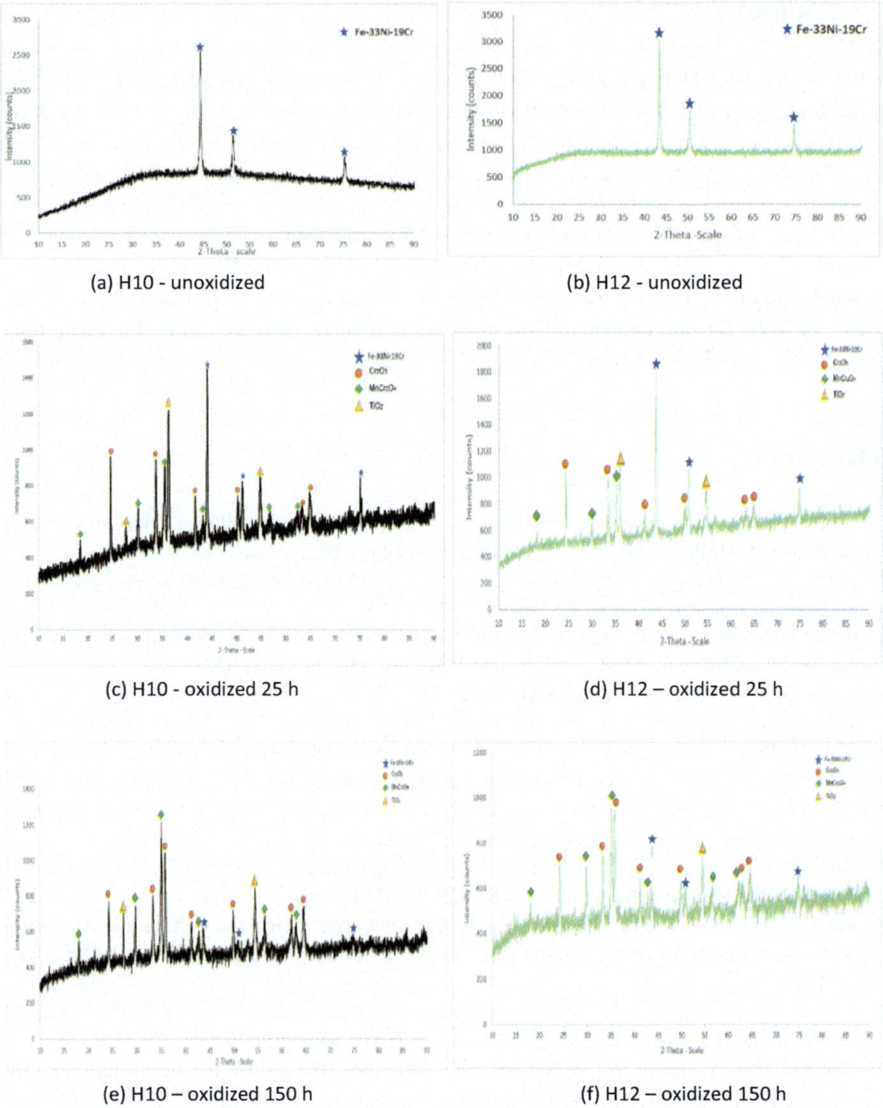

Fig. 3. XRD pattern of heat-treated Fe-Ni-Cr alloy after isothermal oxidation at 700 °C.

the Cr_2O_3 oxide phase [3, 15]. The identification of a Ti-rich fluorite oxide structure suggests the presence of TiO_2 oxide. The existence of Cr-rich and/or Ti-rich oxides helps reduce the evaporation of Cr due to lower Cr vapor pressure [7, 16]. The development of oxides containing Mn and Ti on the alloy demonstrates protective behavior and possesses dense and compact oxide properties [3, 7, 16].

4 Conclusion

Conclusively, the H10 sample, characterized by a fine grain size, displayed a parabolic rate law, suggesting a diffusion-controlled growth rate of the oxide layer and thus demonstrating good oxidation resistance. Conversely, the H12 sample, featuring a coarse grain size, showed a linear rate law, with the weight gain increasing as the exposure time increased. The phase analysis indicated the presence of the base metal phase along with several oxide phases including Cr_2O_3, $MnCr_2O_4$ and TiO_2.

Acknowledgments. The Ministry of Higher Education Malaysia provided support to the authors through the Fundamental Research Grant Scheme (FRGS) with the grant number FRGS/1/2020/TK0/UNIMAP/02/43. The authors wish to express their gratitude for this support.

References

1. Cao, J., Zhang, J., Chen, R., Ye, Y., Hua, Y.: High temperature oxidation behavior of Ni-based superalloy GH202. Mater. Charact. **118**, 122–128 (2016)
2. Wang, X., Szpunar, J.A.: Effects of grain sizes on the oxidation behavior of Ni-based alloy 230 and N. J. Alloy. Compd. **752**, 40–52 (2018)
3. Parimin, N., Hamzah, E.: Role of Mn alloying element on the oxide growth behavior of 800H Nickel-based alloy at 900°C. Solid State Phenom. **336**, 17–22 (2022)
4. Saber, D., Emam, I.S., Abdel-Karim, R.: High temperature cyclic oxidation of Ni based superalloys at different temperatures in air. J. Alloy. Compd. **719**, 133–141 (2017)
5. Col, A., Parry, V., Pascal, C.: Oxidation of a Fe-18Cr-8Ni austenitic stainless steel at 850 °C in O2: Microstructure evolution during breakway oxidation. Corros. Sci. **114**, 17–27 (2017)
6. Nguyen, T.D., Zhang, J., Young, D.J.: Effect of Mn on oxide formation by Fe-Cr and Fe-Cr-Ni alloys in dry and wet CO2 gases at 650°C. Corros. Sci. **112**, 110–127 (2016)
7. Parimin, N., Hamzah, E.: Influence of Ti on Oxide formation during isothermal oxidation of 800H Ni-based alloys. Key Eng. Mater. **929**, 29–34 (2022)
8. Jiang, J., Xiao, G., Wang, Y., Liu, Y.: High temperature oxidation behavior of the wrought Ni-based superalloy GH4037 in the solid and semi-solid state. J. Alloy. Compd. **784**, 394–404 (2019)
9. Zulnuraini, Z., Parimin, N.: Isothermal oxidation behavior of Fe-33Ni-18Cr alloy in different heat treatment temperature. Mater. Sci. Forum **1010**, 46–51 (2020)
10. Cai, D., Mei, Y., Pulin, N., Wenchang, L.: Influence of solution treatment temperature on mechanical properties of a Fe–Ni–Cr alloy. Materials Lettters **57**(24–25), 3805–3809 (2003)
11. Zhang, J.F., Tu, Y.F., Xu, J., Zhang, J.S., Zhang, J.L.: Effect of solid solution treatment on microstructure of Fe-Ni based high strength low thermal expansion alloy. J. Iron Steel **15**(1), 75–78 (2008)
12. Zhang, Z., Webster, R.F., Zhang, J.: Alloying effects of Al, Si and Fe on Ni-20Cr alloy oxidation in water vapour at 650 ∘C. Corros. Sci. **232**, 112050 (2024)
13. Zaiton, N.A.Z., Parimin, N., Anuar, A.I., Hayazi, N.F., Zainal, F.F.: Oxide growth behaviour of 800H and HR-120 series Ni-based alloys on isothermal oxidation. Int. J. Nanoelectron. Mater. **17**(3), 416–421 (2024)
14. Zurek, J., et al.: Growth and adherence of chromia based surface scales on Ni-Base alloys in high and low-pO2 gases. Mater. Sci. Eng., A **477**(1–2), 259–270 (2008)
15. Oquab, D., Xu, N., Monceau, D., Young, D.J.: Subsurface microstructural changes in a cast heat resisting alloy caused by high temperature corrosion. Corros. Sci. **52**(1), 255–262 (2010)

16. Zaiton, N.A.Z., Parimin, N., Hayazi, N.F., Zainal, F.F., Garus, S., Vizureanu, P.: Effect of heat treatment temperature on isothermal oxidation of Ni-based Fe-33Ni-19Cr alloy. Arch. Metall. Mater. **69**(2), 747–752 (2024)

Study of Conductivity and Morphological Properties Using Different Composition of PEDOT: PSS/PEO Film as a Membrane for Lithium-Ion Batteries Recycling

Awatif Hassim[1], Nur Najiha Abd Maliaman[1], Mohamad Arif Kasri[2], Fatin Nur Azmina Mohd Fauzi[1], Suzana Ratim[1(✉)], and Mohd Muzamir Mahat[1(✉)]

[1] Faculty of Applied Sciences, Universiti Teknologi MARA, 40450 Shah Alam, Selangor, Malaysia
{suzanar,mmuzamir}@uitm.edu.my

[2] Kulliyyah of Science, International Islamic University Malaysia, 25200 Kuantan, Pahang, Malaysia

Abstract. Efficient recycling of lithium-ion batteries (LIBs) is crucial for a sustainable future. Electrochemical recycling presents a promising eco-friendly alternative to traditional pyrometallurgy and hydrometallurgy processes. The membrane separator, facilitating efficient ion transport within a battery system is pivotal for the success of electrochemical recycling. This study explores the use of poly (3,4-ethylenedioxythiophene): polystyrene sulfonate (PEDOT: PSS) as a conductive polymer in the membrane, aiming to enhance lithium-ion movement. Membrane will be developed by incorporating PEDOT: PSS into a host polymer, polyethylene oxide (PEO) using the solvent casting method. The conductivity of the separators was measured using electrochemical impedance spectroscopy (EIS), with the 1.0 wt% composition demonstrating the highest conductivity at 6.19×10^{-4} S/cm. Morphological and chemical compositions were further characterized using field emission scanning electron microscopy (FESEM) and Fourier transform infrared spectroscopy (FTIR). These findings hold promise for advancing the recycling efficiency of lithium in Li-ion batteries.

Keywords: membrane · PEDOT: PSS · PEO · electrochemical · lithium recycling

1 Introduction

In recent years, there has been an intensive investigation of solid polymer electrolyte membrane that responds to the electrochemical system of lithium-ion batteries. The polymer electrolyte membrane can be used in electrochemical recycling which simulates the general electrochemical system in the battery. Polymeric electrolyte membranes have gained attention for their potential applications in fuel cells, rechargeable lithium batteries, electrochromic devices, and sensors [1]. Their advantages include good compatibility with lithium metal, no leakage, low self-discharge in batteries, and easy processing.

N. R. Abdul Razak et al. (Eds.): ECOOP 1996, SPPHY 418, pp. 170–176, 2025.
https://doi.org/10.1007/978-981-96-2871-1_24

The recycling of lithium extends beyond what can be done through traditional methods; hydrometallurgy and pyrometallurgy [2]. It is therefore important to improve the membrane separator for better efficiency while using eco-sustainable electrochemical processes. The aim of this study is to help in developing new and sustainable conductive membrane that can help in transporting the lithium-ion in the electrochemical system. Polyethylene oxide (PEO) is a promising solid polymer electrolyte membrane, but its poor ionic conductivity has led to modifications such as the addition of PEDOT: PSS as a conducting polymer (CP) to enhance the electrolyte membrane system's conductivity.

CP offers a promising future as a membrane separator to meet the demand for extracting lithium-ion from waste batteries. The establishment of innovative materials has enabled significant advancements in a variety of sectors, ranging from medicine to battery recycling application endeavors [3]. Most discussed CP are polypyrrole, polyaniline and polythiophenes. Thus, these CP finds extensive application in the fields of organic electronics, sensors, actuators and energy storage devices among other areas that require electrical conduction [4]. CP will be composited with a solid polymer electrolyte membrane to develop a recycling membrane. The most commercially successful conducting polymer is PEDOT: PSS. It can be dispersed in water and certain organic solvents, and it can be easily fabricated into films using conventional solution processing methods like spin-coating [5]. PEDOT: PSS is also a CP with high stability and flexibility as well as tunable conductivity at room temperature, which makes it a promising candidate for the fabrication of conductive membranes [6].

To have a functional recycling membrane, the interpenetrating network (IPN) needs to be developed. IPN indicated a type of polymer structures synthesized by the non-bonded integration of two different networks, which are intricately interlaced and closely curled up together. IPN is produced when one of these components undergoes polymerization or cross-linking in the presence of the other. Consequently, a flexible and interrelated network structure is obtained that combines characteristics from both components [7]. The incorporation of PEO and PEDOT: PSS will create an IPN which will help to facilitate the movement of lithium-ion inside the system. IPN is needed in membranes as it can reduce the crystallinity of the material which will result in high ionic conductivity [8].

2 Experimental Procedure

Sigma-Aldrich manufactured PEO with a molecular weight of 400,000 and also provided the EDOT monomer and the PSS solution. To dissolve the mixture, deionized (DI) water will serve as its solvent. EDOT: PSS appears to polymerize upon the addition of ammonium persulfate (APS) which is also a product from Sigma-Aldrich. To obtain various concentrations of PEO solutions (0,3,7,5,0) DMF will be added as solvent and this was also given by Sigma-Aldrich. After stirring, PVDF powder obtained from Sigma-Aldrich will be directly added to PEO solution while lithium-ion mobility will be enhanced using bis (trifluoromethane) sulfonimide lithium (LiTFSi).

Mixture made up of 0.05 mL of hydrophobic EDOT and 4.5 mL of distilled water, the second component was added to the solution as 0.08 mL of hydrophilic PSS. These steps were undertaken to ensure EDOT: PSS formation involved stirring the mixture at room temperature for 1 hour. After that, ammonium persulfate (APS) which is an oxidizing agent with a mass of 0.55g was introduced into the solution to start EDOT polymerization. The stirring continued for 24 hours leading to the emergence of a dark blue colored solution suggesting successful polymerization of PEDOT. Lastly, another dose 4.5 ml distilled water ordinary tap was added to complete radical cation polymerization in PEDOT: PSS solution and left for another half hour.

To prepare the PEO/PVDF/LiTFSi solution, we will add a mixture of (0, 3, 5, 7, 9) wt% PEO, 3 wt% PVDF and 2 wt% LiTFSi into DMF solvent. The resultant mixture is stirred continuously for at least 5 hours for homogenization purpose. Formulation of membrane with PVDF can enhance the formability of the membrane; this is as a result of PVDF being a polymeric type of electrolyte that has reliable mechanical strength, chemically stable and thermally stable also. Several battery systems have employed PVDF as a binder, separator and polymeric electrolyte [9]. After finding the optimum PEO/PVDF/LiTFSi mixture, next step will be the addition of PEDOT: PSS. PEDOT: PSS will be added into the solution with different ratio of (0.0, 0.1, 0.4, 0.7, 1.0) wt%. The combined solution is now stirred for a period of 24 hours to make sure that it is homogeneous and no agglomeration. Then, about 15 mL of prepared PEDOT: PSS/PEO/PVDF/LiTFSi solution with different ratios of PEDOT: PSS was dropped using a syringe and evenly spread onto the petri dish. The solution was left to dry in the vacuum oven for 24 hours at 80 °C. The membrane was peeled off from the petri dish once it had dried.

The membranes later were characterized by field emission scanning electron microscopy (FESEM) by Apreo 2, fourier transmission infrared (FTIR) by Perkin Elmer 100 infrared spectrum analyzer, and electrochemical impedance spectroscopy (EIS) by HIOKI 3520 LCR Hi- ester.

3 Results and Discussions

3.1 Field Emission Scanning Electron Microscopy (FESEM)

Figure 1 (a), shows the PP0 membrane which contains PEO/PVDF/LiTFSi without the addition of PEDOT: PSS. For polymer film in (a), there are smooth and amorphous regions that are observed on surface. In Fig. 1 (b), the surface of the membrane has some significant irregularities. A critical observation on the surfaces of this polymer is that there are numerous small particles that indicates lithium salt crystals which was labeled, indicating that LiTFSI does not correctly distribute into the acrylate matrix [10]. Figure 1 (c) and (d) show the rougher surface of the membrane are developed. Because of its interactions with the PEO/PVDF matrix, incomplete lithium salt dissolution roughens the film surface, resulting in deficiencies [11]. As shown in Fig. 1 (e), there is possible interpenetrating network (IPN) structure occurred with the addition of 1.0 wt% PEDOT: PSS concentration. The membrane surface indicates a defined IPN that has significantly larger gaps between them. The gaps allows lithium ions to freely migrate through the matrix, which is significant for applications requiring effective diffusion inside materials, such as lithium-ion transport [12].

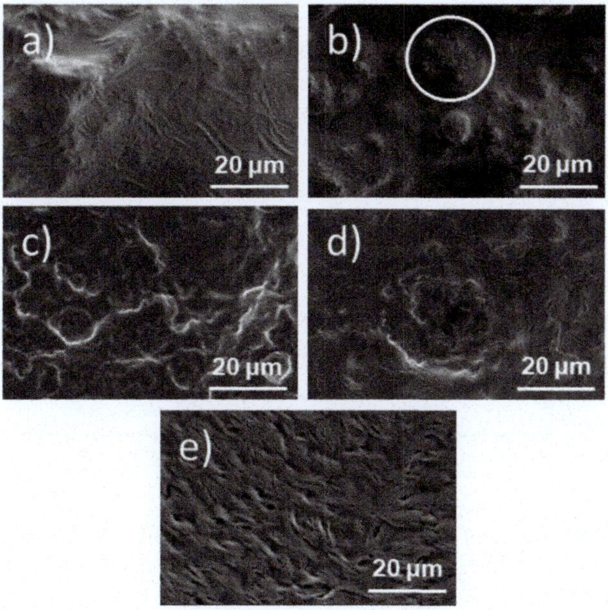

Fig. 1. Image (FESEM) of the membrane with different PEDOT: PSS concentrations a) PP0, b) PP1, c) PP2, d) PP3, and e) PP4

3.2 Fourier Transmission Infrared (FTIR)

In Fig. 2, all the characteristic peaks of PEO are observed, and the band is also detected at 1088 cm^{-1}, which is due to the stretching vibration of C-O-C bond in the ether groups of PEO. PEO also shows peak at 3448 cm^{-1} which correspond to O-H stretching of PEO. The peaks in the spectrum at 664, and 1639 cm^{-1} are attributed to the C-S and C=C, bonds of the thiophene ring and the sulfonic acid group of the PSS. Other LiTFSi peaks observed at 740, 781, 1051, 1193, 1356, and 2876 cm^{-1} correspond to their structure [13]. PVDF peaks at 657, 1254, 1384 and 1436 cm^{-1}. FTIR is a powerful analytical technique used to identify and analyse the molecular composition and structure of substances. It is an effective method for examining changes in structural modifications.

3.3 Electrochemical Impedance Spectroscopy (EIS)

The conductivity measurement of membranes based on PEDOT: PSS was done using Electrochemical Impedance Spectroscopy (EIS) method. It can be seen in Fig. 3 that the conductivity value of a representative membrane changes from 4.76 × 10-4 for 0 wt% (PP0) of PEDOT: PSS to 6.16 × 10-4 for 1.0 wt% (PP4), which indicates that this conductivity increases with increasing concentration. Thus, in terms of this study, it can be concluded that the membrane containing 1.0 wt% addition of PEDOT: PSS are optimal. In general, it can be said that higher concentrations of PEDOT: PSS will result in higher conductivity. A conductive polymer such as PEDOT allows for charge transport through the membrane. Increasing the PEDOT concentration increases the density of conductive pathways in the membrane hence improving its overall conductivity [14]. Previous

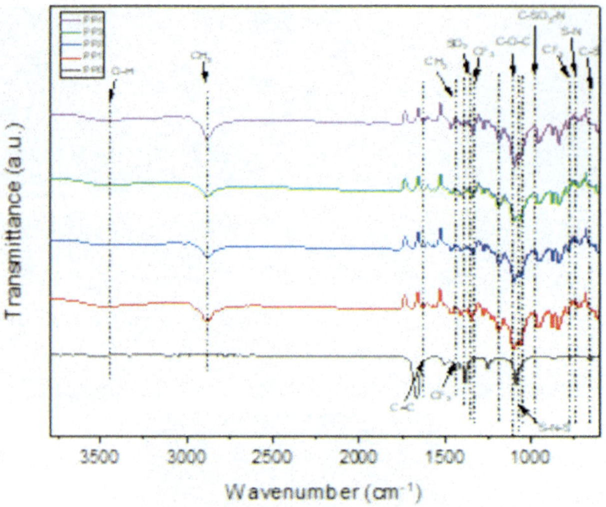

Fig. 2. FTIR spectra of the membrane with different PEDOT: PSS concentration

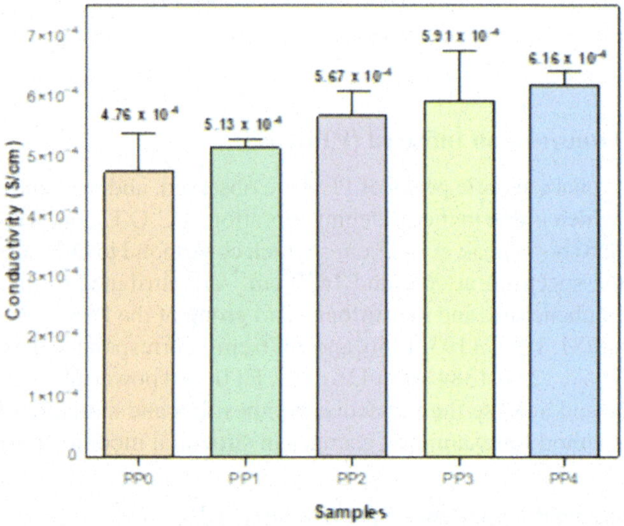

Fig. 3. Conductivity values of PEDOT: PSS PES fabrics at different PEDOT: PSS concentration

study also reported that the highest conductivity was achieved with the addition of 1.0% PEDOT: PSS. Conductivity showed a slight decline with further additions of 1.5% and 2.0% PEDOT: PSS, indicating that 1.0% PEDOT: PSS was the optimal concentration for maximizing conductivity [15].

4 Conclusion

This research delves into creating an eco-friendly recycling membrane for lithium extraction from disposable lithium-ion batteries (LIBs). The investigation focuses on the conductivity and morphological of polymer electrolyte membranes using a mix of PEO and PEDOT: PSS which will create the interpenetrating network (IPN) structure with the help of PVDF and LiTFSi in order to support the mechanical strength and also transportation of lithium-ion. Through analyses such as FESEM, FTIR, and EIS, it has been found that 1.0 wt% PEDOT: PSS is the ideal concentration for obtaining optimum conducting ability. Further addition of PEDOT: PSS concentration into the polymer matrix improves its conductivity as well as morphology thereby enhancing Li-ion diffusion which is important in efficient and sustainable recovery of lithium metal.

Acknowledgments. Authors would like to acknowledge that the study covered in this manuscript was funded by PETRONAS Research Sdn. Bhd. A big thanks to Fakulti Sains Gunaan (FSG) UiTM Shah Alam for accommodating to conduct the research.

References

1. Sanchez-Sanchez, A., del Agua, I., Malliaras, G.G., Mecerreyes, D.: Conductive Poly(3,4-Ethylenedioxythiophene) (PEDOT)-Based Polymers and Their Applications in Bioelectronics, Second Edi. Elsevier Ltd. (2019)
2. Bae, H., Kim, Y.: Materials Advances Technologies of Lithium Recycling from Waste Lithium Ion Batteries : A Review † (2021)
3. Namsheer, K., Rout, C.S.: Conducting polymers: a comprehensive review on recent advances in synthesis, properties and applications. RSC Adv. **11**(10), 5659–5697 (2021)
4. Nezakati, T., Seifalian, A., Tan, A., Seifalian, A.M.: Conductive polymers: Opportunities and challenges in biomedical applications. Chem. Rev. **118**(14), 6766–6843 (2018)
5. Xia, Y., Zhang, H., Ouyang, J.: Highly Conductive PEDOT : PSS Films Prepared Through a Treatment with Zwitterions and Their Application in Polymer Photovoltaic Cells, pp. 9740–9747 (2010)
6. Elschner, A., Kirchmeyer, S., Lovenich, W., Merker, U., Reuter, K.: PEDOT: Principles and Application of an Intrinsically Conducting Polymer (2010)
7. Liu, J., et al.: Synthesis and characterization of interpenetrating polymer networks (IPNs) based on UV curable resin and blocked isocyanate/polyols. Polymer (Guildf) **256**, no. August (2022)
8. Jiang, Y., Guo, N., Dong, F., Xie, H., Liu, J.: Semi-interpenetrating-network PEO-based polymer electrolytes with fast Li-ion conduction for solid Li-metal batteries. J. Energy Storage **73**(PA), 108809 (2023)
9. Ye, Q., et al.: Fabricating a PVDF skin for PEO-based SPE to stabilize the interface both at cathode and anode for Li-ion batteries. J. Energy Chem. **70**, 356–362 (2022)
10. Leš, K., Jordan, C.S.: Ionic conductivity enhancement in solid polymer electrolytes by electrochemical: In situ formation of an interpenetrating network. RSC Adv. **10**(68), 41296–41304 (2020)
11. Li, J., Zhu, K., Wang, J., Yan, K., Liu, J., Yao, Z.: Optimisation of conductivity of PEO/PVDF-based solid polymer electrolytes in all-solid-state Li-ion batteries. Mater. Technol. **00**(00), 1–8 (2020)

12. Fong, K.D., Wang, T., Kim, H.K., Kumar, R.V., Smoukov, S.K.: Semi-interpenetrating polymer networks for enhanced supercapacitor electrodes. ACS Energy Lett. **2**(9), 2014–2020 (2017)
13. Nadiah, N., Subramaniam, R., Kasi, R.: Development and characterization of poly (1-vinylpyrrolidone-co-vinyl acetate) copolymer based polymer electrolytes, **2014** (2014)
14. Kim, N., et al.: Highly Conductive PEDOT : PSS Nanofi brils Induced by Solution-Processed Crystallization, pp. 2268–2272 (2014)
15. Ahmad Ruzaidi, D.A., et al.: Synthesis and characterization of porous, electro-conductive chitosan–gelatin–agar-based pedot: Pss scaffolds for potential use in tissue engineering. Polymers (Basel) **13**(17), 1–21 (2021)

Synthesis of Ferrosilicon Using Sustainable Carbon Materials and Iron Oxide Sources

Saw Wei Qi[1], Anis Nadhirah Ismail[1,2](✉), Rita Mohd Said[1,2], Siti Shuhadah Md Saleh[1,2], and Mohd Hakim Ibrahim[3]

[1] Faculty of Chemical Engineering Technology, Universiti Malaysia Perlis, 02600, Jejawi Arau, Perlis, Malaysia
anisnadhirah@unimap.edu.my

[2] Center of Excellence Geopolymer & Green Technology (CeGeoTech), University Malaysia Perlis (UniMAP), 02600 Arau, Perlis, Malaysia

[3] Research Centre, Mineral and Geoscience Department Malaysia, Jalan Sultan Azlan Shah, 31400 Ipoh, Perak, Malaysia

Abstract. This study investigates ferrosilicon synthesis using coconut char and coke as carbon sources and hematite and EAF slag as iron oxide sources. Coconut char was prepared by coconut shell via pyrolysis process at 450 °C for 2 h in a muffle furnace. From the CHNS result, coconut char exhibits a high carbon content (81.5 wt%), making it an efficient material as reductant. The result corresponds with EDX (C: 92.48%) and XRD (2θ: 23.88° and 43.09°) while also showing a large surface area (215.72 m^2/g) and porosity. The raw material was crushed to less than 63 μm, mixed, and underwent a reduction process for 1300 °C for 2 h. The results indicated coconut char was more effective in facilitating reduction processes than coke. The XRD study identified several phases that formed. At the same time, the SEM/EDX investigations verified the composition and structural properties where the grey metallic colour represents the existence of iron. Among the samples tested, the EAF Slag/Coconut Char/Silica sample achieved the highest reduction percentage (65.63%) which increased about 31.26% from the conventional materials (hematite and coke). The results emphasize that coconut char is sustainable carbon source and EAF slag is a feasible alternative to iron oxide that offers economic benefits through cost savings while also reducing greenhouse gas emissions and promoting environmental health.

Keywords: Ferrosilicon · Activated Carbon · Pyrolysis · Coconut Char · EAF Slag

1 Introduction

Ferrosilicon is a silicon-iron alloy widely used in EAF steel production as a deoxidizer to purify liquid steel and improve steel quality [1]. The synthesis of ferrosilicon has historically relied on reducing iron oxide using conventional carbon sources such as coke, graphite and coal to produce iron alloys [2]. Conventional coke contributes to about 60% or more greenhouse gas emissions from various ferroalloy manufacturing

N. R. Abdul Razak et al. (Eds.): ECOOP 1996, SPPHY 418, pp. 177–185, 2025.
https://doi.org/10.1007/978-981-96-2871-1_25

processes [3]. Similarly, the traditional reliance on iron oxide sources such as hematite for ferrosilicon manufacturing creates environmental issues associated with mining and extraction. Thus, an alternative source of carbon and iron oxide for the production of ferroalloys would be preferred to replace coke and hematite. In the metallurgical field, agricultural waste such as rice husk and palm char are used as carbonaceous material. Biswal et al. (2021) reported that the reduction performance of transformed-spent coffee grounds as carbon material was better than pitch coke; the iron oxide used is EAF slag [4]. Coconut is the fourth most important industry crop in Malaysia, which is also one of the top ten coconut-producing countries in the world. Studies by Hassan et al. (2021) and Farzana et al. (2014) have demonstrated the effectiveness of coconut char in increasing carbon content and enhancing reduction processes [5, 6]. EAF slag is a solid waste by-product from steel production. Research by Ismail et al. (2021) and Kongkarat et al. (2021) supports using EAF slag in ferrosilicon synthesis, showing its efficiency in phase formation and reduction processes [1, 7].

The potential of using EAF slag in combination with coconut char for ferrosilicon synthesis suggesting a gap in understanding how waste materials can be effectively utilized in metallurgical processes. This study focuses on synthesizing ferrosilicon using coconut char and coke as carbon sources and hematite and EAF slag as iron oxide sources to explore sustainable alternatives material. Coconut shells were subjected to a pyrolysis process to produce activated carbon coconut char, followed by the synthesis of ferrosilicon in a muffle furnace at 1300 °C for 2 h.

2 Experimental Procedure

2.1 Sample Preparation and Ferrosilicon Synthesis Study

The coconut shell was collected at Kangar, Perlis area. After that, it was sun-dried and crushed into 2 mm. The coconut shell undergoes pyrolysis in a muffle furnace heating rate of 10 °C/min for 450 °C for 2 h. The breakdown of cellulose, hemicellulose and lignin is optimized at 450 °C for 2 h allows for the pyrolysis reaction to proceed to completion [8, 9]. X-ray diffraction analysis (XRD), Scanning electron microscope (SEM), CHNS Elemental Analysis, and Brunauer–Emmett–Teller (BET) was used to characterize the activated carbon.

The synthesis of ferrosilicon involves using raw materials such as hematite powder (85.9%Fe_2O_3) and EAF slag (43.18% Fe_2O_3) as the iron oxide source, coke (74.4%C) and coconut char (81.5%C) as the carbon source, and silica (99.8%SiO_2) as the silica source. The components were pulverized, sieved to a particle size less than 63μm. Next, using the mass ratio of (1.8 Fe_2O_3: 6.2C: 2.8SiO_2) Hematite/Coke/Silica, (0.8 Fe_2O_3: 6.2C: 2.8SiO_2) EAF slag/Coke/Silica, (1.8 Fe_2O_3: 6.8C: 2.8SiO_2), Hematite/Coconut char/Silica, and (0.8 Fe_2O_3: 6.8C: 2.8SiO_2) EAF slag/Coconut char/Silica were blended in planetary ball mill machine at 250 rpm for 1 h. The mixed samples were placed in a crucible, then sent to a muffle furnace for the synthesis of ferrosilicon. The reduction temperature was 1300 °C for 2 h at the heating rate of 10 °C/min. The reduced samples were analyzed using an X-ray diffractometer and optical microscopy.

3 Results and Discussions

3.1 Structural and Phase Analysis of Raw Materials

Table 1 shows the CHNS elemental analysis for carbon materials. Coconut char and coke differ significantly in their carbon content and purity, with coconut char exhibiting a higher carbon concentration of 81.5 wt% compared to coke's 74.4 wt%. This higher purity is due to the cleaner biomass feedstock of coconut char, resulting in fewer impurities and higher carbon content after pyrolysis [10]. These characteristics make coconut char a more efficient and environmentally friendly alternative to coke in applications such as ferrosilicon synthesis. Table 2 shows the ash analysis of EAF slag.

Table 1. CHNS elemental analysis for carbon materials.

	Carbon (wt%)	Hydrogen (wt%)	Nitrogen (wt%)	Sulphur (wt%)
Coconut char	81.5	3.4	0.2	0.9
Coke	74.4	1.6	0.6	1.4

Table 2. Ash analysis for EAF slag.

	Fe_2O_3	CaO	SiO_2	Al_2O_3
Composition (wt%)	43.18	27.67	16.87	5.77

The X-ray diffraction (XRD) and Scanning Electron Microscopy/Energy Dispersive X-ray Spectroscopy (SEM/EDX) analysis of coconut char and coke are shown in Fig. 1. The XRD found that the coconut shell underwent the pyrolysis process become coconut char exhibited distinct peaks at 23.88° and 43.09°, indicating the breakdown of the cellulose component and the alteration of its structure. Furthermore, the appearance of additional peaks in the non-crystalline area, specifically at around 43.09°, corresponds to previous research indicating the creation of pores and the generation of graphitic structures when coconut shells are subjected to pyrolysis [11]. XRD analysis of the coke contains short range graphite like structure and displays disordered material, giving rise to high background intensity [12]. Sample showed distinct peaks at 21.04°, 26.08°, 36.71°, 45.94°, 50.25°, and 60.00°, indicating the presence of crystalline phases of SiO_2 which aligned with the research conducted by Farzana et al. (2014). Additionally, the research by Xu et al. (2018) on the reactivity and structural evolution of metallurgical coke supports the presence of SiO_2 phases, indicating that the coke undergoes structural changes that enhance its reactivity during high-temperature processes [13].

The microstructure of coconut char underwent a pyrolysis process at 450 °C had a rigid, uneven surface with the formation showing large pore sizes (Fig. 1) due to the influence of the heating process, which causes the decomposition of coconut shells' organic compounds. During pyrolysis of the coconut shell, the volatile matter will escape, leaving pores on the surface of the coconut char [14]. Coconut char shows good combustion reactivity as the porosity increase accessibility of reactive gas to active sites. The EDX spectrum of coconut char obtained 92.48 wt% of carbon while 73.72 wt% for coke. Coke has a significant carbon content in its composition, which could build up at the burning char's retreating surfaces and prevent pore growth. As a result, coke has more delay in oxygen diffusion than coconut char, which leads to decreased combustion performance. The EDX finding was in good agreement with the CHNS results where the carbon content of coconut char higher than coke. The result of the SEM image of the coconut shell and coconut char agrees with the BET result from Table 3, which states that surface area increases as the porosity increases.

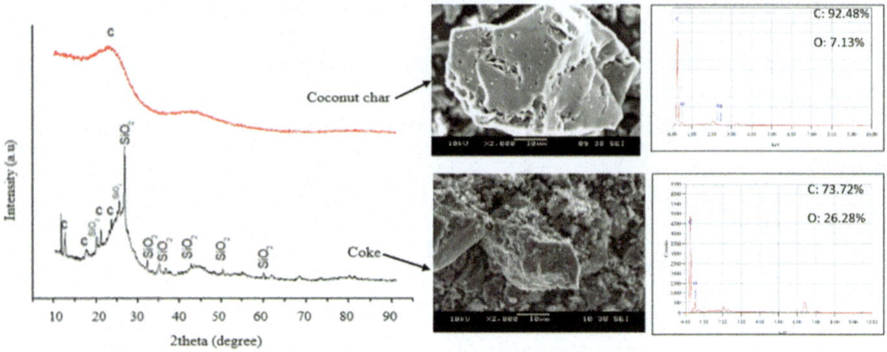

Fig. 1. XRD and SEM/EDX analysis for coconut char and coke.

The BET surface area is a crucial factor in assessing the efficacy of carbon compounds. Table 3 displays the pore surface area results, indicating that coke has a BET surface area of 0.77 m^2/g, while coconut char demonstrates a far larger BET surface area of 215.72 m^2/g. Coconut chars have the highest surface area, which significantly exceeds coke. In addition, Rollinson and Karmakar (2015) investigated the responsiveness of several biomass species and found that the reactivity increases as the specific surface area increases [15]. A greater BET surface area often leads to increased adsorption capacity due to the ability of activated carbon to adsorb a wide range of gases under various conditions [16].

Table 3. Surface area of carbon materials

	Coconut char	Coke
BET surface area (m^2/g)	215.72	0.77

3.2 Phase Transformation

The XRD patterns for reduced samples are combined into Fig. 2 for comparative analysis. The peaks identified for SiO_2, Fe_2O_3, $FeSiO_2$ and Fe_2SiO_4 are summarized in the Table 4.

Several observations were made when comparing the results for coconut char and coke as carbon reductant with EAF slag and hematite as iron oxides. Similar phases were detected in both carbon samples when combined with the same iron oxide material.

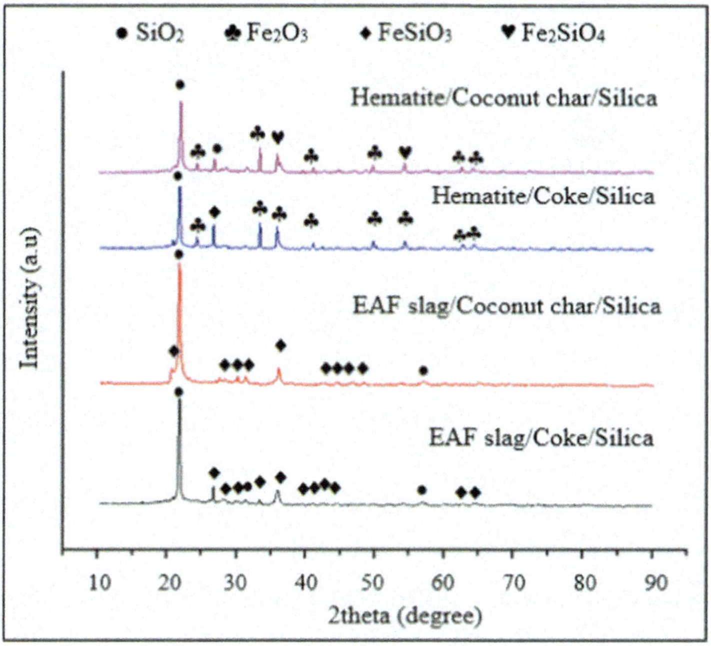

Fig. 2. XRD for reduced samples.

To summarize, hematite, which consists of 96% Fe_2O_3, generates firm Fe_2O_3 peaks when combined with both coconut char and coke. In addition, coke also produces $FeSiO_3$. The primary reaction between EAF slag (containing 43.18% Fe_2O_3) and both coconut char and coke are the formation of $FeSiO_3$. No peaks of Fe_2O_3 are detected in this process. Coconut char, with a carbon content of 81.5%, and coke, with a carbon content of 74.4%, are both efficient in facilitating reduction processes. This makes coconut char a possible substitute for carbon material and EAF slag a feasible substitute for iron oxide in ferrosilicon production.

Apparently, EAF slag contained a variety of oxides (such Fe_2O_3, FeO, CaO and SiO_2) which react at lower temperatures compared to hematite (Fe_2O_3) [17]. The FeO in EAF slag is easier reduced due to lower oxidation state, making the reduction process more efficient than the reduction of Fe_2O_3 (hematite). The presence of CaO and SiO_2 in slag lower the melting point and enhances the fluidity of the slag [18]. A more fluid slag ensures better interaction between the carbon reductant and iron oxides thus improving the reduction kinetics [19].

Table 4. 2θ Angles from XRD Analysis for reduced sample

Samples	Hematite/Coke/Silica	EAF slag/Coke/Silica	Hematite/Coconut char/Silica	EAF slag/Coconut char/Silica
Angle (2θ)				
SiO_2	21.93°	21.79°, 31.31°, 56.85°	22.10°, 26.80°	21.87°, 57.08°
Fe_2O_3	24.31°, 33.38°, 35.80°, 41.07°, 49.74°, 54.37°, 62.70°, 64.28°	–	24.34°, 31.41°, 33.35°, 41.05°, 49.62°, 62.61°, 64.09°	–
Fe_2SiO_4	–	–	35.80°, 54.25°	–
$FeSiO_3$	26.70°	26.73°, 28.20°, 30.03°, 33.29°, 36.06°, 41.03°, 42.46°, 44.51°, 54.07°, 60.08°, 62.83°, 64.33°	–	20.67°, 27.53°, 28.37°, 30.13, 36.08°, 42.49°, 44.61°, 46.85°, 48.50°

3.3 Microstructural and Compositional Analysis

Figure 3 displays SEM image and EDX spectrum, and the composition of elements found in the reduction sample. Iron content corresponds to the composition of iron oxide used. All samples show substantial quantities of silica (SiO_2), which is compatible with the sample's use of silica. The presence of potassium is due to the presence of biomass char or may resulted from impurities in raw materials and the addition of potassium compounds during the coking process [20]. EAF slag/Coconut char/Silica (Fig. 3 (c)) and EAF slag/Coke/Silica samples (Fig. 3 (d)) have significant calcium content, typical of EAF slag from the lime (CaO) and other materials containing calcium during the steelmaking procedure [21].

EAF slag/Coconut char/Silica had the most desirable outcomes compared to others, with the iron (Fe) concentration at 64.01% (Fig. 3 (d)). Coconut char exhibits better reactivity due to its higher carbon content (81.5 wt%) and larger surface area (215.72 m^2/g). The high carbon content provides more available active sites for chemical reactions, particularly in reduction processes that acts as reductant. The larger surface area of coconut char increases the contact area between the carbon and iron oxides thus facilitating faster reactions [17].

3.4 Percentage of Reduction

Figure 4 represents the percentage of sample after reduction. The study found that the EAF Slag/Coconut Char/Silica sample had the highest reduction percentage. The findings suggest that samples containing EAF slag demonstrate a greater reduction percentage compared to samples containing hematite, as EAF slag contained of Fe_2O_3 (43.18%)

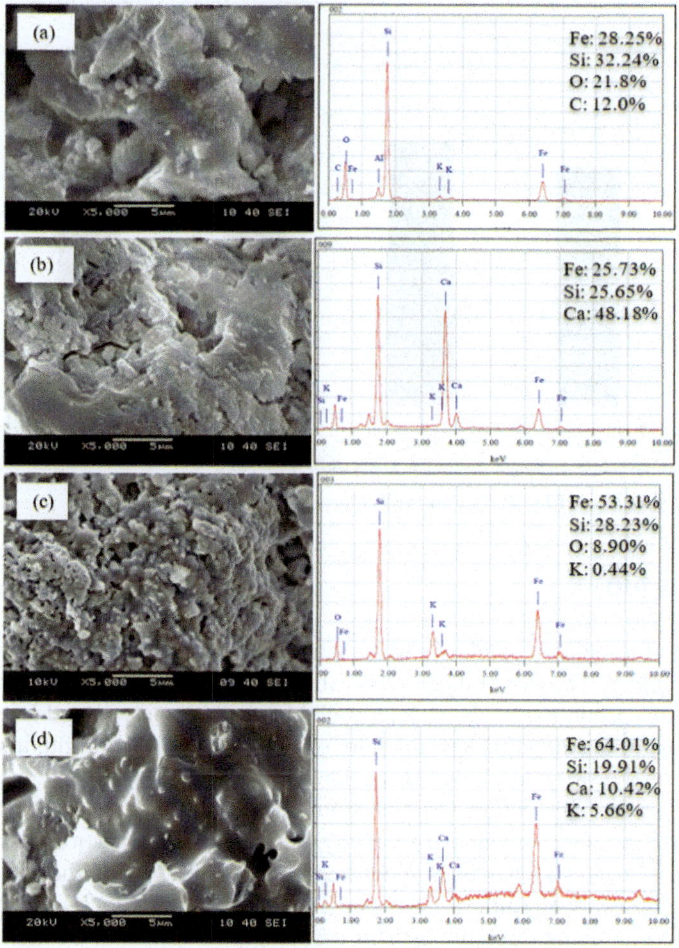

Fig. 3. SEM/EDX analysis for (a) Hematite/Coke/Silica, (b) EAF slag/Coke/Silica, (c) Hematite/Coconut Char/Silica, and (d) EAF Slag/Coconut char/Silica.

and that more efficient reduction process. Furthermore, using coconut char (81.5%) as a carbon source has greater carbon content compared to coke (74.4%), which plays a role in the higher reduction percentages. According to Najmi et al. (2016) and Rashid et al. (2014), respectively, the presence of more carbonaceous material increases the reduction rate and higher carbon content results in increased reduction efficiency [22]. The higher specific surface area of coconut char (215.7 m^2/g) may also contribute to the reduction process since the higher surface area will enhance the oxygen diffusion efficiency thus led to improve the reactivity. The use of wastes (EAF slag and coconut char) in this process demonstrates an increased reduction percentage about 31.26% compared to conventional materials (hematite and coke), which opens significant potential for commercialization and more sustainable resource use in the industry.

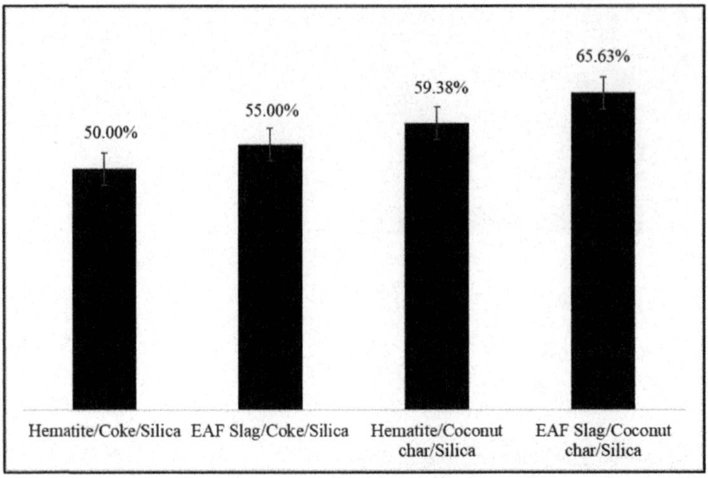

Fig. 4. Percentage of reduction of the samples.

4 Conclusion

In this study, ferrosilicon synthesis was examined using various carbon materials and iron oxide sources. The pyrolysis process exhibited significant increment in the characteristics of the carbon materials as determined by CHNS analysis and EDX spectrum. This higher carbon content makes it a more effective reductant. A rigid, uneven surface with significant porosity developed along the increase in carbon content, making it more reactive. These findings indicate that coconut char is an alternative to coke, providing advantages for the environment and economy due to its higher carbon content and reduction. Similarly, EAF slag has been identified as a possible replacement for traditional iron oxide sources. A mixture of EAF slag and coconut char proved the most efficient method for synthesizing ferrosilicon. This combination resulted in the highest iron content and reduction efficiency, demonstrating the viability of utilizing agricultural waste and industrial by-products as sustainable alternatives in metallurgical operations.

Acknowledgments. The authors would like to deliver their gratitude to Faculty of Chemical Engineering and Technology, University Malaysia Perlis, for supporting the research effort through the materials.

References

1. Kongkarat, S., Boonyaratchinda, M., Chobtham, C.: Formation of ferrosilicon alloy at 1550°C via carbothermic reduction of SiO_2 by coal and graphite: Implication for rice husk ash utilization. Solid State Phenom. **315**, 16–24 (2021)
2. Yunos, N.F.M., Idris, M.A., Munusamy, S.R.R., Perumal, K.: Effect of using palm char and coke as a reductant in production of ferrosilicon. IOP Conf. Ser.: Mater. Sci. Eng. **957**(1), 012031 (2020)

3. Camacho, C.A.M., Torres, J.R.M., Jaegler, A., Gondran, N.: Sustainability metrics for real case applications of the supply chain network design problem: a systematic literature review. J. Clean. Prod. **231**, 600–618 (2019)
4. Biswal, S., Pahlevani, F., Sahajwalla, V.: Synthesis of value-added ferrous material from electric arc furnace (eaf) slag and spent coffee grounds. J. Miner. Metals Mater. Soci. (TMS) **73**(6), 1878–1888 (2021)
5. Hassan, U.F., Sallau, A.A., Ekanem, E.O., Jauro, A., Kolo, A.M.: Effect of carbonization temperature on properties of char from coconut shell. Int. J. Adv. Chem. **9**(1), 34–39 (2021)
6. Farzana, R., Rajarao, R., Sahajwalla, V.: Synthesis of ferrosilicon alloy using waste glass and plastic. Mater. Lett. **116**, 101–103 (2014)
7. Ismail, A.N., Yunos, N.F.D.M., Rajeswari, M.S.R., Idris, M.A.: Production of metallic iron from electric arc furnace (EAF) slag using palm shells char/coke blends. AIP Conf. Proc. **2339**, 020202 (2021)
8. Bielecki, M., Zubkova, V.: Analysis of interactions occurring during the pyrolysis of lignocellulosic biomass. Molecules **28**(2), 1–21 (2023)
9. Jerzak, W., Acha, E., Li, B.: Comprehensive review of biomass pyrolysis: Conventional and advanced technologies, reactor designs, product compositions and yields, and techno-economic analysis. Energies **17**(20) (2024)
10. Buo, T.V., Gray, R.J., Patalsky, R.M.: Reactivity and petrography of cokes for ferrosilicon and silicon production. Int. J. Coal Geol. **43**(1–4), 243–256 (2000)
11. Bakti, A.I., Gareso, P.L.: Characterization of active carbon prepared from coconuts shells using FTIR, XRD and SEM techniques. Al-Biruni/Jurnal Ilmiah Pendidikan Fisika Al-Biruni **7**(1), 33–39 (2018)
12. Yunos, N. F. D. M.: High temperature phenomena occurring during reactions of agricultural wastes in electric arc furnace (2012)
13. Xu, R., Dai, B., Wang, W., Schenk, J., Bhattacharyya, A., Xue, Z.: Gasification reactivity and structure evolution of metallurgical coke under H_2O/CO_2 atmosphere. Energy Fuels **32**(2), 1188–1195 (2018)
14. Nsamba, H.K., Hale, S.E., Cornelissen, G., Bachmann, R.T.: Sustainable technologies for small-scale biochar production—a review. J. Sustain. Bioenergy Syst. **05**(01), 10–31 (2015)
15. Rollinson, A.N., Karmakar, M.K.: On the reactivity of various biomass species with CO_2 using a standardised methodology for fixed-bed gasification. Chem. Eng. Sci. **128**, 82–91 (2015)
16. Skodras, G., Diamantopoulou, I., Zabaniotou, A., Stavropoulos, G., Sakellaropoulos, G.: Enhanced mercury adsorption in activated carbons from biomass materials and waste tires. Fuel Process. Technol. **88**(8), 749–758 (2007)
17. Chang, W.T., et al.: Effect of FeO Content on Foaming and viscosity properties in FeO-CaO-SiO_2-MgO-Al_2O_3 slag system. Metals **11**(2), 1–10 (2021)
18. Hedayati, A., Lestander, T., Rudolfsson, M., Thyrela, M., Ohman, M.: Fate of phosphorus and potassium in single-pellet thermal conversion of forest residues with a focus on the char composition. Biomass Bioenergy (2021)
19. Wang, G., et al.: Influence mechanism of alkali metals on CO_2 gasification properties of metallurgical coke. Chem. Eng. J. **387**, 124093 (2020)
20. Kim, J., Azimi, G.: Valorization of electric arc furnace slag via carbothermic reduction followed by acid baking – water leaching. Resour. Conserv. Recycl. **173**, 105710 (2021)
21. Rashid, R.Z.A., Salleh, H.M., Ani, M.H., Yunus, N.A., Akiyama, T., Purwanto, H.: Reduction of low-grade iron ore pellet using palm kernel shell. Renew. Energy **63**, 617–623 (2014)
22. Wang, Z., et al.: Kinetics analysis of iron oxide reduction by solid carbon in hIsmelt ironmaking Slag. ISIJ Int. **64**(4), 1101–1105 (2024)

Effect of Aluminum Powder Doping in Polyvinylidene Fluoride (PVDF) Electrospun Composites

Muhammad Amirul Farhan Idris[1], Shuhaida Yahud[2], Muhammad Firdaus Mohd Nazeri[1], and Sharifah Shahnaz binti Syed Bakar[1](✉)

[1] Surface Technology Special Interest Group, Faculty of Chemical Engineering and Technology, Universiti Malaysia Perlis, 02600 Jejawi, Arau, Perlis, Malaysia
shahnaz@unimap.edu.my

[2] Faculty of Electronic Engineering Technology, Universiti Malaysia Perlis, Pauh Putra Campus, 02600 Perlis, Arau, Malaysia

Abstract. In this paper, the study has successfully incorporated aluminum (Al) powder into Polyvinylidene Fluoride (PVDF) fibers through electrospinning method. The Al loading of 0.0wt%, 0.25wt%, 0.5wt%, 0.75wt% and 1.0wt% were incorporated into 15wt% PVDF concentration. PVDF was dissolved using DMF using magnetic stirrer before the mixture was electrospun and the yield fiber was collected for further characterization. The resulting fibers exhibited a beaded structure and the presence of agglomerates attributed to the Al powder. The PVDF with 0.75wt% aluminum powder sample showed promising results in reaching the desired ß-phase, while the PVDF with 0.25wt% aluminum powder sample demonstrated the optimum electrical conductivity. The PVDF contained 1.0wt% aluminum powder sample have the largest fiber beads size.

Keywords: Polyvinylidene Fluoride · Aluminum Powder · Electrospinning · ß-phase · Composites

1 Introduction

Polyvinylidene fluoride (PVDF) is a highly versatile polymer known for its excellent properties, including easy processability, chemical resistance, and strong piezoelectric characteristics [1]. It is widely used in various fields such as energy harvesting, sensing, actuation, and biomedical applications [2]. PVDF exhibits different crystal phases, with the polar ß phase showing the best pyroelectric, piezoelectric, and ferroelectric properties. The non-polar α phase is the most stable but lacks electro-activity. PVDF's dielectric and ferroelectric properties stem from the strong dipole created by the arrangement of 〖CF〗 _2 units along the polymer chain. It also possesses hydrophobicity and serves as an excellent dielectric material [3, 4].

Electrospinning, a technique involving the use of a strong electric field to create micro to nanoscale fibers, is commonly used to enhance the ß phase of PVDF [5]. This

N. R. Abdul Razak et al. (Eds.): ECOOP 1996, SPPHY 418, pp. 186–192, 2025.
https://doi.org/10.1007/978-981-96-2871-1_26

process can produce various types of fibers, including core-sheath, composite, multi-component, and porous fibers, with unique properties such as high tensile strength and controllable structure [6]. Conventionally, PVDF in its α phase is non-polar and does not exhibit any piezoelectric effect. However, the β phase of PVDF can be achieved by adjusting electrospinning parameters such as flow rate, voltage, and tip-to-collector distance. Additionally, incorporating fillers can help align fiber orientation, further promoting the formation of the β phase. Electrospinning is particularly suitable for producing PVDF nanofibers for piezoelectric sensors, as it eliminates the need for a separate poling step. The stretching and electric field during electrospinning facilitate the alignment of dipoles, increasing the fiber's ß phase content and visible piezoelectric response [7].

Al powder, widely used in industries like chemicals, pyrotechnics, and aerospace, serves as an important raw material [8]. It is often employed in the production of aluminum-based composite and gradient materials, contributing properties such as thermal and electrical conductivity, response to magnetic fields, heat capacity, and improved molding characteristics [9]. Aluminum powder's narrow size distribution enables the creation of parts with superior mechanical properties. Key considerations for aluminum powder selection include purity, spherical shape for better flowability, narrow fractional composition without large agglomerations, and the absence of moisture and gases [10].

2 Experimental Procedure

15wt% concentration PVDF solution was prepared by dissolving PVDF powder in dimethylformamide (DMF) solution. The solution was stirred using a magnetic stirrer for 2 h until it achieved complete homogeneity, transparency, and visual consistency. To prevent evaporation, aluminium foil was placed on top of the solution. PVDF solution without Al powder acts as a control and named sample A. Subsequently, the PVDF/DMF solution was combined with varying percentages of aluminum powder; namely sample B (0.25wt%), sample C (0.5wt%), sample D (0.75wt%), and sample E (1.00wt%). To ensure proper dispersion and prevent agglomeration of the Al powder, the solution underwent sonication using an ultrasonic homogenizer for 1 h, followed by an additional hour of stirring with a magnetic stirrer.

After the preparation of the PVDF/Al solutions, the electrospinning process was performed. A 10 ml syringe with an inner diameter of 0.8 mm was horizontally placed on the injection holder of a syringe pump, and the PVDF/Al solution was transferred into the syringe. Electrospinning was conducted under controlled conditions, with a flow rate of 1.5 mL/h, an applied voltage of 25 kV, and a distance of 12 cm between the syringe tip and the collector. The electrospun fiber collector consisted of a cylindrical metal rotating drum covered with aluminum foil, positioned perpendicular to the syringe nozzle's tip. Prior to the electrospinning process, the syringe was scrubbed and polished with sandpaper to ensure the needle's level placement on the surface. A crocodile clip was attached to the needle to deliver the voltage from the power source. The PVDF/Al powder solution was allowed to flow from the needle tip by activating the pump machine's power and voltage supply. Aluminum foil was used to collect the resulting PVDF/Al composite fibers.

The morphology, electrical conductivity, and crystalline phase of electrospun PVDF/Al composite fibers were investigated. The composite fiber morphology was examined using the TEM3000 Tabletop Microscope, where the sample was prepared and imaged after sputtering with a platinum coating. Four-point probe testing was conducted to assess electrical conductivity, measuring resistance through current flow and voltage difference. X-ray Diffraction (XRD) analysis with the D2 Phaser machine was employed to determine the crystalline phase of the cpmposite fibers. This comprehensive characterization approach provided valuable insights into the properties of the electrospun PVDF/Al composite fibers, contributing to a better understanding of their structure and performance.

3 Results and Discussions

3.1 Composite Fibers Morphology

The thickness of electrospun PVDF/Al composite fiber was measured using vernier calliper. PVDF/Al fibres range in thickness from 0.03 to 0.07 mm. All samples of 0 wt% Al (A); 0.25 wt% Al (B); 0.50 wt% Al (C); 0.75 wt% Al (D) and 1.00 wt% Al (E) are in paper-like form.

The SEM images in Fig. 1 depicted the electrospun PVDF/Al composite fibers at × 1000 magnification. SEM analysis revealed that the PVDF fibres exhibited a uniform and smooth surface. According to Shehate et al. [11], theoretically an electrospun pure PVDF fibre produce a smooth electrospun fiber. From Fig. 1, it was observed that all the electrospun PVDF/Al composite fibers displayed beaded structures with varying fiber diameters. Sample A which represents pure PVDF fibers without any aluminum powder, displaying a uniform and smooth surface with an average diameter corresponding to the bead formation.

The incorporation of Al powder in Samples B, C, D, and E led to an increase in bead formation and difficulties in dispersing the powder within the PVDF matrix. Larger bead sizes (Samples D and E) exhibited noticeable agglomerations of aluminum, resulting in a non-uniform distribution within the fibers. The presence of Al powder and the increase in bead size caused Samples B, C, D, and E to have larger fiber diameters compared to Sample A. As the Al powder increased, the viscosity of PVDF also increased, which then intensified the surface tension which favor the formation of beads and larger fibers [12]. Table 1 shows the average fiber sizes of electrospun samples.

As can be observed in Table 1, the average fiber diameter of electrospun fibers has increased with increase of filler loading. However, as the filler loading maximized at 1.00 wt%, the fiber size has decreased. As can be seen in the SEM micrograph (Fig. 1), the bead formation becomes significantly large, due to the agglomeration of Al powder onto the bead spots.

3.2 Electrical Conductivity

Figure 2 shows the correlation between the average of resistance and conductance of PVDF filled with various weight percentage of Al powder. It is evident that sample A exhibited the lowest conductance, measuring 2.71 × 10^(−2) Siemens, while sample B had the highest conductance at 3.34 × 10^(−2) Siemens, which also reflecting the

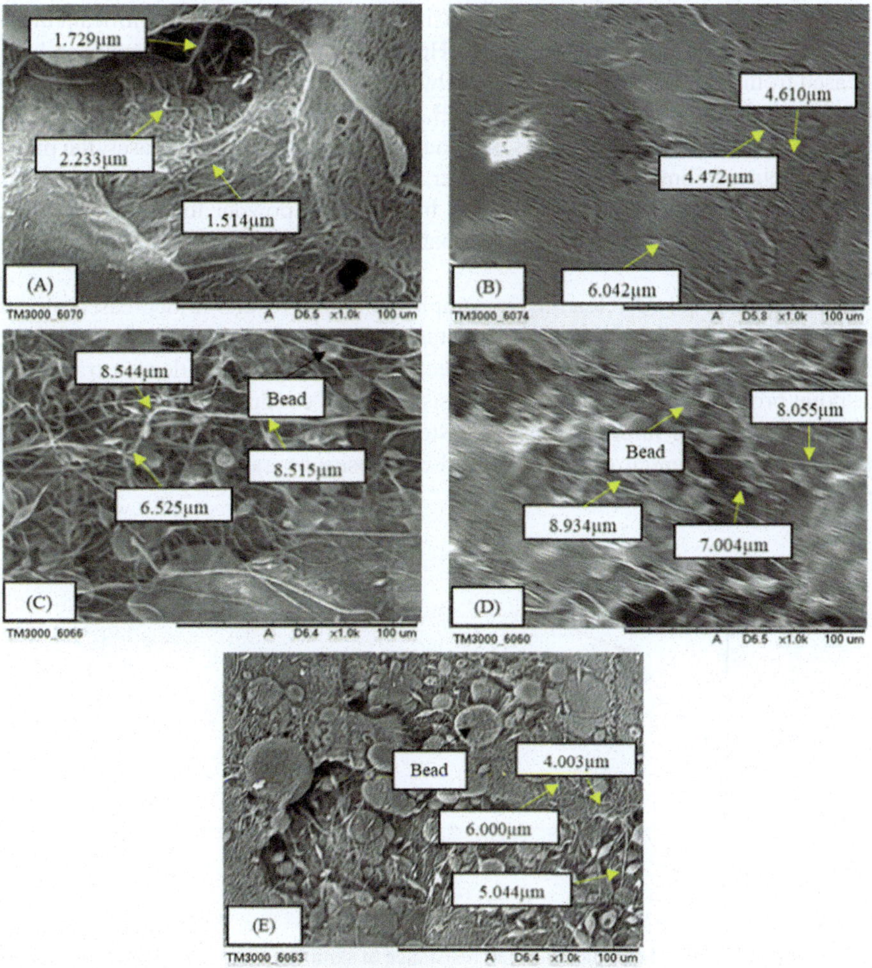

Fig. 1. SEM micrographs of (A) PVDF/0wt% Al powder; (B) PVDF/0.25wt% Al powder; (C) PVDF/0.5wt% Al powder; (D) PVDF/0.75wt% Al powder; and (E) PVDF/1.0wt% Al powder.

Table 1. Example of table

Sample	A	B	C	D	E
Average Fiber Diameter (µm)	1.825	5.041	7.895	8.495	5.016

lowest resistance of 36.99 ϕ . Resistance is a measure of the difficulty of current flow in a material, where lower resistance indicates easier current flow. Conductance, the reciprocal of resistance, represents the ease with which electrons move through the material. Therefore, resistance and conductance are inversely related. Materials with low resistance and high conductance are considered to have high electrical conductivity.

The addition of aluminum powder 0.25wt% in sample B resulted in a significant increase in conductance compared to sample A. However, as the aluminum powder loading increased from 0.25wt% (sample B) to 0.50wt%, 0.75wt% and 1.0wt%, the conductance values showed a slight decrease which are 3.32 × 10^(−2), 3.21 × 10^(−2) and 3.17 × 10^(−2) Siemens, respectively. This is because, as the Al loading increases, the particles come into closer contact with one another, allowing for better electrical conduction. However, when the loading becomes too high, the Al powder tends to agglomerate, forming clusters that block the electron pathways, leading to a decrease in electrical conductivity [13].

Increase in Al powder loading in samples D and E results in decreased electrical performance, evident from the higher resistance and lower conductance values. Overall, the PVDF/Al composite fibers exhibits electrical conductivity, as demonstrated by the four-point probe testing and analysis of resistance and conductance values. The closely grouped conductance values suggest consistent conductive behavior across the samples, highlighting the potential of the PVDF/Al composite fibers for applications requiring electrical conductivity.

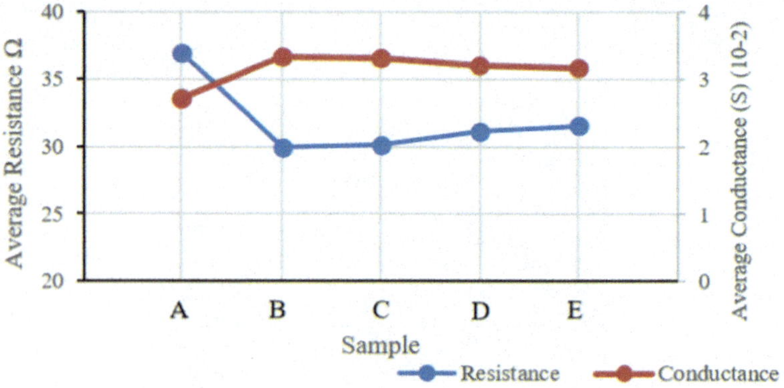

Fig. 2. Average electrical conductance and resistance of samples using four-point probe measurements

3.3 Phase Analysis

Figure 3 shows the diffraction angle of electrospun PVDF/Al composite fibers samples at 25 kV enlightened by X-ray diffraction patterns observed from 5° to 50°. Sample B with 0.25 wt% Al exhibits the most intense peak at 22.44°, indicating a slight alteration in crystalline structure compared to the pure PVDF sample (sample A). This suggests a potential crystallization of the ß-phase due to the addition of Al powder. Sample C (0.5 wt% Al) shows a further shift towards the ß-phase, indicating a significant alteration of the PVDF crystalline structure. Sample D (0.75wt% Al) exhibits a continued shift towards the ß-phase at 20.97°, indicating a higher concentration of this phase and the closest to 20.8° ß-phase peak as reported by Zhao et al. [3]. The formation of ß-phase was aided by Al as a metallic element that promotes the polymer chain to exhibit a TTTT

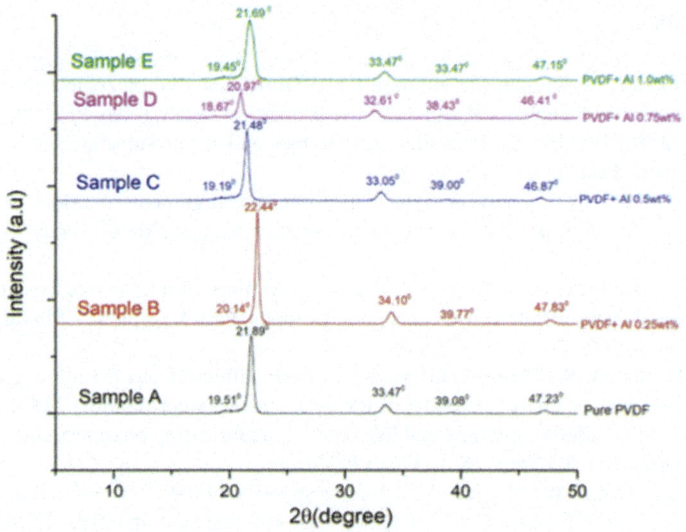

Fig. 3. XRD diffraction plots of PVDF/Al composite fibers.

conformation, which then formed a ß-phase on the PVDF/Al composite fibers [14]. Sample E (1.0 wt% Al) shows a partial transition back towards the α-phase. Overall, the X-ray diffraction results demonstrate the influence of aluminum powder loading on the crystal structure of the PVDF samples, with varying degrees of transition between the α-phase and the ß-phase.

4 Conclusion

We can conclude that the addition of Al powder into electrospun PVDF fibres affected their morphology, microstructure, and electrical properties. SEM analysis revealed that lower Al loading has better dispersion and uniformity which resulted in smaller bead sizes and producing smoother fibres with improved Al powder distribution. As filler loading increased, Al tends to agglomerates and leads to formation of larger bead sizes and rougher surfaces. XRD analysis demonstrated that even a small amount of Al powder could increased the yield of the ß-phase in the PVDF fibres. PVDF with 0.75 wt% Al yield the closest peak of ß-phase. The electrical properties, determined through four-point probe testing, showed that as the weight percentage of aluminum powder increased, the resistance value increased while the conductance value decreased, indicating a decrease in electrical conductivity properties. PVDF with 0.25 wt% Al exhibited the best electrical conductivity among all samples and the addition of more Al into PVDF has resulted in slight decrease in the electrical conductivity.

References

1. Dilip Kulkarni, N., Kumar, M., Kumari, P.: PVDF/RGO based piezoelectric nanocomposite films for enhanced mechanical and dielectric properties. Mater Today Proc. **76**, 81–87 (2023)
2. Sappati, K.K., Bhadra, S.: Piezoelectric polymer and paper substrates: a review. Sensors (Switzerland) **18**(11) (2018)
3. Zhou, J., et al.: Recent advances in dispersion and alignment of fillers in PVDF-based composites for high-performance dielectric energy storage. Mater Today Energy 101208 (2022)
4. Wang, P., et al.: Flexible cellulose/PVDF composite films with improved breakdown strength and energy density for dielectric capacitors. Compos. Part A Appl. Sci. Manuf. **164**, 107325 (2023)
5. Gade, H., Nikam, S., Chase, G.G., Reneker, D.H.: Effect of electrospinning conditions on β-phase and surface charge potential of PVDF fibers. Polymer (Guildf) **228**, 123902 (2021)
6. Tan, S.M., et al.: Electrospinning and its potential in fabricating pharmaceutical dosage form. J. Drug Deliv. Sci. Technol. **76**, 103761 (2022)
7. Singh, R.K., Lye, S.W., Miao, J.: Holistic investigation of the electrospinning parameters for high percentage of β-phase in PVDF nanofibers. Polymer (Guildf) **214**, 123366 (2021)
8. Yang, K., Chen, Y., Xiao, Q., Pang, L.: Influence of venting coefficient on disastrous effects of aluminium powder explosions. Process Saf. Environ. Prot. **156**, 72–88 (2021)
9. Adeniyi, G., Abdulkareem, S.A., Odimayomi, K.P., Emenike, E.C., Iwuozor, K.O.: Production of thermally cured polystyrene composite reinforced with aluminium powder and clay. Environ. Challenges **9**, 100608 (2022)
10. Shinkaryov, S., et al.: Aluminum powder preparation for additive manufacturing using electrostatic classification (2021)
11. Shehata, N., et al.: Static-aligned piezoelectric poly (vinylidene fluoride) electrospun nanofibers/MWCNT composite membrane: facile method. Polymers (Basel) **10**(9), 965 (2018)
12. Fong, H., Chun, I., Reneker, D.H.: Beaded nanofibers formed during electrospinning. Polymer **40**, 4585–4592 (1999)
13. Osman, A.F., Mariatti, M.: Properties of aluminum filled polypropylene composites. Polymers Polymer Compos. **14**(6) (2006)
14. Zhongchen, H., François, R., Maryline, L., Elham, M., Fabien, S.: A review of piezoelectric PVDF film by electrospinning and its applications. Polymers **13**, 174 (2021)

Evaluation of Graphitic Carbon Nitride Loaded Carbon Plate as Photoanode for RR120 Dye Degradation Under the Effect of Supporting Electrolyte

Jai-Xien OrYang[1(✉)], Li-Ngee Ho[1,3], Chang-Chuan Lee[2], Kang-Zheng Khor[1], and Soon-An Ong[3]

[1] Faculty of Chemical Engineering and Technology, Universiti Malaysia Perlis, 02600 Jalan Kangar-Arau, Perlis, Malaysia
oryangjaixien@studentmail.unimap.edu.my

[2] Faculty of Mechanical Engineering and Technology, Universiti Malaysia Perlis (UniMAP), Pauh Putra Campus, 02600 Arau, Perlis, Malaysia

[3] Water Research and Environmental Sustainability Growth, Centre of Excellence (WAREG), Faculty of Civil Engineering and Technology, Universiti Malaysia Perlis (UniMAP), Arau, Perlis, Malaysia

Abstract. This research introduced a photocatalytic fuel cell (PFC) configuration utilising graphitic carbon nitride (GCN) derived from urea through a calcination process and immobilised on a carbon plate for application as the photoanode and platinum-loaded carbon paper as the cathode. The function of the PFC was to treat the organic wastewater containing the azo dye, Reactive Red 120 (RR120) and generate electrical energy simultaneously under irradiation of UVA light. The investigation focused on the effect of supporting electrolytes on the efficiency of PFC in RR120 degradation and power output. Results revealed that 57.31% decolourisation of 10 mg/L RR120 dye was attained in the presence of a supporting electrolyte, compared to a 44.39% removal efficiency without electrolyte over 6 h. Additionally, voltage output increased significantly with the application of supporting electrolyte, reaching a maximum power density of 2.164 mW/m^2 while 0.0852 mW/m^2 for that of non-supporting electrolyte.

Keywords: Graphitic carbon nitride · Photocatalytic fuel cell · Supporting electrolyte

1 Introduction

Wastewater treatment has always been a concern to the world for sustainable purposes and green earth planning, especially in these few years due to the rapid growth of development and water pollution. Photocatalysis is an environmentally friendly and effective method in wastewater treatment. Sulfides and metal oxides are commonly used as photocatalysts for treating organic pollutants, however, they can contribute to environmental

N. R. Abdul Razak et al. (Eds.): ECOOP 1996, SPPHY 418, pp. 193–198, 2025.
https://doi.org/10.1007/978-981-96-2871-1_27

hazards during the production process and non-sustainable [1], so it is crucial to find an alternative material. Recently, the discovery of graphitic carbon nitride (GCN) to be used as a photocatalytic material for wastewater treatment due to its many advantages including proposing utilisations in the water splitting process, organic pollutants deterioration, and photoproduction [2]. GCN has the advantage of high chemical stability in acid and alkaline solutions, and also strong thermal stability under elevated temperatures due to its covalent bond between carbon and nitrogen. Moreover, it has plentiful raw material, a moderate band gap in the visible range, and environmentally friendly properties [3]. Photocatalytic fuel cell (PFC) is one of the effective ways to treat wastewater due to its smart designation which can break down the organic pollutants and together develops electrical energy [4]. By using the power of light energy to initiate the GCN, the combination of PFC and GCN is a valuable design to be investigated for wastewater treatment and sustainable energy production. Therefore, the aim of this research is to synthesise GCN by calcination process and fabricate the GCN-loaded carbon plate for application as a photoanode in PFC to treat Reactive Red 120 (RR120) under ultraviolet light (UVA) irradiation. The synthesised GCN was been characterised through X-ray diffraction (XRD) to confirm the structural and phase analysis. Besides that, the influence of supporting electrolyte for treating RR120 and the performance of PFC were investigated in this study.

2 Experimental Procedure

2.1 Chemicals

A type of azo dye, Reactive Red 120 (RR120) (molecular formula: $C_{44}H_{30}C_{12}N_{14}O_{20}S_6$) was used as the pollutant in this study and supplied by Sigma Aldrich. Urea ($CO(NH_2)_2$) with a molecular weight of 60.06 g/mol was purchased from Bendosen and used as a precursor of GCN. Ethylene glycol ($C_2H_6O_2$) was supplied by HmbG Chemicals to be used in the fabrication of photoanode. Sodium sulphate (Na_2SO_4) was purchased from HmbG Chemicals and used as the supporting electrolyte in the study. The chemicals are analytical grade and employed without further purification.

2.2 Synthesis of GCN

A 50 mL ceramic crucible containing 20 g of urea was weighed and placed in the muffle furnace, which was heated to 520 °C, with a heating rate of 3 °C/min for 3 h. Finally, the sample was cooled to room temperature and ground in the mortar to obtain a fine powder of GCN. X-ray diffraction (XRD) from Bruker (Model: D2 Phaser) with Cu-Kα radiation ($\lambda = 1.54184$ Å) was used to confirm the structural phase of the synthesised GCN and the data were recorded in a scan range, $2\theta = 5° - 80°$ with 0.1 s/step.

2.3 Preparation of GCN-Loaded Carbon Plate (GCN/C) Photoanode

A carbon plate as the substrate for the photoanode was fabricated in the dimensions of $3.0 \times 3.0 \times 1.0$ cm. Ultrasonic cleaning of the carbon plate with deionised water and ethanol was carried out in an ultrasonic bath. 0.4 g of GCN was added with ethylene glycol and deionised water with a ratio of 1:9 and using a magnetic stirrer to homogeneously mix, after that, the mixture was ultrasound for 30 min. The GCN suspension was then loaded on the carbon plate through immobilisation method. The GCN-loaded carbon plate (GCN/C) then be dried in the oven at 60 °C for 2 h and transferred to the muffle furnace at 80 °C, for 10 min for heat treatment.

2.4 PFC Setup and Operation

GCN/C was applied as the photoanode while platinum-loaded carbon paper (Pt/C) was employed as the cathode in a 600 mL beaker with 250 mL 10 mg/L of RR120 to set up a PFC. The two electrodes were linked through an external circuit with a resistivity of 1 kΩ at 1 cm. The irradiation source was an ultraviolet A (UVA) lamp (Philips), while the distance between the lamp and the photoanode was fixed at 5 cm. The PFC system was provided with a flow rate of 0.4 L/min and accomplished at room temperature for 6 h of reaction time.

3 Results and Discussions

3.1 Structural Phase Analysis of Synthesised GCN

The XRD pattern of the synthesised GCN is depicted in Fig. 1. Two diffraction peaks were identified in GCN at $2\theta = 13.0°$ and $27.4°$, which represent the (001) and (002) planes, respectively in agreement with the crystal structures of pure GCN (JCPDS card 87-1526) [5]. The weak diffraction peaks at $2\theta = 13.0°$ with (001) plane of graphitic layered structure, this plane is influenced by hydrogen bonds that help maintain the in-plane structural arrangement of tri-s-triazine units [6]. While the strong diffraction peak at $2\theta = 27.4°$ with (002) plane represents the interlayer stacking of the conjugated aromatic system which manipulated by the weak Van de Waals force [6].

3.2 Influence of Supporting Electrolyte on Electricity Generation in PFC

The electrons travelled through the external circuit from the photoanode to the cathode to produce electricity when the GCN/C photoanode was irradiated under UVA light in the PFC. It can be observed through the voltage output for 6 h as shown in Fig. 2 (a). In the presence of 0.1M Na_2SO_4 as a supporting electrolyte in the PFC system, the voltage output shows a significant difference at 0 h, which is 8.5 mV compared to the PFC system without supporting electrolyte which is 0.2 mV. Throughout the 6 h reaction time, both systems show increasing voltage outputs. Finally, the PFC with supporting electrolyte reached 30.5 mV at 6 h which was 10 times greater than PFC without adding supporting electrolyte having 3.1 mV of voltage output at 6 h. This is due to the enhancement of ionic strength in solution by the supporting electrolyte. The improvement of ionic strength in solution facilitates the electrochemical reaction resulting in higher conductivity by reducing the Ohmic drops, therefore, it allows more current to flow through the PFC system and eventually produce high voltage output [7, 8].

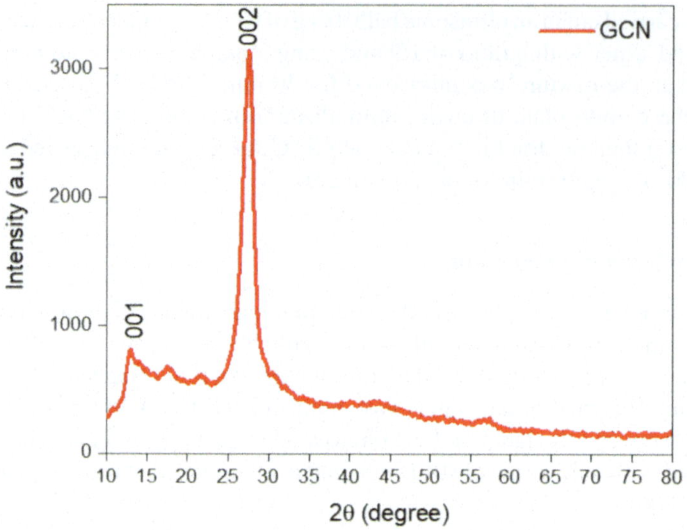

Fig. 1. XRD pattern of synthesised GCN

Figure 2 (b) and (c) illustrate the polarisation curves and power density curves of PFC systems under the influence of supporting electrolyte. The lowest open circuit voltage (V_{oc}), current density short circuit (J_{sc}), and maximum power density (P_{max}) values were recorded in the PFC without a supporting electrolyte under UVA exposure. Upon adding 0.1M Na_2SO_4 as a supporting electrolyte to the PFC under UVA exposure, V_{oc} increased by 1.3 times, from 143 mV to 198 mV compared to the PFC without electrolyte support. Additionally, J_{sc} and P_{max} showed substantial increases in current density and maximum power density, reaching 37.77 mA/m^2 and 2.164 mW/m^2, respectively, compared to 2.10 mA/m^2 and 0.0852 mW/m^2 in the PFC without electrolyte support. These findings indicate that the introduction of a supporting electrolyte significantly influences the power output of the PFC. The rise of V_{oc} and J_{sc} in the existence of Na_2SO_4 may be due to the increase of electron flow to the cathode, Pt/C. Enhanced conductivity due to Na_2SO_4 promotes the movement of photogenerated electrons across solution interfaces, thereby reducing overall internal resistance in the PFC system [9].

3.3 Influence of Supporting Electrolyte on Degradation Efficiency in PFC

Figure 3 discloses the colour removal efficiency of RR120 under UVA irradiation over 6 h duration. In the PFC without a supporting electrolyte, the colour removal efficiency of RR120 was 44.39% after 6 h. However, with the presence of Na_2SO_4 in the PFC, the degradation efficiency increased by approximately 1.3 times, reaching 57.31% after 6 h. This improvement can be attributed to the presence of OH• and superoxide radical anions. The addition of Na_2SO_4 to PFCs enhances the release of photogenerated electrons, which in turn increases the oxidation reaction at the photoanode under UVA light radiation. This facilitates the easier breakdown of the azo bond (-N = N-) due to an increased number of photogenerated holes and improved degradation efficiency [10].

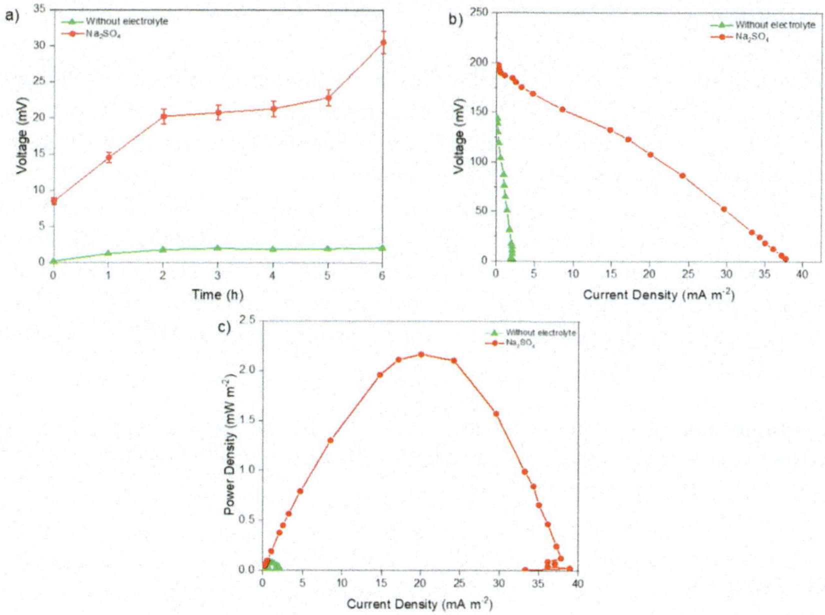

Fig. 2. PFC performance: (a) Voltage output; (b) Polarisation curves; (c) Power density curves, with and without supporting electrolyte under UVA irradiation.

These findings suggest a correlation between dye degradation, conductivity, and voltage output in PFCs, consistent with the findings by Lee et al. (2017) [11].

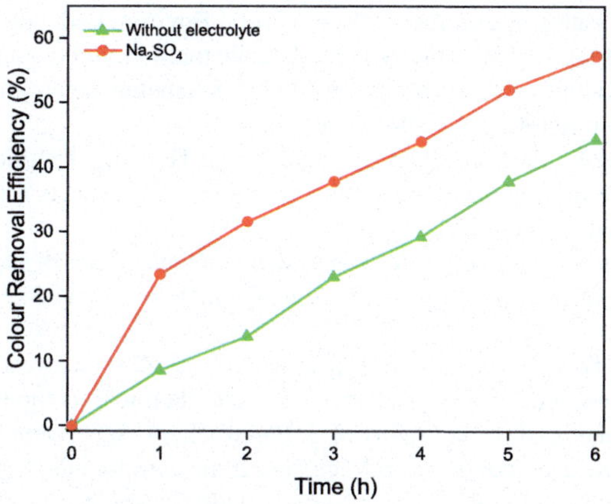

Fig. 3. Colour removal efficiency of PFC at 10 mg/L of RR120 over 6 h irradiation time with and without supporting electrolyte.

4 Conclusion

In conclusion, the supporting electrolyte had a significant impact on the PFC system. Furthermore, the synthesized GCN loaded on carbon plate was successfully performed as a photoanode in the PFC for electricity generation and RR120 azo dye degradation. The presence of supporting electrolyte showed apparent improvement in terms of electricity generation, J_{sc}, P_{max} and dye removal efficiency in the PFC system, which was 1.3 times compared to the PFC without supporting electrolyte. This study attested to the applicability of GCN/C as a photoanode in the PFC. However, there are still areas requiring further investigation, such as low voltage output and colour removal efficiency. Overcoming these challenges is essential for advancing optimised PFC as a sustainable wastewater treatment system in the future.

Acknowledgments. The authors would like to acknowledge the financial support of the Fundamental Research Grant Scheme, grant reference no: FRGS/1/2022/STG05/UNIMAP/02/1, funded by the Ministry of Higher Education Malaysia.

References

1. Xu, J., Antonietti, M., Shalom, M.: Moving graphitic carbon nitride from electrocatalysis and photocatalysis to a potential electrode material for photoelectric devices. Chem. Asian J. **11**, 2499–2512 (2016)
2. An, H., et al.: Photocatalytic degradation of organic dyes using zinc oxide-decorated graphitic carbon nitride composite under visible light. Diam. Relat. Mater. **131**, 109583 (2023)
3. Safaei, J. et al.: Facile fabrication of graphitic carbon nitride, (g-C_3N_4) thin film. J. Alloy. Compd. **769**, 130–135 (2018)
4. Li, M., et al.: Recent advances on photocatalytic fuel cell for environmental applications—the marriage of photocatalysis and fuel cells. Sci. Total Environ. **668**, 966–978 (2019)
5. Kahraman, Z., Kartal, U., Genç, A., Alp, E.: A facile method for boosting the graphitic carbon nitride's photocatalytic activity based on 0D/2D S-scheme heterojunction nanocomposite architecture. Mater. Sci. Eng.: B **304** (2024)
6. Zhang, M., Yang, Y., An, X., Zhao, J., Bao, Y., an Hou, L.: Exfoliation method matters: the microstructure-dependent photoactivity of g-C_3N_4 nanosheets for water purification. J. Hazard. Mater. **424,** 127424 (2022)
7. Xu, L., et al.: Ammonia recovery from wastewater as a fuel: effects of supporting electrolyte on ammonium permeation through a cation-exchange membrane. ACS Omega **7**, 20634–20643 (2022)
8. Yilmaz, A.E., Fil, B.A., Yilmaz, M.T., Bayar, S., Köksal, Z.: The investigation of treatment of pistachio processing industrial wastewater by electrochemical methods in terms of chemical oxygen demand and total phenol removal. J. Electrochem. Sci. Technol. **15**, 178–189 (2024)
9. Xie, S., Ouyang, K.: Degradation of refractory organic compounds by photocatalytic fuel cell with solar responsive WO_3/FTO photoanode and air-breathing cathode. J. Colloid Interface Sci. **500**, 220–227 (2017)
10. Khalik, W.F., et al.: Influence of supporting electrolyte in electricity generation and degradation of organic pollutants in photocatalytic fuel cell. Environ. Sci. Pollut. Res. **23**, 16716–16721 (2016)
11. Lee, S.L., et al.: A highly efficient immobilized ZnO/Zn photoanode for degradation of azo dye Reactive Green 19 in a photocatalytic fuel cell. Chemosphere **166**, 118–125 (2017)

Investigation of Oxidation Rate Law on Fe-Ni-Cr Series Ni-Based Alloy After Isothermal Oxidation

Megat Farhan Nazmi Megat Mohamad Halim[1], Noraziana Parimin[1,2(✉)], Esah Hamzah[3], Nur Farhana Hayazi[1,2], and Siti Sarah Ibrahim[1,2]

[1] Faculty of Chemical Engineering & Technology, Universiti Malaysia Perlis (UniMAP), 02600 Arau, Perlis, Malaysia
noraziana@unimap.edu.my

[2] Surface Technology Special Interest Group, Faculty of Chemical Engineering & Technology, Universiti Malaysia Perlis (UniMAP), 02600 Arau, Perlis, Malaysia

[3] Faculty of Mechanical Engineering, Universiti Teknologi Malaysia (UTM), 81310 Skudai, Johor, Malaysia

Abstract. The study examined the oxidation rate law of Fe-Ni-Cr series Ni-based alloy, which underwent an isothermal oxidation test. Fe-Ni-Cr alloy samples with small grain (SG) and coarse grain (CG) structures were produced through heat treatment at varying temperatures. After the heat treatment, the SG and CG samples were subjected to isothermal oxidation at 900 °C for 300 h. The analysis of the oxidation rate was conducted at 100 h, 200 h, and 300 h exposure durations to observe the behaviour of the oxidized samples. This was achieved by plotting the oxidation kinetics, which involved examining the change in weight per surface area over time. After 100 h of exposure, both SG and CG samples followed a linear rate law, transitioning to a parabolic rate law at 300 h of exposure. SG samples demonstrated superior oxidation behaviour compared to CG samples, exhibiting the shift to a parabolic rate law after only 200 h of exposure. Furthermore, the SG sample displayed higher oxidation resistance as evidenced by its lower parabolic rate constant.

Keywords: Fe-Ni-Cr Alloy · Ni-based Alloy · Oxidation Rate · Isothermal Oxidation · Heat Treatment

1 Introduction

Ni-based alloy belongs to a group of metallic materials that possess impressive qualities such as high-temperature strength, toughness, and resistance to oxidation [1, 2]. It is commonly employed in power generation and heat processing equipment [3]. The Fe-Ni-Cr series studied in this research contains a higher concentration of Cr, which serves as the element for forming oxides [4]. This particular alloy contains 18.9% Cr. Research has shown that a sufficient amount of Cr within the range of 15 to 25% offers the highest level of protection against oxidation by producing a slowly developing surface oxide scale of Cr2O3 during high-temperature operation [5, 6].

N. R. Abdul Razak et al. (Eds.): ECOOP 1996, SPPHY 418, pp. 199–205, 2025.
https://doi.org/10.1007/978-981-96-2871-1_28

The formation of the oxide scale is greatly influenced by the grain structure of the alloy. The formation of a protective oxide scale and good oxidation resistance have been attributed to the fine-grain alloy structure [7, 8]. A parabolic rate law is typically associated with the protective behaviour against oxidation. This type of rate law is based on a diffusion-controlled mechanism for oxide growth, contributing to the formation of an optimal oxide scale layer and excellent oxidation resistance [7]. As part of this investigation, two Fe-Ni-Cr series Ni-based alloys with different grain sizes were subjected to isothermal oxidation testing to analyse the behaviour of the oxidation rate law over prolonged exposure.

2 Experimental Procedure

The alloy composition of Fe-Ni-Cr used in this study was analysed using optical emission spectroscopy (OES), revealing weight percentages of 32.51% Ni, 18.90% Cr, 0.489% Ti, 0.556% Mn, 0.534% Al, 0.315% Si, 0.143% Co, 0.082% Cu, 0.048% Mo, 0.035% W, 0.031% V, 0.036% Nb, with the remaining being Fe. The sample dimensions are 10 x 10 x 3 mm. Following heat treatments at 950 ºC and 1050 ºC for 3 h soaking time and subsequent water quenching, the small grain (SG) and coarse grain (CG) samples exhibited average grain structures of 47.95 μm and 73.05 μm, respectively. The SG and CG samples were exposed to an isothermal oxidation test at 900 ºC for 300 h. Prior to the oxidation test, the surface area and weight of each sample were measured. Weight changes of the samples were recorded at 25 h and 50 h intervals during the test. The oxidation kinetics were calculated by analysing the weight change per surface area over time.

The oxidation rate law was determined using Eq. 1, where x signifies the change in weight per unit area, and t represents time. The value of m in the oxidation rate law designates the type of rate law. When m equals 1 (range: 0.50 to 1.49), it indicates a linear rate law, and m equals 2 (range: 1.50 to 2.49) signifies a parabolic rate law. Equation 2 was employed to establish the oxidation rate constant for linear rate constants, denoted as KL, and Eq. 3 was utilized for the parabolic rate constant, denoted as KP.

$$\log x = 1/m \log t + c \tag{1}$$

$$x = \mathrm{KL}\, t + c \tag{2}$$

$$x2 = \mathrm{KP}\, t + c \tag{3}$$

3 Results and Discussions

The oxidation kinetics are illustrated in Fig. 1, which enables the determination of the oxidation rate law in Fig. 2, and the calculation of the oxidation rate constant in Fig. 3. In Fig. 1, the oxidation kinetics of the SG and CG samples are depicted after 300 h of exposure. Both samples exhibit an escalating pattern of weight gain. Our prior study [7] and another by a different researcher [9] also documented similar findings, demonstrating that greater weight change is associated with prolonged exposure. Initially, at 25 h of exposure, the SG sample displayed a greater weight gain in comparison to the CG sample. This difference can be attributed to the higher grain boundary area present in the SG sample relative to the CG sample. The grain boundary area serves as a pathway for diffusion, particularly for short-circuit diffusion. Consequently, a higher grain boundary area will expand the diffusion path, thereby enhancing the oxidation rate [10, 11]. At the start of the test, a high rate of oxidation leads to the rapid development of an initial oxide scale on the sample surface, providing a protective barrier between the metal surface and the environment. Once the initial oxide layer forms quickly, it will cover the entire surface area of the metal. As the protective oxide layer envelops the metal, the oxidation rate will gradually decrease to reach a stable condition, providing the necessary level of protection. This can be observed after a 50-h exposure when both samples showed similar weight gains. With increased exposure time until 300 h, the formation of the oxide scale decreased slowly, achieving the optimum oxide thickness for enhanced protection.

On the other hand, the behaviour of the CG sample was different. This is because of the lower grain boundary area, which resulted in a slower formation of the oxide scale. The rate then steadily increased to achieve the best possible protection. Referring to Fig. 1 for the CG curve, we can observe that the weight gain increased with longer exposure durations. This trend is attributed to the continuous formation of the oxide scale. To gain a better understanding of the behaviour of oxide scale formation, additional analysis of the data from Fig. 1 was conducted. This analysis focused on determining the oxidation rate law and the oxidation rate constant based on Eqs. 1, 2 and 3, respectively. The results of these analyses are summarized in Table 1.

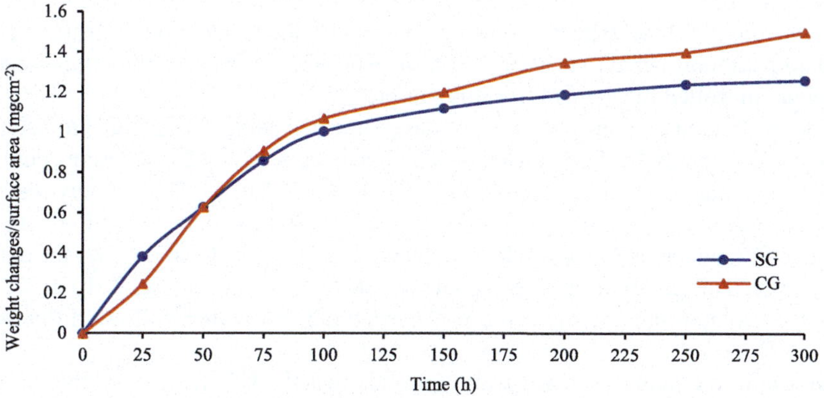

Fig. 1. Oxidation kinetic.

The oxidation rate law of the two samples was further determined at specified time intervals of 100 h, 200 h, and 300 h using Eq. 1 to closely examine the trend in oxidation rate. The primary objective of this analysis was to identify the m value that corresponds to either a linear or parabolic rate law. Figure 2 (a) illustrates the double log plot graph for 100 h, showing m values of 1.41 and 0.92 for the SG and CG samples, respectively. This indicates that both samples exhibited a linear rate law. The m values are summarized in Table 1. In accordance with the weight change data presented in Fig. 1, there is a clear trend of escalating weight gain during the initial 100 h of exposure. The linear rate law suggests that the rate of oxidation remains consistent over time [10, 11]. This law is determined by the steady formation of oxide scale at the interface between the metal and the oxide, resulting in the establishment of a continuous protective oxide scale with a consistent thickness.

Further analysis was conducted using a double log plot at an exposure time of 200 h, as illustrated in Fig. 2 (b). The CG sample maintained a linear rate law (m = 1.26), while the SG sample transitioned to a parabolic rate law with an m value of 1.75. This assessment suggests that the SG sample has reached the limit of diffusion-controlled oxide growth rate due to the presence of an adequate amount of oxide scale and the optimal oxide thickness, which contributes to the alloy's protective behaviour. From Fig. 1, it is evident that the weight gain begins to decrease around 150 h. The reason behind this phenomenon is the larger grain boundary area of the small grain, SG sample, which facilitates rapid outward diffusion of metal ions through a shortcut pathway. Similar observations have been documented by other researchers [8] and our previous research [4, 7], indicating that refine the grain structure improves oxidation resistance. The linear rate law still applies to the coarse grain, CG sample. It suggests that the formation of oxide scale is ongoing. This is due to the coarse grain structure having a smaller grain boundary area, resulting in reduced diffusion paths for outward metal ion diffusion. Consequently, the oxide scale will continue to grow until it reaches the optimal oxide layer for improved protection.

A further examination was conducted at 300 h, as depicted in Fig. 2 (c), which indicated that both SG and CG samples exhibited a parabolic rate constant with m values of 2.11 and 1.50, respectively. The SG sample consistently adhered to the parabolic rate law from 200 h of exposure onwards. However, the CG sample transitioned from linear to parabolic behaviour after 300 h of exposure, suggesting the attainment of a diffusion-controlled mechanism for this sample.

The oxidation rate constant analysis was performed using Eqs. 2 and 3 to determine the linear and parabolic rate constants. This analysis was conducted after confirming the oxidation rate law on the double log plot shown in Fig. 2. The primary discovery focuses on the analysis of the linear rate constant at 100 h, as depicted in Fig. 3 (a), and the parabolic rate constant at 300 h, as depicted in Fig. 3 (d). Furthermore, the analysis of the linear and parabolic rate constants at 200 h is shown in Fig. 3 (b) and (c) for reference. No additional comparison was carried out as the data follows different rate laws.

At 100 h, the analysis of the results reveals that the CG sample exhibits a higher linear rate constant of 3.056 x 10–6 mgcm-2s-1 compared to the SG sample, which has a linear rate constant of 2.333 x 10–6 mgcm-2s-1. This indicates that the CG sample has a greater oxidation rate, signifying a consistent oxide growth rate over time. When the exposure time reaches 200 h, the CG sample continues to follow a linear rate law, but with a decreased linear rate constant of 1.602 x 10–6 mgcm-2s-1, suggesting a decrease in oxidation rate compared to the value at 100 h. Conversely, the SG sample at

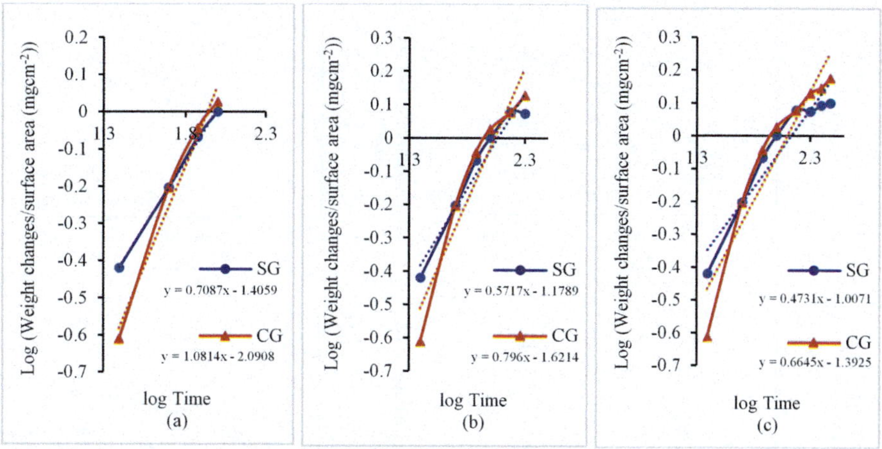

Fig. 2. Oxidation rate law: (a) 100 h, (b) 200 h, and (c) 300 h.

200 h transitions to a parabolic rate law with a parabolic rate constant of 2.028 x 10–6 mg2cm-4s-1.

As the exposure reached 300 h, both SG and CG samples exhibited a parabolic rate law. The parabolic rate constant for the SG sample at 300 h is 1.419 x 10–6 mg2cm-4s-1, a significantly lower value compared to 200 h, indicating a decreasing oxidation rate and optimal protective behaviour. In comparison to the SG sample, the CG sample began to exhibit a shift to parabolic rate law after 300 h, displaying a parabolic rate constant of 2.112 x 10–6 mg2cm-4s-1. This value was higher than that of the SG sample at both 200 and 300 h. This observation indicates that the SG sample has a lower oxidation rate than the CG sample, thus demonstrating good oxidation resistance. Other researchers [8] and our previous studies [4, 7] have also noted the same findings, suggesting that refine the grain structure enhances resistance to oxidation.

The fine grain structure displays good resistance to oxidation and exhibits protective behaviour due to the lower parabolic rate constant. The lower parabolic rate constant suggests that the kinetic oxide growth rate is controlled by diffusion. Initially, a rapid oxide scale forms on the alloy surface to create a continuous oxide layer, providing initial surface protection. With increased exposure time, the thickness of the oxide scale grows, enhancing the protective properties. Once optimal protection is achieved, the formation of the oxide scale reaches a stable state, slowing down the further diffusion process. Our recent study on Fe-40Ni-24Cr alloy [12, 13] and Fe-33Ni-19Cr alloy [4, 13] has documented that the lower parabolic rate constant values indicate the increase in the optimum oxide scale thickness over time.

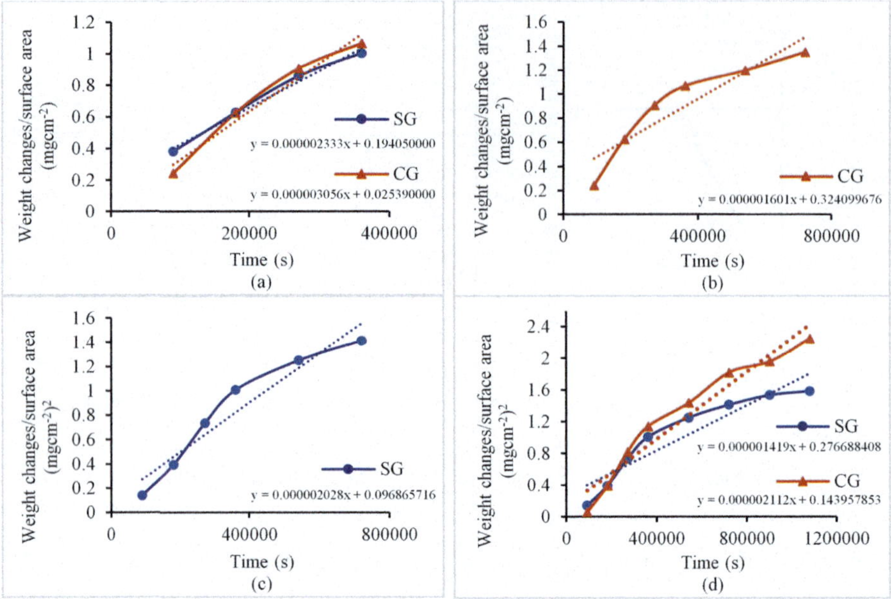

Fig. 3. Oxidation rate constant: (a) 100 h linear rate constant, (b) 200 h linear rate constant, (c) 200 h parabolic rate constant, and (d) 300 h parabolic rate constant.

Table 1. Example of table

Time (h)	SG			CG		
	m	Rate law	Rate constant	m	Rate law	Rate constant
100	1.41	Linear	2.333×10^{-6} $mgcm^{-2}s^{-1}$	0.92	Linear	3.056×10^{-6} $mgcm^{-2}s^{-1}$
200	1.75	Parabolic	2.028×10^{-6} $mg^2cm^{-4}s^{-1}$	1.26	Linear	1.601×10^{-6} $mgcm^{-2}s^{-1}$
300	2.11	Parabolic	1.419×10^{-6} $mg^2cm^{-4}s^{-1}$	1.50	Parabolic	2.112×10^{-6} $mg^2cm^{-4}s^{-1}$

4 Conclusion

Based on our research, we have found that the rate of oxidation is influenced by both the length of exposure and the grain structure of the alloy. During the initial 100 h of the oxidation test, samples with small and coarse grains displayed a linear rate of oxidation, which corresponds to the formation of oxide on the alloy's surface. With increased exposure, the sample with smaller grains demonstrated a shift to a parabolic rate of oxidation, indicating a diffusion-controlled growth of the oxide scale, observed after 200 h of exposure. By the 300-h mark, both small grain and coarse grain samples exhibited a parabolic rate of oxidation, with the small grain sample showing a lower parabolic rate constant, indicating better resistance to oxidation.

Acknowledgments. The study discussed in this manuscript received funding from the Ministry of Higher Education Malaysia through the Fundamental Research Grant Scheme (FRGS) with the grant number FRGS/1/2020/TK0/UNIMAP/02/43. The authors express their gratitude for this support.

References

1. Cao, J., Zhang, J., Chen, R., Ye, Y., Hua, Y.: High temperature oxidation behavior of Ni-based superalloy GH202. Mater Charact **118**, 122–128 (2016)
2. Wang, X., Szpunar, J.A.: Effects of grain sizes on the oxidation behavior of Ni-based alloy 230 and N. J. Alloy. Compd. **752**, 40–52 (2018)
3. Yang, J., Wang, S., Li, Y., Xu, D.: Under-deposit corrosion of Ni-based alloy 825 and Fe-Ni based alloy 800 in supercritical water oxidation environment. Corros. Sci. **167**, 108493 (2020)
4. Zaiton, N.A.Z., Parimin, N., Hayazi, N.F., Zainal, F.F., Garus, S., Vizureanu, P.: Effect of heat treatment temperature on isothermal oxidation of Ni-based Fe-33Ni-19Cr alloy. Arch. Metall. Mater. **69**(2), 747–752 (2024)
5. Oquab, D., Xu, N., Monceau, D., Young, D.J.: Subsurface microstructural changes in a cast heat resisting alloy caused by high temperature corrosion. Corros. Sci. **52**, 255–262 (2010)
6. Zurek, J., et al.: Growth and adherence of chromia based surface scales on Ni-base alloys in high- and low-pO2 gases. Mater. Sci. Eng. A **477**(1–2), 259–270 (2008)
7. Parimin, N., Hamzah, E.: Influence of Ti on oxide formation during isothermal oxidation of 800H Ni-based alloys. Key Eng. Mater. **929**, 29–34 (2022)
8. Tan, L., Ren, X., Sridharan, K., Allen, T.R.: Effect of shot-peening on the oxidation of alloy 800H exposed to supercritical water and cyclic oxidation. Corros. Sci. **50**(7), 2040–2046 (2008)
9. Zhang, Z., Webster, R.F., Zhang, J.: Alloying effects of Al, Si and Fe on Ni-20Cr alloy oxidation in water vapour at 650 ∘C. Corros. Sci. **232**, 112050 (2024)
10. Kofstad, P.: High Temperature Corrosion. Elsevier Applied Science, London (1988)
11. Khanna, A.S.: Introduction to High Temperature Oxidation and Corrosion. ASM International®, Ohio (2002)
12. Anuar, A.I., Parimin, N., Ahad, N.A., Derman, M.N., Garus, S., Vizureanu, P.: High Temperature Isothermal Oxidation at 950°C of Nu-based Fe-40Ni-24Cr alloy. Arch. Metall. Mater. **69**(3), 1191–1198 (2024)
13. Zaiton, N.A.Z., Parimin, N., Anuar, A.I., Hayazi, N.F., Zainal, F.F.: Oxide Growth Behaviour of 800H and HR-120 Series Ni-based Alloys on Isothermal Oxidation. Int. J. Nanoelectron. Mater. **17**(3), 416–421 (2024)

Performance of Graphitic Carbon Nitride Loaded Carbon Plate as the Photocathode in Photocatalytic Fuel Cell for Azo Dye Degradation and Electricity Generation

Kang-Zheng Khor[1(✉)], Li-Ngee Ho[1,2], Nabilah Aminah Lutpi[2], Yunhai Wang[3], Jai-Xien OrYang[1], and Soon-An Ong[2]

[1] Faculty of Chemical Engineering & Technology, Universiti Malaysia Perlis (UniMAP), Arau, Perlis, Malaysia
khorkz@studentmail.unimap.edu.my
[2] Water Research and Environmental Sustainability Growth, Centre of Excellence (WAREG), Faculty of Civil Engineering & Technology, Universiti Malaysia Perlis (UniMAP), Arau, Perlis, Malaysia
[3] Department of Environmental Science and Engineering, Xi'an Jiaotong University, Xi'an 710049, China

Abstract. The photocatalytic fuel cell (PFC) represents a green and sustainable technology capable of treating organic pollutants while simultaneously generating electricity. Graphitic carbon nitride (g-C_3N_4) was synthesized via the calcination method. The synthesized g-C_3N_4 was characterized and confirmed by using X-ray diffraction (XRD) and Fourier-transform infrared (FTIR) spectroscopy. A PFC was assembled with a carbon plate loaded with g-C_3N_4 (g-C_3N_4/C) serving as the photocathode, and another carbon plate loaded with zinc oxide (ZnO/C) acting as the photoanode. In the PFC evaluation, 10 ppm of azo dye Reactive Red 120 (RR120) was applied as the organic pollutant which was irradiated under UV light over 6 h of reaction time. The objective of this study was to investigate the effect of g-C_3N_4 as a cathodic catalyst in the PFC. The results demonstrate that the photocatalytic activity of the PFC was enhanced by the addition of g-C_3N_4 on the carbon plate as a photocathode, leading to improved RR120 color removal and electricity generation. Specifically, the photocathode with g-C_3N_4/C exhibited a color removal efficiency of 98.17% and a maximum power density of 0.8667 mW m^{-2}, surpassing the bare carbon plate photocathode, which achieved a color removal efficiency of 87.49% and a maximum power density of 0.5902 mW m^{-2} after 6 h of UV light irradiation.

Keywords: Graphitic Carbon Nitride · Photocathode · Photocatalytic Fuel Cell

N. R. Abdul Razak et al. (Eds.): ECOOP 1996, SPPHY 418, pp. 206–212, 2025.
https://doi.org/10.1007/978-981-96-2871-1_29

1 Introduction

Recent technological advancements have exacerbated environmental pollution, particularly from wastewater discharged by industries and households [1]. As a result, this pollution has become a significant global issue, especially water pollution. To overcome this, various wastewater treatment methods have been developed. Among these, the photocatalytic fuel cell (PFC) gained considerable attention. This is due to its environmentally friendly, outstanding efficiency, ease of operation, and sustainability [2]. PFCs combine the principles of photocatalysis and fuel cells, enabling the simultaneous degradation of organic pollutants and generating electricity, offering a dual benefit in wastewater treatment. This study focuses on the application of graphitic carbon nitride (g-C_3N_4) loaded on a carbon plate as a photocathode in a PFC system. g-C_3N_4, known for its high photocatalytic activity, stability, and suitable electronic properties, was chosen to enhance the degradation of azo dyes specifically Reactive Red 120 (RR120), which are persistent industrial pollutants [3]. Furthermore, g-$_3N_4$'s narrow band gap (~2.7 eV) and favorable conduction band position make it highly effective for catalytic reactions and electricity generation [4, 5]. By investigating the performance of g-$_3N_4$ in this novel configuration, this research aims to improve the overall efficiency of PFCs and contribute to the evolution of g-$_3N_4$-based photocathodes for potential implementation in both environmental rehabilitation and electricity production.

2 Experimental Procedure

2.1 Chemicals and Materials

The organic pollutant chosen was Reactive Red 120 (RR120), an azo dye manufactured by Sigma Aldrich (St. Louis, MO, USA) with the molecular formula $C_{44}H_{30}Cl_2N_{14}O_{20}S_6$. A 95% pure ethanol, C_2H_5OH which was used for cleaning, and zinc (II) oxide, ZnO, are supplied by HmbG Chemicals. Melamine, as the precursor for the synthesis of g-C_3N_4, $C_3H_6N_6$ was acquired from Sigma Aldrich, which originated from St. Louis, MO, USA. The chemicals used in this research were lab-grade and used without further processing.

2.2 Fabrication of g-C_3N_4 and g-C_3N_4 /C Photocathode

g-C_3N_4 was synthesized via calcining melamine under controlled conditions in a muffle furnace. An alumina crucible with a lid was used in a typical synthesis run for 3 g of melamine. The melamine was heated to 520 °C with 10 °C/min of heating rate and held for 4 h. Finally, the synthesized g-C_3N_4 was left cool to ambient temperature and ground into powder by using mortar.

The photocathode of g-C_3N_4/C was fabricated by the immobilization method. The 0.3 g of synthesized g-C_3N_4 powder was dispersed in the mixture of ethylene glycol and distilled water, magnetically stirring, and ultrasound for 10 min respectively. Next, the solution was covered on the surface of the substrate of the photocathode (a cleaned carbon plate with dimensions of 5.00 cm × 3.00 cm × 0.30 cm) evenly. After that, the immobilized g-C_3N_4/C substrate was dried in an oven (60 °C) for 2 h, followed by a heating attachment in a furnace for 10 min at 200 °C with a 5 °C min^{-1} of heating rate. Lastly, the photocathode g-C_3N_4/C was immersed in distilled water several times to eliminate the loose bound g-C_3N_4 powder, and then let dry at room conditions.

2.3 Materials Characterization

X-ray diffraction (XRD) was used to determine the crystal lattice structure and phases of g-C_3N_4. The XRD model used was a D2 Phaser from Bruker, equipped with Cu-Kα radiation ($\lambda = 1.54184$ Å). In a scanning range of 2θ at 5°–80° with a scanning rate of 0.1 s/step, the XRD data were collected. Fourier transform infrared spectra of the prepared sample of g-C_3N_4 powder were obtained using a spectrophotometer (Perkin Elmer, Spectrum RX1 Spectrometer), by analysis of the sample in the wavelength range of 450–4000 cm^{-1}.

2.4 Evaluation of the g-C_3N_4/C Photocathode in PFC

The sampling was carried out every interval hour for over 6 h of reaction time. A UV-vis spectrophotometer was utilized to evaluate the absorbance at max = 536 nm, within the wavelength ranging from 200–700 mm to examine the removal of color and breakdown of the organic pollutant. The UV-vis spectrophotometer model used was ME-UV1200PC from Mesulab, China.

The dye removal efficiency was calculated by using Eq. (1).

$$\text{Dye Removal efficiency} = \frac{C_0 - C_t}{C_0} \times 100\% \quad (1)$$

where C_t, is the final concentration at reaction time (t) in hours and C_0 is the initial concentration.

The external resistance, R_{ext} was varied to carry out the polarization test from 1 M Ω to 10 Ω with an alterable resistor box, from China (ZX21e). A CD800a digital multimeter originating from Japan (Sanwa) was used to observe the correspondent voltage output during the experiment. Equation (2) was utilized to ascertain the current density based on the voltage produced across the Rext per unit area of the cathode, whereas Eq. (3) was employed to evaluate the power density.

$$\text{Current density} = \frac{V}{R_{ext}A} \quad (2)$$

where V is the correspondent voltage output (mV), and A is the cathode area ($0.0015m^{-2}$).

$$\text{Power density} = \frac{VI}{A} \quad (3)$$

where V is the correspondent voltage output (mV), and I/A is the current density (mA/m^{-2}).

3 Results and Discussions

3.1 Structural Phase Analysis and Molecular Composition and Functional Groups Analysis of g-C_3N_4

Figure 1(a) shows the XRD pattern of g-C_3N_4 powder which calcined at 520 °C for 4h. The diffractogram of the g-C_3N_4 displayed distinct crystalline peaks, with one intense peak standing out at $2\theta = 27.9°$ (JCPDS 87–1526) [6]. The formation of tri-s-triazine units was triggered by the condensation of the precursor, melamine. These units polymerized into plates, which interacted through π-π^* interactions, resulting in the aggregation of a certain number of layers [7]. As a result, a sharp peak appeared at approximately 27.9° is observed in g-C_3N_4. This peak corresponds with the (002) of the $(h\ k\ l)$ plane, which is determined by the aromatic interaction between planes from the π-π^* cooperation of the aromatic heptazine rings [8]. The diffractogram of g-C_3N_4 shows another, considerably broadened and relatively low intense peak located at approximately 13°. This peak is contributed to the (100) plane, from the heptazine rings within a layer spacing [9].

Figure 1(b) shows the FTIR spectra of the g-C_3N_4 powder, which was calcined at 520 °C for 4 h. The peak at 810 cm^{-1} is due to triazine units, and the peaks at 1568 and 1650 cm^{-1} can be attributed to C = N stretching. Moreover, the peaks at 1241,1319 and 1409 cm^{-1} can be assigned to C-N stretching in the aromatic ring. The relatively broad peak in 3100 cm^{-1} could be due to the adsorbed H_2O molecules and end of NH_2 or NH groups [10]. Hence, this analysis suggests that g-C_3N_4 was successfully synthesized and the atomic arrangement of g-C_3N_4 shown in Fig. 1(c).

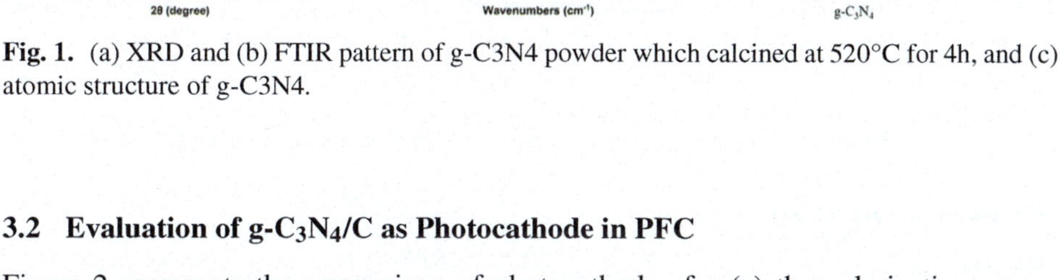

Fig. 1. (a) XRD and (b) FTIR pattern of g-C3N4 powder which calcined at 520°C for 4h, and (c) atomic structure of g-C3N4.

3.2 Evaluation of g-C_3N_4/C as Photocathode in PFC

Figure 2 represents the comparison of photocathodes for (a) the polarization curves and (b) the power density curves of the PFC system. Significant differences in electricity generation are observed between the two photocathodes. In Fig. 2(a), the cell voltage (V_{oc}) decreases as the current density (J_{sc}) increases because of losses of ohmic in the performing cell [12]. The V_{oc} increases from 146.9 mV to 244 mV when the photocathode changes from a bare carbon plate to g-C_3N_4 loaded carbon plate. Similarly, in Fig. 2(b), the J_{sc} and P_{max} increase from 21.56 mA m^{-2} and 0.59 mW m^{-2} to

26.22 mA m^{-2} and 0.87 mW m^{-2}, respectively, when the photocathode changes from a bare carbon plate to g-C_3N_4. These outcomes suggest that g-C_3N_4 as a photocathode generates higher electricity. This phenomenon may be due to the g-C_3N_4 exhibiting an efficient hole-scavenging in the PFC, which suggested that g-C_3N_4 is efficient for accepting the electrons for oxygen reduction reaction (ORR) activity [12].

Figure 2(c) shows the color removal efficiency of RR120 over irradiation time with different photocathodes in PFC. Generally, after 6 h of light irradiation, the color removal efficiency of RR120 for g-C_3N_4 as photocathode exhibited a higher decolorization efficiency, 98.17% compared to the bare carbon plate as photocathode, which is 87.49%. This suggested that g-C_3N_4 as a photocathode improved the color removal efficiency of RR120 compared to bare carbon plate under the same conditions. This phenomenon could be attributed to g-C_3N_4 possessing a high surface area (28.68 m^2 g^{-1}) and a porous structure compared to the bare carbon plate (0.1 m^2 g^{-1}) shown in Table 1, providing many active sites for the surface assimilation and reaction of dye molecules [13]. This increases the contact area between the photocatalyst and the dye, promoting more efficient degradation. Additionally, the nitrogen-rich structure of g-C_3N_4 can form coordination bonds with dye molecules, enhancing their adsorption onto the photocatalyst surface [14]. Moreover, g-$_3N_4$ has a band gap of approximately 2.7 eV, enabling efficient light absorption and generating electron-hole pairs. The excited electrons can then participate in oxygen reduction, which is crucial for ORR [15]. Thus, g-$_3N_4$'s structural and electronic properties contribute to its superior photocatalytic performance compared to bare carbon plate.

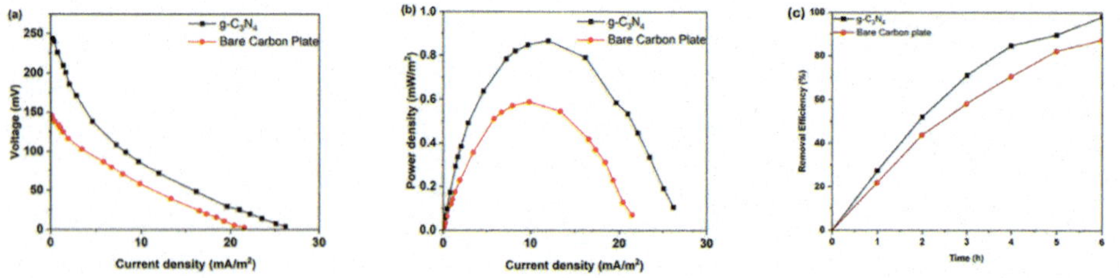

Fig. 2. PFC performance: (a) Polarization curves, (b) power density curves (c) color removal efficiency of RR120 over irradiation time of PFC system with different photocathodes.

Table 1: Specific surface area (S_{BET}) of photocathodes materials.

Photocathode materials	**S_{BET} (m^2 g^{-1})**
g-C_3N_4	28.68
Carbon plate	0.10

4 Conclusion

This research demonstrates the capability of g-C_3N_4 as an effective cathodic catalyst in PFCs for both azo dye degradation and electricity generation. The results revealed that g-C_3N_4/C as a photocathode significantly improved the performance of the PFC. The g-C_3N_4/C photocathode generated higher electricity, with a maximum power density of 0.8667 mW m^{-2}, surpassing the bare carbon plate's 0.5902 mW m^{-2}. This increased electricity generation is likely due to g-C_3N_4's efficient hole-scavenging ability and enhanced ORR activity. Additionally, the g-C_3N_4/C photocathode achieved a color removal efficiency of 98.17%, compared to 87.49% for the bare carbon plate. This improvement is attributed to g-C_3N_4's high surface area, porous structure, and nitrogen-rich composition, which enhance dye molecule adsorption and degradation. Overall, this research highlights the effectiveness of g-C_3N_4 as a photocathode in PFCs, demonstrating its potential for environmental and energy applications. The findings contribute to the understanding of g-C_3N_4-based photocathodes and their role in improving PFC efficiency, paving the way for future advancements in wastewater treatment and sustainable energy generation.

Acknowledgment. The authors want to acknowledge and thank to the financial support provided by the Ministry of Higher Education Malaysia for this study, under grant reference no. FRGS/1/2022/STG05/UNIMAP/02/1.

References

1. Lellis, B., Fávaro-Polonio, C.Z., Pamphile, J.A., Polonio, J.C.: Effects of textile dyes on health and the environment and bioremediation potential of living organisms. Biotechnology Research and Innovation. **3**, 275–290 (2019). https://doi.org/10.1016/j.biori.2019.09.001
2. Jia, M., et al.: Integrating N and F co-doped TiO_2 nanotubes with ZIF-8 as photoelectrode for enhanced photo-electrocatalytic degradation of sulfamethazine. Chem. Eng. J. **388**, 124388 (2020). https://doi.org/10.1016/j.cej.2020.124388
3. Anastopoulos, I., Ahmed, M.J., Ojukwu, V.E., Danish, M., Stylianou, M., Ighalo, J.O.: A comprehensive review on adsorption of Reactive Red 120 dye using various adsorbents. J. Mol. Liq. **394**, 123719 (2024). https://doi.org/10.1016/j.molliq.2023.123719
4. Li, D., et al.: Synthesis of a carbon dots modified g-C_3N_4/SnO_2 Z-scheme photocatalyst with superior photocatalytic activity for PPCPs degradation under visible light irradiation. J. Hazard. Mater. **401**, 123257 (2021). https://doi.org/10.1016/j.jhazmat.2020.123257
5. Taghipour, S., Ataie-Ashtiani, B., Hosseini, S.M., Yeung, K.L.: Graphitic carbon nitride-based composites for photocatalytic abatement of emerging pollutants. In: Nanostructured Carbon Nitrides for Sustainable Energy and Environmental Applications. pp. 175–214. Elsevier (2022)
6. Alizadeh, T., Nayeri, S., Hamidi, N.: Graphitic carbon nitride (g-C_3N_4)/graphite nanocomposite as an extraordinarily sensitive sensor for sub-micromolar detection of oxalic acid in biological samples. RSC Adv. **9**, 13096–13103 (2019). https://doi.org/10.1039/C9RA00982E
7. Fina, F., Callear, S.K., Carins, G.M., Irvine, J.T.S.: Structural investigation of graphitic carbon Nitride via XRD and neutron diffraction. Chem. Mater. **27**, 2612–2618 (2015). https://doi.org/10.1021/acs.chemmater.5b00411

8. Zandipak, R., Bahramifar, N., Torabi, M., Calero, M., Muñoz-Batista, M.J., Solís, R.R.: Synergistic effect of graphitic-like carbon nitride and sulfur-based thiazole-linked organic polymer heterostructures for boosting the photocatalytic degradation of pharmaceuticals in water. Chem. Eng. J. **494**, 152843 (2024). https://doi.org/10.1016/j.cej.2024.152843
9. Sunasee, S., et al.: Sonophotocatalytic degradation of bisphenol A and its intermediates with graphitic carbon nitride. Environ. Sci. Pollut. Res. **26**, 1082–1093 (2019). https://doi.org/10.1007/s11356-017-8729-7
10. Fahimirad, B., Asghari, A., Rajabi, M.: Magnetic graphitic carbon nitride nanoparticles covalently modified with an ethylenediamine for dispersive solid-phase extraction of lead (II) and cadmium (II) prior to their quantitation by FAAS. Microchim. Acta **184**, 3027–3035 (2017). https://doi.org/10.1007/s00604-017-2273-5
11. Wang, A., Wang, C., Fu, L., Wong-Ng, W., Lan, Y.: Recent Advances of Graphitic Carbon Nitride-Based Structures and Applications in Catalyst, Sensing, Imaging, and LEDs. Nanomicro Lett. **9**, 47 (2017). https://doi.org/10.1007/s40820-017-0148-2
12. Ong, Y.-P., et al.: A highly sustainable hydrothermal synthesized MnO2 as cathodic catalyst in solar photocatalytic fuel cell. Chemosphere **263**, 128212 (2021). https://doi.org/10.1016/j.chemosphere.2020.128212
13. Hadi, P., Yeung, K.Y., Barford, J., An, K.J., McKay, G.: Significance of "effective" surface area of activated carbons on elucidating the adsorption mechanism of large dye molecules. J. Environ. Chem. Eng. **3**, 1029–1037 (2015). https://doi.org/10.1016/j.jece.2015.03.005
14. Wu, J., et al.: Nitrogen-rich covalent triazine frameworks for high-efficient removal of anion dyes and the synergistic adsorption of cationic dyes. Chemosphere **272**, 129622 (2021). https://doi.org/10.1016/j.chemosphere.2021.129622
15. Lee, Y.-J., Jeong, Y.J., Cho, I.S., Park, S.-J., Lee, C.-G., Alvarez, P.J.J.: Facile synthesis of N vacancy g-C3N4 using Mg-induced defect on the amine groups for enhanced photocatalytic •OH generation. J. Hazard. Mater. **449**, 131046 (2023). https://doi.org/10.1016/j.jhazmat.2023.131046

Benzoxazine Resin Derived from Cardanol as an Efficient Additive in EPDM Rubber Compounds

Cao Xuan Viet[1,2](✉), Phan Quoc Phu[1,2], and Nguyen Thi Le Thanh[1,2]

[1] Ho Chi Minh City University of Technology (HCMUT), 268 Ly Thuong Kiet Street, District 10, Ho Chi Minh City, Vietnam
caoxuanviet@hcmut.edu.vn

[2] Vietnam National University Ho Chi Minh City, Linh Trung Ward, Thu Duc District, Ho Chi Minh City, Vietnam

Abstract. This preliminary work present the synthesis and successful ulilization of benzoxazine from natural resourse as reinforcing resins in Ethylene propylene diene monomer (EPDM) rubber compounds. Benzoxazine resin were obtained from the use of cost-effective and renewable starting material such as cardanol (agricultural by-product of cashew nut processing), hexamethylene tetramine, and formalin. The benzoxazine resin were incorporated in EPDM/silica mixture to investigate its effect on the physical and mechanical of the rubber compounds. The results show that the addition of benzoxazine significantly improve the tensile properties and thermal properties to some extent. This study provides good perspective in the utilization of benzoxazine as reinforcing agent as well as processing aids to replace traditional phenolic resins for rubber applications.

Keywords: Benzoxazine · Ethylene Propylene Diene Monomer

1 Introduction

Ethylene propylene diene monomer (EPDM) is known as an outstanding synthetic rubber thanks to its superior features compared to other rubbers such as high temperature resistance, weather resistance, ozone resistance and good chemical resistance. However, EPDM compounds shows poor tensile strength due to the low degree of unsaturation and the unsaturated site is not the polymer backbone [1, 2].

Typically, the properties of EPDM compounds are improved through vulcanization by creating a crosslinked structure using sulfur or organic peroxide as the vulcanizing agent [3]. Filler reinforcement such as calcium carbonate, clay, carbon black or silica, has also been used as an alternative method for enhancing the physical and chemical properties of the EPDM matrix as well as reducing its production cost [4, 5]. EPDM rubber is capable to be extended with fillers and plasticizers to a very high level in comparison with the other rubbers. Other appproach is using reactive monomers or polymers

N. R. Abdul Razak et al. (Eds.): ECOOP 1996, SPPHY 418, pp. 213–219, 2025.
https://doi.org/10.1007/978-981-96-2871-1_30

to improve the properties of rubber. Phenolic resin has been widely explored as crosslinking agents and modifiers in various types of rubbers such as EPDM, NBR, chloroprene, etc., bringing remarkable improvements in physical and mechanical properties [6–8].

Benzoxazine resin is derived from the phenolic resin family, it has a phenol ring network structure similar to traditional phenolic resins. Benzoxazine resin has more outstanding properties but still retains the properties of traditional phenolic resins synthesized through coagulation reactions between phenol derivatives, primary amine derivatives and aldehyde-Mannich reactions [9, 10]. Exhibiting excellent mechanical properties, PBz has been used as a reinforcing agent for rubber applications [11]. Cardanol ($C_{21}H_{36\text{-}2n}O$) is a major constituent of the cashew nut shell liquid (CNSL) which is an agricultural by-product abundantly available in many parts of the world, particularly in ASEAN countries and Vietnam. Cardanol based benzoxazine resin with flexibility in molecular structure due to the presence of a long hydrocarbon will offer a better processability [12, 13].

The main objective of this preliminary study is to investigate the effect of benzoxazine resin synthesized from cardanol as an effficient additive to improve the mechanical properties of silica filled EPDM compounds.

2 Experimental Procedure

Cardanol obtained from Son Chau (Vietnam) Co., Ltd. ($\rho = 0.9363$ g/cm^3 (in 25°C), water content <0.1%, hydroxyl value 184.4 mg KOH/g). Formalin, citric acid and hexamethylenetetramine (urotropin) are from Xilong (China), methanol provided by Fisher Chemical (USA). Other chemicals (chloroform, natri hydroxide, natri sulfate) are from Chemsol (Vietnam).

EPDM rubber (EPT 3072EM) purchased from Mitsui Chemical. Vulcanizing chemicals such as stearic acid, zinc oxide, 2-mercaptobenzothiazole (MBT), tetramethylthiuram disulfide (TMTD), sulphur and silica (Tokusil 185G), PEG 4000 were obtained from local suppliers.

In first step, cardanol, formalin (formaldehyde solution), and citric acid are combined in a molar ratio of 1:0.7:0.01. Cardanol was gradually heated to 100 °C ± 5 °C for 30 min. After that, citric acid was added in 5 ml of methanol in a beaker, and then formalin was added to the mixture. After completely adding the formalin and acid catalyst mixture for 1 h, the mixture was heated, and the temperature was maintained at 120 °C ± 5 °C. The pH was checked about every 30 min. The initial pH of the mixture was 6.5, and the reaction was finished when the pH dropped to 4. The mixture was then cooled to room temperature. The progress of the reaction was monitored by checking the pH, and it was finally confirmed by FTIR analysis. The sample obtained (Ca-F) was in reddish-brown vicous liquid form.

In second step, Ca-F and hexamethylenetetramine (urotropin) are combined in a molar ratio of 1:0.1. Ca-F is added to the flask and heating at 120 °C ± 5 °C, for 30 min. Dissolve urotropin in methanol, with the minimum amount of methanol. The reaction temperature was reduced to 80 ± 5 °C. Adding slowly the dissolved solution into CA-F mixture through the neck of the flask for 15 min. After complete addition of urotropin, the temperature was raised and maintained at 100 ± 5 °C for 15 min. The progress of the reaction was checked of pH and the benzoxazine product (Ca-Bz) finally confirmed by FTIR.

Next, preparation of silica filled EPDM compounds. All ingredients (except the curing agents) were firstly blended in internal mixer and followed by further mixing with accelerators and sulfur on the two rolls mill. After that, the sample was stabilized for 24 h prior to Rheometer measurement to determine the curing parameters. Compounds were pressed on a hydraulic press with temperature of 170 °C, pressure of 50kgf/cm^2, and curing time determined from the vulcanization curve. Samples were then cut into shapes for various testing and characterization.

3 Results and Discussions

Figure 1(a) depicts the FT-IR spectrum of cardanol, Ca-F, and Ca-Bz. The spectrum of cardanol shows a signal peak at 3354 cm^{-1}, which is characteristic of the stretching vibration of the -OH group present in cardanol. In the Ca-F spectrum, the appearance of a peak at 3354 cm^{-1} shows the stretching vibration of the -OH group, and the peak at 1108 cm^{-1} shows the stretching vibration of the C–O bonding of CH_2OH. There are also other characteristic peaks at 3078 cm^{-1} (Ar C=C–H stretching), 3009 cm^{-1} (C=C–H stretching of the aromatic ring), 2926 and 2854 cm^{-1} (asymmetric and symmetric CH_2-CH_3 stretching), 1588 cm^{-1} (C=C stretching of the hydrocarbon chain), 1266 cm^{-1} and 1154 cm^{-1} (phenolic C–OH stretching), and 779 cm^{-1} (C–H bonding of the aromatic ring). The Ca-Bz spectrum has a reduced intensity of the -OH group at 3341–3350 cm^{-1}, and it has lost the peak at 1108 cm^{-1}, which is characteristic of the C–O in CH_2OH. The Ca-Bz spectrum reveals characteristic absorptions for the oxazine ring at 1350 cm^{-1} (C–N, stretching vibration), 1264 cm^{-1} (C–O–C, asymmetric stretching vibration), 1118 cm^{-1} (C–N–C, asymmetric stretching vibration), and 990 cm^{-1} (O–CH_2–N, stretching vibration), which are present in the oxazine ring.

The variation in mechanical properties with the addition of Ca-Bz to EPDM/Silica compound is shown in and Fig. 1(b), with σ_u: tensile strength at break (MPa), ε: elongation at break (%), M100: stress at 100% elongation (MPa), M300: stress at 300% elongation (MPa).

The tear strength of EPDM/silica compound was improved by the addition of Ca-Bz. The EPDM/silica compounds without Ca-Bz exhibits the lowest tear strength (17.99 N/mm), while the EPDM-10phr Ca-Bz sample demonstrates the highest tear strength (26.73 N/mm), representing a 48.58% increase. Subsequently, with an increase in Bz content to 15phr, there is a slight decrease in tear strength (EPDM-15Bz), reaching 23.61 N/mm.

It can be observed from Table 1 and Fig. 1(b) that the tensile strength of EPDM samples significantly increases by adding the benzoxazine resin. Tensile strength of EPDM sample with 10phr Ca-Bz shows an remarkable increase by 332.75% compared to the control sample.

These results might be explained by formation of the crosslinking network of Ca-Bz resin in the vulcanized EPDM network. Under stress, the molecular chains of polybenzoxazine rearrange along the direction of deformation, contributing to increased tensile strength and tear strength. On the other hand, tensile strength decreases for the NR-15Bz

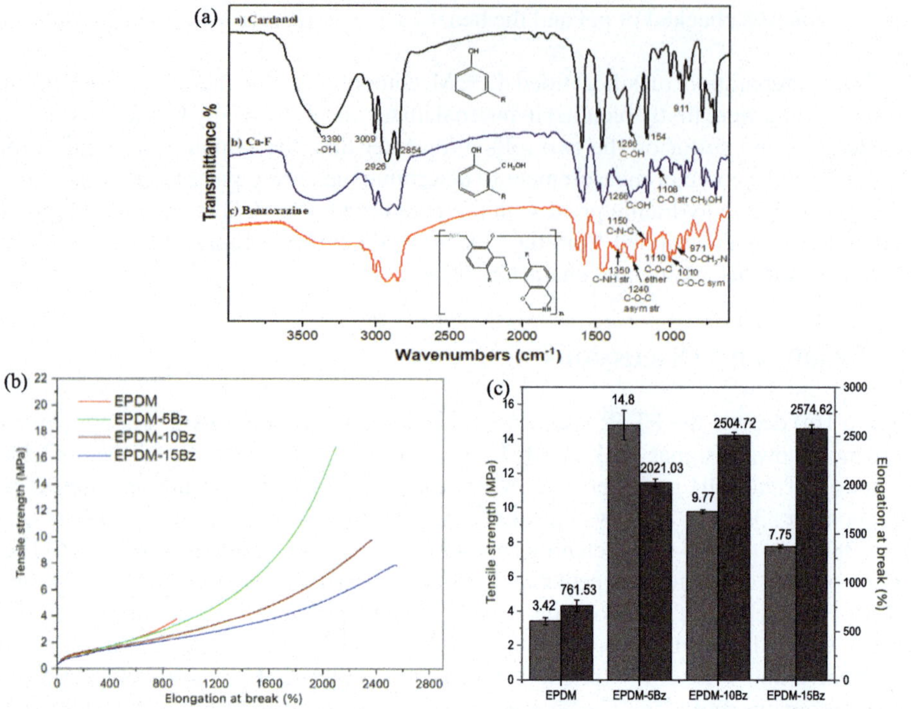

Fig. 1. (a) FT-IR spectra of Cardanol, Ca-F, BZ-CAFU, (b) Stress-strain curve of the compounds with different Ca-Bz contents, (c) Tensile strength and elongation at break of EPDM compounds at different Ca-Bz contents

sample because as the Ca-Bz content increases, only a portion interacts and binds with rubber, while another part distributes unevenly in the rubber network. Additionally, the long hydrocarbon chains of Ca-Bz contribute to the spatial effect, potentially hindering some crosslink reactions between rubber and sulfur.

Elongation at break also increases with increasing Ca-Bz content Fig. 1(c). Particularly, elongation at break increases from 761.5% for the control sample without Ca-Bz to 2504.7% for the sample containing 10 phr Bz. This is explained by the intertwining and cross-linking of long hydrocarbon chains of Ca-Bz in the vulcanized rubber network. In addition, the rubber chains can easily slide on the long hydrocarbon chains of Ca-Bz, enhancing the flexibility of the rubber molecular chains, thereby increasing the elongation at break.

In general based on Table 2, tensile strength and elongation at break of EPDM rubber slightly reduces after thermal aging. However, it is noticeable that for samples containing BZ-CAFU, the decrease in elongation at break is less pronounced compared to the sample without Ca-Bz. This can be attributed to the heat resistance of Ca-Bz resin network.

Based on the analysis of the tensile fractured surface of EPDM samples by SEM in Fig. 2, it can be observed the presence of silica particles on the sample surface in both cases. However, when Ca-Bz is added to the compound, there is an enhancement

Table 1. The mechanical properties of silica filled EPDM compouds with different Ca-Bz benzoxazine content

Sample	Hardness (Shore A)	Tear strength (N/mm)	Tensile strength (σ_u) (MPa)	M100 (MPa)	M300 (MPa)	Elongation at break (ε) (%)
EPDM	55	17.99 ± 1.60	3.42 ± 0.21	1.06 ± 0.01	1.56 ±. 0.02	761.5 ± 58.87
EPDM-5Bz	50	24.53 ± 0.60	14.8± 0.85	1.00 ± 0.01	1.53 ± 0.01	2021.0 ± 38.57
EPDM-10Bz	43	26.73 ± 2.0	9.77 ± 0.11	0.98 ± 0.02	1.30 ±0.03	2504.7 ± 34.5
EPDM-15Bz	38	23.61 ± 0.84	7.75 ± 0.10	0.921 ± 0.04	1.30 ± 0.02	2574.6 ± 42.74

Table 2. Tensile strength and elongation at break of EPDM compounds before and after thermal aging (at 100° C for 72 hours according to ASTM D573)

Sample	Tensile strength (σ_u) (a)	Tensile strength (σ_u) (MPa) after aging	Elongation at break (ε) (%)	Elongation at break (ε) (%) after aging
EPDM	3.42 ± 0.21	3.43 ± 0.23	761.5 ± 58.87	572.44 ± 22.77
EPDM-5Bz	14.8± 0.85	10.69 ± 0.77	2021.0 ± 38.57	1424. 95 ± 39.97
EPDM-10Bz	9.77 ± 0.11	12.82 ± 0.29	2504.7 ± 34.5	2304.74 ± 44.02
EPDM-15Bz	7.75 ± 0.10	10.17 ± 0.34	2574.6 ± 42.74	2148.16 ± 40,45

in interaction between polymer matrix and silica filler particles displayed by the better filler dispersion as well as bonding between the phases.

Figure 3 shows the thermal degradation and stability of the sample with and without the benzoxazine resin. It can be seen that the char yield and the decompostion of EPDM compounds with 5 phr Ca-Bz are higher the those of unmodified compounds. This results confirms the excellent heat resistance of benzoxazine which contributes to the better thermal properties of EPDM compounds.

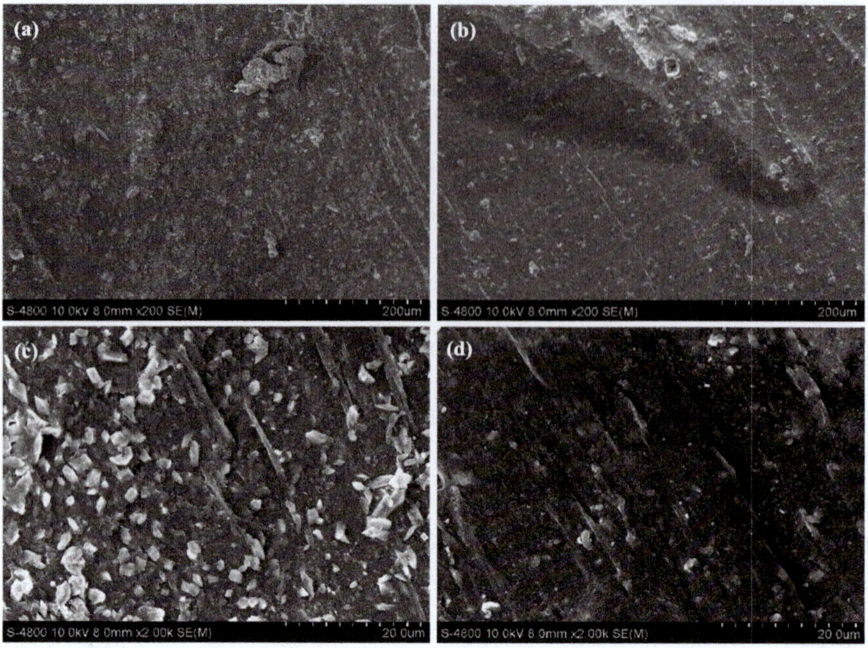

Fig. 2. SEM images of EPDM/silica compound (a & c) and EPDM/silica compound with 5phr Ca-Bz (b & d).

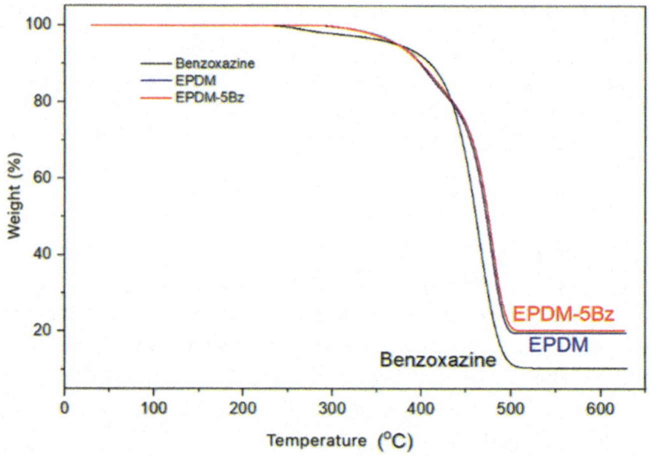

Fig. 3. Thermogravimetric analysis (TGA) results of Ca-Bz benzoxazine resin, EPDM/silica compound without and with 5 phr Ca-Bz.

4 Conclusion

This study introduces an approach to enhance the performance of silica filled EPDM compounds by using the efficient additive benzoxazine resin derived from inexpensive, environment friendly sources. The tensile strength, tear strength, and thermal aging resistance of EPDM samples increased with the addition of BZ-CAFU, especially at low Ca-Bz content. The interlocking network of Ca-Bz resin within the EPDM network migh responsible for the enhancement. Results from the tensile measurement, TGA and SEM of the cured compounds provides the possibility to employ benzoxazines precursors to reinforce rubber compounds.

Acknowledgments. This research is funded by Vietnam National University Ho Chi Minh City (VNU-HCM) under grant number C2023-20-17. We acknowledge the support of time and facilities from Ho Chi Minh City University of Technology (HCMUT), VNU-HCM for this study.

References

1. Hofmann, W.: Rubber Technology Handbook. Hanser Publishers, Munic Vienna, New York (1989)
2. Morton, M.: Rubber Technology, 3rd edn. Van Nostrand Reinhold, New York (1987)
3. Restrepo-Zapata, N.C., Osswald, T.A., Hernández-Ortiz, J.P.: Vulcanization of EPDM rubber compounds with and without blowing agents: identification of reaction events and TTT-diagram using DSC data. Polym. Eng. Sci. **55**, 2073–2088 (2015)
4. Su, J., Li, C.H.: Effect of carbon nanotube content on properties of ethylene propylene diene rubber/$CaCO_3$ carbon nanotube composites. Adv. Mater. Res. **1142**, 206–210 (2017)
5. Ashok, N., Balachandran, M.: Effect of nanoclay and nanosilica on carbon black reinforced EPDM/CIIR blends for nuclear applications. Mater. Res. Exp. **6**, 125364 (2020)
6. Kumar, S.R., Asseref, P., Dhanasekaran, J., Mohan, S.K.: A new approach with prepregs for reinforcing nitrile rubber with phenolic and Benzoxazine resins. RSC Adv. **4**(24), 12526–12533 (2014)
7. Ye, L., Zhang, Y., Wang, Z.: Mechanical properties and microstructure of acrylonitrile–butadiene rubber vulcanizates reinforced by in situ polymerized phenolic resin. J. Appl. Polym. Sci. **105**(6), 3853–3857 (2007)
8. Ye, L., Zhang, Y.: In-situ polymerized phenolic resin for reinforcement of chloroprene and ethylene–propylene rubber vulcanizates. J. Appl. Polym. Sci. **111**(3), 1177–1184 (2009)
9. Yagci, Y., Kiskan, B., Ghosh, N.N.: Recent advancement on polybenzoxazine—a newly developed high performance thermoset. J. Polym. Sci. Part A: Polym. Chem. **47**(21), 5565–5576 (2009)
10. Takeichi, T., Kawauchi, T., Agag, T.: High performance polybenzoxazines as a novel type of phenolic resin. Polym. J. **40**(12), 1121–1131 (2008)
11. Trejo-Machin, A., et al.: Synthesis of novel Benzoxazines containing sulfur and their application in rubber compounds. Polymers **13**(8), 1262 (2021)
12. Voirin, C., et al.: Functionalization of cardanol: towards biobased polymers and additives. Polym. Chem. **5**(9), 3142–3162 (2014)
13. Viet, C.X., Trinh, N.H., Uy Lan, D.N., Takeichi, T.: Synthesis and characterization of monofunctional benzoxazine from cardanol. Vietnam J. Sci. Technol. **55**(1B), 160–168 (2017)

Effect of Cesium Tungsten Bronze Nanoparticle Concentration on Optical and Thermal Properties of Cs_xWO_3/PMMA Composite

Luu Tuan Anh[1,3](✉), Tran Phuc Hoang[1,3], Truong Vinh Phuoc[1,3], and Nguyen Thi Minh Nguyet[2,3]

[1] Faculty of Materials Technology, Ho Chi Minh City University of Technology (HCMUT), 268 Ly Thuong Kiet Str., Dist. 10, Ho Chi Minh City, Vietnam
luutuananh@hcmut.edu.vn

[2] VNU-HCM Key Laboratory for Material Technologies, Ho Chi Minh City University of Technology (HCMUT), 268 Ly Thuong Kiet Street, Ward 14, District 10, Ho Chi Minh City, Vietnam

[3] Vietnam National University Ho Chi Minh City, Linh Trung Ward, Thu Duc Dist., Ho Chi Minh City, Vietnam

Abstract. Cesium tungsten bronze (Cs_xWO_3) materials, which have strong near-infrared (NIR) absorption and excellent photothermal conversion properties, have attracted worldwide attention for their applications in energy-saving window glasses. In this work, Cs_xWO_3 particles modified with polyvinylpyrrolidone (PVP) as a surfactant are successfully synthesized by a solvothermal method using cheap sodium tungstate ($Na_2WO_4.2H_2O$) and cesium chloride (CsCl) as raw materials. The results indicate that the Cs_xWO_3 content in the coatings affects the NIR-shielding rate and the optical band gap of the coatings. The NIR-shielding rate of the prepared Cs_xWO_3/PMMA coatings attains more than 70% when visible light transmittance reaches more than 72%. Results also demonstrate the excellent heat insulation ability of the Cs_xWO_3/PMMA coatings through the simulated thermal insulation test.

Keywords: Cs_xWO_3 · Solvothermal · Near-infrared shielding · Photothermal conversion · Thermal insulation

1 Introduction

About 98% of the solar radiation incident on the Earth's surface is distributed in the wavelength range of 200–2500 nm, of which ultraviolet radiation (UV - from 200 nm to 380 nm) accounts for about 5%, visible light (Vis - from 380 nm to 780 nm) accounts for about 43% and near-infrared radiation (NIR - from 780 nm to 2500 nm) accounts for about 52% [1]. Most of the solar energy that supplies the Earth's surface is in the form of infrared radiation, which then passes through the glass surface into buildings, raising the temperature in the indoor space and causing the aging and deterioration of materials.

N. R. Abdul Razak et al. (Eds.): ECOOP 1996, SPPHY 418, pp. 220–226, 2025.
https://doi.org/10.1007/978-981-96-2871-1_31

The development of thermal insulating window glasses has garnered significant interest recently in reducing building energy consumption. Low-E glass, which has been around for over 40 years, is widely used as transparent heat-shielding glass in commercial applications [2]. The Low-E glass typically uses a multilayer silver coating with a low solar thermal emissivity and relatively high visible light transmittance. The thermal insulation mechanism of Low-E glasses is based on the ability to strongly reflect infrared light from the sun, which contributes to the urban heat island effect and heat reflection-induced degradation of surrounding building structures. Many thermal insulating coatings using heat-absorbing particles have been fabricated, such as antimony-doped tin oxides (ATO) [3], indium tin oxide (ITO) [4], rare-earth hexaborides (RB6, where R is a rare-earth element such as La, Pr, Ce, etc.) [5, 6], etc, ... to address the urban heat island phenomenon caused by heat-reflective glasses. However, these particles have shortcomings that limit their practical use in energy-saving materials [7]. Both ATO and ITO exhibit relatively high transmittance in the wavelength range of 780 – 1500 nm, so it can be seen that they can only shield the NIR with wavelength from 1500 – 2500 nm. Rare-earth hexaboride compounds have NIR blocking capability, but their NIR blocking efficiency starts to degrade at 1200 nm.

In this work, Cs_xWO_3 nanoparticles are synthesized using PVP as a nontoxic polymeric surfactant by solvothermal method. Then, we utilize PMMA as the matrix and as-prepared Cs_xWO_3 powder as a NIR-shielding additive to prepare transparent heat insulation Cs_xWO_3/PMMA composite films by blending method. PVP is a nonionic surfactant exhibiting hydrophilic and lipophilic properties due to the high polar lactam group in pyrrolidone and the pyrrole ring with long chains containing nonpolar methylene, respectively [8]. This surfactant plays a significant role in the morphology regulation of nanocrystals and can control crystal growth and particle size. The effect of different Cs_xWO_3 contents on the optical and thermal properties of composite films is evaluated.

2 Experimental Procedures

2.1 Synthesis of Cs_xWO_3 Powders

Cs_xWO_3 powders were prepared by solvothermal reaction using sodium tungstate ($Na_2WO_4.H_2O$) as a W source and cesium chloride (CsCl) as a Cs source. First, 2.474g $Na_2WO_4.2H_2O$ was added into 30 ml deionized water under magnetic stirring to obtain a uniform sodium tungstate aqueous solution with a 0,25 mol/L concentration. Then, a certain amount of 2 mol/L diluted hydrochloric acid was added into a solution of sodium tungstate dropwise under continuous magnetic stirring to create a yellow flocculent gel-like tungstic acid ($WO_3.H_2O$) suspension. The suspension was then filtrated, washed with water five times, and redispersed in 35 ml ethylene glycol ($C_2H_6O_2$) acting as the solvent and reducing agent to obtain a gel-like tungstic acid/ethylene glycol dispersion solution. After that, 0.421g CsCl was dissolved in 5 ml deionized water to obtain a CsCl aqueous solution. The precursor solution in solvothermal synthesis was prepared by mixing tungstic acid/ethylene glycol dispersion, CsCl aqueous solution, and 0,5g PVP under 30-min magnetic stirring. Next, the precursor solution was transferred into a Teflon-lined autoclave with 100 mL internal volume, and a solvothermal reaction was conducted at 200 °C for 24 h. The obtained dark-blue precipitate was then collected by centrifugation and washing with deionized water and ethanol three times, respectively. Finally, Cs_xWO_3 powders were obtained by drying the dark-blue solid form of the precipitate at 80 °C for 4 h.

2.2 Fabrication of Cs_xWO_3/PMMA Coatings

Cs_xWO_3/PMMA coatings were fabricated by a simple blending process using PMMA as a film-forming agent, as-prepared Cs_xWO_3 particles as NIR-shielding additives, and propylene glycol methyl ether acetate (PGMEA) as solvent. First, PMMA powders were dissolved in PGMEA to form a homogeneous dispersion with 8.5 wt%. After that, a certain amount of Cs_xWO_3 powders was added into the above dispersion with continuous magnetic stirring for 1 h to obtain a uniform coating solution. Finally, the Cs_xWO_3/PMMA coatings were prepared on the glass substrate by a dip-coating method (for microscope slide) and flow-coating method (for large-sized glass panels).

2.3 Characterization

The phase composition and crystallographic structure of the Cs_xWO_3 particles were determined by X-ray diffraction analysis (XRD, PANalytical Empyrean, England) with Cu Kα radiation with a wavelength of 1.540598 Å. The morphology of the Cs_xWO_3 particles was investigated by scanning electron microscopy (SEM, JSM - IT200, Japan). Energy dispersive X-ray spectroscopy (EDS) was used to identify the elemental composition. The size distribution of Cs_xWO_3 particles in the solution and the dispersibility in the solvent were characterized by dynamic light scattering analysis (DLS, Horiba SZ-100, Japan). The transmittance spectra and absorbance spectra of the as-prepared films were characterized by UV–Vis-NIR spectrometer (Agilent Cary 6000i, USA). Heat-insulation performance was investigated by measuring the change in inner temperature of the simulated thermal insulation box in real time using an electronic thermometer with one-sided window glass being irradiated by a 300 W infrared lamp. Thermographic image analysis using a thermal temperature gun (Fluke TiS40, USA) also evaluated temperature changes on the coating surface.

Based on UV-Vis-NIR transmittance measurements, the NIR-shielding ability of composite films can be quantitatively assessed using the SETS (solar energy transmittance selectivity) value, calculated according to Eq. (1) [9]:

$$SETS = \frac{1}{2}\left(1 + \frac{\int_{380}^{780} E(\lambda)T(\lambda)d\lambda}{\int_{380}^{780} E(\lambda)} - \frac{\int_{780}^{2500} E(\lambda)T(\lambda)d\lambda}{\int_{780}^{2500} E(\lambda)}\right) \tag{1}$$

In this equation, E(λ) is the solar irradiance spectrum. T(λ) is the transmittance spectrum. SETS represents the solar energy transmittance deviation. The value of SETS ranges from 0 to 1. The minimal value of ‘0’ represents a material with visible or ultraviolet light-shielding ability. The coating with no light absorption performance is represented by the middle value of “0.5”. A higher SETS value indicates that more visible light is allowed to go through while a greater amount of NIR light is blocked, making the material more appropriate for use in NIR-shielding films.

3 Results and Discussions

Figure 1a presents the XRD pattern of the Cs_xWO_3 nanocrystal prepared by solvothermal method. The pattern exhibits characteristic peaks with strong intensities at 2θ values of approximately 23.55^0, 27.43^0, 27.96^0, and 33.91^0, corresponding to the (002), (102), (200), and (112) diffraction planes, respectively. These peaks agreed well with the hexagonal diffraction peaks of $Cs_{0.32}WO_3$ (JCPDS 83–1334) with the space group of P63/mcm. Additionally, the pattern containing those different peaks indicates that the Cs_xWO_3 powder is polycrystalline, and almost no diffraction peak from impurities is observed, indicating that the as-prepared Cs_xWO_3 sample has relatively high purity. The average crystallite size of $Cs_{0.32}WO_3$ sample is determined using the Williamson-Hall (W-H) method given by the relation (2) [10]:

$$\beta_{hkl}\cos\theta = \frac{k\lambda}{D} + 4\varepsilon\sin\theta \tag{2}$$

where λ is the wavelength of the X-ray wavelength, θ is Bragg angle, D is the average crystallite size, β is the full width at half-maximum (FWHM) of the Bragg peak, and ε is the lattice strain. The average crystallite size of the Cs_xWO_3 sample is around 11.4 nm.

Figure 1b shows the SEM image of the as-prepared Cs_xWO_3 sample. It can be seen that the sample mainly consists of uniformly distributed nanorod particles with average size of about 97.41 nm in length and 45.18 nm in diameter. The rod shape of Cs_xWO_3 nanoparticles is formed due to the role of the capping agent of PVP surfactant. The capping effect occurs mainly during the evolution of crystalline seeds. The PVP ligands can be chemically adsorbed onto the specific surface of Cs_xWO_3 crystals, inhibit the crystal growth on that surface, and promote the crystal growth of the other surfaces in the remaining dimension. The chemical composition analysis based on the EDS spectrum in Fig. 1c and the element mapping in Fig. 1d indicates that uniformly distributed Cs, W, and O elements exist in the selected area of the sample, and there are no other elements except Cs, W, and O in the samples. The atomic ratio of Cs/W is 0.3197.

The FTIR measurement is conducted to detect the various characteristic functional groups associated with the synthesized nanoparticles Cs_xWO_3 and clarify whether PVP is completely decomposed during solvothermal reaction. The FTIR spectrum was recorded in the range of 400–4000 cm^{-1} as shown in Fig. 1e. It is inferred that the as-prepared Cs_xWO_3 sample has absorption peaks at wavenumbers of 3421 cm^{-1}, 2925 cm^{-1}, 2846 cm^{-1}, 1648 cm^{-1}, 964 cm^{-1} and 618 cm^{-1}. The band around 618 cm^{-1} is related to the stretching vibration for O–W–O in WO_3, and the small peak at 964 cm^{-1} is assigned as the stretching vibration of short W = O bonds of $WO_3.H_2O$. The absorption peak at 1648 cm^{-1} corresponds to the stretching vibration of pyrrolidone C = O from the lactam group in the side chains of PVP. Other small peaks at 1284 cm^{-1}, 1421 cm^{-1}, and 1458 cm^{-1} are due to the C–N stretching vibrations and the $-CH_2$ absorption of PVP, and the absorption peak at 1371 cm^{-1} is due to the C–H bond in PVP. Moreover, absorption bands at 2846 cm^{-1} and 2925 cm^{-1} are ascribed to the symmetric stretching vibrations of CH_2 of the side chains of PVP and asymmetric stretching vibrations of CH_2 of the pyrrole ring, respectively. A large and broad peak located at 3421 cm^{-1} is due to –OH symmetric stretching.

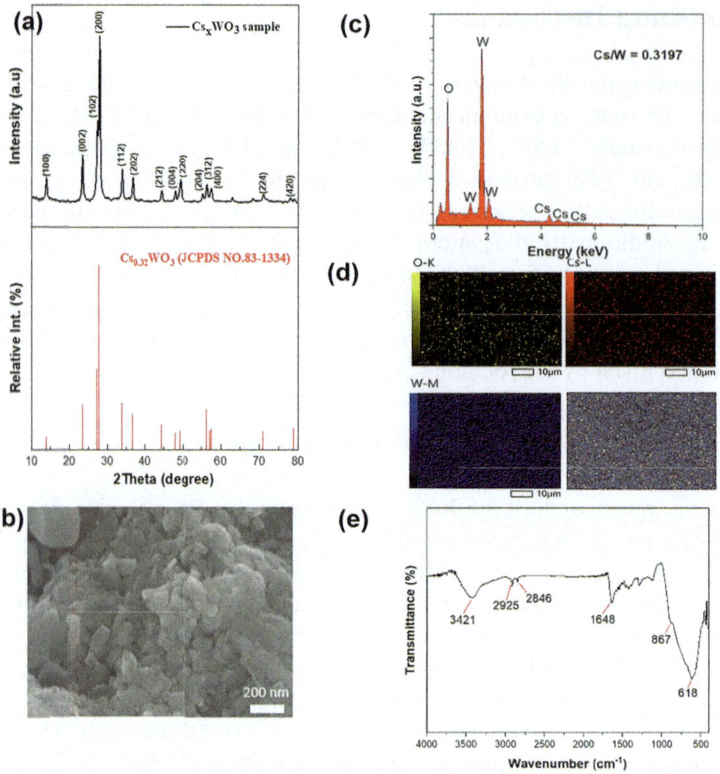

Fig. 1. (**a**) XRD pattern, (**b**) SEM image, (**c**) EDS spectrum, (**d**) Elemental mapping, (**e**) FTIR spectrum of the Cs_xWO_3 nanoparticles synthesized by solvothermal method with PVP surfactant

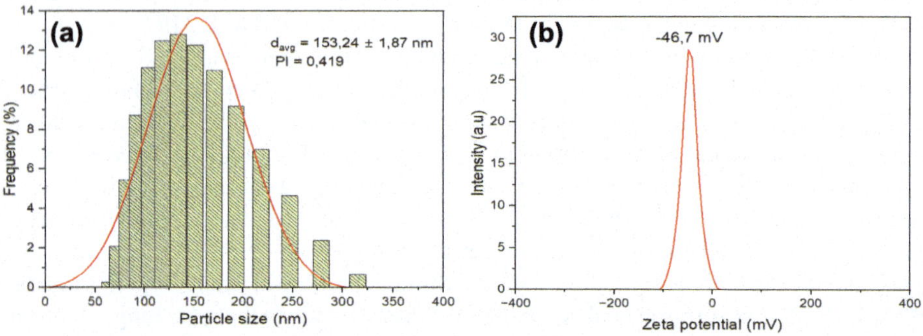

Fig. 2. (**a**) Particle size distribution (DLS), (**b**) Zeta potential distribution of Cs_xWO_3 nanoparticles in PGMEA dispersant

The average particle size (hydrodynamic size) of Cs_xWO_3 particles dispersed in PGMEA solvent was measured using the DLS method, and the results are shown in

Fig. 2a. The particle size distribution indicates that the average particle size of the synthesized Cs_xWO_3 in the dispersed system is about 153.24 ± 1.87 nm, corresponding to the particle size value determined at the peak of the Gaussian corrected curve. The stability of the Cs_xWO_3 particles in PGMEA dispersant was evaluated by the zeta potential measurement based on the dynamic light scattering method. The average zeta potential value of the dispersion system is −46.7 mV (as shown in Fig. 2b). This result demonstrates that the dispersion system has high stability and the Cs_xWO_3 particles are well dispersed in the PGMEA solvent.

Table 1. The calculated optical indices of glasses coated with composite films

Sample	T_{Vis}[a]	T_{NIR}[b]	SETS
Pure PMMA	0.93	0.98	0.477
1 wt%	0.81	0.65	0.581
3 wt%	0.70	0.32	0.689
5 wt%	0.65	0.18	0.735
7 wt%	0.48	0.01	0.733

[a]The average visible light transmittance (380–780 nm): $T_{vis} = \frac{\int_{380}^{780} T(\lambda)d(\lambda)}{(780-380)\times 100}$

[b]The average NIR transmittance (780–2500 nm): $T_{NIR} = \frac{\int_{780}^{2500} T(\lambda)d(\lambda)}{(2500-780)\times 100}$

In order to compare the optical properties of these coating films quantitatively, some optical indices are listed in Table 1, including the average visible light transmittance (T_{Vis}), the average NIR transmittance (T_{NIR}), and the SETS value (Eq. (1)). It can be seen that the SETS values of the Cs_xWO_3/PMMA film with Cs_xWO_3 content of 5 wt% and 7 wt% is 0.735 and 0.733, respectively, which is much higher than the SETS value of the pure PMMA film (0.477).

4 Conclusions

In this work, Cs_xWO_3 nanoparticles with excellent NIR-shielding ability were successfully synthesized by solvothermal process, in which the rod-like shape of particles was controlled via PVP structure-directing surfactant. Near-infrared shielding polymer nanocomposite Cs_xWO_3/PMMA films were fabricated using PMMA as the matrix dispersed with as-prepared Cs_xWO_3 nanoparticles. The influences of the Cs_xWO_3 content of polymer composite films on the optical and thermal properties were further studied. It was found that the Cs_xWO_3/PMMA composite film at a concentration of 7 wt% of Cs_xWO_3 particles exhibits the most efficient NIR-shielding performance (with about 100% NIR-shielding rate) as well as a comparatively high visible transmittance (76.6%) and a qualified SETS value (0.733). The measurement of temperature control effect in the model box indicated that the 7 wt% Cs_xWO_3/PMMA-coated glass significantly reduced the inner temperature compared to the pure PMMA-coated glass.

Acknowledgments. We acknowledge the support of time and facilities from Ho Chi Minh City University of Technology (HCMUT), VNU-HCM for this study.

References

1. Su, K., Mao, Z., Yang, Z., Zhang, J.: J. Vinyl Addit. Technol. **45**(4), 356–366 (2021)
2. Karlsson, J., Roos, A.: Thin solid films, **373**, 345–348 (2001)
3. Gao, R., Zhang, J., Liu, H.: Advanced Functional Materials **28**(5) (2018). Singapore: Springer
4. Pattathil, P., Giannuzzi, R., Manca, M.: Nano Energy **29**, 242–251 (2016)
5. Kimura, S., Nanba, T., Kunii, S., Suzuki, T., Kasuya, T.: Sol. State Commun. **75**(4), 717–720 (1990)
6. Takeda, H., Kuno, H., Adachi, K.: J. Am. Ceram. Soc. **91**(8), 2897–2902 (2008)
7. Chao, L., Bao, L., Wei, W., Tegus, O.: Solar Ener. **180**, 10–17 (2019)
8. T. Song et al, Nanoscale, 3895–3910 (2021)
9. Y. Wang et al, Nanoscale Adv, 3177–3183 (2021)
10. W. H. Hall, Proc. Phy. Soc., 741–743 (1949)

Effect of Doping Elements on the Structure of Argyrodite-Type Li_6PS_5Cl Solid Electrolyte Using PXRD

Phung Khoi Nguyen[1,2], Nguyen Manh Duc[1,2], Nguyen Thi Thanh Mai[1,2], and Nguyen Huu Huy Phuc[1,2(✉)]

[1] Faculty of Materials Technology, Ho Chi Minh City University of Technology (HCMUT), 268 Ly Thuong Kiet Str., Dist. 10, Ho Chi Minh City, Vietnam
nhhphuc@hcmut.edu.vn

[2] Vietnam National University Ho Chi Minh City, Linh Trung Ward, Thu Duc Dist., Ho Chi Minh City, Vietnam

Abstract. This paper reported the preparation of Argyrodite-type Li_6PS_5Cl solid electrolytes doped with multivalent cations, using a mechanical milling synthesis method. The purpose of this treatment is to increase the ionic conductivity of the solid electrolyte. The synthesized samples were analyzed by X-ray diffraction (XRD) and electrochemical impedance spectroscopy (EIS) methods. From the data obtained, combined with the formulas and calculation equations, we can analyze the change in the crystal lattice of solid electrolytes type Argyrodite Li_6PS_5Cl when doped, and clarify movement of Li ions in these solid electrolytes.

Keywords: Argyrodite Li_6PS_5Cl · Multivalent Cations

1 Introduction

All solid-state batteries (ASSBs) are considered the next generation of energy battery technology and will compete directly with lithium batteries for dominance in the current market. Solid electrolytes with high ionic conductivity at 25 °C are important components of solid-state Li-ion batteries [1]. Because the S^{2-} ion has larger ionic radius and higher polarity than the O^{2-} ion, sulfide-based solid electrolytes typically show higher ionic conductivity than most oxide-based solid electrolytes. Furthermore, sulfide-based electrolytes typically have lower melting temperatures and require lower synthesis temperatures than oxides and phosphates. In current solid-state battery research, solid-state Li-ion batteries using sulfide-based solid electrolytes are a potential candidate because of their high energy density, thermal stability, and easy cell design [2]. A number of sulfide-based electrolytes have been developed that have Li^+ conductivities equivalent to or even higher than many liquid electrolytes. Among them, Argyrodite-type Li_6PS_5Cl is a promising sulfide-based solid electrolyte material due to its low cost and high Li^+ conductivity at room temperature [3].

N. R. Abdul Razak et al. (Eds.): ECOOP 1996, SPPHY 418, pp. 227–234, 2025.
https://doi.org/10.1007/978-981-96-2871-1_32

Argyrodite-type Li_6PS_5Cl can have its ionic conductivity improved by lithium vacancy method, or doping method. The method used in this report is to dope multivalent cations in place of Li^+. Argyrodite-type Li_6PS_5Cl is synthesized by mechanical milling synthesis method. This technique allows the production of small-sized powder materials by grinding the materials, based on the impact of the materials with hardened steel balls that are centrifuged in a closed chamber at high speeds. This process takes place in a closed environment, which can be evacuated and filled with inert gases to ensure safety and performance. The component powders are ground using the above method at high rotation speed in an inert gas environment. The crystal structure of the solid electrolyte samples was characterized by XRD.

2 Experimental Procedure

Chemicals used include: Li_2S (99.9%), P_2S_5 (99%), LiCl (99.9%) combined with one of the following substances (depending on the type of cation that needs to be doped): $CaCl_2$ (99.9%), $BaCl_2$ (99.9%), $MgCl_2$ (99.9%), YCl_3 (99.9%), $AlCl_3$ (99.9%), $ZnCl_2$ (99.9%). These chemicals were obtained from Macklin and used as received without further purification.

All synthesis experiments were performed in a dry Argon atmosphere. Starting ingredients were put into the mortar with an appropriate amount and mixed for about 15 min. This mixture was then transferred into a zirconia pot (45 ml), along with about 30 g of zirconia balls (diameter 5 mm). The pot was planetary ball milled at 550 rpm for 20 h, using a FRITSCH Pulverisette device. Then, the samples were recovered and characterized without a further heat treatment step.

After synthesis, the prepared samples were loaded into an air-tide sample holder. The crystal structure of prepared samples was characterized by an XRD measuring device (X8, Bruker). Data were analyzed using Origin software.

For resistivity measurement, the samples were prepared by uniaxially cold pressing the powder at a pressure of 330 MPa to form pellets, with diameter of 10 mm [4]. DC conductivities were carried out using blocking electrodes (the stainless steel (SUS) rods were employed as the electrodes). The prepared cells were then applied one voltage (DC) and the current was measured to determine the dominant mobile ions in the sample. AC impedance spectroscopy measurements were carried out using a potentiostat (PGSTAT302N, Autolab, Herisau, Switzerland) in the frequency range from 10 Hz to 9 MHz.

3 Results and Discussions

Figure 1(a) and 1(b) showed the XRD patterns of $Li_{6-2x}Ca_xPS_5Cl$ and Li_6PS_5Cl solid electrolyte doped with Ba, Y and Zn. All the peaks could be assigned to argyrodite phase (JCPDS 34-0688). To investigate the effect of multivalent cation substitution on the crystal structure of Li_6PS_5Cl, the lattice constant, plane distance and micro strain were calculated using the XRD data. The observed peaks were indexed to [111], [200],

[220], [311], [222], [422], [511], [440], [600] and [620] planes. The lattice parameters were obtained at each diffraction angle θ, using the Bragg equation [5].

$$n\lambda = 2d_{hkl}\sin\theta \;\&\; d_{hkl} = \frac{a}{\sqrt{h^2 + k^2 + l^2}} \rightarrow a_{(Bragg)} = \frac{\lambda}{2\sin\theta}\sqrt{h^2 + k^2 + l^2}$$

The Nelson–Riley equation was employed to determine lattice parameter a_0

$$a_{(Nelson\text{ - }Riley)} \approx \frac{1}{2}\left[\frac{\cos\theta^2}{\sin\theta} + \frac{\cos\theta^2}{\theta}\right]$$

a_0 was determined from the linear fit equation, which was derived from a plot with $a_{(Nelson\text{-}Riley)}$ as the *x*-axis, and $a_{(Bragg)}$ as the *y*-axis.

The Williamson–Hall (WH) analysis is a simple method to estimate the lattice strain (ε) [6]

$$\beta_{hkl}\cos\theta = 4\varepsilon\sin\theta + \frac{k\lambda}{d}$$

where β_{hkl} is the full width at half maximum (FWHM) of the diffraction peak and θ is the Bragg angle. The equation above has the form y = ax + b, so the value of ε can be derived from the slope value of the regression line which was obtained by plotting $\beta_{hkl}\cos\theta$ against $\sin\theta$ ($\varepsilon = a/4$).

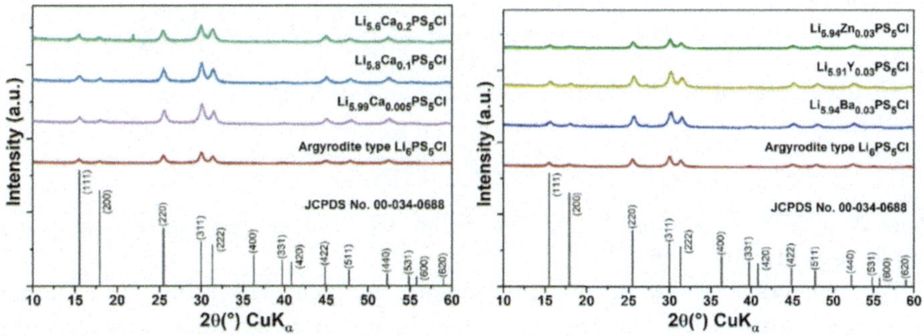

Fig. 1. (a) XRD patterns of Li6-2xCaxPS5Cl solid electrolytes, (b) XRD patterns of Li6PS5Cl solid electrolyte doped with Ba, Y and Zn.

The values of lattice constant a0, plane distance (d220, d311, d222) and micro strain (ε) were shown in Table 1. The variation in these values can be explained in terms of attractive and repulsive interactions in the crystal lattice. In crystals, cations are surrounded by anions and vice versa; Therefore, electrostatic interactions among oppositely charged ions help strengthen the crystal structure.

Substitution of Li ions with multivalent cation resulted in the formation of vacancies. The attraction is due to the cation interacting with surrounding oppositely charged anions. Repulsion occurs in holes, when anions of the same sign push each other away. The lattice constant a_0 of Li_6PS_5Cl was 9.84 Å and in good agreement with the reported value 9.8397 Å of Li_6PS_5Cl prepared by mechanochemical method [7].

Table 1. The structural parameters of the prepared samples

Samples	Lattice parameters (Å)	d_{220} (Å)	d_{311} (Å)	d_{222} (Å)	Micro strain
Li_6PS_5Cl	9.840	3.488	2.974	2.851	0.003275
$Li_{5.99}Ca_{0.005}PS_5Cl$	9.849	3.475	2.966	2.843	0.003525
$Li_{5.8}Ca_{0.1}PS_5Cl$	9.838	3.482	2.970	2.846	0.003675
$Li_{5.6}Ca_{0.2}PS_5Cl$	9.823	3.493	2.978	2.851	0.004
$Li_{5.94}Ba_{0.03}PS_5Cl$	9.836	3.466	2.959	2.837	0.00345
$Li_{5.91}Y_{0.03}PS_5Cl$	9.842	3.464	2.958	2.836	0.00477
$Li_{5.94}Zn_{0.03}PS_5Cl$	9.846	3.477	2.966	2.845	0.00467

Most of the values of plane distance (d_{220}, d_{311}, d_{222}) of the substituted samples were smaller than Li_6PS_5Cl. The micro strain represents the displacement of unit cells about their normal positions. All of the prepared samples showed a positive value of lattice strain and the substituted samples exhibited higher strain than Li_6PS_5Cl. Therefore, those results suggested that multivalence cation was successfully incorporated into the crystal structure of Li_6PS_5Cl.

The complex ion conductivity (σ^*) with 2 parts: real (σ') and imaginary (σ''), is determined from the complex impedance (Z^*):

$$\sigma^* = \sigma' + j\sigma'' = \frac{1}{AZ'} + j\frac{1}{AZ''} \sigma' = \frac{1}{Z'} \cdot \frac{1}{A} \& \sigma'' = \frac{1}{Z''} \cdot \frac{1}{A}$$

So, to calculate ionic conductivity, we have the formula:

$$\sigma' = \frac{1}{Z'} \cdot \frac{0.06}{0.785}$$

where A = 0.785 (cm^2) and l = 0.06 (cm) are the contact area and thickness of the sample, respectively.

The ionic conductivities of prepared samples are compared in Fig. 2(c). The ionic conductivity at room temperature of the doped substances increases compared to the original Argyrodite sample. Among solid electrolytes, samples doped with aluminum (Al) have the highest ionic conductivity (at 25 °C).

The frequency dependence of the real part of conductivity (σ') of Li_6PS_5Cl and $Li_{5.91}Al_{0.03}PS_5Cl$ obtained from 10 Hz to 9 MHz at different temperatures are shown in Fig. 2(a) and 2(b), respectively. Those are also called conductivity isotherms. When the temperature increases, the ionic conductivity of the sample also increases. The isotherms of all samples consist of three regions: the electrode polarization region of the material at low frequency region, the region of frequency-independent conductance (σ_{DC}) at intermediate frequency region, and the current region induced current (σ_{AC}) at high frequency region. The isotherms are analyzed using Jonscher power law to understand the ion dynamics of the prepared electrolytes. The conductivity at the intermediate frequency region followed the Jonscher power law equation:

$$\sigma = \sigma_{DC} + A\omega^k \rightarrow \sigma - \sigma_{DC} = A\omega^k$$

$$\rightarrow \log_{10}(\sigma - \sigma_{DC}) = k \log_{10}\omega + \log_{10}A$$

where σ_{DC} is the DC conductivity, A is a pre-factor, and k is the frequency exponent in the range $0 < n < 1$ [8]. The equation above has the form $y = ax + b$. So values of k and A at each temperature can be obtained by fitting the conductivity spectra using Eq. (2). The values of A and k are shown in Table 2.

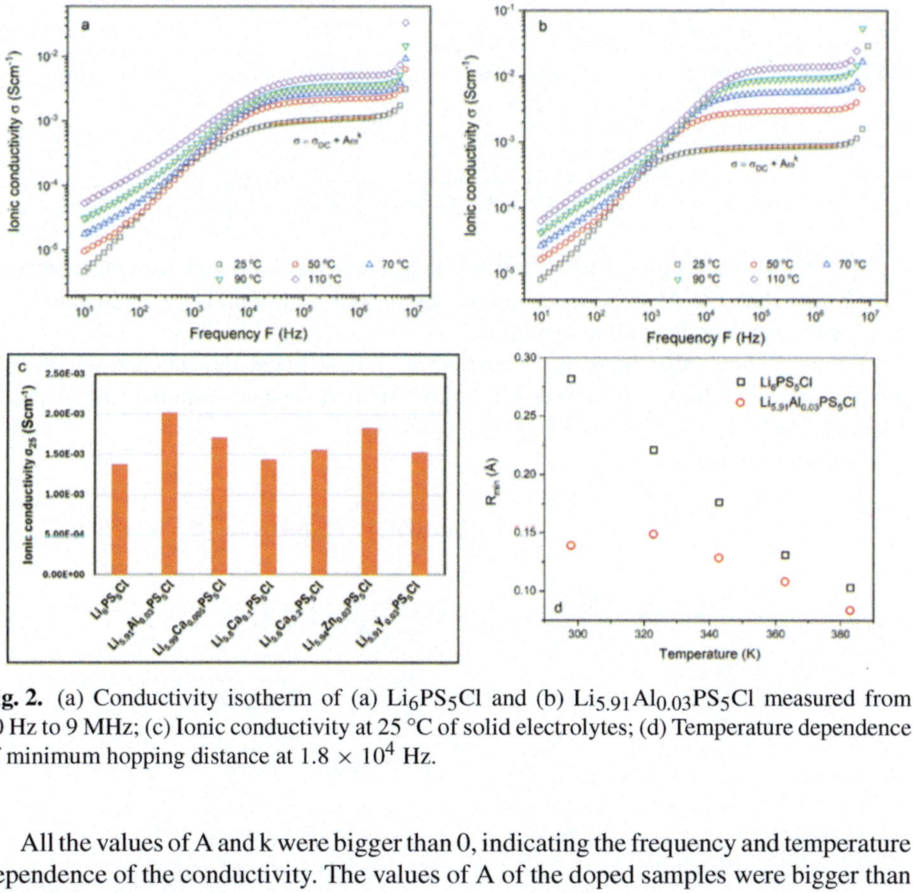

Fig. 2. (a) Conductivity isotherm of (a) Li_6PS_5Cl and (b) $Li_{5.91}Al_{0.03}PS_5Cl$ measured from 10 Hz to 9 MHz; (c) Ionic conductivity at 25 °C of solid electrolytes; (d) Temperature dependence of minimum hopping distance at 1.8×10^4 Hz.

All the values of A and k were bigger than 0, indicating the frequency and temperature dependence of the conductivity. The values of A of the doped samples were bigger than that of Li_6PS_5Cl, suggesting that multivalence cation doping enhanced the frequency dependence conductivity.

The dependence of the exponential coefficient k on temperature can be used to determine the origin of the conduction mechanism. Table 2 shows that k decreases as temperature increases. This behavior proposed that correlated barrier hopping (CBH) conduction is the main conduction mechanism [9]. In the CBH model, charge carriers move between localized sites, over the Coulomb barrier separating two defect sites. The minimum (shortest) hopping distance is calculated using the following equation:

$$R_{min} = \frac{2e^2}{\pi\, \varepsilon_0\, \varepsilon\, W_m} = \frac{2e^2}{\pi\left(55.263e^2\right)\varepsilon\, \frac{6k_B T}{1-k}}$$

Table 2. Values of A and k according to temperature levels

T (K)	Li_6PS_5Cl		$Li_{5.91}Al_{0.03}PS_5Cl$	
	k	A	k	A
298	0.6355	1.30497E-08	0.6105	1.70216E-08
323	0.6223	3.24788E-08	0.4464	5.22516E-07
343	0.5841	1.03395E-07	0.4252	1.5787E-06
363	0.5368	2.82553E-07	0.3906	6.85804E-06
383	0.4475	3.5148E-06	0.3768	1.3605E-05

$$= \frac{2}{\pi\ 55.263\ \varepsilon\ \frac{6\times 8.617\times 10^{-5}\times T}{1-k}}\ (\mu m)$$

Figure 2(d) shows that at 1.8×10^4 Hz, R_{min} decreases with increasing temperature. Along with that, the Al ion-doped sample has shorter minimum hopping distance than the Argyrodite sample at all temperatures.

Figure 3 shows that log(σ_{DC}) satisfied an almost linear dependence on inversed temperature; therefore, it followed Arrhenius equation. The activation energy $E_{a,DC}$ of Li_6PS_5Cl and $Li_{5.91}Al_{0.03}PS_5Cl$ was then calculated:

Arrhenius equation:

$$\sigma DC = \sigma 0\, e^{\frac{-E_a}{RT}} \rightarrow \ln(\sigma DC) = \ln(\sigma 0) - \frac{E_a}{RT}$$

$$\ln(x) = \frac{\log(x)}{\log(e)} = \log(x)\frac{1}{\log(e)} = \log(x) \times 2.3$$

$$\rightarrow \log(\sigma DC) \times 2.3 = \log(\sigma 0) \times 2.3 - \frac{E_a}{RT} = \rightarrow \log(\sigma DC)$$

$$= -\frac{E_a}{R \times 2,3} \times \frac{1}{T} + \log(\sigma 0) \rightarrow E_{a,DC} = -(a \times 2.3 \times R)$$

Li_6PS_5Cl: $E_{a,\,DC} = -(-0.75 \times 2.3 \times 8.32) = 15\ (kJ\ mol^{-1})$
$Li_{5.91}Al_{0.03}PS_5Cl$: $E_{a,\,DC} = -(-1.5 \times 2.3 \times 8.32) = 29\ (kJ\ mol^{-1})$

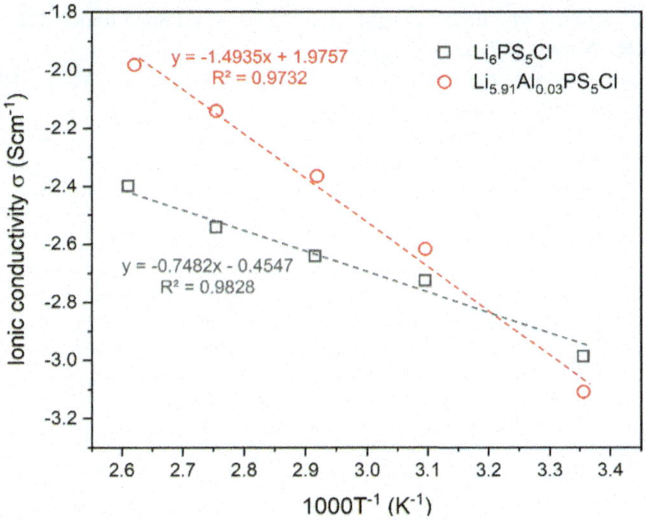

Fig. 3. Temperature dependence of ionic conductivity of Li_6PS_5Cl and $Li_{5.91}Al_{0.03}PS_5Cl$.

4 Conclusion

Argyrodite-type Li_6PS_5Cl solid electrolyte doped with multivalent cations (Mg^{2+}, Ba^{2+}, Zn^{2+}, Al^{3+}, Y^{3+}) was successfully prepared by mechanochemical synthesis. The lattice constant (a_0), plane distance (d_{hkl}) and micro strain (ε) showed that the doping elements were incorporated into the crystal structure of Li_6PS_5Cl. The plane distances (d_{220}, d_{311}, d_{222}) of most doped samples are smaller than Li_6PS_5Cl. All samples have positive micro strain, and most doped samples have higher strain than Li_6PS_5Cl. The doped samples had higher ionic conductivity than Li_6PS_5Cl at 25 °C. $Li_{5.91}Al_{0.03}PS_5Cl$ has the highest σ_{DC} ionic conductivity at 25 °C among the samples. The movement of charge carriers at the grain boundaries changed from long-range diffusion in Li_6PS_5Cl to short-range diffusion in doped samples.

Acknowledgments. We acknowledge Ho Chi Minh City University of Technology (HCMUT), VNU-HCM for supporting this study.

References

1. Zhou, L., Minafra, N., Zeier, W.G., Nazar, L.F.: Acc. Chem. Res. **54**, 2717–2728 (2021)
2. Kato, Y., et al.: Nat. Energy **1**, 16030 (2016)
3. Yu, C., Zhao, F., Luo, J., Zhang, L., Sun, X.: Nano Energy **83**, 105858 (2021)
4. Phuc, N.H.H., Totani, M., Morikawa, K., Muto, H., Matsuda, A.: Solid State Ionics **288**, 240–243 (2016)
5. Jubeer, E.M., Manthrammel, M.A., Subha, P.A., Shkir, M., Biju, K.P., AlFaify, S.A.: Sci. Reports **13** (2023)
6. Tran, A.T., Tran, V.T., Nguyet, N.T.M., Luong, A.T.Q., Van Le, T., Phuc, N.H.H.: ACS Omega **8**, 36253–36260 (2023)

7. Boulineau, S., Courty, M., Tarascon, J.M., Viallet, V.: Solid State Ionics **221**, 1–5 (2012)
8. Jonscher, A.K.: Nature **267**, 673–679 (1977)
9. Imran, Z., Rafiq, M.A., Ahmad, M., Rasool, K., Batool, S.S., Hasan, M.M.: AIP Adv. **3** (2013)

Effect of NaOH Concentration on the Synthesis of Calcium Silicate Hydrate (CSH) from PV Solar Waste Glass and Calcium Hydroxide

Tran Ngo Quan[1,2,3,4], Bui Thi Thao Nguyen[1,2], Huynh Dung[1,2], Takaaki Wajima[5], and Pham Trung Kien[1,2,3,4](✉)

[1] Ho Chi Minh City University of Technology (HCMUT), 268 Ly Thuong Kiet Street, District 10, Ho Chi Minh City, Vietnam
phamtrungkien@hcmut.edu.vn
[2] Vietnam National University Ho Chi Minh City, Linh Trung Ward, Thu Duc District, Ho Chi Minh City, Vietnam
[3] Polymer Research Center, Ho Chi Minh City University of Technology (HCMUT), 268 Ly Thuong Kiet Street, District 10, Ho Chi Minh City, Vietnam
[4] VNU-HCM Key Laboratory for Material Technologies, Ho Chi Minh City University of Technology (HCMUT), 268 Ly Thuong Kiet Street, District 10, Ho Chi Minh City, Vietnam
[5] Department of Materials Science, Graduate School of Engineering, Chiba University, 1-33 Yayoi-Cho, Inage-Ku, Chiba 263-8522, Japan

Abstract. Calcium Silicate Hydrate (CSH) mineral has long been highlighted as a material with many potential applications, especially in the fields of construction and environmental protection, thanks to its special physical and mechanical properties such as durability, high surface area, low cost. In this study, synthesizing CSH by taking advantage of available waste from glass products and Ca(OH)2, acting as the main raw material to form CSH minerals in the hydrothermal environment. The role of NaOH medium was selected to study with different NaOH concentraion such as 4M, 8M and 12M. The result show that CSH can be form when the NaOH concentration is 8M or 12M, indicating the alklaine activator process. The needle-like CSH crystal have length of 2μm have potential be used in thermal insulator and heavy metal removal.

1 Introduction

One of the global challenges contributing to the increased impact of the carbon footprint is the demand for natural resource extraction and the issue of landfilling and disposal of waste glass [1]. Today, glass recycling is becoming increasingly popular worldwide, with a commitment to developing processing technologies and adhering to environmental policies. In this context, it is essential to develop glass recycling strategies, as glass is a material that does not degrade under environmental influences and remains stable over a long period.

More and more research on synthesie Xonotlite mineral [$Ca_6Si_6O_{17}(OH)_2$] as indicating in the research paper [2–8]. Brieftly, xonolite can be synthesized under suitable

N. R. Abdul Razak et al. (Eds.): ECOOP 1996, SPPHY 418, pp. 235–239, 2025.
https://doi.org/10.1007/978-981-96-2871-1_33

hydrothermal process conditions and is widely used in various fields such as insulation, construction materials, paper production, artificial wood, and biology etc. as indicating in research paper. Therefore, the study of Xonotlite synthesis is crucial, especially from agricultural and industrial waste sources, to address environmental issues. Glass is also a material used for synthesizing Xonotlite mineral [9]. Thus, this research aims to utilize industrial glass waste as a raw material to develop a Xonotlite synthesis process. This approach not only helps improve the environment by addressing glass waste but also creates a high-value product from the waste material. Previous studies on Xonotlite synthesis have mainly focused on using rice husk ash (an agricultural byproduct), but these are limited due to seasonal dependency and its inconsistent availability. Therefore, replacing the input material with industrial glass waste can overcome these limitations and promises to bring economic benefits. In this research, the effect of NaOH medium concentration was investigated on the formation of Calcium Silicate Hydrate (CSH),

2 Materials and Methods

2.1 Preparation of CSH Pellet from Wasted Glass Powder and $Ca(OH)_2$

Firstly, the wasted glass cullet powder from PV-solar glass and $Ca(OH)_2$ (Xiaolong, China) were mix with Ca/Si molar ratio of 1.0. The glass powder and $Ca(OH)_2$ samples are analyzed by microstructure such as XRD, SEM to evaluate the structure and properties. A small amount of water is added to increase adhesion to support the ability to shape CSH minerals before hydrothermal treatment. The mixture of wasted glass powder and $Ca(OH)_2$ is compressed into pellets using an 8mm diameter stainless steel mold.

2.2 Synthesis of CSH Using Hyrothermal Reaction Using NaOH Medium

The resulting pellet sample is then placed as steam hydrothermal synthesis in a NaOH environment at 180 °C for 96 h. The concentration of NaOH is vary from 4M, 8M and 12M to investigate its role on the formation of CSH. Following hydrothermal autoclaving, the sample is dried at 90 °C for 24 h to ensure complete water removal from the CSH sample. After synthesis, the CSH sample undergoes analysis methods to evaluate its microstructure, phase composition, and mineral composition before and after the hydrothermal treatment. The flowchart process to synthesis CSH was shown in Fig. 1.

2.3 Materials Characterization

Fourier-transform infrared spectroscopy (FT-IR) method by Thermo Scientific Nicolet 6700 FTIR is used to analyze the chemical bonds in CSH materials scanning wavenumber in range 400 – 4000 cm^{-1}.

X-Ray diffraction analysis (XRD) is used to analyze the phase composition of CSH before and after synthesizing by Bruker D8 Advance scanning from 5 to 80°, operation at 40kV and 40mA.

The surface microstructure and chemical composition analysis methods used are Scanning electron microscope (SEM) and Energy-dispersive X-ray spectroscopy (EDX) on Hitachi S-4800 at 10kV.

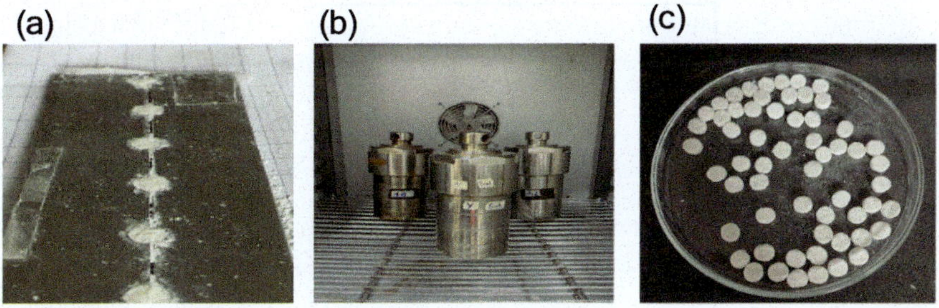

Fig. 1. Sample preparation: (a) Sample making with Ca/Si 1.0, (b) Steam hydrothermal at 180 °C/96h with different NaOH concentraion 4M, 8M and 12M and (c) CSH sample

3 Results and Discussions

The chemical composition of glass powder and CSH were udentify by XRF analysis in Table 1. The mass ratio between Ca and Si is equivalent with Ca/Si 1.0. XRF has proven that the synthesis process is stable without loss. The mass ratio Ca/Si ~ 1 also proves the possibility of forming specific CSH minerals by hydrothermal synthesis such as Xonotlite, Tobermorite.

Table 1. Chemical composition of glass powder and $Ca(OH)_2$

.Materials	SiO2	Na2O	CaO
Glass	69.3	14.9	12.7
$Ca(OH)^2$	N/A	N/A	74.6

Figure 2 shown XRD patterns of CSH sample before and after steam hydrothermal treatment at 180oC for 96 h using NaOH medium at different concentration such as 4M, 8M and 12M. The peak of xonotlite can be detected upon hydrothermal treatment at 180oC for 96 h event in NaOH 0M (pure water) or 12M at 2theta of 32degree. However, when we use NaOH 8M, the single peak of xonolite at 32degree show sharply, thus indicating the formation of xonotlie condition.

Figure 3 show the SEM images of Ca/Si sample upon hydrothermal treatment at 180 °C/96h using NaOH medium at (a) 4M; (b) 8M and (c) 12M. At concentration of 4M NaOH, we cannot observed the needle-like structure. However when increasing the NaOH concentration from 4M to 8 and 12M, the needle-like structure can be obtain as show by yellow arrow. Thus, the formation of xonotlite is strongly effected by the NaOH concentration.

Through the surface microstructure results, we can see the possibility of using NaOH medium as alkaline activator. This research should be considered in the context of the experimental conditions and the potential applications of the findings. The main objective was to investigate the role of NaOH on the formation of CSH made form wasted glass and Ca(OH)2 upon hydrothermal treatment at 180 °C/96h. As mentioned in introduction

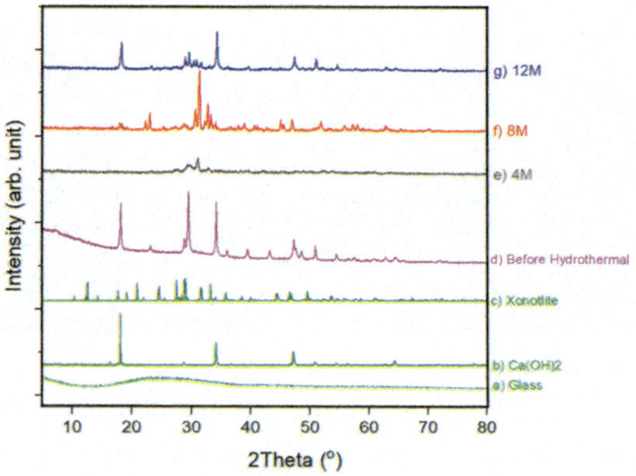

Fig. 2. XRD patterns of CSH at different synthesis concentration: (a) Glass powder; (b) Ca(OH)2; (c) Xonolite standard; mixture with Ca/Si 1.0 hydrothermal treatment at 180 °C/96h using NaOH with (d)0M; (e) 4M; (f) 8M and (g) 12M

Fig. 3. SEM images of Ca/Si sample after hydrothermal treatment at 180oC/96 h using NaOH medium at (a) 4M; (b)8M and (c)12M

section, wasted glass powder can be consider as main source to supply Si, while $Ca(OH)_2$ is the main source to supply Ca. Upon mixing wasted glass powder with $Ca(OH)_2$ with Ca/Si 1.0, the Ca/Si green pellet can be used for steam hydrothermal treatment at 180 °C/96h. When increasing the concentration of NaOH, the appear of CSH needle like crystal seem more precise. Based on the needle-like shape morphology of CSH, it can be interlocked to enhance the linkage chain between Ca-Si. The main CSH mineral forming after hydrothermal reaction is Xonotlite as indicated in Fig. 3f and 3g with the needle-like morphology. It need to emphasized that the needle-like crystal size of xonotlite with diameter of 0.5μm and length of 2μm can be used as thermal insulator.

This research shown the potential to recycle both wasted glass powder and $Ca(OH)_2$ to form CSH mineral toward the heat insulator in industry.

4 Conclusions

The steam hydrothermal process in 4,8 and 12M NaOH of CSH minerals from glass powder and $Ca(OH)_2$ has demonstrated the ability to synthesize waste source materials into CSH minerals with potential application value in many fields. The increasing of NaOH concentration activate the needle-like crystal forming. The next research direction is proposed to take advantage of different waste sources such as fly ash, steel slag, etc. to synthesize CSH minerals. The study can be extended to evaluate the porosity of CSH minerals such as adsorption applications.

Acknowledgments. We acknowledge Ho Chi Minh City University of Technology (HCMUT), VNU-HCM for supporting this study.

References

1. Yuefeng, M., et al.: Influences of leaching on the composition, structure and morphology of calcium silicate hydrate (C–S–H) with different Ca/Si ratios. J. Build. Eng. **58**, 105017 (2024)
2. Dambrauskas, T., et al.: Effect of intercalated metal ions on the specific surface area and porosity of dibasic calcium silicate hydrate. Surf. Interfaces **33**, 102243 (2022)
3. Fang, Q., et al.: Recycling of blast furnace slag to prepare calcium silicate hydrate by mechanical-chemical co-activation and its application to calcium silicate fireproof board. Process. Saf. Environ. Prot. **165**, 1–12 (2022)
4. Suji, W., Young, C.: Synthesis of calcium silicate hydrate nanoparticles and their effect on cement hydration and compressive strength. Constr. Build. Mater. **407**, 133559 (2023)
5. Andreas, P., Luc, N., Andre, N., Christophe, L., Helmut, C.: Influence of polymers on the nucleation of calcium silicate hydrates. Cem. Concr. Res. **174**, 107329 (2023)
6. Bohong, Z., Wenyu, L., Hongyang, M., Jie, H.: In situ monitoring of the hydration of calcium silicate minerals in cement with a remote fiber-optic Raman probe. Cement Concr. Compos. **142**, 105214 (2023)
7. Yiru, Y., et al.: Kinetics of Al uptake in synthetic calcium silicate hydrate (C-S-H). Cem. Concr. Res. **172**, 107250 (2023)
8. Gedeike, I., Baltakys, K., Eisinas, A.: Thermal stability of synthetic high basicity calcium silicate hydrates substituted with Cr3+ ions. Mater. Charact. **193**, 112231 (2022)
9. Shengrui, S., Ya, T., Jiayi, L., Jiahui, K., Yangqiao, L.: Fly ash derived calcium silicate hydrate as a highly efficient and fast adsorbent for Cu(ii) ions: role of copolymer functionalization. RSC Adv. **12**(35), 22843–22852 (2022)

Formation of $MgAl_2O_4$ Spinel from $MgO-Al_2O_3$ Composite Particle by Single Step Reaction Sintering

Tran Anh Tu[1,2,3] and Nguyen Huu Huy Phuc[1,2,3](✉)

[1] Faculty of Materials Technology, Ho Chi Minh City University of Technology (HCMUT), 268 Ly Thuong Kiet Str., Dist. 10, Ho Chi Minh City, Vietnam
nhhphuc@hcmut.edu.vn

[2] Vietnam National University Ho Chi Minh City, Linh Trung Ward, Thu Duc Dist., Ho Chi Minh City, Vietnam

[3] National Key Laboratory of Polymer and Composite Materials, 268 Ly Thuong Kiet, District 10, Ho Chi Minh City, Vietnam

Abstract. This paper reported the preparation of $MgAl_2O_4$ spinel using one step sintering of $MgO-Al_2O_3$ composite powder in ambient air at 1450 °C, 1500 °C and 1600 °C. Rietveld refinement of XRD data showed the excellent agreement between experimental and calculated patterns. The Halder-Wagner-Langford's equation was employed to derive crystal size and micro strain of the prepared samples from XRD data. In addition, the transparent $MgAl_2O_4$ spinel pellet was successfully fabricated using spark plasma sintering.

Keywords: $MgAl_2O_4$ Spinel · Halder-Wagner-Langford's · Spark Plasma Sintering

1 Introduction

Magnesium aluminate spinel ($MgAl_2O_4$, MAS) is widely used in metallurgy of steel, cement, glass and ceramics because of its high melting temperature (2135 °C), low coefficient of thermal expansion (8.0×10^{-6} K^{-1}), high hardness (16 GPa), resistance to thermal shock and erosion [1]. $MgAl_2O_4$ can be synthesized by two main preparation methods, which are solid state sintering and electro melting. Solid state sintering is more cost-effective and energy efficient as well as better mechanical and refractory properties. The basic principle for MAS fabrication is to sinter the compressed raw powder under controlled high temperature with or without pressure. MgO and Al_2O_3 are the common raw materials for synthesizing MAS by solid state reaction. Conventional solid annealing methods require two-stage annealing to form homogeneous and high-density spinel phases [2]. It was shown that MAS could be obtained by one step solid state reaction using $MgO-Al_2O_3$ composite particle [3].

N. R. Abdul Razak et al. (Eds.): ECOOP 1996, SPPHY 418, pp. 240–245, 2025.
https://doi.org/10.1007/978-981-96-2871-1_34

Transparent $MgAl_2O_4$ ceramic is considered one of the important materials for vis-IR transparency due to its wide bandgap (7.8 eV) [4, 5]. Furthermore, its high thermo-mechanical strength, rigidity and thermal shock resistance make it suitable for a variety of optical applications including transparent armor, infrared sensor windows and domes [6]. It is difficult to fabricate near-perfect density $MgAl_2O_4$ ceramics due to its low oxygen diffusion capability and residual holes of less than 0.01% in the microstructure that can make it transparent. Transparent $MgAl_2O_4$ ceramics with high optical properties from the visible to infrared wavelength range can be fabricated through advanced ceramic powder processing technology with a special sintering method. The commonly used sintering methods for transparent MAS fabrication are hot-press (HP), hot-isostatic pressing (HIP) and spark plasma sintering (SPS) [7, 8].

This study aimed to fabricate $MgAl_2O_4$ spinel from MgO-Al_2O_3 composite particle using conventional sintering and spark plasma sintering methods. The MAS pellet fabricated by SPS method exhibited relatively high optical transparent of 70% in visible range. The crystal structure of the sintered MAS pellets was characterized by XRD. Rietveld refinement was employed to analyze the structural parameters from XRD data.

2 Experimental Procedure

Al_2O_3 (reagent-grade) and heavily sintered MgO (reagent-grade) were purchased from Wako. Poly(sodium 4-styrenesulfonate) (PSS, Mw 70000) and poly(diallyl dimethylammonium chloride) (PDDA, Mw 400000–500000) were purchased from Sigma-Aldrich. LiF (99.99%) was obtained from Mackin. The chemicals were used as received without further purification.

MgO-Al_2O_3 (1:1 molar ratio) composite powder was fabricated using our previously reported procedure [3]. The prepared powder was made into pellets by pressing 0.3 g into a 10 mm stainless steel mold at 30 MPa. Afterward, the green bodies were sintered at different temperatures and times in ambient air.

The transparent MAS pellets were fabricated using SPS method. LiF was employed as additive. The prepared composite powder was manually mixed with LiF (about 1% weight) using an agate mortar. Then, about 0.2g of the mixture was loaded into the graphite mold (10 mm in diameter) for SPS process (Fuji Electronic Industrial Co., Ltd.).

The crystal structure of the sintered MAS pellets was characterized with XRD (Bruker X8). The data was analyzed using Fullprof software. The optical transmittance of the MAS pellet was investigated using UV-Vis spectroscopy (V-670, Jasco).

3 Results and Discussions

Figure 1(a) showed the XRD patterns of the MAS samples sintered in ambient air at different temperature and duration. The peaks of starting materials were not observed. All the peaks could be assigned to $MgAl_2O_4$ spinel (JCPDS card 01 – 070 – 5187). The observed peaks located at $2q \approx 19.03°$, 31.33°, 36.91°, 38.62°, 44.89°, 55.77°, 59.47°, 65.37°, 68.77° were indexed to [111], [220], [113], [222], [040], [422], [511], [440] and [315] planes. Rietveld refinement was performed using Fullprof program to analyze the crystal structure of the prepared samples. The crystal structure of all the sample belonged to Fd-3m:1 space group. Figure 1(b) showed the refined XRD pattern of the sample sintered at 1450 °C for 6h. The refined results ($R_p = 8.13\%$, $R_{wp} = 10.6\%$ and $c^2 = 0.772$) exhibited the excellent agreement between the experimental and calculated pattern.

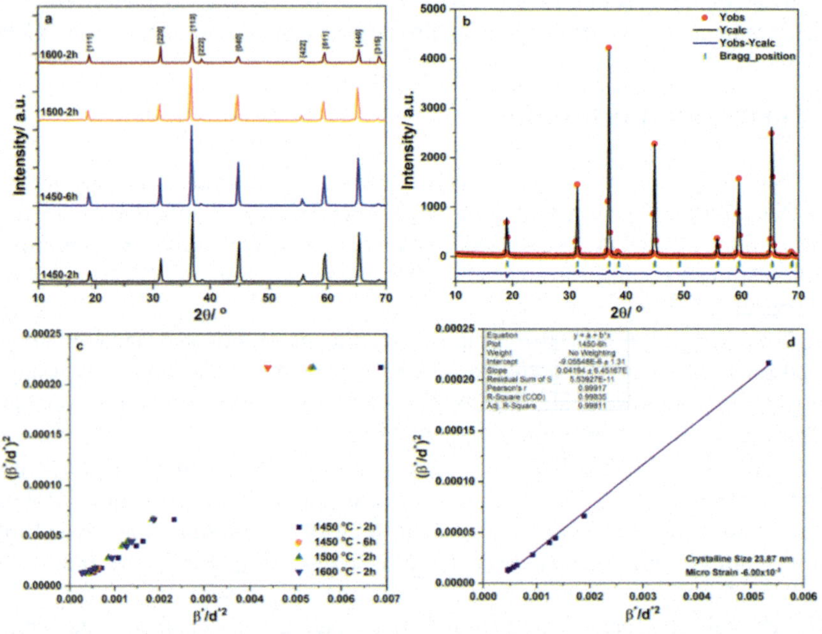

Fig. 1. (a) XRD patterns of the MAS samples sintered in ambient air at different temperature and duration, (b) Refined XRD pattern of the sample sintered at 1450 °C for 6h, (c) Harder-Wagner-Langford plots of the MAS samples sintered in ambient air at different temperature and duration, (d) Fitting result of the WHL plot of the sample sintered at 1450 °C for 6h.

The lattice parameter of the samples obtained from the refinement are summarized in Table 1. The mean crystallite size and micro strain of the prepared MAS was deduced from the Halder-Wagner-Langfor''s (HWL) plot technique applied to the XRD data. The HWL equation relates the FWHM of peaks, β, with the mean crystallize size, "T", and the micro strain of a grain, ε, as follow:

$$\left(\frac{\beta^*}{d^*}\right)^2 = \frac{1}{T}\left(\frac{\beta^*}{d^2}\right) + \left(\frac{\varepsilon}{2}\right)^2$$

$$\beta^* = \frac{\beta}{\lambda}\cos\theta, d^* = \frac{2}{\lambda}\sin\theta$$

where l is the X-ray wavelength (0.15406 nm). Figure 1(b) showed the HWL plots of the prepared samples. Based on the linear fit equation, the crystallize size and micro strain could be obtained. Figure 1d showed the fitting results of the sample sintered at 1450 °C for 6h. The crystalline size and micro strain obtained from HWL plots were illustrated in Table 1.

Table 1. The structural parameters of the prepared samples

Samples	Lattice parameters (Å)	Crystalline size (nm)	Micro strain (%)
1450-2h	8.0692	31.15	-4.47×10^{-3}
1450-6h	8.0680	23.87	-6.00×10^{-3}
1500-2h	8.0665	24.33	-4.90×10^{-3}
1600-2h	8.0709	30.58	$+2.00 \times 10^{-3}$

One of the most important physical properties of $MgAl_2O_4$ spinel is its transparency. The transparency is only observed at high relative density spinel (>98.5%) because the existent of voids will lead to the light scattering effect. In this study, the transparent spinel was obtained using SPS (Fig. 2(a)). The composite particles were prepared via ESA method was mixed with LiF (1 wt %) then loaded into the graphite dice; the SPS chamber was then evacuated to pressure of about 20Pa following by pressing the powder at 50MPa. The graphite dice was then heated to about 1100 °C at heating rate of ~ 100 K/min from room temperature, then ~ 20K/min from 1200 to 1350 °C (the current was 500A). The applied force was about 5MPa during this heating step. After temperature reached its max value, the applied force was slowly increased to 30MPa and hold for 20 min prior to fast cooling down.

Figure 2(b) and (c) showed the optical images of the transparent samples employed $MgO-Al_2O_3$ composite and conventional $MgO-Al_2O_3$ mixture, respectively. The beads milling pellet still had un-reacted or porous part accumulated at the boundary of the pellet. However, cloudy parts were still observed in the pellet using composite powder. The optical transparency measurement showed that the pellet using composite powder was superior to the one using conventional mixed powder (Fig. 2(d)).

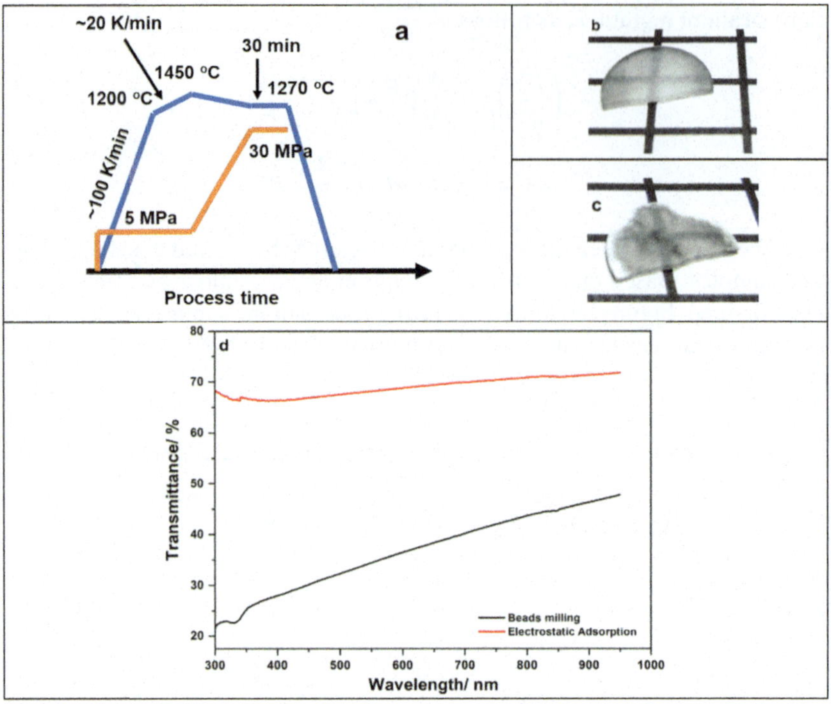

Fig. 2. (a) The SPS process for fabrication of transparent MAS pellet, (b) and (c) optical images of the transparent samples employed $MgO\text{-}Al_2O_3$ composite and conventional $MgO—Al_2O_3$ mixture, respectively, (d) the UV-Vis curves of the transparent samples employed $MgO\text{-}Al_2O_3$ composite (Electrostatic adsorption) and conventional $MgO—Al_2O_3$ (Beads milling).

4 Conclusion

$MgAl_2O_4$ spinel was successfully prepared by conventional sintering $MgO\text{-}Al_2O_3$ composite powder in ambient air at different temperature and time. Rietveld refinement of XRD data showed that the prepared samples had cubic structure belonging to Fd-3m:1 space group. The lattice parameter reduced when sintering temperature increased from 1450 to 1500 °C or sintering time increased from 2 to 6h. The micro strain obtained from Halder-Wagner-Langford's equation indicated that the sample sintered at 1600 °C had positive strain value while the others had negative strain value, indicating the expansion of the unit cell of sample sintered at 1600 °C. In addition, the transparent $MgAl_2O_4$ spinel pellet was successfully fabricated using spark plasma sintering.

Acknowledgements. We acknowledge Ho Chi Minh City University of Technology (HCMUT), VNU-HCM for supporting this study.

References

1. Ganesh, I.: Int. Mater. Rev. **58**, 63–112 (2013)
2. Ganesh, S.B.I., Saha, B.P., Johnson, R., Rajeshwari, K., Sengupta, R., Ramana Rao, M.V., Mahajan, Y.R.: Ceram. Inter. **28**, 245–253 (2002)
3. Tran, A.T., Tran, V.T., Nguyet, N.T.M., Luong, A.T., Le, T.V., Phuc, N.H.H.: ACS Omega **8**, 36253–36260 (2023)
4. Biswas, P., Chakravarty, D., Suresh, M.B., Johnson, R., Mohan, M.K.: Ceram. Inter. **42**, 17920–17923 (2016)
5. Jung, W.K., et al.: Ceram. Inter. **50**, 5007–5012 (2024)
6. Gajdowski, C., et al.: Ceram. Inter. **48**, 18199–18211 (2022)
7. Shi, Z., Zhao, Q., Guo, B., Ji, T., Wang, H.: Mater. Des. **193**, 108858 (2020)
8. Xiao, Z., et al.: Mater. Sci. Eng. R. Rep. **139**, 100518 (2020)

Influence of Biopolymer and Weighting Agent on Properties of Water-Based Drilling Mud

L. N. H. Nam[1,2(✉)], D. P. Hau[1,2], H. T. Quang[1,2], and T. N. T. Tam[1,2]

[1] Faculty of Geology and Petroleum Engineering, Ho Chi Minh City University of Technology, 268 Ly Thuong Kiet, District 10, Ho Chi Minh City, Vietnam
lnhnam@hcmut.edu.vn

[2] Vietnam National Univeristy Ho Chi Minh City, Linh Trung Ward, Thu Duc City, Ho Chi Minh City, Vietnam

Abstract. Drilling fluid plays a crucial role in drilling operations, significantly impacting performance by managing bottom hole pressure, ensuring wellbore stability, and maintaining hole cleanliness. Without effective drilling fluid, issues such as poor well control, unstable wellbores, inadequate transmission capacity, excessive torque, pipe sticking, drag, and damage would occur. This study investigates the enhancement of water-based bentonite drilling mud properties by adding calcium carbonate (CaCO3) and xanthan gum polymer. Various concentrations of calcium carbonate (2, 4, and 6 percentage by weight (%wt)) and xanthan gum (0.1, 0.2, 0.3, and 0.4 wt%) were tested to evaluate their effects on rheology and filtration properties. The plastic viscosity of fluids with calcium carbonate varied slightly from 6 to 7.5 cp, while density increased significantly, peaking at 8.93 ppg at 6 wt%. The filtration rate showed minimal change, decreasing slightly from 16.8 ml to 16.4 ml, and cake thickness remained constant. Xanthan gum significantly increased gel strength from 0.1 wt% to 0.3 wt% (52 to 72 lb/100ft^2 at 10 min), with negligible changes beyond 0.3 wt%. Filtration volume decreased notably from 12.2 ml at 0.1 wt% to 10.2 ml at 0.4 wt%, and cake thickness reduced from 2 mm to 1.5 mm, highlighting xanthan gum's effectiveness in controlling fluid loss.

1 Introduction

In response to the rising demand for hydrocarbon energy, the oil and gas industry has intensified efforts to discover new resources in uncharted regions and deeper geological formations. Drilling the wellbore is both the initial and most costly phase in the oil and gas industry. Drilling expenses account for approximately 25% of the overall oil-field exploitation costs, with a significant portion directed towards exploration and well development. Drilling fluids alone can constitute up to 15 to 18% of the total well drilling expenses [1, 2]. Drilling fluids play crucial roles in oil well drilling processes, including transporting broken formation rock from the bottom to the surface, maintaining the hole wall, lubricating and cooling the bottom hole assembly [3]. To address challenges encountered during drilling, various methods have been developed to design suitable drilling muds. These drilling muds should be easy to use, affordable, and economically

N. R. Abdul Razak et al. (Eds.): ECOOP 1996, SPPHY 418, pp. 246–250, 2025.
https://doi.org/10.1007/978-981-96-2871-1_35

sustainable. They are specifically formulated to reduce damage and ensure the practicality and cost-effectiveness of rotary drilling in oil and gas reservoirs. Consequently, drilling muds are treated with various additives, such as different polymers and chemicals, to meet specific well requirements, including rheology, fluid loss control, and mud weight. Calcium carbonate is an important material used to control mud density due to its high molecular weight [4]. Additionally, xanthan gum has been used as a fluid-structure builder [5–7], primarily to improve rheology and filtration properties.

This study conducted a series of experiments to assess the effectiveness of calcium carbonate and xanthan gum in controlling filtration loss and enhancing the properties of mud samples at different concentrations. The findings indicate that both calcium carbonate and xanthan gum reduced fluid loss and had a positive effect on rheology.

2 Method and Analysis

In this study, bentonite powder, tap water, Calcium Carbonate ($CaCO_3$), and Xanthan Gum (XG) were utilized as drilling fluid additives. The water used in the experiment was sourced from the water system at the university campus. The bentonite, as depicted in Fig. 1a, was prepared in our laboratory, while the $CaCO_3$, shown in Fig. 1b, was procured from Xilong Scientific Co., Ltd, boasting a purity of 99%. In Fig. 1c, the Xanthan gum is manufactured by Deosen Biochemical (ordos) Company.

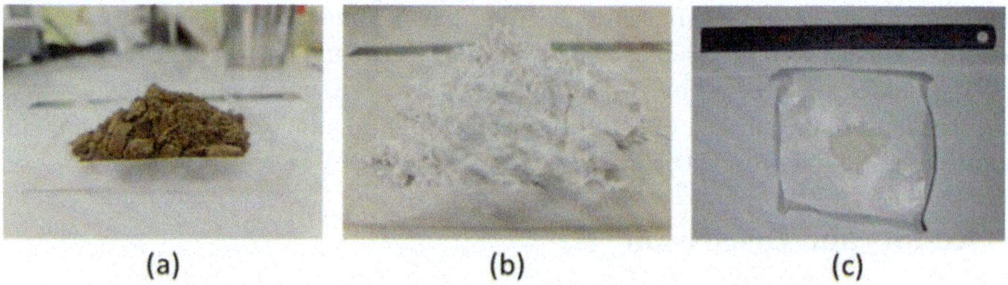

Fig. 1. (a) Bentonite powder; (b) calcium carbonate; (c) Xanthan Gum.

In this section, several fluid samples were prepared to assess the impact of various additives on their rheological properties. These fluids will be examined through laboratory experiments, with each material tested at different concentrations to observe and document their effects.

First three fluid will be containing constant concentration of bentonite and 3 different concentration of calcium carbonate to mainly evaluate changes in density and other rheologies. The fluids after that will also have constant concentration of bentonite but with 4 different concentrations of xanthan gum to determine the change in viscosity and fluid loss control. Table. 1 presents the initial three fluids, which utilize 2 wt% to 6 wt% calcium carbonate, and fluids 4 through 7, which contain 0.1 wt% to 0.4 wt% xanthan gum. These fluids will illustrate the impact of each additive on the properties of the drilling fluid. Procedure: First, bentonite powder is mixed with water using a Hamilton Beach mixer. Next, calcium carbonate and of xanthan gum are gradually added. This

Table 1. Selected variable parameters and levels.

Additives	Sample						
	1	2	3	4	5	6	7
Tap water, (%wt)	93	91	89	94.9	94.8	94.7	94.6
Bentonite, (%wt)	5	5	5	5	5	5	5
CaCO3, (%wt)	2	4	6	0	0	0	0
Xanthan Gum(%wt)	0	0	0	0.1	0.2	0.3	0.4

method ensures that all additives are evenly incorporated. The fluid is then stirred for 30 min at a constant room temperature of 20 °C. Afterward, rheology and filtration characteristics are measured in the subsequent section.

The rheological properties of the drilling fluid were evaluated using the API 13B standard procedure for field testing. An OFITE rotational viscometer (USA) was employed to measure plastic viscosity (PV), yield stress (YP), and gel strength across seven different fluid samples (as shown in Fig. 2b). These measurements were calculated by obtaining the dial reading at rotational speeds of 600, 300, 200, 100, 60, 20, 6 and 3 rpm. Also, gel strength was measured at 10 s and 10 min. The PV and YP parameters were calculated.

The filtrate of drilling mud samples was examined using an API filter press (as shown in Fig. 2c). The liquid volume of filtrate, which indicates the simulated leakage from the fluid to the formation, was measured. A filter cake, composed of solid particles in the sample, forms to help mitigate excessive fluid loss. The mud sample was poured into a cylinder holder and pressed by 100 psi $\pm$ 5 psi. Once pressure is applied, the test period starts to measure the amount of filtrate collected after 30 min. Report the filtrate volume in milliliters (to the nearest 0.1 ml).

3 Results and Discussion

All of the laboratory experiment results will be demonstrated in Table 2. Including OFITE viscometer reading, gel strength with plastic viscosity, yield point and apparent viscosity are calculated, stirring time, density in ppg and filtration volume recorded throughout out 30 min, and filter cake thickness in milimeter.

Figure 2 illustrates the relationship between the concentration of calcium carbonate ($CaCO_3$) in a drilling mud and its performance characteristics, including Plastic Viscosity (PV), Yield Point (YP), and Gel Strength (GS) at 10 s and 10 min. As the $CaCO_3$ concentration increases from 2% to 6%, the PV, YP, and GS all exhibit a gradual upward trend, indicating that higher $CaCO_3$ content enhances mud viscosity, flow resistance, and particle suspension properties. Figure 3 shows how increasing $CaCO_3$ concentration in a drilling mud impacts fluid loss and filter cake thickness. As the $CaCO_3$ concentration rises from 2% to 6%, the fluid loss gradually decreases, declining from around 16.8 ml to approximately 16.4 ml. However, the filter cake thickness remains consistent at around 2 mm across all $CaCO_3$ concentrations. These findings suggest that adding $CaCO_3$ to the drilling mud can effectively reduce fluid loss while maintaining a relatively stable filter cake thickness.

Table 2. Experiment results for 7 samples

Parameter	Sample						
	1	2	3	4	5	6	7
Density (g/cc)	1.04	1.06	1.07	1.02	1.02	1.02	1.02
Reading @ 600RPM	43	47	52.5	70	105	130	158
Reading @ 300RPM	37	40	45	60	94	117	139.5
PV (cP)	6	7	7.5	10	11	13	18.5
YP (cP)	31	33	37.5	50	83	104	112
Gel strength @ 10s	30	36	40	42	51	65	66
Gel strength @ 10min	35	38	42	52	64	72	71
Fluid loss after 30 min (ml)	16.8	16.6	16.4	12.2	12	10.8	10.2
Filter cake thichness (mm)	2	2	2	2	1.7	1.7	1.5

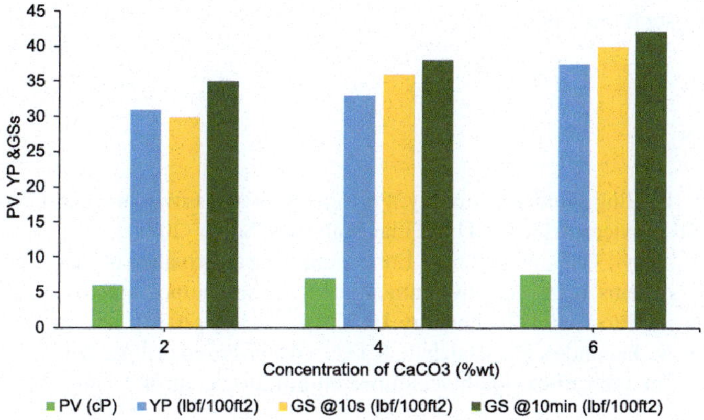

Fig. 2. Effect of $CaCO_3$ on Gel strength values.

4 Conclusions

In summary, adding Xanthan Gum and Calcium Carbonate to bentonite mud has shown encouraging results in improving rheological characteristics and lowering volume fluid loss. Increasing concentrations of Calcium Carbonate and Xanthan Gum resulted in higher plastic viscosity and increased fluid density, enhancing the fluid's effectiveness as a weighting agent for drilling tasks. Moreover, Xanthan Gum contributed to reducing the volume of fluid loss and forming a thinner filter cake. Therefore, Calcium Carbonate and Xanthan Gum emerge as viable additives for enhancing drilling fluid performance.

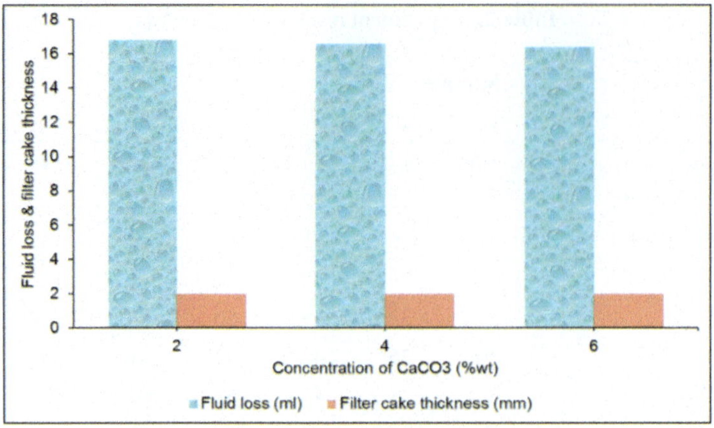

Fig. 3. Effect of $CaCO_3$ on filtration properties.

Acknowledgments. We acknowledge Ho Chi Minh City University of Technology (HCMUT), VNU-HCM for supporting this study.

References

1. Khodja, M.: Drilling Fluid Technology: Performances and Environmental Considerations. In Products and Services; from R&D to Final Solutions, Sciyo (2010)
2. Akpan, E.U., Enyi, G.C., Nasr, G.G.: Enhancing the performance of xanthan gum in water-based mud systems using an environmentally friendly biopolymer. Journal of Petroleum Exploration and Production Technology **10**(5), 1933–1948 (2020)
3. Da Luz, R.C.S., Fagundes, F.P., Balaban, R.D.C.: Water-based drilling fluids: The contribution of xanthan gum and carboxymethylcellulose on filtration control. Chem. Pap. **71**(12), 2365–2373 (2017)
4. Hudson, T.: Div of Dow Chemical USA Coffey, and D. C. Div of Dow Chemical USA Sauer Jr, Fluid-Loss Control Through the Use of a Liquid-Thickened Completion and Workover Brine (1983)
5. Carico, R.D., Bagshaw, F.R., Div Merck, K.: Description and Use of Polymers Used in Drilling, Workovers, and Completions (SPE 7747)
6. Talabani, S., U. of Oklahoma, Hatzignatiou, D.G., Chukwu, G.A.: Comprehensive Description and Evaluation of Polymers as Drilling Fluids (1993)
7. Lucas, E.F., Mansur, C.R.E., Spinelli, L., Queirós, Y.G.C.: Polymer science applied to petroleum production. Pure Appl. Chem. **81**, 473–494 (2009)

Mechanochemical Synthesis of AlFeCoNiTi and AlFeCuZnTi High Entropy Alloys

Vu Nguyen Duy Anh[1,2], Ngo Gia An[1,2], Do Thi Ngoc Anh[1,2], and Nguyen Huu Huy Phuc[1,2](✉)

[1] Faculty of Materials Technology, Ho Chi Minh City University of Technology (HCMUT), 268 Ly Thuong Kiet Str., Dist. 10, Ho Chi Minh City, Vietnam
nhhphuc@hcmut.edu.vn

[2] Vietnam National University Ho Chi Minh City, Linh Trung Ward, Thu Duc Dist, Ho Chi Minh City, Vietnam

Abstract. The high entropy alloy AlFeCoNiTi and AlFeCuZnTi were prepared by high energy ball milling of elemental powders in toluene at 250 rpm for 3h, 20h, 30h, 50h and 100h. The XRD data results revealed that both alloys exhibit the formation of BCC and FCC phases. The Scherrer equation was employed to derive crystal size of the prepared samples from XRD data.

1 Introduction

The basic goal and motivation for the field of materials science is to develop advanced materials with appropriate properties during this period [1]. Additionally, there is an essential need for new materials to realize new energy technologies [2]. A new concept for material design has been developed, offering the ability to influence the phase stability of solid solutions through precise control of configurational entropy. This development originates from the discovery of high-entropy alloys (HEAs) [1]. Interest in high-entropy materials due to their unique structural characteristics, chemical composition, and outstanding properties such as high hardness, excellent thermal stability, and remarkable resistance to oxidation and surface corrosion has been increasingly attracting attention in the fields of environmental science and renewable energy technology [1].

The current study aimed at the preparation of AlFeCoNiTi and AlFeCuZnTi HEAs by mechanical alloying. The methods proposed in this study were successful in the production of HEAs from constituting elements with different ranges of melting point, namely, Al, Co, Cu, Fe, Ni, Ti, and Zn. Besides, the phase analysis and the structural evolution of prepared HEAs was investigated in order to study effect of preparation method and phase formation process of high-entropy alloys.

2 Experimental

Al, Co, Cu, Fe, Ni, Ti, and Zn powder with purity of higher than 99% (Macklin, China) were used at the starting materials of ball milling process for synthesizing equiatomic alloys.

N. R. Abdul Razak et al. (Eds.): ECOOP 1996, SPPHY 418, pp. 251–255, 2025.
https://doi.org/10.1007/978-981-96-2871-1_36

Both Al-Fe-Co-Ni-Ti and Al-Fe-Cu-Zn-Ti powders are mixed together in a 1:1 molar ratio and manually pre-ground before being milled using a ball mill at a speed of 300 rpm with a ball-to-powder ratio of 50:1 in 10 ml of toluene for the times specified above.

Both AlFeCoNiTi and AlFeCuZnTi were characterized with XRD (Bruker X8) with Cu-Kα radiation (wavelength of $\lambda = 1.54$ Å). The data was analyzed using Origin-Lab Learning Edition software. Field emission scanning electron microscopy (FESEM) and energy-dispersive spectroscopy (EDS) were used for morphological and elemental characterization of the powders, respectively.

3 Results and Discussions

3.1 Results

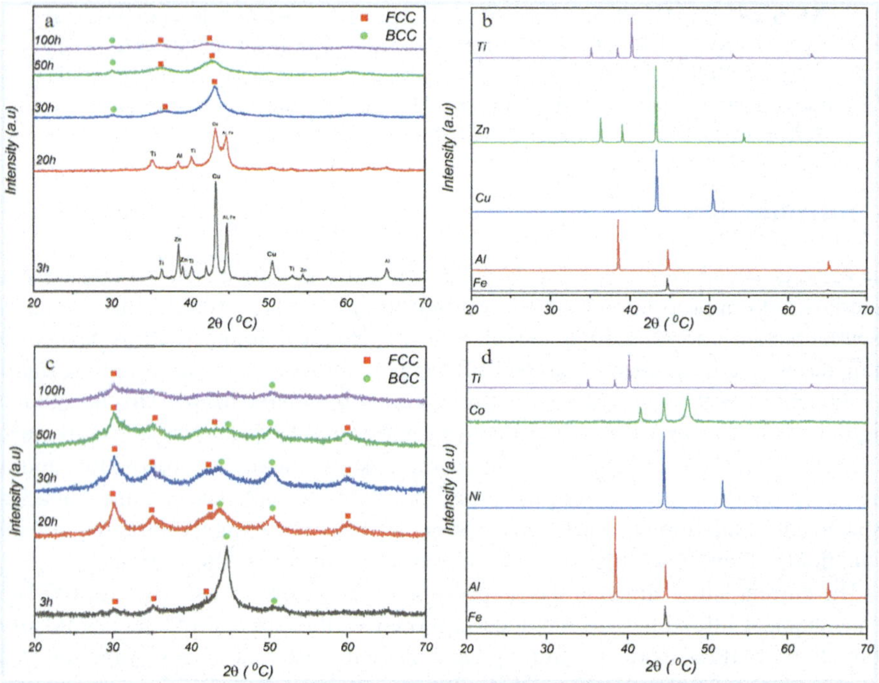

Fig. 1. a), d) XRD patterns of mechanically alloyed AlFeCuZnTi and AlFeCoNiTi powder at different milling times, respectively; b), d) XRD patterns of metals prepared for the synthesis of high-entropy alloys.

Figure 1a showed XRD patterns of mechanically alloyed AlFeCuZnTi powder at different milling times. At the initial stage of grinding (3 h), peaks corresponding to the constituent metals are observed. As the grinding time increases from 3 h to 20 h, the intensity of the peaks decreases, and the peaks become broader. Specifically, the Zn peaks exhibit a faster decrease in intensity compared to the peaks of other elements. This is related to the melting point of each constituent element of the alloy. Elements with lower melting points tend to experience a more significant decrease in peak intensity. In the term of AlFeCoNiTi powder, after 3 h of grinding, the width of the diffraction peaks increased compared to those of the pure metals at Fig. 1c&d. The diffraction peaks tend to shift towards lower 2θ, suggesting that the powder sample has undergone compressive stress during ball milling. In addition, there is a disappearance of the diffraction peaks of the original elements, indicating the formation of a solid solution, i.e., the beginning of alloying. The positions of the diffraction peaks of the powder sample further confirm the formation of a solid solution. Moreover, during high-energy planetary ball milling, after 100 h of grinding, the peaks are fully broadened, indicating a decrease in the particle size of the elemental components.

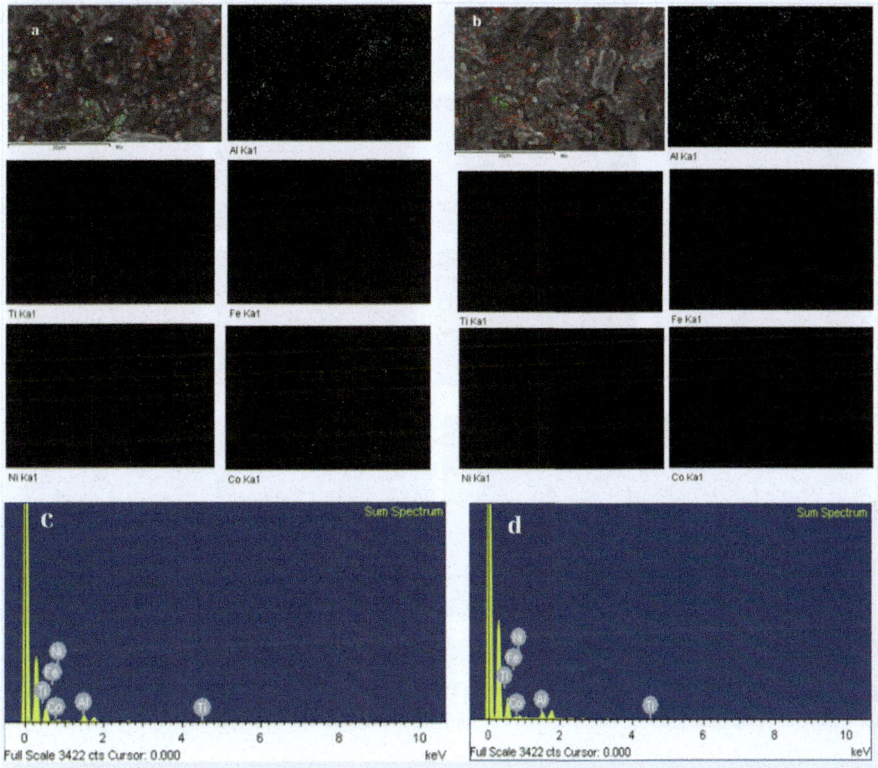

Fig. 2. a) & b) the EDS maps of AlFeCoNiTi alloy at milling time 0.5h and 1h, respectively; b) & c) EDS spot analysis of AlFeCoNiTi alloy at milling time of 0.5h and 1h, respectively.

SEM-EDS analysis confirmed the presence of all the original elements used to prepare the high-entropy alloy powder, indicating the reaction between the metals before the formation of the high-entropy alloy in Fig. 2c & d. After 0.5 h of ball milling, the fabricated AlCoFeNiTi alloy exhibits an uneven distribution of elements. A significant amount of Ti is observed in the sample, which can be attributed to its high melting point of 1941K. Additionally, the presence of a large amount of Al can be explained by the diffusion of Co, Fe, and Ni into the Al lattice (Fig. 2a). Compared to the sample milled for 0.5 h, the EDS mapping at 1h (Fig. 2b) hour shows a significant decrease in the distribution of Fe, Co, and Ni elements. This is attributed to their diffusion into the Al lattice. The distribution of Ti remains relatively uniform due to its higher melting point compared to the other three elements in the AlCoFeNiTi high-entropy alloy. This observation aligns with the initial prediction based on the XRD analysis of the AlCoFeNiTi alloy, where the disappearance sequence of XRD diffraction peaks for each element was correlated with their melting temperatures.

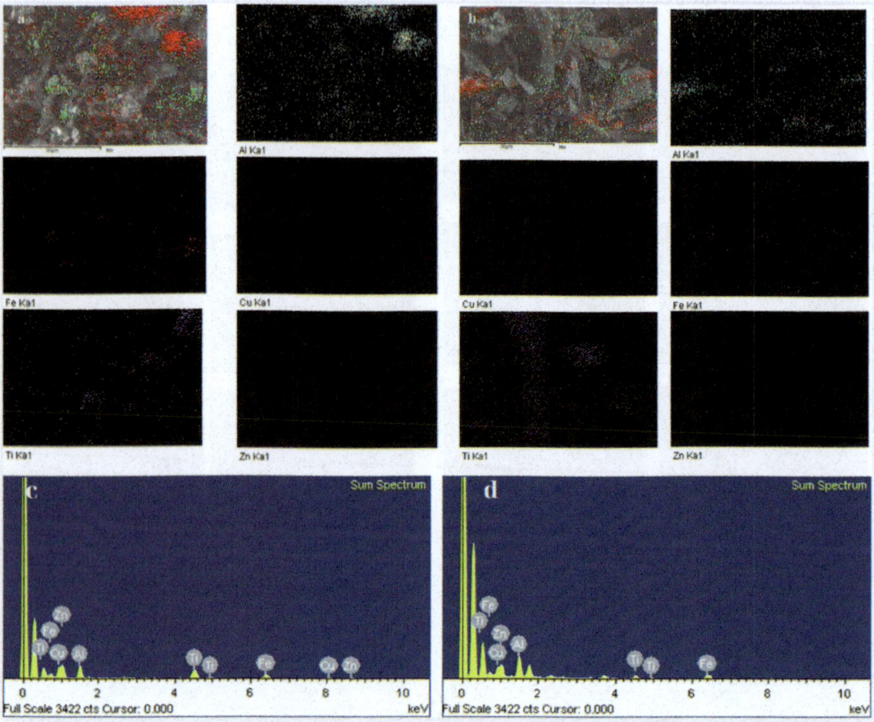

Fig. 3. a) & b) the EDS maps of AlFeCuZnTi alloy at milling time 0.5h and 1h, respectively; b) & c) EDS spot analysis of AlFeCuZnTi alloy at milling time of 0.5h and 1h, respectively.

Figure 3a showed that elemental distribution of Al, Cu, Fe, Zn, and Ti after 0.5 h of ball milling: The fabricated AlFeCuZnTi alloy exhibits a relatively uniform distribution of elements. Due to the high melting points of Ti and Fe, a significant amount of these elements is observed in the sample. Notably, the presence of Al can be attributed to the diffusion of Cu and Zn into the Al lattice. Similar to the EDS mapping at 0.5 h, there is no significant change in the elemental distribution at 1 h (Fig. 3b). This is attributed to the high melting points of Ti and Fe, which hinder the formation of the alloy.

4 Conclusions

Both high entropy alloy samples were successfully synthesized via high-energy ball milling for different durations. X-ray diffraction analysis revealed the presence of peaks indicative of the formation of a dual-phase microstructure comprising both FCC and BCC phases. The formation of these solid solution phases is attributed to the difference in atomic sizes.

Acknowledgments. We acknowledge Ho Chi Minh City University of Technology (HCMUT), VNU-HCM for supporting this study.

References

1. Sarkar, A., et al.: High-entropy oxides: fundamental aspects and electrochemical properties. Adv. Mater. **31**(26), 1806236 (2019)
2. Ma, Y., et al.: High-entropy energy materials: challenges and new opportunities. Energy Environ. Sci. **14**(5), 2883–2905 (2021)

Predicting Loss of Circulation During Drilling Using Decision Trees and Ensemble Learning Algorithms

T. N. T. Tran(✉), Q. T. Hoang, N. N. H. Le, and H. T. Nguyen

Ho Chi Minh City University of Technology (HCMUT), VNU-HCM, 268 Ly Thuong Kiet, District 10, Ho Chi Minh City, Vietnam
trantam2512@hcmut.edu.vn

Abstract. Drilling fluid is a crucial factor in the drilling operation. The loss of drilling fluid during drilling is inevitable. Attempts to compute based on physical modeling are not feasible, and experimental correlations are of-ten imprecise in calculating the amount of drilling fluid lost. Currently, machine learning is the most effective method to solve the above problem. In this study, the authors will perform the prediction of loss of circulation during drilling using Decision Tree and ensemble learning algorithms including Random Forest, Extra Tree, Gradient Boosting, and XGBoost. The data used in the study were obtained from the Marun oil field in Iran. The prediction results of the algorithms are compared based on statistical parameters such as Mean Absolute Error (*MAE*), Root Mean Squared Error (*RSME*), and Coefficient of Determination (R^2), and it is found that the Extra Tree algorithm has the most accurate prediction results with the lowest *MAE* and *RMSE* (3.6284 and 10.2039), and the highest R^2 (0.9891) in the five algorithms mentioned.

1 Introduction

Drilling fluid is a mixture of oil and/or water and clay that use in oil-well drilling to cool and lubricate the bit, flush rock cuttings to the surface and stability the formation. One of the common problems while drilling is loss of circulation. Because during fluid plays such an important role, it is necessary to control the loss of circulation. The best approach is forecasting to prevent loss of drilling fluid. There are two common approaches for prediction: physical modeling and data-driven. The first approach is relatively difficult and specific to each formation. A few research efforts can be listed here as [1–3]. The second approach has been most common in recent years. The explosion of artificial intelligence and machine learning has given researchers very powerful tools to solve engineering problems and challenges in general and petroleum engineering in particular. There are so many studies regarding the use of machine learning and deep learning techniques for predicting loss of circulation during drilling. The earliest can be mentioned in some studies using ANNs to predict [4]. Later, developments in machine learning techniques have given researchers many effective methods to more accurately predict loss of circulation. These studies use supervised machine learning algorithms

N. R. Abdul Razak et al. (Eds.): ECOOP 1996, SPPHY 418, pp. 256–260, 2025.
https://doi.org/10.1007/978-981-96-2871-1_37

such as artificial neural networks, support vector machines, particle swarm optimization, decision trees, convolutional neural network and hybrid machine learning algorithms [5–10].

In this study, the authors extend predictive methods based on recently developed supervised learning algorithms such as decision trees, random forest, extra trees, gradient boosting, and XGBoost. Research data collected from the Marun field in Iran [11]. Using statistical parameters including Mean Absolute Error (*MAE*), Root Mean Squared Error (*RSME*), and Coefficient of Determination (R^2), the estimated results are then compared to evaluate and select the optimal method.

2 Experimental Procedure

Basically, the Decision Trees (DT) algorithm is not ensemble learning, but rather the foundation of ensemble learning algorithms such as Random Forest (RF), Extra Trees (ET), Gradient Boosting (GB), and XGBoost (XGB). Ensemble learning is a general meta-approach for improving predictive performance by combining predictions from multiple models. It has three main methods are bagging, boosting, and stacking.

Bagging is a combined term of bootstrapping and aggregating. Bagging generates multiple subsets of data from the original dataset through random sampling. Then, it trains a base learner (such as decision trees) on each bootstrap subset independently. After training the base learners, predictions are aggregated to make a final prediction. For regression tasks, the final result is averaging the predictions from all base learners. For classification tasks, the final prediction is often determined by majority voting among the base learners.

Boosting is a technique that aims to create the strong learner by sequentially training weak learners (for instance, decision trees). It starts with a first weak learner trained on the entire training dataset by initializing weights for all data points. The next weak learners will be trained for updating weights through the loss function to find the best strong learner. With regression tasks, the loss function is typically the Mean Squared Error (ME). While with classification tasks, the popular loss function is Log Loss.

Stacking, also known as stacked generalization, is a machine learning technique that combines predictions from multiple base models (e.g., decision trees, random forest, extra trees, gradient boosting, XGBoost) to create a meta-model for improved accuracy. Base models are trained on a loss function like Mean Squared Error (MSE) for regression or Log Loss for classification. Decision trees identify attributes using Information Gain, while random forests and extra trees build multiple trees, with random forest using subsamples and extra trees using the full dataset. Gradient boosting and XGBoost train models sequentially, compensating for previous errors to refine predictions, making XGBoost highly efficient.

3 Results and Discussions

Using processed training and testing sets with the scikit-learn library in Python, predictions of the loss of circulation rate values were made. With the Decision Tree Regressor function in scikit-learn, prediction results were obtained and compared with measured results, as shown in Fig. 1(a). Random Forest (RF). Using the Random Forest Regressor function in the scikit-learn library, we have the prediction results compared with the measured results, as shown in Fig. 1(b).

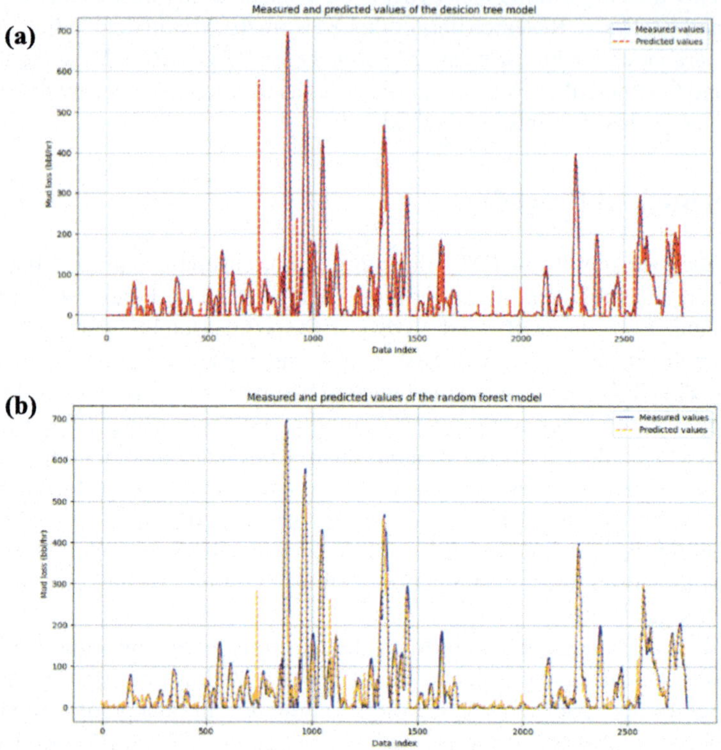

Fig. 1. (a) Measured and predicted values of the decision tree model and (b) Measured and predicted values of the random forest model

Extra Trees (ET). Using the Extra Trees Regressor function in the scikit-learn library, we have the prediction results compared with the measured results, as shown in Fig. 2. The prediction results were obtained using the Gradient Boosting Regressor function in the scikit-learn library and were compared with the measured results. As shown in Fig. 2, the Extra Trees algorithm produced the most accurate predictions, yielding the lowest MAE and RMSE (3.6284 and 10.2039, respectively) and the highest R^2 value (0.9891) among the five learning algorithms.

The comparison reveals how the dataset is handled by each algorithm, with improvements shown by ensemble methods (Random Forest and Extra Trees) over single decision trees due to their ability to reduce variance. Extra Trees is noted as the most accurate, having the lowest error rates (MAE and RMSE) and the highest R2 score, indicating that it best fits this specific data for precise predictions.

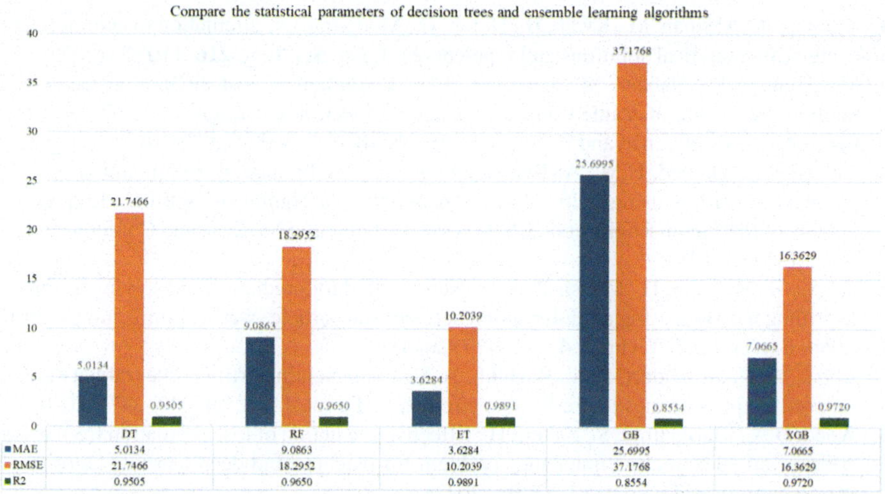

Fig. 2. Statistical parameters of decision trees and ensemble learning algorithms

4 Conclusion

This study predicts loss of circulation while drilling based on Decision Trees (DT) and four ensemble learning algorithms, including Random Forest (RF), Extra Trees (ET), Gradient Boosting (GB), and XGBoost (XGB). The research data is taken from open literature in the Marun oil field, Iran. Here are the key points about this work:

- First is that optimal model training depends heavily on data preprocessing and feature selection. As shown in the study, from the original 19 features and 2820 data records, after analysis, only 16 features and 2783 data records were selected to build the model.
- With the estimation and comparison of different methods for the regression task, the study has shown the great advantages of the Extra Trees algorithm. ET model has the the highest R^2 (0.9891) and the lowest *MAE* and *RMSE* (3.6284 and 10.2039, respectively). The highest value of R^2 and lowest values of *MAE* and *RMSE* prove that the ET is best algorithm in this study.

This research can also be applied to estimate other real data sets and extended by applying deep learning or hybrid machine learning algorithms.

Acknowledgements. The authors would like to thank Ho Chi Minh City University of Technology (HCMUT), VNU-HCM with the support of time and facilities for this study.

References

1. Albattat, R., Hoteit, H.: Modeling Lost-Circulation into Fractured Formation in rock drilling operations. Drilling Technology (2021)
2. Feng, Y., Gray, K.E.: Modeling lost circulation through drilling-induced fractures. SPE J. **23**, 205–223 (2018)

3. Albattat, R., AlSinan, M., Kwak, H., Hoteit, H.: Modeling lost-circulation in natural fractures using semi-analytical solutions and type-curves. J. Pet. Sci. Eng. **216**, 110770 (2022)
4. Moazzeni, A.R., Nabaei, M., Jegarluei, S.G.: Prediction of lost circulation using virtual intelligence in one of Iranian oilfields. Society of Petroleum Engineers - Nigeria Annual International Conference and Exhibition 2010, NAICE, 1, 393–400 (2010)
5. Abbas, A.K., Hamed, H.M., Al-Bazzaz, W., Abbas, H.: Predicting the amount of lost circulation while drilling using artificial neural networks: An example of Southern Iraq oil fields. Society of Petroleum Engineers - SPE Gas Oil Technology Showcase and Conference, GOTS 2019, 1–14 (2019)
6. Toreifi, H., Rostami, H., Manshad, A.K.: New method for prediction and solving the problem of drilling fluid loss using modular neural network and particle swarm optimization algorithm. J. Pet. Explor. Prod. Technol. **4**, 371–379 (2014)
7. Abbas, A.K., Al-haideri, N.A., Bashikh, A.A.: Implementing artificial neural networks and support vector machines to predict lost circulation. Egypt. J. Pet. **28**, 339–347 (2019)
8. Wood, D.A., Mardanirad, S., Zakeri, H.: Effective prediction of lost circulation from multiple drilling variables: a class imbalance problem for machine and deep learning algorithms. J. Pet. Explor. Prod. Technol. **12**, 83–98 (2022)
9. Jafarizadeh, F., et al.: A new robust predictive model for lost circulation rate using convolutional neural network: a case study from Marun Oilfield. Petroleum (2022)
10. Sabah, M., Mehrad, M., Ashrafi, S.B., Wood, D.A., Fathi, S.: Hybrid machine learning algorithms to enhance lost-circulation prediction and management in the Marun oil field. J. Pet. Sci. Eng. **198**, 108125 (2021)
11. Marun field. https://en.wikipedia.org/wiki/Marun_Field

Rheological Properties of Two-Stage Concrete (TSC) Mortar with Varying Silica Fume and Superplasticizer Additives

Nguyen Ngoc Tri Huynh[1,2], Nguyen Hoang Thien Khoi[1,2], and Nguyen Khanh Son[1,2,3](✉)

[1] Department of Silicate Materials, Faculty of Materials Technology, Ho Chi Minh City University of Technology (HCMUT), 268 Ly Thuong Kiet Street, District 10, Ho Chi Minh City, Vietnam
ksnguyen@hcmut.edu.vn

[2] Vietnam National University Ho Chi Minh City, Linh Trung Ward, Thu Duc City, Ho Chi Minh City, Vietnam

[3] Centre of Excellence Geopolymer and Green Technology (CEGeoGTech), Universiti Malaysia Perlis (UniMAP), 01000 Kangar, Perlis, Malaysia

Abstract. Control over the flowability of cement mortar in the two-stage concreting method is crucial for ensuring the desired performance and durability of the final concrete structure. Thus, a comprehensive study using an advanced vane-type coaxial rheometer, which precisely measures the material's flow characteristics, is essential alongside regular flow cone measurements. In this study, TSC cement mortar was mixed with varying amounts of silica fume (SF) and superplasticizer (SPs) to investigate their impact on rheological properties. The results indicated that adding SF increased the viscosity and yield stress while adding SPs reduced these properties. The variations emphasize the significance of meticulous selection and proportioning of supplementary cementitious materials (SCMs) and admixtures to attain the desired rheological properties of cement mortar. Refining the mixture using the insights obtained from this study makes it feasible to improve the workability, strength, and durability of the concrete, ultimately leading to more efficient construction practices.

1 Introduction

Two-Stage Concrete (TSC), also known as Preplaced Aggregate Concrete (PAC), is a special concreting method that differs from conventional concrete. In TSC, coarse aggregates are pre-placed in the formwork, and grout mortar is injected into the formwork to fill the voids between the pre-placed aggregates. This unique placement technique offers several technological and sustainability advantages in modern concrete technology. However, the quality control of TSC heavily depends on the properties of the grout mortar used and its ability to flow around the pre-placed aggregate particles and effectively fill voids. According to the guideline ACI 304.1R, the flowability of grout mortar can be controlled using the Flow Cone test (ASTM C939), which measures the time of

N. R. Abdul Razak et al. (Eds.): ECOOP 1996, SPPHY 418, pp. 261–271, 2025.
https://doi.org/10.1007/978-981-96-2871-1_38

discharge of a sample of grout through an orifice [1–3]. Mechanical resistances are also controlled after 28 days of curing time. The properties of the mortar and concrete samples at both early and later ages are usually determined by the mixture of materials used. In this context, it is crucial to investigate how different components affect the rheological properties of grout mortar. Chemical and mineral admixtures, such as superplasticizers (SPs), fly ash (FA), and silica fume (SF), can significantly improve the overall properties of concrete. Understanding the effects of these components on the flowability and other rheological properties of grout mortar is essential for ensuring high-quality mortar and optimizing the casting technology.

The rheological properties of fresh mortar and concrete are characterized by their complexity. However, understanding these properties is important for controlling the quality of the mortar and concrete after mixing, as well as optimizing the casting technology. The complexity of the rheological properties is due to the heterogeneous mixtures and time-dependent properties of cement mortar and concrete at an early age. Immediately after mixing, the cement particles are attracted to each other, forming a flocculated structure. To initiate the flowability of the cement mortar, a sufficiently large shear stress is required to break the bonds between these cement particles in their initial state, resulting in the yield stress of the cement mortar or Bingham fluid model [4–6]. Besides, cement mortars and concrete exhibit shear-thinning or thickening behaviors, which result from the partial recovery of the bonding state between the cement particles after shearing. This is represented by a decrease over time in the material's viscosity as the shear stress or shear rate increases, or by an increase over time in the material's viscosity as the shear stress or shear rate decreases [7]. The non-linear relationship between shear stress and shear rate can be described using the three-parameter Herschel-Bulkley model (τ_0, K, n). However, due to its simplicity, the two-parameter Bingham model has been widely used in many studies on the rheological behavior of cement mortar and concrete (Fig. 1).

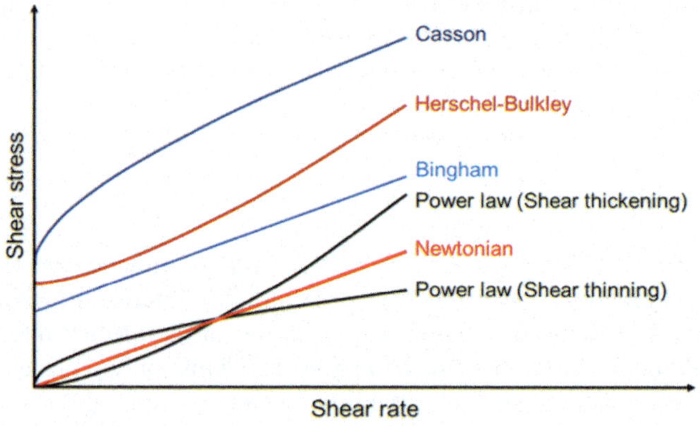

Fig. 1. Basic constitutive relationships for fluid flow [8–11].

The following section investigates the rheological properties of fresh mortar mixtures using a vane-type rheometer developed in our laboratory based on a simple rod stirrer apparatus. This study utilized the Bingham model to fit the torque data obtained from the vane's rotation speed to estimate the grout's viscosity and yield stress. To evaluate the reliability of the rheometer, the variations in viscosity, yield stress, and flow time were correlated through a cone's orifice (Flow Cones test ASTM C939). Subsequently, experiments were conducted using mortar mixes containing FA, SF, and SPs to analyze the effects of these ingredients on the mortar's rheological properties and optimize the material mixture. In the conclusion section, the results obtained were highlighted, which can provide valuable insights into optimizing mix designs for TSC, resulting in improved mechanical properties and rheological performance.

2 Methodology

2.1 Materials and Mortar Mixing Procedure

The grout mortar was formulated with a mixture of cement, river sand as fine aggregate, tap water, and a superplasticizer admixture to achieve optimal performance. The river sand used in the mortar was meticulously prepared. It was first washed with water to remove impurities, then dried in an oven to eliminate moisture, and finally subjected to sieve analysis. This careful process resulted in a fineness modulus of 1.6, indicating a significant presence of fine particles.

To enhance the workability and reduce the mixture's water content, Sika® ViscoCrete®-8700 M, a high-performance superplasticizer, was incorporated at a dosage of 1.2% by weight of cement. This admixture effectively reduced the water content by approximately 40% by weight, ensuring a more workable and durable mortar. The composition of one batch of the control mixture was carefully measured to include 1000 g of cement, 309 g of river sand, 240 g of water, and 15 g of superplasticizer.

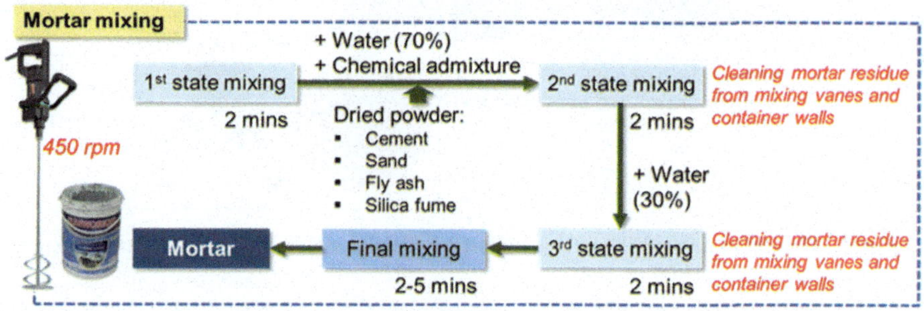

Fig. 2. Handheld mixer Eibenstock EHR 15,1 SB and different steps of mortar mixing.

Given the high cement content, thorough mixing was crucial to achieve a homogeneous, workable mortar. Therefore, a handheld mixer with a maximum rotation speed of up to 450 rpm was employed at the final mixing stage, facilitating the efficient dispersion of cement and improving the overall flowability of the mixture compared to the planetary mixer (three other mixing stages). The total mixing time for one batch was approximately 10 min, ensuring that all components were well-integrated and the desired consistency was achieved. This meticulous formulation and preparation process (Fig. 2) guaranteed the quality and performance of the grout mortar. It provided a reliable baseline for further experimentation and optimization in creating high-performance concrete structures.

2.2 Rheological Measurement and Estimation of Viscosity, Yield Stress

This study used a vane-type coaxial rheometer in conjunction with a customized rod stirrer apparatus from Heidolph Instruments GmbH & Co. KG. The rheometer comprised four additional components: a stainless steel rack, a wand, a container, and a six-bladed brass propeller, as depicted in Fig. 3. Advanced software was employed to control the stirring speed and record torque values. The outer cylinder of the rheometer was a container with a 10 cm diameter, while the rotating propeller formed the inner cylinder, consistent with the schematic assembly of a coaxial system. A plastic liner with a sharp surface was introduced to prevent adhesion between the cement mortar and the container wall, which also facilitated cleaning after each experiment. The six-bladed propeller (Fig. 4a), meticulously designed and manufactured, had dimensions of 5 cm in diameter, 5 cm in height, and 1 mm blade thickness. It was affixed to the steel wand using a threaded system.

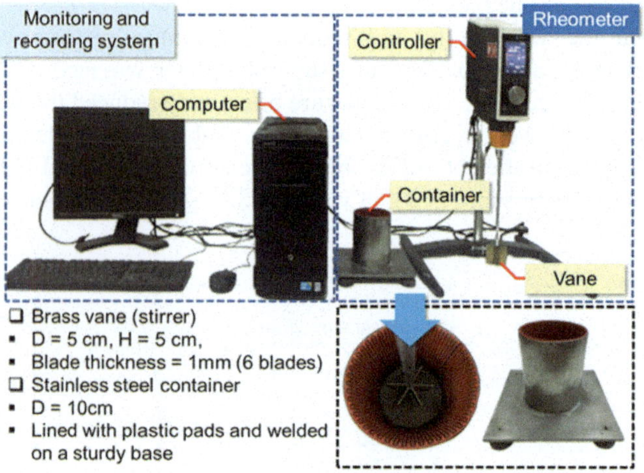

Fig. 3. Handheld mixer Eibenstock EHR 15,1 SB and different steps of mortar mixing.

During operation, the stirrer rotor and propeller rotated, and the torque value was recorded every second based on the set speed. As the propeller applied torque, it sheared the mortar, causing internal stress within the fresh mortar material. This setup enabled precise measurement of the rheological properties of the mortar, providing valuable data for the study.

Before commencing any rheometer measurements, initiating and rotating the stirrer at a speed ranging from 300 to 400 rpm for 30 min is essential, which is imperative to ensure the machine's smooth operation and diminish any effects caused by engine inertia and noise in the data. The stirrer should be firmly fixed to the support, and the entire setup must remain stable throughout the experiment to avoid measurement errors. It has been observed that air bubbles in the mortar immediately after mixing can significantly impact the results. Thus, measurements should only be taken after stabilizing the mortar for at least 15 min. Moreover, the control program for the propeller's rotation over time should be considered a critical parameter, which varies depending on the type of fluid being studied, such as cement mortar. This study examined rotation programs (profile) with three stages. These stages include (i) initial acceleration to break the initial high viscous state of the mortar; (ii) constant rotational speed after the mortar loses its structure and becomes flowable, and (iii) final deceleration to allow the mortar to reconstruct its plastic and viscous state. These rotation programs were selected after carefully reviewing previous studies in the literature, which are depicted in Fig. 4b. The rotation profile comprises three stages, namely, (I) ramping up from 40 to 80 rpm over 10 s, (II) maintaining a constant speed of 80 rpm for 30 s, and (III) ramping down from 80 to 40 rpm in a ladder shape over 1 min and 30 s. Each step is spaced at 5 rpm and held for 10 s before settling at 40 rpm for 20 s (Fig. 5).

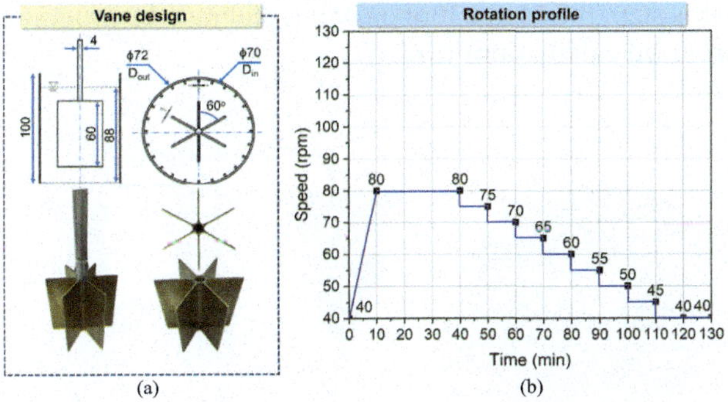

Fig. 4. Vane design and rotation profile with three stage

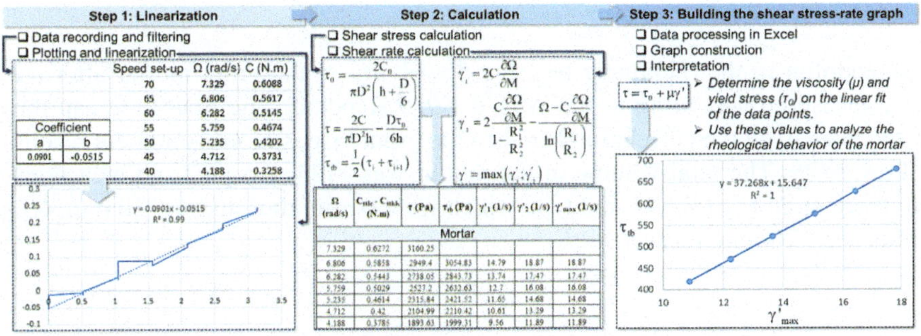

Fig. 5. Data processing for calculating and plotting shear stress-strain diagram.

In the early stages of the experiment, the torque values versus rotation speed exhibit a continuous increase until the initial state of the vicious mortar is disrupted. The critical shear stress, or yield threshold τ_0, can be calculated by determining the line intersecting the torque axis from this nearly linear relationship, utilizing the formula that accounts for the influence of the fluid's vortex motion around the propeller and at the bottom of the container, which is fixed at the height of 60 mm. The shear stress τ and the shear strain velocity γ' must be related to estimating the viscosity value, assuming that the cement mortar behaves rheologically according to the Bingham model. The formula for calculating shear stress on the cylindrical section is $\tau = 2T/\pi r2$ where T is the measured torque, r is the cylindrical section's radius, and τ is the shear stress.

The relationship between the initial torque and the shear strain rate γ' depends on the fluid flow regime in the early stages of the test. Therefore, two shear strain rate values are calculated: the first γ' corresponding to the partial shear force applied to the mortar at the beginning of the measurement and the second γ' when the mortar has been fully sheared. The higher of these two values is chosen for viscosity calculations, which is obtained by selecting the maximum value of γ' between the first and second stages of the experiment. The shear stress and shear strain velocity values on the graph can be used to estimate the viscosity μ and the yield threshold τ_0 using the Bingham equation $\tau_0 = \pi + \mu\gamma'$. In this step, the Marco tool on MS Excel may be programmed to perform quick and synchronized calculations (Fig. 6).

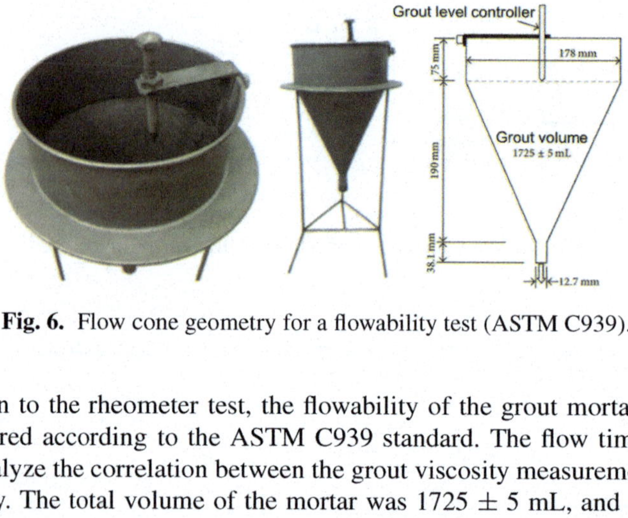

Fig. 6. Flow cone geometry for a flowability test (ASTM C939).

In addition to the rheometer test, the flowability of the grout mortar after mixing is also measured according to the ASTM C939 standard. The flow time can thus be utilized to analyze the correlation between the grout viscosity measurement results and the flowability. The total volume of the mortar was 1725 ± 5 mL, and measurements were taken at t_0, t_{30}, t_{60}, t_{90}, and t_{120}, which were 0 min after mixing, 30 min, 60 min, 90 min, and 120 min, respectively.

3 Results and Discussions

3.1 Flowability of Cement Mortar

After 90 min, the mortar exhibits significantly decreased flow, even stopping the flow, while it still flows adequately at 30 and 60 min, satisfying the requirements for two-stage concrete. The two-stage concrete mortar is expected to have a flow time of less than 90 s. For high-strength TSC, achieving a grout efflux time of 35–40 ± 2 s is recommended as per ACI 304.1 [12, 13]. This decline in flowability over time is further illustrated in Fig. 7, where the diameter through which the grout flows diminishes as the retention time is extended.

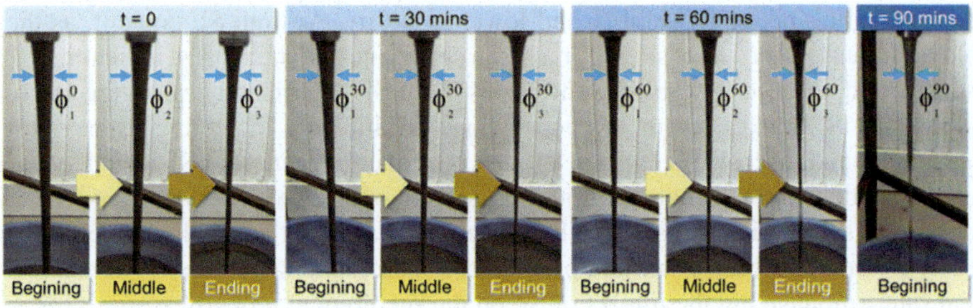

Fig. 7. Flowability of mortar through the flow cone test after different periods after mixing.

As shown in Fig. 7, the diameter through which the grout flows diminishes as the retention time is extended. This relationship is illustrated by the progressive decrease in flow diameter, denoted as ($\phi_1^i > \phi_2^i > \phi_3^i$), and similarly by ($\phi_1^0 > \phi_1^{30} > \phi_1^{60} > \phi_1^{90}$). These notations indicate that the flow diameter decreases at successive intervals as the retention time increases. At intervals of 30, 60, and 90 min, the mortar attains greater density due to the cement's ongoing hydration process. During hydration, cement particles react with water to form a hardened matrix, resulting in a progressive increase in the mortar's density and a corresponding reduction in its flowability. This densification process enhances the bond strength between the particles, gradually surpassing gravity's influence. Gravity is typically the primary force that causes fluid bonds to disintegrate and enables the mortar to flow freely from the base of a cone or similar apparatus used for measuring flow diameter.

As the retention time progresses, the increasing density and bond strength restricts the flow of the mortar, leading to a smaller flow diameter. This phenomenon is critical for understanding the workability and performance of the mortar over time. Initially, the mortar exhibits higher flowability, making it easier to handle and place. However, as the hydration process advances, the mortar becomes denser and less flowable, eventually reaching a point where it transitions from a fluid to a semi-solid state. This behavior has significant implications for construction practices, particularly regarding timing and application. For instance, in scenarios requiring extended working times, such as large-scale pours or intricate formwork, it is crucial to account for the decreasing flowability over time. Contractors and engineers must plan and execute the mixing and placing of mortar within a specific timeframe to ensure optimal workability and performance.

3.2 Effect of Varying SF on the Flowability of Mortar

Figure 8a shows how different amounts of SF affect the rheological properties of mortar samples at various retention times while the SPs content remains constant. The addition of SF increases the mortar's viscosity (corresponding slope values: 9.3, 13.5, 21.4, 36.2 Pa.s), making it less flowable, as indicated by the steep slopes in the graph (Fig. 8a). The relationship between shear stress and shear rate (Fig. 8b) reveals that higher SF content leads to more significant yield stress (corresponding stress: 4.1, 5.3, 12.1, 16.8 Pa). At low SF content, mortars with low SF content exhibit lower shear stress at given shear rates, indicating better flowability and lower viscosity. At high SF content, as the SF content increases, the shear stress for a given shear rate also increases, indicating higher viscosity and reduced flowability. The inclusion of SF significantly raises the viscosity; adding 10% SF makes it almost non-flowing, significantly impacting its workability. This heightened viscosity is attributed to the pozzolanic reaction of SF, which enhances the packing density of the cementitious matrix, reduces free water content, and increases the mixture's overall stiffness, which the steeper slopes in the graph can demonstrate. Higher SF content leads to a more pronounced increase in shear stress with increasing shear rate, emphasizing the impact of SF on the rheological properties.

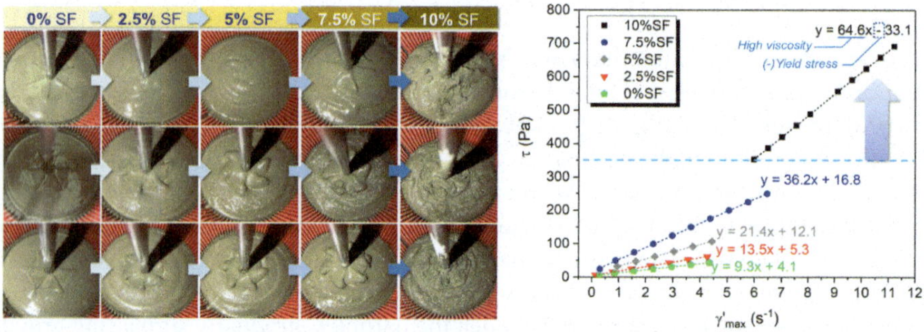

(a) (b)

Fig. 8. (a) State of mortar samples while measuring rheology at different retention times with constant SPs content and varying amounts of SF. (b) Corresponding relationship between shear stress and shear rate.

3.3 Effect of Varying SPs on the Flowability of Mortar

Figure 9a presents the impact of SPs on mortar rheology while maintaining a constant amount of silica fume. Without SPs, the mortar has very limited flowability and high viscosity, even with a minimal SPs content of 0.25%. At this low SPs content, the behavior of the mortar remains similar, indicating that such a small amount of SPs does not significantly improve workability or reduce viscosity. As the SPs content increases, the mortar flow becomes more accessible. At 0.5% SPs content, there is a noticeable improvement in flowability, suggesting that moderate amounts of SPs can enhance the workability of the mortar. At 1% SPs content, the mortar exhibits significantly reduced shear stress and very low viscosity, making it much easier to handle, mold, and place, even in complex forms. This improvement in flowability is attributed to the effective dispersion of cement particles facilitated by the SPs, reducing inter-particle friction and enhancing the lubricating effect within the mortar matrix.

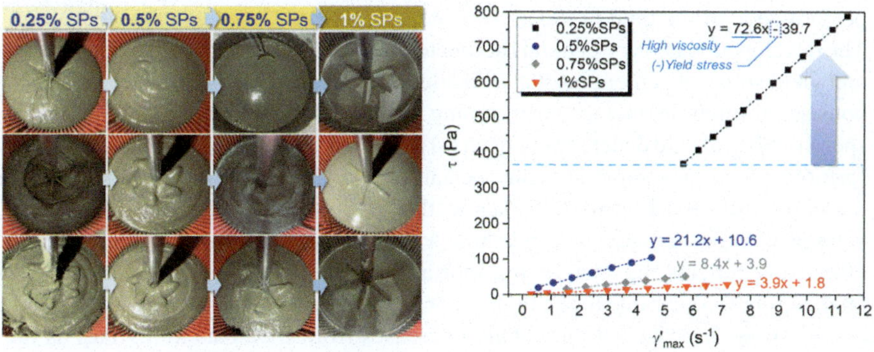

(a) (b)

Fig. 9. (a) State of mortar samples while measuring rheology at different retention times with constant SF content and varying amounts of SPs. (b) Corresponding relationship between shear stress and shear rate.

Figure 9b presents the corresponding relationship between shear stress and shear rate for mortar samples with varying superplasticizer amounts and a constant amount of silica fume. At low shear rates, the initial shear stress is relatively high for mixtures with low SPs content, which indicates that without sufficient SPs, the mortar has a higher resistance to flow, even at low shear rates. As the amount of SPs in the mixture increases, the initial shear stress decreases (corresponding yeild values: 10.6, 3.9, 1.8 Pa), which means the mortar becomes more flowable. This reduction in shear stress is due to the SPs reducing the internal friction between the particles, enhancing the lubricating effect, and allowing the mortar to flow more easily. The slope of the shear stress-shear rate curve represents the mixture's viscosity. With increasing SPs content, the slope becomes less steep, indicating a reduction in viscosity (corresponding slope values: 21.2, 8.4, 3.9 Pa.s), suggesting that the mortar requires less force to achieve the same shear rate, enhancing its workability. The critical shear stress (yield point) is significantly lower for mortars with higher SP content. This result means that the force required to initiate flow is reduced, which is beneficial for applications requiring high flowability and ease of placement. A higher SPs content is beneficial in high-flowability applications, such as self-consolidating concrete or intricate formworks. For applications requiring higher structural integrity and less deformation under stress, a balance between SPs content and other components must be achieved to ensure optimal performance.

4 Conclusions

Based on the study conducted, several conclusions can be drawn that enhance our understanding of the rheological properties of cement-concrete mortar and their implications for practical applications:

- Viscosity and yield threshold are critical indicators of the working properties during the early stages of setting and hardening of cement mortar. These properties accurately reflect the mortar's workability, providing valuable insights for adjusting mix design, construction conditions, and casting technology to ensure the highest strength and performance of the final product.
- The developed vane-type coaxial rheometer is capable of measuring both the viscosity and yield stress of cement mortar with adequate reliability. The tested rotation profile consists of three distinct stages: ramping up from 40 to 80 rpm, maintaining a fixed speed of 80 rpm, and decelerating from 80 to 40 rpm in a ladder-like fashion to measure formulated mortar samples. On the one hand, the results reveal that viscosity gradually increases from 9.3 Pa.s to 36.2 Pa.s as the SF content varies from 0% to 7.5% by weight, with corresponding yield stress values ranging from 4.1 Pa to 16.8 Pa. This effectively reflects the state of the mortar as it becomes thicker, denser, and more challenging to shear due to the higher content of superfine SF. On the other hand, the results show that viscosity gradually decreases from 21.2 Pa.s to 3.9 Pa.s as the SP content varies from 0.5% to 1% by weight, with corresponding yield stress values ranging from 10.6 Pa to 1.8 Pa. This effectively reflects the state of the mortar as it becomes thinner, more flowable, and easier to shear due to the higher content of the superplasticizer.
- There is a strong correlation between the flow time and rheological parameters of the mortar, indicating that these parameters are reliable predictors of the material's behavior. The resulting viscosity and yield stress show higher sensitivity to changes in the mortar's state, providing a comprehensive understanding of its rheological characteristics throughout the setting and hardening process.

Future research will explore the effects of various supplementary cementitious materials, admixtures, and environmental conditions on cement-concrete mortar's rheological behavior and performance.

Acknowledgments. We acknowledge Ho Chi Minh City University of Technology (HCMUT), VNU-HCM for supporting this study.

References

1. da Silva, O., Gomes, P.C.C., Morais, K.A.M.: Analytical method for determining the saturation point of superplasticizers in cement pastes using the Marsh cone. Cerâmica **65**, 252–260 (2019)
2. Chairunnisa, N., Fardheny, A.F.: The study of flowability and the compressive strength of grout/mortar proportions for pre-placed concrete aggregate (PAC). In: MATEC Web of Conferences. p. 04010. EDP Sciences (2019)

3. Agulló, L., Toralles-Carbonari, B., Gettu, R., Aguado, A.: Fluidity of cement pastes with mineral admixtures and superplasticizer—A study based on the Marsh cone test. Mat. Struct. **32**, 479–485 (1999). https://doi.org/10.1007/BF02481631
4. Weng, Y., Li, M., Tan, M.J., Qian, S.: Design 3D printing cementitious materials via Fuller Thompson theory and Marson-Percy model. Constr. Build. Mater. **163**, 600–610 (2018). https://doi.org/10.1016/j.conbuildmat.2017.12.112
5. Yahia, A., Khayat, K.H.: Analytical models for estimating yield stress of high-performance pseudoplastic grout. Cem. Concr. Res. **31**, 731–738 (2001)
6. Güllü, H.: Comparison of rheological models for jet grout cement mixtures with various stabilizers. Constr. Build. Mater. **127**, 220–236 (2016)
7. Brahme, A.: Comprehensive biomedical physics. Newnes (2014)
8. Sochi, T.: Pore-scale modeling of non-Newtonian flow in porous media. arXiv preprint. arXiv:1011.0760 (2010)
9. Mardaru, A., Djelveh, G., Isopencu, G., Jinescu, C., Mares, A.M., Ghimiciu, G.: Characterizing foams flow obtained by continuous foaming operation. Univ. Polit. Bucharest Sci. Bull. **73**, 117–132 (2011)
10. Wang, X., Ban, X., He, R., Wu, D., Liu, X., Xu, Y.: Fluid-solid boundary handling using pairwise interaction model for non-Newtonian fluid. Symmetry **10**, 94 (2018)
11. Leusheva, E., Brovkina, N., Morenov, V.: Investigation of non-linear rheological characteristics of barite-free drilling fluids. Fluids **6**, 327 (2021). https://doi.org/10.3390/fluids6090327
12. Najjar, M.F., Soliman, A.M., Nehdi, M.L.: Critical overview of two-stage concrete: Properties and applications. Constr. Build. Mater. **62**, 47–58 (2014)
13. Stodola, P.R., et al.: Guide for the use of preplaced aggregate concrete for structural and mass concrete applications. ACI Mater. J. **88**, 650–668 (1992)

Structural Characterization and Li-ion Dynamic in Multivalence Cation Doped Li_2S Using XRD and AC Impedance Spectroscopy

Nguyen Manh Duc[1,2], Nguyen Thi Thanh Mai[1,2], Phung Khoi Nguyen[1,2], and Nguyen Huu Huy Phuc[1,2](✉)

[1] Faculty of Materials Technology, Ho Chi Minh City University of Technology (HCMUT), 268 Ly Thuong Kiet Str., Dist. 10, Ho Chi Minh City, Vietnam
nhhphuc@hcmut.edu.vn

[2] Vietnam National University, Linh Trung Ward, Thu Duc Dist., Ho Chi Minh City, Vietnam

Abstract. Doping other ions into the Li_2S crystal structure has been researched and tested to increase the ionic conductivity of Li_2S. From there, it can be applied to lithium-sulfur batteries. From there, I deeply researched the XRD and EIS results of solid solutions of Li_2S with Al_2S_3, AlI_3, CaS, and MgS to see the structural changes of Li_2S when doped and the movement of Li ions. I received the results that when doping AlI_3, the solid solution brought a large ionic conductivity compared to the remaining solid solutions. The activation energy of the solid solution has decreased by 3.5 - 4 times (depending on the doping concentration) compared to the original Li_2S. I have also shown the movement model of Li ions in the crystal structure and calculated the smallest ion jump. Furthermore, during my research, I also discovered that the electron conductivity of Li_2S increased when doping with AlI_3. The bandgap energy has decreased obviously after doping, in theory the doped Li_2S can become a semiconductor. The above results can support future research to turn Li_2S into a semiconductor and apply it to lithium-sulfur battery systems.

1 Introduction

Nowadays, traditional lithium batteries are significant in life. They are used in so many fields such as renewable energy storage, electric vehicles (EVs), hybrid electric vehicles (HEVs),... On the path of improving and developing lithium batteries, many types of lithium batteries have been created such as lithium iron phosphate batteries, Lithium Manganese Oxide, Lithium Titanate batteries. Among them, the battery with the highest specific capacity and specific energy is the lithium-sulfur battery (LSB) [1]. It is composed of sulfur at the cathode, lithium metal or lithium oxide at the anode and the electrolyte does not contain water. However, lithium metal has been found to tend to deposit in the form of dendrites or moss which can penetrate the separator and cause a short circuit, leading to explosion or spontaneous combustion. So Li_2S was used to replace lithium metal at the anode to overcome this weakness, but Li_2S is an insulator with high bandgap energy.

N. R. Abdul Razak et al. (Eds.): ECOOP 1996, SPPHY 418, pp. 272–276, 2025.
https://doi.org/10.1007/978-981-96-2871-1_39

2 Experimental Procedure

First, from data XRD, we got value of Bragg angle (θ). we used the equation:

$$n\lambda = 2d_{hkl} \sin\theta \tag{1}$$

and

$$d_{hkl} = \frac{a}{\sqrt{h^2 + k^2 + l^2}} \tag{2}$$

with Z' is resistor value, we is sample thickness (0.06 cm) and A is sample cross-section (0.785 cm^2). We plotted the dependence of ionic conductivity on frequency and found that the conductivity is divided into two components. One is in the medium frequency region when the shape of the conductivity here is almost horizontal, it can be concluded that this is the frequency-independent conductance region (σ_{DC}). Second, in the low and high frequency regions, the shape of the conductance is a straight line that tends to increase with frequency, this is the region of frequency-dependent conductance (σ_{AC}).

3 Results and Discussions

As can be seen in Fig. 1, the network parameter a_0 tends to increase as the x value increases. This is because when the two ions Al^{3+} and I^- begin to enter the crystal lattice of Li_2S, the crystal size expands, increasing lattice parameters. Then, because the charge of Al^{3+} ion is much larger than that of Li, when increasing the x value, the electrostatic attraction of Al^{3+} ion and surrounding negative ions is stronger than that of Li^+ ion, causing the lattice parameter to decrease. However, because the radius of iodine is very large (206 pm) compared to the radius of other ions (Al^{3+} is 67.5 pm; Li^{2+} is 73 pm; S^{2-} is 170 pm) and the large electronegativity of I^- pushed other negative ions around it, causing the lattice size to continue to increase as x increased, and the lattice parameters also increased. It can be seen that the network distortion tends to increase as the x value increases (at x = 7, the result is no longer accurate). This is due to the excessive influence of I^- ions on the crystal lattice as mentioned above. And the number of deflection lines tends to increase greatly compared to the original substance. It can be concluded that there has been movement of Al^{3+} ions into the crystal lattice and pushed Li^+ ions out of their original positions.

With the results as shown in Fig. 2, the activation energy after doping AlI_3 is much lower than the original Li_2S with $E_a = 141.83$ kJ.mol^{-1} [4]. From there it can be concluded that the ionic conductivity of the solid solution has increased significantly with this method.

a)

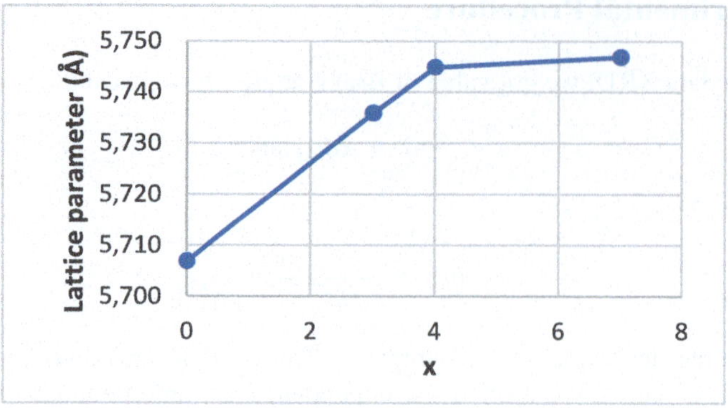

b)

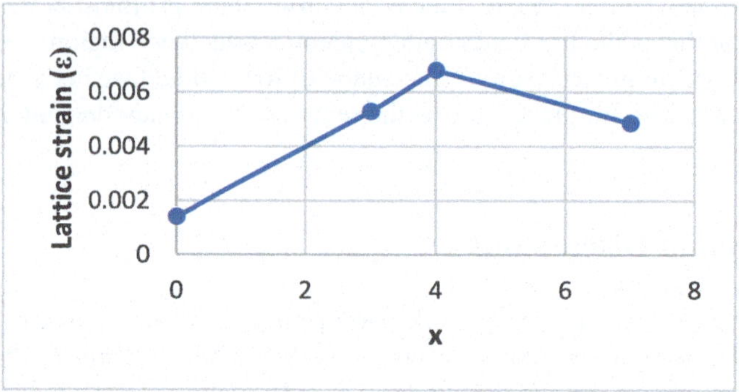

Fig. 1. a) XRD results of (100-x)Li_2S − xAlI_3 solid solution; b) The influence of x value on crystal lattice parameter; c) The influence of x value on network deformation.

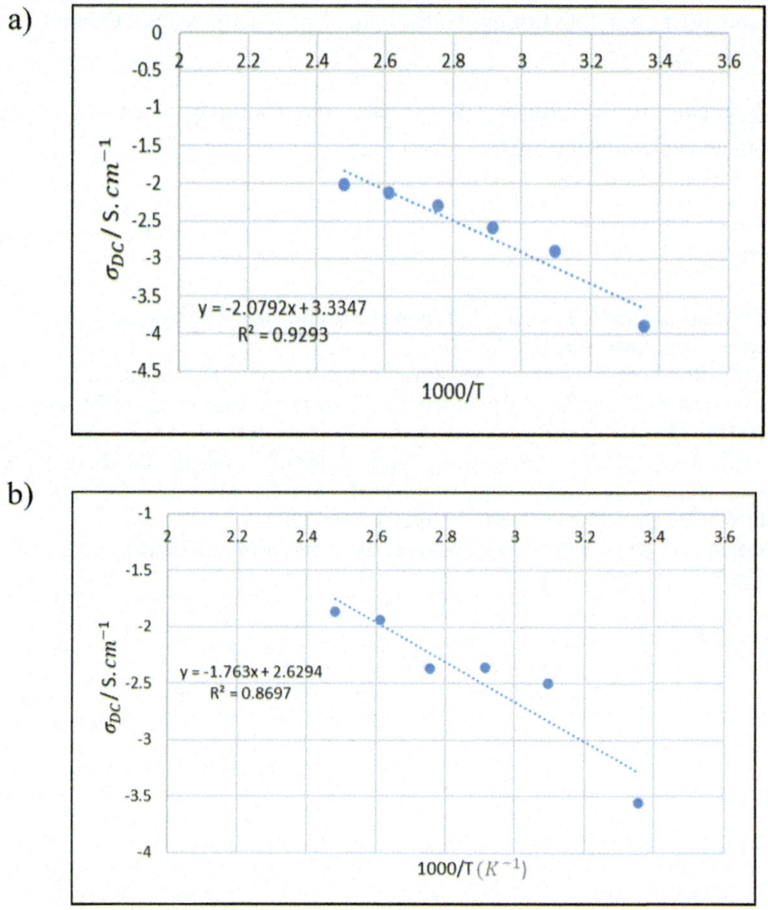

Fig. 2. a) Dependence of ionic conductivity on temperature at x = 2.5; b) Dependence of ionic conductivity on temperature at x = 3.

4 Conclusion

From the presented content and results, I can draw the conclusion that doping AlI_3 into the Li_2S crystal lattice will help improve the ionic conductivity of this material. It can be seen most clearly that the final product has a much lower activation energy than the original Li_2S. In addition, I also confirmed the movement model of Li^+ ions in the crystal lattice structure and calculated the smallest jump of ions in doped materials to serve the next research.

However, in this topic, it is not possible to conclude which AlI_3 doping concentration is best for increasing ionic conductivity because we have not calculated the concentration of Li^+ ions moving inside the crystal structure.

This topic can be developed further if this material is applied to the battery system as an electrode and continues to research the remaining components of the battery to produce the most complete lithium sulfur battery product possible. Commercialization in the future.

Acknowledgment. We acknowledge Ho Chi Minh City University of Technology (HCMUT), VNU-HCM for supporting this study.

References

1. Kato, Y., Hori, S., Saito, T., et al.: High-power all-solid-state batteries using sulfide superionic conductors. Nat. Energy **1**, 16030 (2016)
2. Imran, Z., Rafiq, M.A., Ahmad, M., Rasool, K., Batool, S.S., Hasan, M.M.: Temperature dependent transport and dielectric properties of cadmium titanate nanofiber mats. AIP Adv. **3**(3), 032146 (2013)
3. Tran, A.T., Luong, A.T.Q., Viet, C.X., Nguyen, H.H.P.: Ionic conductivity and relaxation dynamics in amorphous Li2.94Mn0.06/nPS4 (M= Mg, Ca, Al) and Li2.94CaPS3.94Xy0.06/y (X= I, N) solid electrolytes. Ceramics International (2024)
4. Moradabadi, A., Kaghazchi, P.: Thermodynamics and kinetics of defects in Li2S. Appl. Phys. Lett. **108**(21), 213906 (2016)

Green Materials

Characterization of Pineapple Leaves Reinforced Cement Based Composites

Rosmamuhamadani Ramli[1](✉), Alif Akhyar Nor Azlan[1], Syaiful Osman[1], Mohd Nazarudin Zakaria[2], and Sabrina M. Yahaya[3]

[1] Material Science Technology Programme, Faculty of Applied Sciences, Universiti Teknologi MARA, 40450 Shah Alam, Selangor, Malaysia
rosma614@uitm.edu.my

[2] Eco-Technology Programme, Faculty of Applied Sciences, Universiti Teknnologi MARA, 40450 Shah Alam, Selangor, Malaysia

[3] Applied Chemistry Programme, Faculty of Applied Sciences, Universiti Teknnologi MARA, 40450 Shah Alam, Selangor, Malaysia

Abstract. The cement-based composite material known as concrete is a brittle material. After the formation of a fracture, the material is unable to absorb force and breaks immediately. Cracks can be caused by microcracks or shear stress in the interfacial transition zone between the aggregates, which is the weakest region in the matrix. The strength of concrete is determined by the link between the particles in the concrete matrix. Cracks can also arise as a result of shear stress. As a result of the widespread usage of concrete in construction, this may present a potential risk to safety. In the event that the concrete is not strong enough to absorb tension, it may result in potentially hazardous circumstances, such as accidents that occur close to the building. The use of pineapple leaf fiber (PALF) into the concrete recipe will result in an improvement in its workability. In the course of this investigation, a variety of PLAF concentrations-0, 5, 10, and 15 wt.% were combined with water, cement, and sand. The purpose of the characterisation was to ascertain the water absorptivity of PALF cement-based composites as well as their mechanical parameters, which included compressive and flexural strength. In order to examine the microstructures of the cement composites, the machines that were utilized were the Universal Testing Strength (UTS) and the Scanning Electron Microscopy (SEM). The composite material that was based on cement was produced by combining water, cement, sand, and PALF. The fiber content of the composite material was composed of 0, 5, 10, and 15 wt.%. According to the findings that were obtained, the incorporation of natural fibers resulted in an increase in the flexural toughness of the composite material, which was 0.611 MPa, and the rate of water absorption by the material was 29.9%. Nevertheless, the compressive stress was reduced to 0.977 MPa as a result of the improper distribution of the fiber and the large length of the fiber. Through the utilization of PALF, which is both biodegradable and renewable, the composite reduces carbon emissions and reduces the amount of waste produced by agriculture, making it a more environmentally friendly option for the building industry.

Keywords: Cement-based composite · Pineapple leaf fibre · Mechanical properties · Microstructure

N. R. Abdul Razak et al. (Eds.): ECOOP 1996, SPPHY 418, pp. 279–287, 2025.
https://doi.org/10.1007/978-981-96-2871-1_40

1 Introduction

Recently, there has been a renewed interest in utilizing natural plant fibers as substitutes for traditional reinforcements in cementitious composites [1]. The growing apprehension of environmental and sustainability issues has led to heightened interest in the development of biocomposites. Consequently, numerous manufacturing sectors, such as automotive, construction, packaging, and biomedical industries, have redirected their attention towards the utilization of eco-friendly products, leading to an increased demand for high-performance materials derived from plant fibers recently [2]. Natural plant fibers enhance the ecological sustainability of engineering practices due to their renewability, biodegradability, energy efficiency, and non-toxic nature [1]. The manufacture of eco-friendly items derived from natural resources not only captivates material researchers but also provides farmers with income-generating opportunities [3].

PALF has long been utilized to enhance composites, yielding significant results; yet, its application in civil engineering remains inadequate due to the absence of standards. Recent studies have commenced exploring the use of pineapple to enhance concrete reinforcement. The stress-strain curves of the composites were notably altered when short and thin PALF were utilized for reinforcement, and extensive additional study on composites demonstrated substantial enhancements attributable to the advantages of pineapple leaf fiber.

Concrete, which is a product of cement-based composite, is brittle; tension cannot be absorbed once a fracture has formed in the concrete, and the concrete fails immediately after the crack has formed. It is the link between the aggregates in the concrete matrix that controls the strength of the concrete, and the crack that appears is the result of a microcrack or shear stress in the interfacial transition zone between the aggregates. This zone is the weakest location in the concrete matrix. Given that concrete is often utilized in the construction process, this can result in a potentially hazardous situation. In the event that the concrete that is utilized is brittle and unable to absorb tension, it may result in undesirable outcomes, such as an accident taking place in the vicinity of the building. Consequently, PALF, which is a natural fiber, can be incorporated into concrete to create a novel composite material. The incorporation of fiber into concrete will have an effect on the workability performance of the material. Additionally, the incorporation of PALF will prevent cracks that are brought on by the shrinking of plastic and drying shrinkage. In addition to this, they reduce the permeability of the concrete, which results in less water leaking out.

The World Commission on Environment and Development characterizes sustainability as the capacity to satisfy current needs without compromising the ability of future generations to achieve their own objectives. A fundamental issue facing society is the growing global human population and the consequent demand on the construction environment. Synthetic and natural fibres are frequently utilized in the construction of cement-based composites in contemporary applications. Numerous studies in the published literature indicate that the incorporation of synthetic fibers can enhance the tensile strength, toughness, ductility, crack resistance, and impact resistance of cement composites. The production of synthetic fibres requires substantial fossil energy, and the trade-off between performance and cost of these fibres remains unresolved. Natural plant fibers offer a sustainable solution to the aforementioned challenges due to their abundant availability, contribution to environmental preservation, energy efficiency, and biodegradability [1].

This research focuses on PALF as the primary subject of study regarding natural fibres. This study aimed to examine the properties and microstructures of PALF reinforced with cement-based composites. This research will utilize the equipment such as SEM and UTS.

2 Methodology

2.1 Pineapple Leaf Fibre (PALF)

PALF was obtained through two methods: extraction from the pineapple leaf and online purchase. The fibers undergo alkaline treatment to enhance fiber/matrix adhesion by increasing the surface roughness of the fiber. The fibers were subjected to a 1% sodium hydroxide (NaOH) solution at room temperature for a duration of 3 h. PALF was subsequently cleaned to eliminate any residual NaOH traces. Subsequently, PALF was exposed to sunlight until it was dried for concrete preparation.

2.2 Fabrication of PALF with Cement-Based Composite

The preparation of the cement-based composite will commence with the mixing of water, cement, soil, and PALF as outlined in Table 1.

Table 1. Percent composition for fabrication of PALF cement-based composite

Samples (% wt. Fibre)	Cement (wt.%)	Sand (wt.%)	PALF (wt.%)
0	100	100	0
5	100	95	5
10	100	90	10
15	100	85	15

The mixture is subsequently poured into a mold and allowed to dry until it attains a hardness comparable to that of brick. The samples for the compressive tests measure (100 × 100 × 20) mm, whereas the samples for the flexural test measure (38 × 112 × 20) mm.

2.3 Physical Characterization-Water Absorption Test

This test assesses the material's capacity for water absorption, which is crucial for evaluating its durability and resistance to weathering and moisture-related damage. The test was conducted in accordance with ASTM C-1585 [4]. The weight of samples was measured prior to immersion in tap water at room temperature for a duration of three days. After three days, the samples were removed from the water and dried to eliminate excess moisture. The weight of samples submerged in tap water was measured. The percentage of water absorption was determined using Eq. 1:

$$\text{Water Absorption \%} = \frac{(\text{Wet Weight} - \text{Dry Weight})}{(\text{Dry Weight})} \times 100 \quad (1)$$

2.4 Mechanical Characterization

2.4.1 Compression Test

This test aimed to assess the concrete's capacity to withstand the load of a building or structure. The test was performed utilizing a UTS in conjunction with ASTM C-109 [5]. The samples were positioned between the compression platens of the testing machine. The samples were aligned to ensure axial and even load application. The test was performed at a speed of 0.01 mm/sec.

2.4.2 Flexural Test

This test aimed to assess the concrete's resistance to flexural stress by measuring the force required to bend the samples, in accordance with ASTM C293/C293M [6]. The samples were positioned horizontally on the supports of the testing machine. The sample's center was accurately aligned with the loading head of the testing machine, and the two supports were equidistant from one another. The test was performed at a speed of 0.01 mm/sec.

2.5 Microstructures of PALF Cement-Based Composite

The SEM was employed to examine the microstructures of the PALF cement-based composite. The specimen was extracted from the sample for the flexural test. The sample was affixed to double-sided tape mounted on a stub. Subsequently, to enhance the conductivity of the sample for SEM analysis, gold coating is applied using a sputter coater.

Figure 1 illustrates the methodology employed in this project, presented in a flow chart format.

3 Results and Discussion

3.1 Water Absorption Properties of PALF Cement-Based Composite

Table 2(a) presents the percentage of water absorption for the samples, while Fig. 2(a) illustrates the corresponding graph of water absorption percentages for the samples. The sample exhibiting the greatest water absorption rate is 15 wt.% fibre, recorded at 29.9%. The sample exhibiting the lowest water absorption rate is the one with 0 wt.% fiber, recording a value of 5.47%. The sample with 15 wt.% fiber content exhibits the highest water absorption rate, attributable to the increased PALF content in the concrete relative to other samples. The hydrophilic nature of natural fibers accounts for water absorption in composites; thus, a higher percentage of fibers leads to increased water absorption. PALF is a natural fiber characterized by its ability to absorb significant quantities of water due to its hydrophilic properties. PALF can serve as effective reinforcements for polymer and cement matrices due to its superior properties.

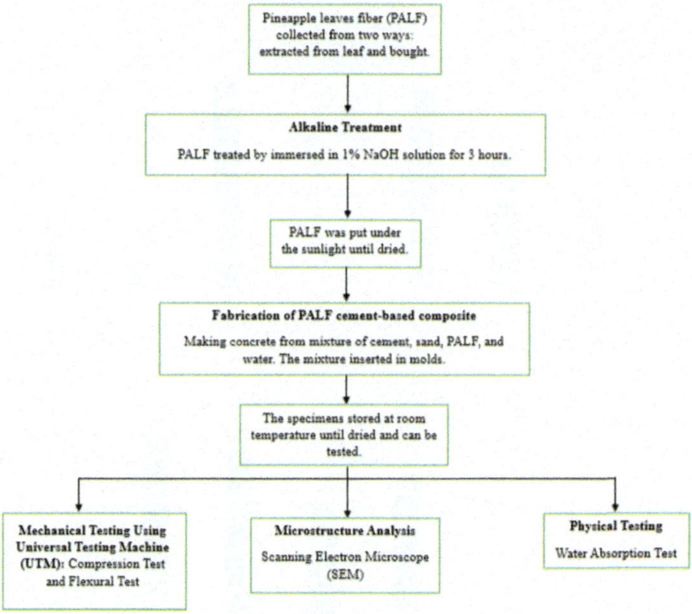

Fig. 1. Flowchart of methodology used in this project.

3.2 Compression Properties

Table 2(b) presents the peak compressive stress (MPa) results for cement composites, while Fig. 2(b) illustrates the corresponding graph of these peak compressive stress values (MPa) for the composites. Cement with 0 wt.% fiber content exhibited the highest compressive stress, measuring 791 MPa. The minimum compressive stress observed at peak occurred at 15 wt.% fiber, registering a value of 0.977 MPa. According to [7], the addition of fibre to concrete enhances mechanical properties, including compressive strength. In this particular case, the compressive strength of the PALF cement-based composite decreases as the fibre content increases. This results from the extended dimensions of the fiber and inadequate distribution during the mixing process.

According to [7], the effect of fibre length on mechanical characteristics is inversely proportional. Short fibre-reinforced composites possess a greater number of end points, facilitating the penetration of cement hydration products. The loss of flexibility in the fibres is accelerated. Furthermore, increased fibre length correlates with enhanced water absorption, as the incorporation of long coir fibres into the mixture diminishes workability and elevates void spaces. Shorter fibers are proposed to be well-arranged, leading to their separation from one another, which results in increased density and strength. Longer fibers, conversely, result in fiber clustering, which contributes to voids, reduced density, and diminished strength. The long dimensions of the fibres in these samples influence the compression strength results. As noted in [8], an increase in fibre content, coupled with a constant or reduced amount of cement, results in a diminished quantity of cement available for hydration and binding the matrix and fibre. Consequently, this will lead to a reduction in the compression strength of the concrete.

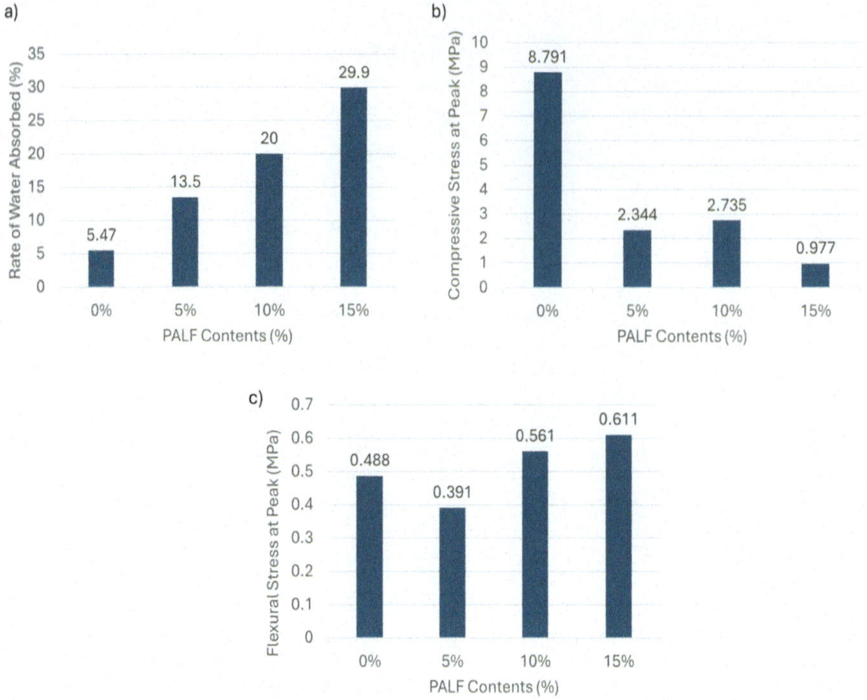

Fig. 2. **(a)** Rate of water absorbed (%) versus cement reinforced with different fibre contents, **(b)** Compressive stress at peak (MPa) versus cement reinforced with different fibre contents, **(c)** Flexural stress at peak (MPa) versus cement reinforced with different fibre contents.

Table 2. (a) Rate of water absorbed (%) by sample

Sample (wt.% fibre)	Rate of water absorbed (%)
0	5.47
5	13.5
10	20.0
15	29.9

Table 2. (b) Sample and its compressive stress at peak

Sample (wt.% fibre)	Compressive stress at peak (MPa)
0	8.791
5	2.344
10	2.735
15	0.977

3.3 Flexural Properties

Table 2(c) presents the flexural stress at peak (MPa) for the samples, while Fig. 2(c) illustrates the corresponding graph of these results. Sample 15 wt.% fibre exhibited the highest peak flexural stress, recorded at 0.611 MPa. Sample 5 wt.% fibre exhibits the lowest peak flexural stress, recorded at 0.391 MPa. This results from the inherent behaviour of the PALF. The incorporation of natural fibre enhanced flexural strength while preserving adequate flexural ductility. The incorporation of fibres significantly improved the flexural strength of the concrete. The incorporation of fibers into the concrete matrix enhances its ductility, in contrast to the brittleness of plain concrete. The fibers enhance the concrete's resistance to sudden failure and improve its energy absorption and post-cracking performance. The uniform distribution of fibers is a key factor contributing to the enhancement of flexural strength values [9]. When distributed uniformly, they create a reinforcing network that efficiently distributes external loads. The homogeneous distribution of the fibres enhances their ability to withstand applied stresses, thereby reducing the likelihood of localized stress concentrations and the initiation of cracks.

Table 2. (c) Sample and its flexural stress at peak

Sample (wt.% fibre)	Flexural stress at peak (MPa)
0	0.488
5	0.391
10	0.561
15	0.611

3.4 Microstructures Analysis

SEM was employed to examine the microstructures of the samples at a magnification of 150X, as illustrated in Figs. 3(a) and (b). Figure 3(a) illustrates the microstructures of 5 wt.% PALF, exhibiting a flexural stress of 0.391 MPa. Figure 3(b) illustrates the microstructures of the 10 wt.% PALF sample, which exhibits a greater flexural stress of 0.611 MPa compared to the 5 wt.% PALF sample. This analysis aimed to examine the impact of PALF on the structural integrity of concrete and the associated flexural stress described previously.

Figure 3(a) illustrates the presence of a limited number of cracks and some fibers at the outset. The low fiber content has obscured the concrete. The formation of cracks is attributed to the extended length of the fibers, which leads to clustering within the structures. Consequently, it results in several voids and cracks that negatively impact the flexural test outcomes of the 5 wt.% fibre, which are inferior to those of the 10 wt.% fibre.

Figure 3(b) illustrates that the overall structures were obscured by the PALF. The number of cracks was lower in comparison to the 5 wt.% PALF sample. Ultimately, the application of 10 wt.% PALF, combined with 150X magnification, has effectively addressed the cracks. The flexural test results indicate that the 10 wt.% fibre exhibited higher values than the 5 wt.% fibre (Fig. 2c). The cracks formed as a result of the extended length of the fibers, which led to clustering within the structures. However, the 10 wt.% fiber content yielded the highest results compared to the 5 wt.% content. The natural behavior exhibiting flexural strength influences the results of flexural stress, indicating that samples with higher fiber content demonstrate increased flexural stress.

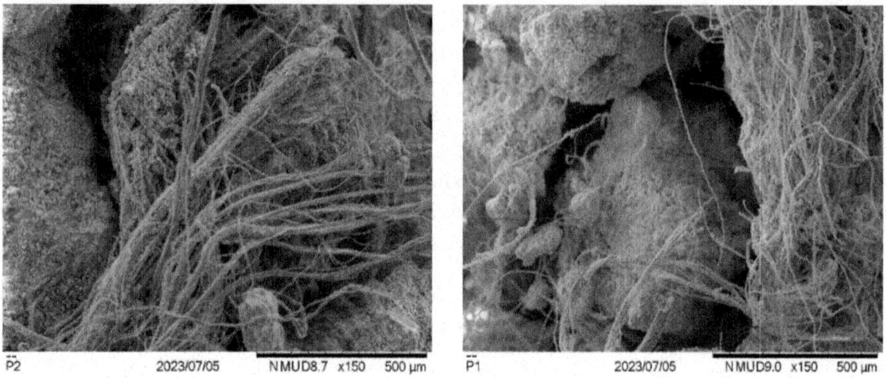

Fig. 3. Microstructures of cement composited reinforced with 5 and 10 wt.% fibre contents at 150X magnifications.

4 Conclusions

PALF, as an eco-friendly product, can be produced from leaves, which serve as a renewable and biodegradable resource. Utilizing these leaves as reinforcement in cement composites enhances sustainability. This reduces dependence on non-renewable resources and aids in decreasing the carbon footprint linked to conventional building materials. The water absorption test conducted with 15 wt.% fiber content yielded the highest rate of 29.9% for the physical properties. The findings indicate that PALF functions effectively as a water-absorbing fiber owing to its hydrophilic properties. The mechanical properties indicate that 0 wt.% fibre content exhibited the highest compressive stress, measuring 8.791 MPa, in contrast to 15 wt.% fibre content, which recorded a value of 0.611 MPa. PALF serves as an effective reinforcement to endure the stress exerted on concrete during flexural testing. In compression tests, the strength of PALF is ineffective due to the non-uniform distribution of fibers within the concrete, resulting in reduced energy for binding the matrix and the fibers. The analysis of microstructures indicates that the 10 wt.% fiber sample effectively covered the major cracks in the concrete, demonstrating that PALF serves as a reinforcement in the concrete matrix.

Acknowledgments. The author would like to thank the Faculty of Applied Sciences, Universiti Teknologi MARA, Shah Alam for the facilities and guidance.

References

1. Zhao. K., Xue, S., Zhang, P., Tian, Y., Li, P.: Application of natural plant fibers in cement-based composites and the influence on mechanical properties and mass transport. Materials **12**(21), 1–16 (2019)
2. Thiagamani, S.M.K., Nagarajan, R., Jawaid, M., Anumakonda, V., Siengchin, S.: Utilization of chemically treated municipal solid waste (spent coffee bean powder) as reinforcement in cellulose matrix for packaging applications. Waste Manag. **69**, 445–454 (2017)
3. Krishnasamy, S., et al.: Effect of fibre loading and $Ca(OH)_2$ treatment on thermal, mechanical, and physical properties of pineapple leaf fibre/polyester reinforced composites. Mater. Res. Express **6**(8), 085545 (2019)
4. ASTM C1585-20: Standard test method for measurement of rate of absorption of water by hydraulic-cement concretes (2020)
5. ASTM C109/C109M-20: Standard test method for compressive strength of hydraulic cement mortars (Using 2-in. or [50-mm] Cube Specimens) (2020)
6. ASTM C293/C293M-16: Standard test method for flexural strength of concrete (Using simple beam with center-point loading) (2016)
7. Zuraida, A., Norshahida, S., Sopyan, I., Zahurin, H.: Effect of fiber length variations on mechanical and physical properties of coir fiber reinforced cement-albumen composite (CFRCC). IIUM Eng. J. **12**(1), 63–76 (2011)
8. Babafemi, A.J., Kolawole, J.T., Olalusi, O.B.: Mechanical and durability properties of coir fibre reinforced concrete. J. Eng. Sci. Technol. **14**(3), 1482–1496 (2019)
9. More, F.M.D.S., Subramanian, S.S.: Impact of fibres on the mechanical and durable behaviour of fibre-reinforced concrete. Buildings **12**(9), 1436 (2022)

Enhanced Synthesis Method for Graphene Oxide

N. M. N. Azira[1(✉)], M. A. Idris[1], Dewi Suriyani Che Halin[1], M. Darus[2], and N. F. M. Yunos[2]

[1] Faculty of Chemical Engineering & Technology, Universiti Malaysia Perlis, Taman Muhibbah, Kompleks Pengajian Jejawi 2, 02600 Arau, Perlis, Malaysia
noorazira@unimap.edu.my

[2] Faculty of Mechanical Engineering & Technology, Universiti Malaysia Perlis, Kampus Alam UniMAP, Pauh Putra, 02600 Arau, Perlis, Malaysia

Abstract. Graphene oxide (GO) possesses unique properties that have garnered significant interest for numerous industrial applications. In this study, GO was efficiently synthesized via an environmentally friendly improvement of the Hummers method, directly from graphene oxidation. The synthesized products were analyzed using Raman spectroscopy, Fourier transform infrared spectroscopy (FTIR), X-ray powder diffraction (XRD), and scanning electron microscopy (SEM). Raman spectroscopy and FTIR analysis confirmed the formation of chemical bonds such as -OH, C=O, C=C, and C-O-C with graphene. The results indicated the oxidation of graphite, where oxygen atoms were introduced between the graphite layers. XRD analysis revealed a characteristic peak at $2\theta=11.9^{\circ}$, confirming the formation of GO. Raman spectra showed a prominent G-band at 1551 cm^{-1} and D-band at 1330 cm^{-1}. SEM images illustrated closely stacked sheets for graphite, whereas GO exhibited exfoliated sheets that appeared more separated and dispersed. Therefore, the improved Hummers method successfully synthesized graphite into graphene oxide (GO), showcasing promising materials for diverse applications including biomedical, energy storage, and notably, photocatalysis.

Keywords: Graphite · Graphene Oxide · Improved Hummer Method

1 Introduction

Carbon, with an atomic number of 6, exists as a solid at room temperature and appears in various natural allotropes including graphite, fullerenes, diamond, carbon nanospheres, and carbon nanotubes. These allotropes vary depending on synthesis conditions or formation processes [1–3]. Graphene, a two-dimensional (2D) lattice of carbon atoms featuring sp2 hybridization, forms a single layer of just one atom thick [4]. Each carbon atom covalently bonds with adjacent carbon atoms, creating a densely packed honeycomb crystal lattice composed of two overlapping triangular networks [5–8]. Initially identified in 1960, graphene was successfully isolated as a single layer using the Scotch® tape method by Andre Geim and Konstantin Novoselov in 2004 [3].

N. R. Abdul Razak et al. (Eds.): ECOOP 1996, SPPHY 418, pp. 288–294, 2025.
https://doi.org/10.1007/978-981-96-2871-1_41

The two main approaches to synthesizing graphene: bottom-up and top-down methods [8]. Crystalline growth on Silicon Carbide and chemical vapor deposition (CVD) are examples of bottom-up approaches, utilising a carbon-containing gas source to produce graphene [9]. However, optimizing CVD synthesis remains crucial to reduce preparation time and minimize the use of chemical oxidants in producing this intermediate product [10]. The method chosen to fabricate graphene significantly influences its properties and potential applications [4, 5, 11]. Therefore, graphene fabrication continues to be a highly active area of research [12]. It should be noted that no single graphene synthesis method consistently produces graphene with optimal properties for all potential applications. Hence, for this study, the top-down approach has been specifically selected for its photocatalytic application.

In the top-down approach, graphite is used as a precursor, and its layers are separated or broken down to form graphene sheets. Techniques in this approach include micromechanical cleavage, exfoliation of graphite intercalation compounds (GICs), carbon nanotube unzipping, and graphene oxide (GO) exfoliation through methods such as solvent-based exfoliation [13]. Before obtaining graphene oxide, it is crucial to synthesize graphite oxide as it serves as the precursor for GO. The synthesis of graphite oxide involves using a strong acid concentration as an oxidation agent, which is an exothermic process [14].

Graphite oxide (GrO) is a collection of functionalized graphene sheets primarily consisting of oxygen, carbon, and hydrogen atoms. Its structure closely resembles graphite but contains oxygenated groups that increase the spacing between the graphene layers [15, 16]. Well-known methods for synthesizing GrO include the Brodie, Staudenmaier, Hummers and Offeman methods, and their modified versions like the Modified Hummers or Improved Hummers methods [4, 17, 18].

Graphene oxide (GO) consists of a graphene layer decorated with oxygen-based functionalities, including carbonyl (C=O), hydroxyl (OH), and alkoxy (C-O-C) groups in a hexagonal arrangement. Researchers typically produce GO by oxidizing pristine graphene following established frameworks [19]. The size of graphene oxide and the distribution of oxygen domains are influenced by the carbon source, typically graphite, and the specific synthesis process used.

Recently, researchers have adopted a "primitive" approach involving the exfoliation and oxidation of graphite in aqueous solutions using strong oxidizing agents, typically sulfuric acid (H_2SO_4) [11, 20]. The three primary reaction pathways: mechanical production of graphene followed by oxidation, ultrasonic treatment for exfoliation in aqueous media, and simultaneous oxidation and exfoliation in a strong acidic environment similar to the Brodie-Staudenmaier-Hummers method. These pathways collectively yield graphene oxide (GO), each imparting distinct structural property such as variations in structure and reactive sites [19].

In this study, the improved Hummers method is chosen for synthesizing graphite oxide (GrO) due to its advantages: minimal defects in the basal plane, higher yield of oxidized graphite, comparable conductivity to other methods, and high performance [16, 21]. Importantly, this method is environmentally friendly and does not produce toxic gases during preparation. The exfoliation process occurs mechanically during product centrifugation [22]. This paper explores recent advancements in graphene preparation, production, placement, and handling specifically for photocatalytic applications, particularly in fabricating titanium oxide-graphene oxide (TiO_2-GO) composites.

2 Experimental Procedure

To synthesize graphene oxide (GO), graphite powder was treated using the improved Hummers method. The experimental procedure began with the initial oxidation of graphite. First, a mixture of sulfuric acid (H_2SO_4), 1.5 g of graphite powder, and potassium permanganate ($KMnO_4$, 9.0 g) was prepared. The mixture was continuously stirred at a constant speed for 6 h. Ice water and hydrogen peroxide (H_2O_4) were added to the mixture. Next, the solution was centrifuged for 7 min to yield GO. Subsequently, hydrochloric acid (HCl) was used to decant and remove the supernatant. The washing process was repeated multiple times using distilled water (DI) to achieve a pH of 4–5. Finally, the resulting material was dried in an oven overnight.

Raman spectroscopy was used to confirm the formation of graphene oxide (GO). The spectra were recorded using a Renishaw Raman Microscope (WITEC, Germany). X-ray diffraction (XRD) analysis was employed to investigate the crystal structures of the carbonaceous materials. For microstructural examination, scanning electron microscopy (SEM) coupled with energy-dispersive X-ray spectroscopy (EDS) was performed using a Hitachi S-4800 field-emission scanning electron microscope (FESEM).

3 Results and Discussions

Graphene oxide is synthesized with an oxidizing agent, and graphite is placed in concentrated acid. The improved Hummers' method is acknowledged for its safety and efficiency in oxidizing graphite. Furthermore, it is widely accepted alongside its several modified iterations. According to Lerf et al., the graphene adorned on both surfaces and along its edges with oxygen functional groups for each GO sheet. The edges of carboxyl groups enable GO to achieve electrostatic stabilization, allowing the formation of colloidal suspensions in alcohols, water, and specific organic solvents, without requiring surfactants [26]. Ultrasonic agitation or rapid heating can assist in separating graphite oxide into individual sheets, although excessive ultrasonication may reduce lateral dimensions [27].

When graphite undergoes oxidation, it changes into a brown-colored thick liquid mixture containing graphite oxide, separated sheets, and some graphite particles that remain unoxidized, alongside oxidizing agents. During oxidation, the salts and ions can be eliminated through iterative methods such as centrifugation, sedimentation, or dialysis to purify the GO suspension. Further centrifugation helps settle non-oxidized graphite particles and thick graphite oxide sheets, facilitating the isolation of single-layer GO suspensions. Density-gradient centrifugation can also produce uniformly dispersed GO flakes depending on their lateral dimensions [5].

Raman spectroscopy was employed to analyze whether carbonaceous materials are in ordered or disordered crystal structures. Figure 1(a) displays the Raman spectra for both carbonaceous materials, used as the precursor, and graphene synthesized through the modified Hummers method. In the spectra, three distinct Raman-active bands are clearly observed. These include the graphite's lattice, where the G band indicate in-phase vibration. Alongside the relatively faint disorder at 1551 cm^{-1} refer to the graphite's edge structure. The D band represent peak at 1330 cm^{-1}. Additionally, a peak at 2655 cm^{-1} is also found. The G and D peaks in the first-order, which are 1551 cm^{-1} and 1330 cm^{-1}, arise from sp^2 carbon vibrations [28]. For the G peak represent the optical E_{2g} phonons located at the Brillouin zone center. It's found in both rings and chains arising from sp^2 carbon pairs that the bond stretched.

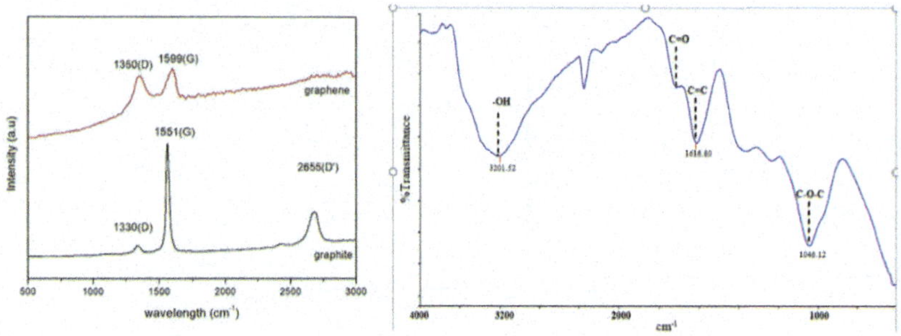

(a) (b)

Fig. 1. (a) Raman spectrum for graphite and graphene after synthesis and (b) FTIR spectrum of graphene oxide (GO).

An aromatic ring corresponds to the D peak, indicating the defect. The degree of disorder is represented by its intensity. The 2D peak at 2655 cm^{-1}, is associated with the number of graphene layers (N). It is emerging from dual resonance transitions, producing two phonons with opposite momentum. The 2D peak remains active even in the absence of defects. The presence of weak and broad 2D peaks further suggests disorder. The strong D peak around ~1330 cm^{-1}, compared to the G peak around ~1551 cm^{-1}, shows substantial structural disorder in GO and the broad bandwidths [17, 27].

According to Ferrari et al., leveraging the second-order 2D characteristic in Raman spectra facilitated the number of graphene layers. Malard et al. explored the theoretical basis for Raman-scattering mechanism, in which the double-resonance, accounts for the key feature observed in graphene's Raman spectra [30]. As a result, by deconvoluting the 2D peak characteristic of each graphene type, researchers revealed that monolayer graphene used as a filler in composites also displays a symmetric single peak[11].

The hydrophilicity of the sample is indicated by analysing the FTIR spectrum shown in Fig. 1(b). The spectrum must have a high-frequency of broad peak extending from 2900 to 3600 cm^{-1}. Besides, the appearance peak at 1616 cm^{-1} is attributed to the OH group's bending and stretching vibrations that water molecules attached to GO. Furthermore, the CH_2 groups are attributed by peaks observed at 2352 cm^{-1} and 2180 cm^{-1}. Moreover, it represents stretching vibrations that symmetric and anti-symmetric. The 1630 cm^{-1} and 1046 cm^{-1} peaks represent C=C and C-O-C bonds stretching vibrations. These carboxylic acid and carbonyl groups confirm an oxidation process of graphite [27]. The existence of these groups containing oxygen, especially surface hydroxyl groups, facilitates hydrogen bonding between graphene and water molecules. It can enhance the GO hydrophilic characteristic.

GO confirmed by a clear diffraction peak is evident at $2\theta = 11.9°$ for GO's XRD spectrum, indicating an interlayer distance of 0.77 nm (Fig. 2). In contrast, graphite exhibits a peak at $2\theta = 27°$. Using the Scherrer equation, the stacking height of GO is calculated to be 9.1nm, with approximately 12 layers. The GO peak aligns with the [011] interlayer spacing of 7.43 Å as reported in studies by Yu et al. and Liu et al. [30, 31].

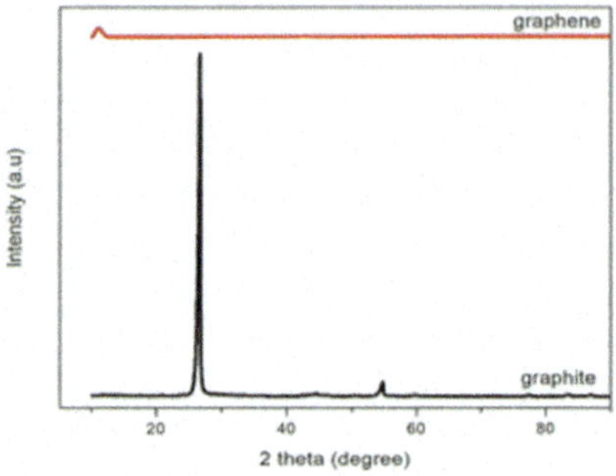

Fig. 2. XRD spectrum for graphite and graphene.

The SEM images shown in Fig. 3 highlight clear morphological differences between graphite and graphene oxide. Graphite exhibits closely stacked sheets, whereas graphene oxide displays exfoliated sheets that are more separated and dispersed. This exfoliation process transforming graphite into graphene oxide is consistent with observations reported by Paulchamy et al. [32].

Fig. 3. SEM morphology at 1000X magnification for (a) graphite and (b) GO.

4 Conclusion

In conclusion, graphene oxide was successfully synthesized by enhancing the Hummers method, thereby improving the oxidation process of graphite. Raman spectroscopy confirmed the presence of sp^2 carbon pairs in rings and chains. FTIR analysis further supported the oxidation of graphite by forming strong oxidant-induced bonds such as -OH, C=O, C=C, and C-O-C, resulting from the introducing oxygen atoms into the graphene layers. The primary objective was to create carbon materials with significant surface oxidation for fabricating composites suitable for photocatalytic applications, a goal emphasized by the FTIR analysis highlighting the presence of -OH bonds. X-ray diffraction (XRD) measurements revealed a characteristic 2θ value of 11.9° for graphene oxide, corresponding to an interlayer spacing of 0.77 nm. SEM microstructure images depicted uniformly distributed graphene particles. These findings demonstrate that the synthesized graphene oxide possesses the necessary attributes for developing TiO_2-GO composites to enhance photocatalytic activity.

Acknowledgments. The research was supported by Fundamental Research Grant Scheme, FRGS 2021: 9003-00919 (FRGS/1/2021/STG05/UNIMAP/02/6). The authors would like to acknowledge the Faculty of Chemical Engineering & Technology, Management & Innovation Centre (RMIC) of Universiti Malaysia Perlis.

References

1. Chung, D.D.L.: A review of exfoliated graphite. J. Mater. Sci. **51**(1), 554–568 (2015)
2. Dideikin, A.T., Vul', A.Y.: Graphene oxide and derivatives: the place in graphene family. Front. Phys. **6** (2019)
3. Drahansky, M., et al.: Graphene—A Wonder Material for Scientist and Engineer, Introductory Chapter: Brief Scientific Description to Carbon Allotropes—Technological Perspective (2016)
4. Mbayachi, V.B., Ndayiragije, E., Sammani, T., Taj, S., Mbuta, E.R., Ullah Khan, A.: Graphene synthesis, characterization and its applications: a review. Results Chem. **3**, 100163 (2021)
5. Moosa, A.A., Abed, M.S.: Graphene preparation and graphite exfoliation. Turkish J. Chem. **45**(3), 493–519 (2021)
6. Almeida, N.A., et al.: TiO_2-graphene oxide Immobilized In P(VDF-Trfe) electrospun membranes with enhanced visible-light-induced photocatalytic performance. J. Mater. Sci. **51**(14), 6974–6986 (2016)
7. Guerrero-Fajardo, C.A., Giraldo, L., Moreno-Piraján, J.C.: Graphene oxide: study of pore size distribution and surface chemistry using immersion calorimetry. Nanomaterials **10**(8), 1–22 (2020)
8. Trikkaliotis, D.G., Christoforidis, A.K., Mitropoulos, A.C., Kyzas, G.Z.: Graphene oxide synthesis, properties and characterization techniques: a comprehensive review. Chem Eng. **5**(3), 64 (2021)
9. Zhang, Q., et al.: Advanced fabrication of chemically bonded graphene/TiO2 continuous fibers with enhanced broadband photocatalytic properties and involved mechanisms exploration. Sci. Rep. **6** (2016)
10. Bhuyan, M.S.A., Uddin, M.N., Islam, M.M., Bipasha, F.A., Hossain, S.S.: Synthesis of graphene. Int. Nano Lett. **6**(2), 65–83 (2016)

11. Lavin-Lopez, M.P., et al.: Synthesis and characterization of graphene: Influence of synthesis variables. Phys. Chem. Chem. Phys. **16**(7), 2962–2970 (2014)
12. Brownson, D.A.C., Kampouris, D.K., Banks, C.E.: An overview of graphene in energy production and storage applications. J. Power. Sources **196**(11), 4873–4885 (2011)
13. Kigozi, M., Koech, R.K., Kingsle, O., Ojeaga, I., Tebandeke, E., Kasozi, G.N., Onwualu, A.P.: Synthesis and characterization of graphene oxide from locally mined graphite flakes and its supercapacitor applications. Results Mater. **7** (2020)
14. Jiříčková, A., Jankovský, O., Sofer, Z., Sedmidubský, D.: Synthesis and applications of Graphene Oxide. Materials **15**(3), 920 (2022)
15. Kumar, S., et al.: A review on properties and environmental applications of graphene and its derivative-based composites. Catalysts **13**(1), 111 (2023)
16. Boulanger, N., et al.: High Surface Area '3D Graphene Oxide' for Enhanced Sorption of Radionuclides. Adv. Mater. Interf. **9**(18) (2022)
17. Zhu, Y., et al.: Graphene and graphene oxide: Synthesis, properties, and applications. Adv. Mater. **22**(35), 3906–3924 (2010)
18. Singh, V., Joung, D., Zhai, L., Das, S., Khondaker, S.I., Seal, S.: Graphene based materials: Past, present and future. Prog. Mater. Sci. **56**(8), 1178–1271 (2011)
19. Pendolino, F., Armata, N.: Synthesis, characterization and models of graphene oxide. SpringerBriefs Applied Science and Technology, (9783319604282), 5–21 (2017)
20. Bonaccorso, F., et al.: Graphene, related two-dimensional crystals, and hybrid systems for energy conversion and storage. Science **347**(6217) (2015)
21. Chen, X., Qu, Z., Liu, Z., Ren, G.: Mechanism of oxidization of graphite to graphene oxide by the hummers method. ACS Omega **7**(27), 23503–23510 (2022)
22. Singh, J., Jindal, N., Kumar, V., Singh, K.: Role of green chemistry in synthesis and modification of graphene oxide and its application: a review study. Chem. Phys. Impact **6** (2023)
23. Cai, C.-J., Xu, M.-W., Bao, S.-J., Ji, C.-C., Lu, Z.-J., Jia, D.-Z.: A green and facile route for constructing flower-shaped TiO_2 nanocrystals assembled on graphene oxide sheets for enhanced photocatalytic activity. Nanotechnology **24**(27) (2013)
24. Ndagijimana, P., et al.: A review on activated carbon/graphene composite-based materials: synthesis and applications. J. Clean Prod. **417** (2023)
25. Bonaccorso, F., Lombardo, A., Hasan, T., Sun, Z., Colombo, L., Ferrari, A.C.: Production and processing of graphene and 2d crystals. Mater. Today **15**(12), 564–589 (2012)
26. Lerf, A., He, H., Forster, M., Klinowski, J.: Structure of graphite oxide revisited. J. Phys. Chem. B **102**(23), 4477–4482 (1998)
27. L. Shahriary and A. a. Athawale, Graphene Oxide Synthesized by using Modified Hummers Approach. Int. J. Renew. Energy Environ. Eng. **02**(01), 58–63 (2014)
28. Ferrari, A.C., et al.: Raman spectrum of graphene and graphene layers. Phys. Rev. Lett. **97**(18), 1–4 (2006)
29. Malard, L.M., Pimenta, M.A., Dresselhaus, G., Dresselhaus, M.S.: Raman spectroscopy in graphene. Phys. Rep. **473**(5–6), 51–87 (2009)
30. Yu, H., Xiao, P., Tian, J., Wang, F., Yu, J.: Phenylamine-functionalized rGO/TiO2 photocatalysts: spatially separated adsorption sites and tunable photocatalytic selectivity. ACS Appl. Mater. Interf. **8**(43), 29470–29477 (2016)
31. Liu, J., Bai, H., Wang, Y., Liu, Z., Zhang, X., Sun, D.D.: Self-assembling tio2 nanorods on large graphene oxide sheets at a two-phase interface and their anti-recombination in photocatalytic applications. Adv. Func. Mater. **20**(23), 4175–4181 (2010)
32. Paulchamy, B., Arthi, G., Bd, L.: A simple approach to stepwise synthesis of Graphene Oxide. J. Nanomed. Nanotechnol. **6**(1), 2–5 (2015)

Exploring Irradiation as a Means to Toughen and Improve the Impact Resistance of Dolomite Filled PP/r-PET Composites

Ismail Ibrahim[1](✉), Aiswaria Ramalingam[1], Siti Salwa Mohammad Shirajuddin[2], Mohd Fairul Sharin Abdul Razak[1], Mohd Hafiz Zainol[1], and Jatuporn Kaew-On[3]

[1] Faculty of Chemical Engineering & Technology, University Malaysia Perlis, 01000 Kangar, Perlis, Malaysia
ismailibrahim@unimap.edu.my
[2] Agency Nucklear Malaysia, 43000 Kajang, Selangor, Malaysia
[3] Mechanical Engineering Program, Faculty of Engineering, Thaksin University of Thailand, Khao Rup Chang, Thailand

Abstract. This study investigates the effects of irradiation on the mechanical properties of dolomite-filled polypropylene (PP)/recycled polyethylene terephthalate (r-PET) composites across varying filler ratios. Composites with PP/r-PET ratios of 100/0, 90/10, 80/20, 70/30, and 60/40, each containing 3% dolomite, were subjected to irradiation doses of 0, 30, and 90 kGy. Tensile and impact tests, along with scanning electron microscopy (SEM) for surface morphology analysis, were employed to investigate performance changes post-irradiation. It is expected that the composite with a 90/10 PP/r-PET ratio at a 30 kGy dose will demonstrate the highest tensile and impact strengths, while increased irradiation doses are predicted to reduce mechanical properties across other ratios. For raw PP, elongation at break is anticipated to reach its peak post-irradiation, with the 90 kGy dose likely yielding the highest Young's modulus. Morphologically, non-irradiated samples are expected to exhibit fewer microcracks and retain greater structural integrity. Overall, the study aims to elucidate how irradiation influences the enhancement or reduction of mechanical properties in these composites, dependent on composition and irradiation dose.

Keywords: Polypropylene (PP) · recycled Polyethylene Terephthalate (r-PET) · Dolomite · Irradiation

1 Introduction

Polymer matrix composites are increasingly replacing metal components in industries like electronics, aerospace, and automotive due to their affordability, ease of production, and high specific strength and modulus [1, 2]. This technology is integral to sustainable materials design, as it incorporates recycled or low-cost polymers while maintaining performance [3]. Reinforcing polymers with fibers, fillers, nanoparticles, and nanotubes

N. R. Abdul Razak et al. (Eds.): ECOOP 1996, SPPHY 418, pp. 295–302, 2025.
https://doi.org/10.1007/978-981-96-2871-1_42

enhances their strength, thermal stability, and electrical conductivity, making them suitable for high-performance applications [4]. Recent research has focused on improving common polymers like polypropylene (PP) and polyethylene terephthalate (PET) through reinforcements to overcome limitations in properties such as thermal stability, flame retardancy, and stiffness [5]. For instance, dolomite—a cost-effective mineral—is widely used to improve the thermal stability and stiffness of PP and recycled PET (r-PET) blends [6]. However, a key knowledge gap exists in understanding how irradiation, especially electron beam irradiation, impacts the structure and performance of these composites. Studies indicate that irradiation can alter polymer matrices at the molecular level, potentially enhancing tensile strength and impact resistance or, in some cases, leading to degradation depending on composite formulation [7, 8]. Techniques like scanning electron microscopy (SEM) are essential for analyzing these microstructural changes, offering insights into tailoring the material for greater durability and targeted performance [9]. This study investigates the effects of irradiation on dolomite-filled PP/r-PET composites, addressing the knowledge gap in how irradiation influences mechanical and thermal properties. By optimizing PP/r-PET blends with dolomite under various irradiation doses, this research advances the development of durable, sustainable composites designed for applications in high-demand industries, supporting both ecological and economic goals.

2 Experimental Procedure

2.1 Film Preparation

The PP, r-PET pellets, and dolomite powder were manually mixed in an extruder at temperatures between 225–245 °C, with a total compounding time of 20 min. The blend preparation occurred for samples intended for both irradiated and non-irradiated conditions. After extrusion, the composite was hot-pressed at 225–245 °C for 20 min. Electron-beam irradiation was then applied to three sets of dolomite-filled PP/r-PET composites at doses of 0, 30, and 90 kGy.

PP, r-PET, dolomite was prepared based on the weight percentage shown in Table 1.

Table 1. Formulation of different weight percentage of dolomite filled PP/r-PET composite

Code (PP/r-PET/Dolomite)	PP (wt %)	r-PET (wt %)	Dolomite (wt %)
100/0/3	100	0	3
90/10/3	90	10	3
80/20/3	80	20	3
70/30/3	70	30	3
60/40/3	60	40	3

Mechanical properties were assessed using tensile and impact tests. The Izod impact test was performed using an impact machine according to ASTM D256-D, and tensile testing was conducted with a Model 5569 universal testing machine following ASTM D638 standards. The effect of irradiation on the composites was further examined through scanning electron microscopy (SEM) to evaluate surface structure and morphology. Overall, the composite is anticipated to exhibit enhanced mechanical properties following irradiation.

3 Results and Discussions

3.1 Tensile Strength

Based on Fig. 1(A), the dolomite-filled PP/r-PET composite with a 90/10/3 ratio exhibits the highest tensile strength among all categories, reaching 47.865 MPa in irradiated samples, suggesting a robust resistance to tensile forces. This property aligns with findings in related studies, where similar filler ratios have shown enhanced durability, making this composite suitable for applications demanding strength and longevity [1, 5]. The 90/10/3 irradiated composite demonstrates the ability to withstand substantial forces without significant deformation, likely due to optimal interfacial adhesion and the reinforcing effect of r-PET and dolomite. In contrast, raw PP samples have the second-highest tensile strength at 47.851 MPa, indicating competitive tensile properties but slightly lower resistance compared to the 90/10/3 irradiated composite. This is consistent with literature noting that pure PP generally offers high tensile strength but can benefit from fillers for improved performance [2, 6]. Meanwhile, the lowest tensile strength was observed in the 60/40/3 PP/r-PET/dolomite composite, especially under a 90 kGy irradiation dose, where it recorded 8.364 MPa. Though comparatively lower in tensile capacity, this ratio may still be advantageous in applications prioritizing flexibility, ease of processing, or cost-effectiveness over mechanical strength [3, 7]. These findings align with prior research on the effects of irradiation on polymer composites, which reports modest strength increases at optimal doses but a decline with higher irradiation, likely due to polymer chain degradation [4]. The critical comparison with previous studies underscores the effectiveness of the 90/10/3 ratio, positioning it as a high-performance composite configuration for load-bearing applications.

Figure 1(B) illustrates the Young's modulus across various PP/r-PET ratios and irradiation doses, with raw PP irradiated at 90 kGy demonstrating the highest modulus at 8.44 MPa. This indicates that a 90 kGy dose significantly enhances the stiffness and rigidity of the composite, aligning with studies that report increased molecular cross-linking and structural reinforcement in polymer matrices under high-dose irradiation [1, 5]. For all composites at 90 kGy, the Young's modulus is markedly higher than for those exposed to 0 or 30 kGy, underscoring that high-dose irradiation yields improved resistance to deformation. The 90/10/3 PP/r-PET/dolomite composite also exhibits elevated stiffness with Young's modulus values of 5.28 MPa and 5.89 MPa at 30 and 90 kGy doses, respectively, further supporting the effectiveness of this irradiation level in enhancing rigidity, which is similarly observed in comparable studies [2, 7]. Conversely, the 80/20/3 composite shows the lowest modulus at 1.31 MPa under irradiation, indicating that higher r-PET ratios may reduce stiffness under similar conditions. The non-irradiated samples

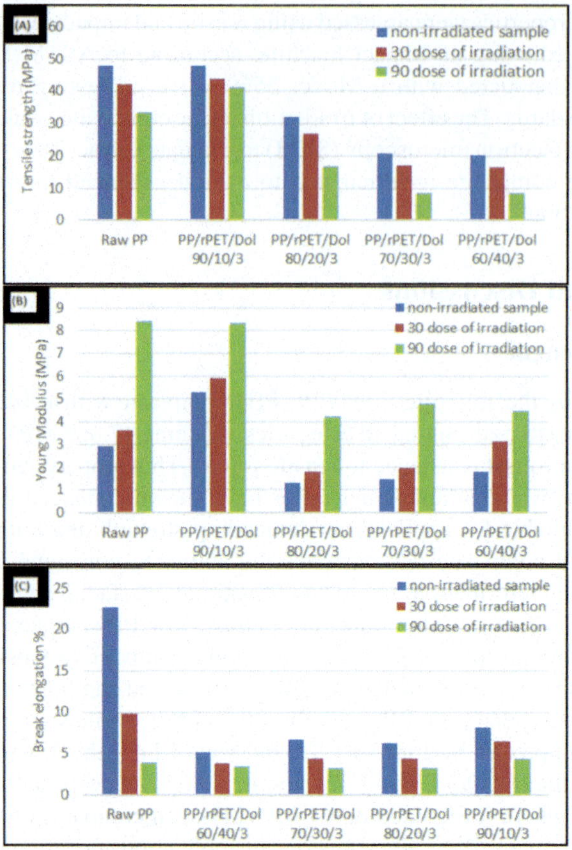

Fig. 1. (A) Tensile strength of dolomite-filled PP/r-PET composites across different filler ratios and irradiation doses; (B) Young's modulus of dolomite-filled PP/r-PET composites for varying ratios and irradiation doses; (C) Elongation at break of dolomite-filled PP/r-PET composites at different filler ratios and irradiation doses.

in each composite ratio recorded the lowest modulus, highlighting the limited stiffness in unmodified polymer structures and their tendency towards flexibility [3]. Interestingly, the 30 kGy irradiation dose produced intermediate Young's modulus values across the samples, suggesting moderate enhancement of rigidity without reaching the peak stiffness associated with 90 kGy. These findings correspond with established observations that moderate doses can improve structural properties to an extent but fall short of the stiffening effect seen at higher irradiation levels [4].

Meanwhile, Fig. 1(C) presents an analysis of elongation at break across the samples, highlighting their flexibility and response under stress. The non-irradiated raw PP sample exhibits the highest elongation at break at 22.73%, indicating superior flexibility and fatigue resistance, which allows it to withstand repeated cycles of deformation. This aligns with prior studies that emphasize the resilience of non-irradiated PP in dynamic applications [7]. Following this, raw PP exposed to a 30 kGy dose shows a notable decrease to 9.83%, reflecting moderate flexibility and stretching capacity before failure. This intermediate dose likely promotes some degree of cross-linking without extensive chain scission, maintaining flexibility while improving durability, a trend similarly reported in other irradiated polymer research [8]. Conversely, the lowest elongation values are observed at higher irradiation levels, with the 70/30/3 PP/r-PET/dolomite composite recording 3.2% elongation and the 80/20/3 composite showing 3.23%, both at a 90 kGy dose. These low values imply that high doses of irradiation induce brittleness by increasing chain scission, a phenomenon also noted in comparable studies on radiation-induced brittleness [9]. The results confirm that non-irradiated samples generally exhibit higher elongation compared to irradiated counterparts, supporting the view that irradiation at higher doses compromises flexibility due to chain breakdown and reduced molecular weight. The observed patterns reveal a clear trade-off: lower doses, such as 30 kGy, can enhance flexibility and toughness due to cross-linking, while 90 kGy doses cause chain scission, yielding stiffer but more brittle materials. This relationship between radiation dose and mechanical behavior provides critical insights for tailoring PP/r-PET composites for applications balancing flexibility and structural resilience.

3.2 Impact Strength

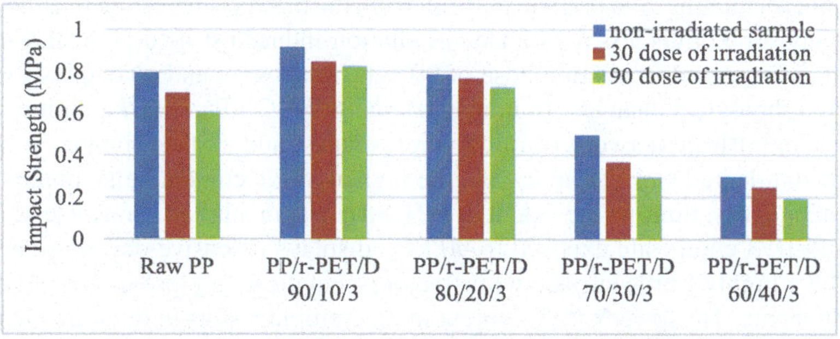

Fig. 2. Impact strength for dolomite filled PP/r-PET composite for different ratio and irradiation dose.

Based on Fig. 2, non-irradiated PP/r-PET/Dolomite composite with the ratio of 90/10/3 has the highest impact strength which is 0.92 MPa. 30 dose irradiation from the same ratio which is 90/10/3 ratio of PP/r-PET/Dolomite has the second highest impact strength which is 0.85 MPa. The third highest is also from the same ratio but 90 kGy dose of irradiation which is 0.83 MPa. The impact strength was improved from the cross-linking. The impact strength of blends decreases with the increasing of r-PET content and increase in irradiation dose. This may be the results of the blend's mechanical properties being adversely affected by the lower molecular weight and increased chain scission [10]. The lowest impact strength recorded was 0.2 MPa by PP/r-PET/Dolomite composite with the ratio of 60/40/3 and 90 kGy irradiation dose. Second lowest impact strength was 0.25 MPa from the same ratio but with 30 dose irradiation. 0.3 MPa impact strength was shown by PP/r-PET/Dolomite composite of ratio 70/30/3 non- irradiated sample and 60/40/3 for 90 kGy irradiation dose. This graph shows a pattern where the impact strength for each ratio become less as irradiation dose increases.

3.3 Scanning Electron Microscopy

The SEM micrographs in Fig. 3, captured at 5000x magnification, provide detailed insights into the structural and morphological changes in polypropylene (PP) and dolomite-filled PP/r-PET composites with different filler ratios and irradiation doses. Image (a) shows the non-irradiated raw PP sample, presenting a smooth and uniform surface that signifies its undisturbed, homogeneous texture. In image (b), the non-irradiated PP/r-PET/Dolomite composite with a 60/40/3 ratio displays a relatively smooth morphology with slight ductile characteristics and some fibrillation, which indicates good interfacial bonding among components in the absence of radiation.

In the irradiated samples, notable structural transformations occur due to radiation exposure. Image (c) shows the 60/40/3 PP/r-PET/Dolomite composite after a 30 kGy dose. Here, surface roughness, microcracks, and microvoid formations begin to appear, reflecting moderate degradation. The higher r-PET content, however, appears to enhance stability and limit the extent of surface deterioration compared to pure PP. Image (d) presents the 90/10/3 PP/r-PET/Dolomite composite also irradiated at 30 kGy. In this sample, degradation is more significant: the lower r-PET content results in increased brittleness, microcracking, and particle pull-out from the matrix, suggesting that the reduced r-PET proportion offers less resistance to radiation-induced structural breakdown.

With an increased irradiation dose of 90 kGy, images (e) and (f) display even more severe morphological changes. In image (e), the 60/40/3 PP/r-PET/Dolomite composite shows an intricate network of microvoids, cracks, and surface roughness, pointing to significant degradation. Despite this, the higher r-PET content helps maintain some structural integrity, limiting the extent of PP matrix fragmentation. Conversely, in image (f), the 90/10/3 composite exposed to 90 kGy displays extensive degradation, characterized by a highly brittle surface with fractures, cavities, and loss of cohesion among the components. The lower r-PET content in this sample results in reduced stabilization against radiation, allowing more pronounced breakdown of the PP matrix.

Overall, these SEM images indicate that higher r-PET content in the PP/r-PET/Dolomite composite enhances the material's resistance to radiation-induced degradation by stabilizing the matrix and limiting structural breakdown, particularly under higher irradiation doses. The comparative analysis suggests that r-PET plays a crucial role in preserving composite integrity under irradiation stress, which is most evident in samples with a 60/40/3 composition compared to those with a 90/10/3 ratio [11–15].

4 Conclusion

The study titled "Exploring Irradiation as a Means to Toughen and Improve the Impact Resistance of Dolomite-Filled PP/r-PET Composites" investigated the effects of irradiation on the mechanical properties of these composites. The incorporation of recycled PET aimed to reduce plastic waste, while dolomite served as a cost-effective filler. The

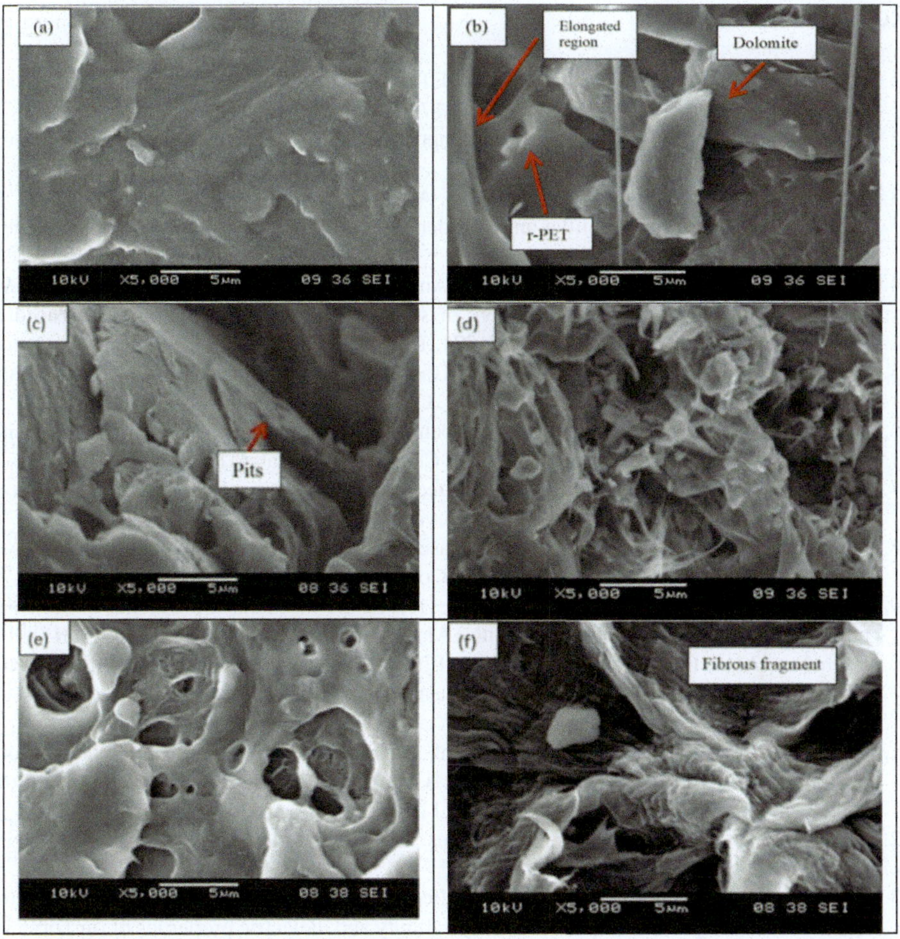

Fig. 3. SEM micrographs at 5000x magnification of various samples: (a) non-irradiated raw PP, (b) non-irradiated PP/r-PET/Dolomite composite with a 60/40/3 ratio, (c) PP/r-PET/Dolomite composite with a 60/40/3 ratio irradiated at 30 kGy, (d) PP/r-PET/Dolomite composite with a 90/10/3 ratio irradiated at 30 kGy, (e) PP/r-PET/Dolomite composite with a 60/40/3 ratio irradiated at 90 kGy, and (f) PP/r-PET/Dolomite composite with a 90/10/3 ratio irradiated at 90 kGy.

research utilized tensile and impact tests, supported by SEM for structural characterization. Results demonstrated that irradiation led to only modest improvements in tensile and impact strength, with the composite ratio itself proving more influential. The highest mechanical strength was achieved in irradiated samples with a PP/r-PET/dolomite ratio of 90/10/3. However, increased irradiation doses resulted in a decline in mechanical strength. SEM analysis offered valuable morphological insights, confirming that the composite displayed favorable structural characteristics.

References

1. Nonato, R.C., Bonse, B.C.: A study of PP/PET composites: factorial design, mechanical and thermal properties. Polym. Testing **56**, 167–173 (2016)
2. Subramanian, M.N.: Polymer blends and composites: chemistry and technology. John Wiley & Sons. (2017)
3. Ahmad Fauzi, A.A., Osman, A.F., Alrashdi, A.A., Mustafa, Z., Abdul Halim, K.A.: On the use of dolomite as a mineral filler and co-filler in the field of polymer composites: a review. Polymers **14**(14), 2843 (2022)
4. Szajerski, P., Celinska, J., Gasiorowski, A., Anyszka, R., Walendziak, R., Lewandowski, M.: Radiation induced strength enhancement of sulfur polymer concrete composites based on waste and residue fillers. J. Clean. Prod. **271**, 122563 (2020)
5. Ku, H., Wang, H., Pattarachaiyakoop, N., Trada, M.: A review on the tensile properties of natural fiber reinforced polymer composites. Compos. B Eng. **42**(4), 856–873 (2011)
6. Ahmed, A., et al.: A review on the tensile behavior of fiber-reinforced polymer composites under varying strain rates and temperatures. Constr. Build. Mater. **294**, 123565 (2021)
7. Arrigo, R., Malucelli, G., Mantia, F.P.L.: Effect of the elongational flow on the morphology and properties of polymer systems: a brief review. Polymers **13**(20), 3529 (2021)
8. Atnurkar, V., Schuster, J., Shaik, Y.P.: Increased elongation at breaking point with improved mechanical characteristics in PLA. Open J. Compos. Mater. **13**(2), 13–28 (2023)
9. Jose, A.S., Athijayamani, A., Jani, S.P.: A review on the mechanical properties of bio waste particulate reinforced polymer composites. Mater. Today Proc. **37**, 1757–1760 (2021)
10. Bachchan, A.A., Das, P.P., Chaudhary, V.: Effect of moisture absorption on the properties of natural fiber reinforced polymer composites: a review. Mater. Today Proc. **49**, 3403–3408 (2022)
11. Ahmadlouydarab, M., Chamkouri, M., Chamkouri, H.: Compatibilization of immiscible polymer blends (R-PET/PP) by adding PP-g-MA as compatibilizer: Analysis of phase morphology and mechanical properties. Polym. Bull. **77**(11), 5753–5766 (2020)
12. Ashfaq, A., et al.: Polymerization reactions and modifications of polymers by ionizing radiation. Polymers **12**(12), 2877 (2020)
13. Suljovrujic, E., Stojanovic, Z., Dudic, D., Milicevic, D.: Radiation, thermo-oxidative and storage induced changes in microstructure, crystallinity and dielectric properties of (un) oriented isotactic polypropylene. Polym. Degrad. Stab. **188**, 109564 (2021)
14. Martinez-Barrera, G., Avila-Cordoba, L., Urena-Nunez, F., Martínez, M.A., Álvarez-Rabanal, F.P., Gencel, O.: Waste polyethylene terephthalate flakes modified by gamma rays and its use as aggregate in concrete. Constr. Build. Mater. **268**, 121057 (2021)
15. More, C.V., Alsayed, Z., Badawi, M.S., Thabet, A.A., Pawar, P.P.: Polymeric composite materials for radiation shielding: a review. Environ. Chem. Lett. **19**, 2057–2090 (2021)

Effect of Annealing Process on the Structural Properties of Perlis Dolomite Powder

Nur Hasnidah Ahmad Shukeri[1](✉), Syed Nuzul Fadzli Syed Adam[1], Hasmaliza Mohamad[2], and Heah Cheng Yong[1]

[1] Faculty of Mechanical Engineering Technology, Universiti Malaysia Perlis, Kampus Tetap Pauh Putra, 02600 Arau, Perlis, Malaysia
hasnidahnur@gmail.com

[2] School of Materials and Mineral Resources Engineering, Universiti Sains Malaysia (USM), 14300 Transkrian, Penang, Malaysia

Abstract. Dolomite, $CaMg(CO_3)_2$ is a carbonate mineral that found naturally in the sedimentary rocks which have high potential as raw material for bioceramic production. However, raw dolomite has a certain composition of organic compounds, carbonates and moisture. Thus, annealing process was required for dolomite in order to decarbonise the carbon, remove impurities and improve its purity. Since not many scientific findings on the heat treatment were obtained from previous researchers specifically for Perlis dolomite, this study aims to investigate the effect of annealing process on the structural properties of Perlis dolomite powder through XRF, XRD and FTIR analysis and determine its suitability for bioceramic production. From the results, it was found that the annealing process on the Perlis dolomite has transformed the carbonate compounds into CaO and MgO compounds and promoted the heat-treated dolomite as the promising precursors in the bioceramic production.

Keywords: Perlis dolomite · annealing process · bioceramic production

1 Introduction

Up to the present moment, carbonate mineral has been extensively explored in most of the reservoirs to acquire its profits for economic importance through the utilization in many fields. In Perlis, dolomite is the second most prevalent carbonate that comes after calcite in the sedimentary carbonate rocks. It is naturally found in the dolomitic rocks called dolostone [1] and can be found easily in abundance in the vicinity of Chuping and Kangar [2]. It mostly known as 'Batu Reput' by the locals [2, 3] and practically used for soil conditioning and fertilizers making in the agricultural, as well as in road pavement and building materials in construction industries [3, 4].

Technically, dolomite with chemical formula of $CaMg(CO_3)_2$ is a double carbonate mineral that is mainly formed by calcium oxide (CaO) and magnesium oxide (MgO). Abdullah et al. [5] has stated that the CaO and MgO compounds in the dolomite most of the time are almost equivalent yet occasionally, one element which is CaO could be

N. R. Abdul Razak et al. (Eds.): ECOOP 1996, SPPHY 418, pp. 303–310, 2025.
https://doi.org/10.1007/978-981-96-2871-1_43

greater than the other. The ideal chemical composition of dolomite is 30.4 wt. % of CaO, 21.7 wt. % of MgO and 47.9 wt. % of carbon dioxide (CO_2) where the percentages indicated the oxides [6]. However, the physical appearance and the properties of the dolomite from different sources may be different according to their structural, textural, and chemical properties [7].

Annealing is a heat treatment process carried out at a significantly low temperature below its melting point to alter the initial properties of the material while decarbonize the carbon content and eliminate any impurities substances. The dolomite mineral could achieve complete decomposition directly and produce CaO and MgO at 1000 °C [8]. In addition, the decomposition of dolomite also can be divided into two phases where there are contradictions in the applied temperature. According to Jablonski et al. [9], the first decomposition of the dolomite happened at low temperature where it forms MgO and $CaCO_3$ while releasing CO_2 gases. Meanwhile, the second decomposition of the $CaCO_3$ created the CaO and again releases the CO_2 residues. Thus, dolomite powder requires annealing process prior to the sintering process during bioceramic samples preparation to avoid samples cracking due to gas burn off. The thermal reactions during the annealing process of Perlis dolomite is expected to occur accordance or similar with decomposition reaction mechanism shown in Eqs. 1–5 [10]:

$$MgCO_3 \rightarrow MgO + CO_2 \quad \Delta 311.2\ ^\circ C \tag{1}$$

$$CaMg(CO_3)_2 \rightarrow MgO + CaCO_3 + CO_2 \quad \Delta 315.5\ ^\circ C \tag{2}$$

$$CaMg(CO_3)_2 \rightarrow CaCO_3 + MgCO_3 \quad \Delta 331.9\ ^\circ C \tag{3}$$

$$CaMg(CO_3)_2 \rightarrow CaO + MgO + 2CO_2 \quad \Delta 538.2\ ^\circ C \tag{4}$$

$$CaCO_3 \rightarrow CaO + CO_2 \quad \Delta 845.8\ ^\circ C \tag{5}$$

Based on the previous study, the heat treatment on the dolomite has significantly changed its structural properties depending on the initial source of the dolomite powder. However, only limited studies have been carried out for dolomite sources from Perlis Dolomite. Besides, the scientific knowledge and discussion on the heat-treated Perlis dolomite for the preparation of bioceramic material are still unclear. Therefore, this study aims to investigate the effect of heat treatment on the Perlis dolomite powder by identifying its structural properties after the annealing process. The characterization of samples includes chemical composition, phase transformation and functional groups and then relates the results with the potential of this Perlis dolomite into bioceramic class of material.

2 Experimental Procedures

In this study, a fine dolomite mineral Perlis (Perlis Dolomite) supplied from the local quarry at Chuping, Perlis was used as it is without any subsequent process. In order to investigate the effect of the annealing process, the dolomite was heated at 1000 °C for 4 h with 10 °C/min heating rate. This temperature was selected based on literature and from the author's own lab work finding.

Chemical compositions of the raw and heat-treated Perlis dolomite were determined by the analysis via X-Ray Fluorescent (XRF) by Philips PANalytical Minipal 4 machine. The phase transformations were identified by X-ray diffraction (XRD) using Test Instrument Machine Bruker D2 Phaser, German (2010). The powder samples were scanned for 10 min with the angle diffraction of step-scanning was set within 10° to 80° (2θ). The obtained diffractogram were then analyzed by X'pert Highscore Plus software. Fourier transform infrared spectroscopy (FTIR) analysis was performed by using Perkin Elmer machine to identify the characteristic bands of the functional groups that exist in the samples where the samples were mixed together with potassium bromide powder (KBr).

3 Results and Discussions

3.1 Chemical Composition Analysis

The chemical compositions of raw and heat-treated dolomite analyzed by XRF is shown in Fig. 1. It showed that the raw Perlis dolomite was mainly composed of CaO, and MgO with 65.07%, and 23.1% respectively. Whereas, Na_2O, Ag_2O and Al_2O_3 were also found in the range of 1.0% to 8.0%. There are also other minor compounds which are found to be below 1.0% such as Sc_2O_3, SiO_2, SnO_2 and others. It was found that the percentages for all chemical compositions were slightly changed after the annealing process. The amount of CaO has decreased from 65.07% to 54.55% because CaO is more reactive with CO_2 and tends to easily remove along with the impurities and volatile substances during annealing process. Meanwhile, the amount of MgO and Na_2O have increased about 32% and 50% respectively due to less volatility because of high thermal stability and remain stable during the annealing process. The percentages of the minor compounds also have mostly reduced. From this result, it has shown that the annealing process has significant effect on the chemical composition of Perlis dolomite as it has highlighted the main compounds of the powder yet small amount of the impurities were still available in the heat-treated dolomite. By the way, the raw and heat-treated Perlis dolomite is free from hazardous compounds, non-toxicity and safe for bioceramic productions [11]. Besides, the main compounds in the dolomite (CaO and MgO) were suitable for the bioceramic production as it is able to contribute significant roles in the biomedical applications. Generally, the CaO is the main source of calcium in the human bone cancellous, teeth and skull [12]. Meanwhile, the MgO helps the bone growth modulation, repair and regeneration as well as promote osteoblasts proliferation and mineralize the calcified tissue also bone metabolism activity [12].

3.2 Phase Transformation Analysis

Figure 2 illustrates the phase transformation of raw and heat-treated dolomite. Based on Fig. 2 (a), it shows that the XRD pattern of raw Perlis dolomite was mainly dominated by calcium magnesium carbonate, $CaMg(CO_3)_2$ phase (JCPDS 01-081-8227) with a crystal system of hexagonal structure where the most intensified peak positioned at 31.148°. Nearly entire peaks in the diffractogram of Perlis dolomite which meets the similarity

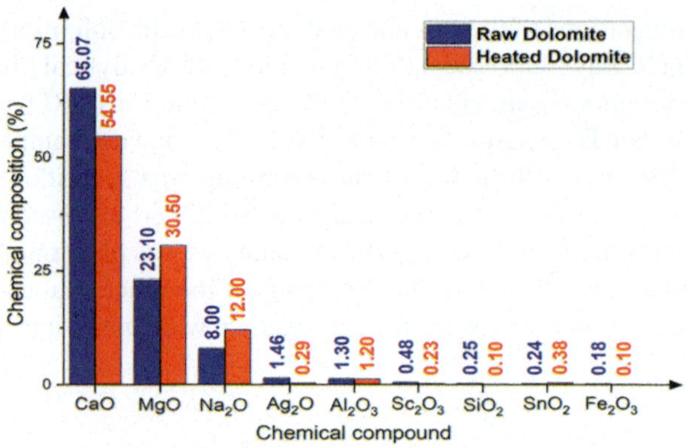

Fig. 1. Chemical compositions of main precursors by XRF analysis.

with previous studies [5, 13]. Moreover, the minority of the peaks in the diffractogram of raw Perlis dolomite were attributed to the phases of calcium magnesium silicate, $CaMgSi_2O_6$ (JCPDS 04-008-8060) and sodium aluminium silicate, $NaAlSi_3O_8$ (JCPDS 04-021-1621) where the crystal structure of the phases were monoclinic and anorthic, respectively.

Figure 2 (b) shows the heat-treated dolomite that appears to have changed the XRD patterns due to phase transformation that happened after the annealing process. The phases of $CaMg(CO_3)_2$ from raw dolomite has decomposed into several single phases where most of the significant peaks were assigned by cubic structures of CaO (JCPDS 04-005-4398) and MgO (JCPDS 04-016-2776) phases. This result proved the dominant contents of the heat-treated dolomite which meets the agreement with the results from Abdullah et al. [5]. Furthermore, there were also spotted some other phases in the heat-treated dolomite powder including sodium oxide, NaO_2 (JCPDS 04-005-4449), calcium silicon oxide, Ca_3SiO (JCPDS 04-021-0148), aluminium, Al (JCPDS 04-014-0129) and silica, SiO_2 (JCPDS 04-015-7161). These phases which formed in the heat-treated dolomite are well correlated with the earlier discussed results obtained from XRF analysis.

According to the phase analysis, it can be reported that the annealing process at 1000 °C is particularly required to enable transformation of initial phases of raw dolomite, $CaMg(CO_3)_2$ into CaO and MgO phases in the heat-treated Perlis dolomite powder due to high thermal energy that breaks the bonds between the chemical atoms in the dolomite structures [14]. This statement can be supported by the findings from Abdullah et al. [5] where the phases showed significant changes at 800 °C and 1000 °C while no apparent changes were found at 400 °C and 600 °C. Thus, the phase transformation of this Perlis dolomite is found be similar and follow the thermal decomposition and reaction mechanism shown previously in Eqs. 1–5. Therefore, the produced phases of CaO and MgO after the annealing process has put the potential and good perspective for the production of bioceramic based material especially for calcium and magnesium based bioceramic. These phases have been proven to be important and contribute in vital roles in biomedical application especially in the practices of bone engineering tissues [15–17].

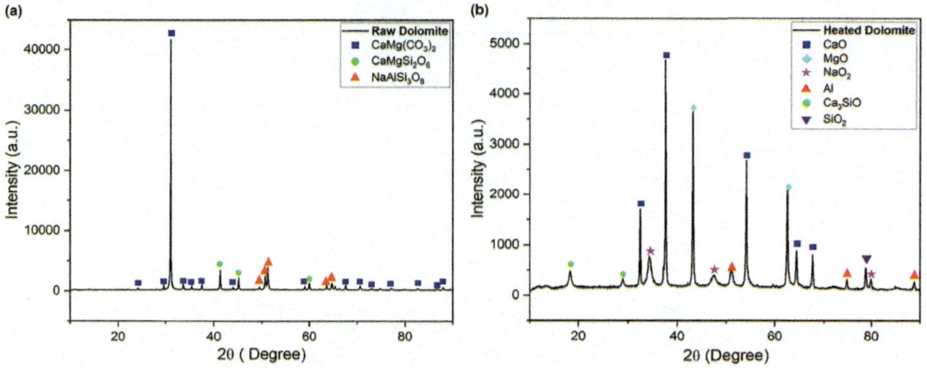

Fig. 2. Diffractogram of (a) raw dolomite and (b) heat-treated dolomite.

3.3 Functional Group Analysis

Figure 3 presents the infrared spectral band of raw and heat-treated Perlis dolomite. For raw dolomite, a strong and broad asymmetric stretching band around 1421 cm^{-1}, sharp bending bands at 875 cm^{-1} and 726 cm^{-1} has proved the presence of complex carbonate mineral group proportional [18, 19] which containing CaO and MgO as its main absorption components in the dolomite [5]. The deep bands at 668 cm^{-1} and 3642 cm^{-1} indicated the presence of CaOH bonds. Meanwhile, the weak bands at 875 cm^{-1}, 1005 cm^{-1} and 1080 cm^{-1} represented the vibration bond of Si–O which proved the silicate existence as the minor impurities in the raw dolomite as reported in the chemical composition analysis [3]. Furthermore, the band at 2895 cm^{-1} attributed as organic matter while the bands at 3400 cm^{-1} and 3875 cm^{-1} were identified as O–H stretching bond due to the hydrated carbonates in the dolomite [18].

Based on the spectrum of heat-treated dolomite, the FTIR analysis after the annealing process does not show apparent changes but there were several transmission bands that have reduced, disappeared and also formed new peaks. The complex bonds of raw dolomite at 1421 cm^{-1} turns into a narrow band after the annealing process indicates the decomposition of dolomite, $CaMg(CO_3)_2$ into formation of CaO and MgO. The strong and extreme bands at 3642 cm^{-1} and 3694 cm^{-1} appeared after the heat treatment process, particularly due to the formation of $Ca(OH)_2$ from the hydroxyl bonds [5, 19]. The weak bands within 875–1080 cm^{-1} have combined together to create a strong band indicating the existence of silica, Si–O bond in vibration mode as the impurities in the dolomite mineral [5]. The reduced intensity of the bands at 2895 cm^{-1} and 2510 cm^{-1} indicated the residue of C–H stretching bond from organic matter and O–H stretching bonds from the hydroxyl groups in the heat-treated dolomite respectively. The minor presence of hydrogen bonds can be fully decomposed with high temperature applied during the sintering process for the bioceramic production.

From this result, it shows that the raw dolomite proved the changes occurred on the functional groups of Perlis dolomite after the thermal treatment process. The annealing process has broken the bonds between the functional group of the dolomite matrix by decomposing the carbonate groups and creating CaO and MgO bonds individually. Besides, it also showed the presence of hydroxyl groups in the heat-treated dolomite as

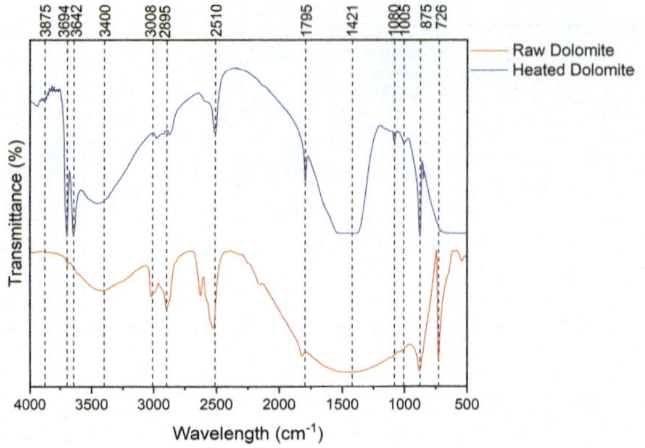

Fig. 3. FTIR spectrum of raw and heat-treated dolomite.

it was crucial in improving the absorption performance especially in the bioceramic production. This is because the hydrogen bonds have significant roles in the interaction with the biological tissues to provide biocompatibility reactions, accelerate cell proliferation and tissue growth in the bone tissue engineering application. Thus, the Perlis dolomite with applied heat treatment has the potential as the main precursors and suitable to be used in the production of bioceramic.

4 Conclusion

The investigation on the effect annealing process of the Perlis dolomite was successfully achieved. It can be summarized that the heat treatment on the raw dolomite at 1000 °C has decomposed the carbonates and formed CaO and MgO. The heat-treated dolomite has verified its suitability as main precursor for bioceramic productions and able to contribute vital roles in the bone tissue engineering applications according to its favourable properties that has been analyzed via XRF, XRD and FTIR. Hence, the annealing process is necessary and shown significant impacts on the Perlis dolomite powder. It should be explored widely for the bioceramic production in order to maximize its potentials associated with biological analysis to confirm its bioactivity performances.

Acknowledgments. Gratitude was expressed to the Faculty of Mechanical Engineering & Technology, Faculty of Chemical Engineering & Technology of Universiti Malaysia Perlis (UniMAP) and School of Materials and Mineral Resources Engineering, Universiti Sains Malaysia (USM) for providing the facilities of labs, machines and testing equipment.

References

1. Resio, L.C.: Characterization of Argentine dolostones with potential application in the manufacture of refractory materials. Open Ceram. **15** (2023)
2. Azurahanim, C., Albert, E.L., Gee Een, L., Zarin, M.A., Mohd Yusoff, M.Z.: Effect of thermal treatment on physical properties of malaysian dolomitic limestone. Int. J. Electroact. Mater. (2022)
3. Mohammed, M.A.A., Salmiaton, A., Wan Azlina, W.A.K.G., Mohamad Amran, M.S., Taufiq-Yap, Y.H.: Preparation and characterization of Malaysian dolomites as a tar cracking catalyst in biomass gasification process. J. Energy **2013** (2013)
4. Hussin, K., Shamsul, J., Ruzaidi, C.M., Sobri, M.I., Nazry, M.S., Nizar, K.: The development of artificial marble from dolomite (Batu Reput). In: 2006 TMS Fall Extraction and Processing Division: Sohn International Symposium, vol. 6, November 2014, pp. 617–621 (2006)
5. Abdullah, S.F.A., Saleh, S.S.M., Mohammad, N.F., Idris, M.S., Saliu, H.R.: Effect of thermal treatment on natural dolomite. J. Phys. Conf. Ser. **2080**(1) (2021)
6. Resio, L.C.: Dolomite thermal behaviour: a proposal to establish a definitive decomposition mechanism in a convective air atmosphere. Open Ceram. **15** (2023). https://doi.org/10.1016/j.oceram.2023.100405
7. Daye, M., Higgins, J., Bosak, T.: Formation of ordered dolomite in anaerobic photosynthetic biofilms. Geology **47**(6) (2019). https://doi.org/10.1130/G45821.1
8. Vaganov, V., Kireev, A., Avdeev, S., Šahmenko, G., Šinka, M.: Prospects for effective use of dolomite in concrete compositions. Constr. Sci. **19**(1) (2017). https://doi.org/10.1515/cons-2016-0008
9. Jablonski, M.O.M.: Thermal behavior of natural dolomite, November 2014, pp. 2239–2248 (2015). https://doi.org/10.1007/s10973-014-4301-6
10. Vasilevich, S.V., Malko, M.V., Degterov, D.V., Asadchyi, A.N.: Influence of dolomite particle size on the kinetics of its thermal decomposition process. Proc. Natl. Acad. Sci. Belarus, Phys. Ser, **67**(2) (2022). https://doi.org/10.29235/1561-8358-2022-67-2-204-213
11. Dias, G.F., Alves, F.B.T., de Andrade, A.V.C., Ferreira, A.M., Dos Santos, F.A.: Toxicity assessment and chemical properties of dolomite for use in dentistry. Rev. Odontol. da Univ. Cid. São Paulo **30**(1) (2018). https://doi.org/10.26843/ro_unicidv3012018p33-46
12. Fiume, E., Tulyaganov, D., Ubertalli, G., Verné, E., Baino, F.: Dolomite-foamed bioactive silicate scaffolds for bone tissue repair. Materials (Basel) **13**(3), 628 (2020). https://doi.org/10.3390/ma13030628
13. Sompech, S., Dasri, T., Thaomola, S.: Preparation and characterization of amorphous silica and calcium oxide from agricultural wastes. Orient. J. Chem. **32**(4) (2016). https://doi.org/10.13005/ojc/320418
14. Ramli, M., Maisarah, M.R., Saidi Murniana, N., Idris, N.: Aceh local dolomite modified by Alkali metals as a low-cost solid inorganic catalyst for biodiesel synthesis. In: 2020 IOP Conference Series: Earth and Environmental Science, vol. 1034, no. 1 (2022). https://doi.org/10.1088/1755-1315/1034/1/012027
15. Zadehnajar P., et al.: Recent advances on akermanite calcium-silicate ceramic for biomedical applications. Int. J. Appl. Ceram. Technol. **18**(6) (2021). https://doi.org/10.1111/ijac.13814
16. Choudhary, R., Koppala, S., Swamiappan, S.: Bioactivity studies of calcium magnesium silicate prepared from eggshell waste by sol-gel combustion synthesis. J. Asian Ceram. Soc. **3**(2), 173–177 (2015). https://doi.org/10.1016/j.jascer.2015.01.002
17. Collin, M.S., et al.: Solution combustion synthesis of functional diopside, akermanite, and merwinite bioceramics: excellent biomineralization, mechanical strength, and antibacterial ability. Mater. Today Commun. **27**, October 2020 (2021). https://doi.org/10.1016/j.mtcomm.2021.102365.

18. Harrati, A., et al.: Akermanite-based ceramics from Moroccan dolomite and perlite: characterization and in vitro bioactivity assessment. Open Ceram. **10**, 100276 (2022). https://doi.org/10.1016/j.oceram.2022.100276
19. Shahraki, B.K., Mehrabi, B., Dabiri, R.: Thermal behavior of Zefreh dolomite mine (Central Iran). J. Min. Metall. Sect. B Metall. **45**(1), 35–44 (2009). https://doi.org/10.2298/JMMB0901035S

Fabrication of Asymmetric Polylactic Acid Membrane Incorporated with Titanium Dioxide for Methylene Blue Dye Removal

Amira Mohd Nasib(✉), Gayathri Arumugam, Mohamad Syahmie Mohamad Rasidi, and Siti Kartini Enche Ab Rahim

Faculty of Chemical Engineering Technology, Universiti Malaysia Perlis, 02600 Arau, JejawiPerlis, Malaysia
amira@unimap.edu.my

Abstract. Asymmetric flat sheet PLA membranes were fabricated via non-solvent induced phase separation (NIPS). The polymer solutions were made by dissolving 16 wt. % of PLA polymer and predetermined TiO_2 (0, 1, 2, and 3 wt.%) in dimethylacetamide solvent. The prepared membranes were characterized through several characterizations to examine the effects of varying TiO_2 compositions on the membrane properties like membrane porosity, hydrophilicity and functional group. The membrane's performance was tested in terms of water flux, permeability and methylene blue dye removal ability. The results show that the PLA membrane was successfully synthesized using NIPS method and the varying compositions of TiO_2 at 3 wt. % have a remarkable outcome on the porosity and hydrophobicity of the membrane. The water flux and permeability of the membrane showed increasing trend from 0.6795 to 1.2307 L/m^2.hr and 0.3397 to 0.6153 L/m^2.hr.bar, respectively with increasing compositions of TiO_2. Lastly, MB dye removal test was carried out at 1 bar and with composition of 0.01 mg/mL and able to remove up to 96.5%.

Keywords: Polylactic acid · Membrane · Titanium Dioxide · Methylene Blue Removal

1 Introduction

In today's industrialized nations, water contamination by a variety of harmful metals, chemicals, and microorganisms is common [1]. The demand of clean water increases as the population rises along with various industrial and agricultural activities. Despite fulfilling human's daily needs and activities, a large amount of water is being released into the environment as wastewater. To protect the environment from harmful chemicals and toxins that are contained in the water and to meet the demand, wastewater treatment is very crucial. Thus, membrane technology has been introduced for the proper management of wastewater and to provide all living things on Earth with a clean source of water [2]. There are around 100,000 dyes commercially available, with a production

N. R. Abdul Razak et al. (Eds.): ECOOP 1996, SPPHY 418, pp. 311–318, 2025.
https://doi.org/10.1007/978-981-96-2871-1_44

capacity of around 700,000t per year. Dyes are widely utilized in a wide range of industries, including textiles, leather, paper, paint, and cosmetics. Large amounts of dyes were spilled into the aquatic environment during the application. Most dye molecules include an aromatic ring, making them carcinogenic, non- biodegradable, highly poisonous, and mutagenic to humans and aquatic life [3].

Hir et al. [4] used a phase inversion technique make a series of polyethersulfone-titanium dioxide (PES/TiO_2) photocatalyst membrane film. In the above study, these nanocomposite membranes were utilized to study the degradation of methyl orange dye (MO). The porosity of the film exhibited changes in membrane morphology corresponding to different TiO_2 content. The porosity of the neat PES membrane was 0.04%, whereas for the composite membrane with 13 wt.% TiO_2, it increased to 0.22%. However, at 15 wt.% TiO_2 loading, the porosity slightly decreased to 0.20% due to the presence of aggregated particles disrupting polymer chain packing, limiting mobility, and resulting in a less porous structure. Moreover, the reduction in porosity could be attributed to the blockage of pores by the TiO_2 particles [4]. Since the research was done at wide range of TiO_2 from 0 to 15 wt. %, its crucial to know at which percent of TiO_2 that gives the optimum performance.

Another research done was Sakarkar et al. [5] modified a polyvinylidene fluoride (PVDF) membrane coated with a polyvinyl alcohol (PVA)/titanium dioxide (TiO_2) solution. The researchers varied the PVA content while maintaining a fixed TiO_2 content of 1 wt.%. According to the results, at 3 wt. % of PVA content in the coating, the dyes' maximum photodegradation can been seen, whereby 44.4% for reactive blue, 47.8% for methyl orange, and 45.8% for rhodamine B. However, when the PVA content exceeded 3 wt%, there was a tendency for TiO_2 nanoparticles to agglomerate in the coating, leading to a reduction in their photocatalytic efficiency [5]. Since, PVDF is fossil-based polymer, its needed to find other greener polymer that can replace the material with another biodegradable material like PLA.

Thus, this study was focused on the effect of different compositions of TiO_2 on the PLA membranes characterization and performance. The PLA membrane was fabricated with the incorporation of TiO2 through the phase inversion technique, known as non-solvent induced phase separation (NIPS). To investigate the membrane characteristics, the membrane porosity, FTIR, and water contact angle of the prepared PLA membranes assisted with TiO_2, were evaluated. Meanwhile, the membrane performance was tested by measuring the water flux, permeability and dye rejection.

2 Experimental Procedure

2.1 Materials and Chemicals

Polylactic acid (PLA) was obtained from Sigma Aldrich from Fisher Scientific and N,N-dimethylacemide (DMAc) ($\geq$ 95.0% purity) solvent which used in the polymer solution preparation was purchased from Merck Milipore Corporation. Distilled water acted as strong non solvent, and methanol (as weak non-solvent) was supplied from SouceL HmBG Chemicals Malaysia. Methylene blue dye was purchased from Merck Milipore Corporation and used as to the dye rejection during membrane performance test.

2.2 Membrane Fabrication

16 wt.% of PLA polymer was dissolved in DMAc solvent with additional of predetermined weight TiO_2. The solution was agitated continuously at 200 rpm in a water bath shaker until it became clear and homogenous. The temperature of the mixture was maintained at 70 °C throughout the stirring process. After that, a uniform polymer solution was allowed to degas at ambient temperature until a transparent solution was obtained.

To cast a flat sheet membrane, the non-solvent induced phase inversion separation (NIPS) method was applied. With a 400 ± 10 μm knife gap, a polymer solution was cast onto a glass plate using a hand-casting knife. The step followed by exposing the casted film to the ambient air for 30 s (evaporation period). The glass plate with the casted film on top was immersed into a coagulation bath consists of 80% methanol and 20 wt.% pure distilled water immediately for 24 h after the evaporation period come to an end. The polymer solution medium consists of different composition of TiO_2 (Table 1). Finally, the fabricated membrane was removed from the coagulation bath and dried for 24 h at room temperature.

Table 1. Composition by (wt.%) of prepared polymer solution

Samples	Polylactic Acid (PLA) (wt.%)	Titanium Oxide (TiO_2) (wt.%)	Dimethylacetamide (DMAc) (wt.%)
M1	16	0	84
M2	15	1	84
M3	14	2	84
M4	13	3	84

2.3 Membrane Characterization

The gravimetric method was used to determine the membrane porosity. The membrane was soaked in isopropanol for 10 min. The excess of the alcohol on the membrane surface was dried by using a filter paper. The membrane weight was measured before and after isopropanol absorption by the membrane. The porosity was calculated using Eq. 1.

$$\varepsilon = \frac{\frac{W_w - W_d}{\rho_i}}{\frac{W_w - W_d}{\rho_i} + \frac{W_d}{\rho_{PLA}}} \quad (1)$$

where ε is the porosity of the membrane (%), w_w and w_d are the weight of wet and dry membrane (g), respectively. ρ_{PLA} is the density of PLA membrane (g/m^3) and ρ_i is the density of isopropyl alcohol (g/m^3).

The contact angle measures of the hydrophilicity or hydrophobicity of the surface membranes. The sessile drop method was implemented to determine the contact angle of water on the membrane surface. Firstly, 3 μl of distilled water at ambient temperature was placed on top of the membrane surface. A clear high-quality picture was taken, and the contour of the water droplet was analysed using the Image-J software. Every sample's average water contact angle was determined by replicate three measurements at various locations of top membrane surface.

Fourier-transform infrared spectroscopy (FTIR) spectroscopy was used to characterize the functional groups on the PLA membrane surface. This analysis used to determine the functional groups and chemical bonds or structures of the membranes obtaining an infrared absorption spectrum using a Perkin Elmer FT-IR spectrometer Model L128044. A total of 5 scans were performed for each of the membrane samples. The membranes were placed under the scope of the FTIR equipment. A spectrum of 4000 to 400 cm^{-1} in the mid-infrared region (IR) range with a resolution of 1 cm^{-1} was collected.

2.4 Membrane Performance

Flux was known as the volume of water that passes through the membrane per unit of surface area and time of the membrane. Samples were taken in triplicate. Membrane flux (J) and permeability (P) were calculated using Eq. 2 and Eq. 3, respectively.

$$J = \frac{\Delta V}{A_m \Delta T} \quad (2)$$

$$P = \frac{J}{\Delta P} \quad (3)$$

where ΔV is the permeate volume (L), *Am* is the effective membrane area (m^2), Δt is the time (h) and Δp is the difference in transmembrane pressure (bar).

Dye removal test was carried out at the transmembrane pressure of 2 bar for 50 ppm of methylene blue dye solution using a dead-end permeation cell. The concentration of the dye removal after filtration was measured by using the absorbance tested in UV-Vis. The final dye concentration (Ce) was determined and the dye removal (%) was calculated using Eq. 4 where Ci is initial dye concentrations and Ce is final (equilibrium) dye concentration (mg/L).

$$Dye\ removal\ (\%) = \frac{C_i - C_e}{C_i} \times 100 \quad (4)$$

3 Result and Discussion

3.1 Membrane Characterization

The porosity of each membrane was reported in Fig. 1(a). The difference of porosity can be explained by the variation of the composition of TiO_2 in the membrane. The average porosity of membrane was increased with the increased of TiO_2 composition (Fig. 1(a)). At 3 wt.% of TiO_2, the membrane exhibited the highest porosity at 74.8%, followed closely by 2 wt.% of TiO_2, which achieved a porosity of 73.8%. For 1 wt.% TiO_2, the porosity was recorded at 68.1%, while the lowest porosity, at 64.7%, occurred in the absence of TiO_2.

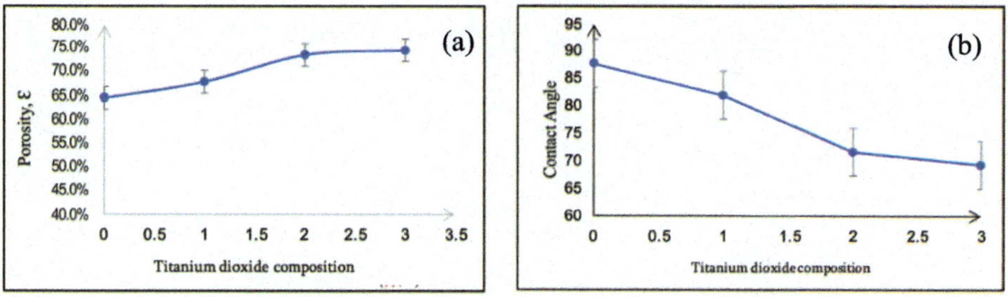

Fig. 1. (**a**) Effect of titanium dioxide (TiO2) composition on PVDF-co-PTFE membrane porosity; (**b**) Effect of titanium dioxide (TiO2) composition on PLA membrane hydrophobicity.

TiO2 composition used in the preparation of the membrane is one of the parameters that influence membrane porosity. The nanoparticles such as TiO2 are hydrophilic, adding more nanoparticles and increasing the content during the phase inversion process causes a greater penetration of water into the casting solution. Moreover, nanoparticles also prevent solvent molecules from interacting with polymer molecules, whereby the solvent molecules able to diffuse more freely from the polymer matrix into the coagulation bath. As a result, the solidified membrane poses larger average pores and more porosity [6].

The water contact angle was common parameter used to describe the hydrophilicity of the membranes surface is shown in Fig. 1(b). The membrane water contact angle decreases with increasing composition of nanoparticles such as TiO_2, making the membrane surface hydrophilic [7, 8]. The average water contact angle for M1 with 0% TiO_2 was 88.0°, M2 with 1 wt. % TiO_2 obtained an average water contact angle of 82.2°, M3 with 2 wt. % TiO_2 obtained an average water contact angle of 71.8° while M4 with 3 wt.% TiO_2 obtained the lowest average water contact angle of 69.5°.

Figure 2 shows the FTIR of PLA membrane with different TiO_2 compositions, whereby FTIR confirms that TiO_2 is present in the M2, M3 and M4. FTIR peak wave numbers of membranes 1, 2, 3 and 4 were close to each other which shows that membranes prepared by different TiO_2 compositions have no drastic effect on the functional groups that exist. The peaks at 1756 cm^{-1} from PLA are the skeletal vibrations of the C = O. A broad adsorption peak at 1585 cm^{-1} in M2, M3 and M4 membranes is attributed to the TiO_2 of the –OH group's stretching vibration. The peaks become obvious as the composition of TiO_2 increases from M1 to M4 [7].

3.2 Membrane Performance

The effect of different compositions of TiO_2 on the water flux and permeability can be seen from Fig. 3(a, b) shows the MB dye removal performance. In general, as the composition of TiO_2 increased from 0 wt.% to 3 wt.%, the water flux increased from 0.6795 to 1.2307 L/m^2.hr whereas the water permeability increased from 0.3397 to 0.6153 L/m^2.hr.bar.

The results obtained were similar to the research conducted by Sotto et al. (2012) where the water flux and water permeability of polyethersulfone (PES) and polysulfone

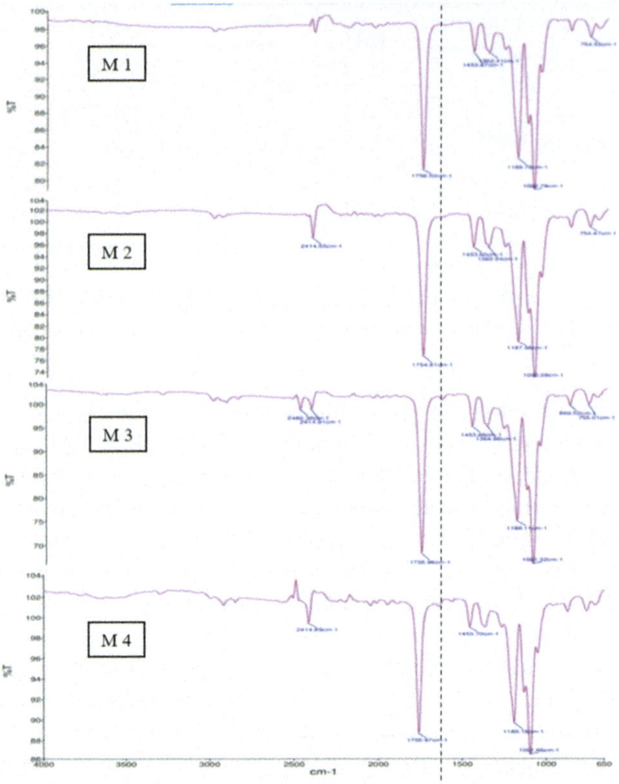

Fig. 2. FTIR spectra of PLA membrane prepared with different titanium dioxide (TiO2) compositions.

(PSF) membranes increased when 0.125 wt.% of TiO_2 were added. This condition is due to their increased porosity and their attraction to water, nanoparticles have an impact on the permeation properties of materials [9].

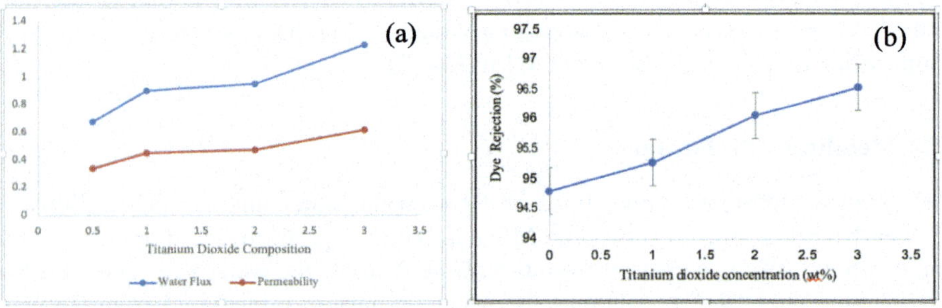

Fig. 3. (**a**) Effect of titanium dioxide (TiO2) composition on water flux and permeability; (**b**) Effect of titanium dioxide (TiO2) composition on dye rejection.

The percentage of dye rejection of PLA membranes were summarized in Fig. 3(b). MB dye rejection improved from 94.8 to 96.5% when the composition of TiO_2 increased from 0 wt.% to 3 wt.%. The results obtained in this study were parallel with the research conducted on polyethyleneimine-based nanofiltration membrane by Kebria et al. [8] where the SiO_2 nanoparticle content was raised to 0.05 and 0.1 wt.%, a developing trend in dye rejection was noticed [8]. This is most likely caused by the roughness of the membrane surface, in addition to the fact that the gel layer that forms on the membrane's upper surface contributes significantly to high SiO_2 nanoparticle contents, which resulted in decreased dye rejection at low SiO_2 nanoparticle contents and increased dye rejection at high and low SiO_2 nanoparticle contents, respectively.

4 Conclusion

In this study, the PLA membranes were successfully fabricated using phase inversion method. Different compositions of TiO_2, in PLA membrane matrix able to alter the membrane porosity and contact angle. The porosity of all prepared membranes started to increase, and the water contact angle started to decrease after TiO_2 composition increased from 0 wt.% to 3 wt.%. FTIR confirms that TiO_2 was present in the membranes. The water flux and permeability increased with increasing TiO_2 compositions from 0 wt.% to 3 wt.%. The MB dye rejection rate improved from 94.8 to 96.5%. This is most likely due to the membrane surface's roughness on the membrane's top surface also contributes significantly to high TiO_2 compositions, which increased dye rejection.

Acknowledgments. The author would like to acknowledge the financial support from Faculty of Chemical Engineering & Technology, Universiti Malaysia Perlis (UniMAP).

References

1. Vedula, S.S., Yadav, G.D.: Wastewater treatment containing methylene blue dye as pollutant using adsorption by chitosan lignin membrane: Development of membrane, characterization and kinetics of adsorption. J. Indian Chem. Soc. **99**(1), 100263 (2022)
2. Othman, N.H., Alias, N.H., Fuzil, N.S., et al.: A review on the use of membrane technology systems in developing countries. Membranes **12**(1) (2022)
3. Cai, Z., Sun, Y., Liu, W., Pan, F., Sun, P., Fu, J.: An overview of nanomaterials applied for removing dyes from wastewater. Environ. Sci. Pollut. Res. **24**(19), 15882–15904 (2017)
4. Hir, Z.a.M., Moradihamedani, P., Abdullah, A. H., Mohamed, M.A.: Immobilization of TiO 2 into polyethersulfone matrix as hybrid film photocatalyst for effective degradation of methyl orange dye. Mat. Sci. Semicond. Process. **57**, 157–165 (2017)
5. Sakarkar, S., Muthukumaran, S., Jegatheesan, V.: Tailoring the effects of titanium dioxide (TIO2) and polyvinyl alcohol (PVA) in the separation and antifouling performance of Thin-Film composite polyvinylidene fluoride (PVDF) membrane. Membranes **11**(4), 241 (2021)
6. Tibi, F., Park, S.J., Kim, J.: Improvement of membrane distillation using PVDF membrane incorporated with TiO2 modified by silane and optimization of fabricating conditions. Membranes (2021)

7. Sakarkar, S., Muthukumaran, S., Jegatheesan, V.: Polyvinylidene fluoride and titanium dioxide ultrafiltration photocatalytic membrane: Fabrication, morphology, and its application in textile wastewater treatment. J. Environ. Eng. **146**(7). https://doi.org/10.1061/(asce)ee.1943-7870.0001716 (2020)
8. Kebria, M.R.S., Jahanshahi, M., Rahimpour, A.: SiO_2 modified polyethyleneimine-based nanofiltration membranes for dye removal from aqueous and organic solutions. Desalination **367**, 255–264 (2015)
9. Dali Youcef, L., Belaroui, L.S., Lopez-Galindo, A.: Adsorption of a cationic methylene blue dye on an Algerian palygorskite. Appl. Clay Sci. **179**, 105145 (2019)

Preparation and Characterization of Carbon from Rice Husk by Pyrolysis Method

Faizul Che Pa(✉), Ruhiyuddin Mohd Zaki, and Khairel Rafezi Ahmad

Faculty of Chemical Engineering & Technology, Universiti Malaysia Perlis, 02600 Jejawi, Arau, Perlis, Malaysia
faizul@unimap.edu.my

Abstract. Rice husk, the outermost coating of rice grains, is a substantial byproduct of rice milling in Malaysia, a major rice producer. This research aims to produce and analyze carbon derived from rice husk using the pyrolysis method with various pyrolysis temperatures (250, 350, 450, and 550 °C) and soaking times (3 and 4 h). The study seeks to examine the physical and chemical characteristics of the biochar that is produced. Scanning electron microscopy (SEM) and energy-dispersive X-ray spectroscopy (EDS) provide information on biochar's morphology and chemical composition. These techniques reveal that higher temperatures result in more porosity and increased exposure to silica. Fourier Transform Infrared spectroscopy has been used to detect and characterize functional groups. A distinct peak at 450 °C is seen for the Si-O-Si functional group. The increase in porosity may be ascribed to the elimination of volatile compounds and the developing micropores at higher temperatures. With an increase in the pyrolysis temperature, a greater amount of volatile matter was eliminated, resulting in the formation of larger pores of the biochar.

Keywords: Rice Husk · Carbon · Biochar

1 Introduction

Biochar, derived from the agricultural biomass and rice milling sectors, has many benefits, including a distinctive pore structure, a substantial specific surface area, intricate surface-active functional groups, and chemically stable characteristics [1]. The substance can immobilize or remove pollutants from soil, water, and the air. It is extensively used in environmental applications, passivation, and the elimination and adsorption of contaminants such as heavy metals.

The high carbon content (35.2–44.7%) and volatile matter concentration (70.2–78.5%) encourage the pyrolysis of rice husk to produce biochar [2]. Microwave-assisted hydrothermal carbonization yielded the maximum amount of biochar from rice husk (57.9%), but wet pyrolysis generated the highest carbon content (71.2%) in biochar [3].

Increasing the pyrolysis temperature increases carbon content, pH, and electrical conductivity, whereas decreasing the temperature has the opposite effect [4]. The main components of rice husk biochar consist of carbon (C), hydrogen (H), oxygen (O),

N. R. Abdul Razak et al. (Eds.): ECOOP 1996, SPPHY 418, pp. 319–326, 2025.
https://doi.org/10.1007/978-981-96-2871-1_45

nitrogen (N), phosphorus (P), potassium (K), and silicon (Si). A significant positive correlation exists between the pyrolysis temperature and the carbon (C) content in rice husk biochar. In contrast, a significant negative correlation exists between the pyrolysis temperature and the hydrogen (H) content. Increasing the pyrolysis temperature leads to a decrease in the hydrogen (H) content of rice husk biochar, while the carbon (C) content is comparable to that of the rice husk [5]. The pyrolysis temperature is the most crucial process parameter since the amount of biochar produced is inversely linked to the carbon content generated at different temperatures.

2 Experimental Procedure

The rice husk has been pyrolyzed by placing the sample into 50 mL of the covered porcelain with a sample weight of 5–7 g. Second, the porcelain will be wrapped tightly in aluminium foil to prevent air contact and prevent oxygen from entering the porcelain and then placed the sample in the furnace using soaking times, which is three and four hours at various pyrolysis temperatures (250, 350, 450, and 550 °C). Then, the remaining biochar in the porcelain will be collected, ground, and sieved through a 355-micron mesh. The surface functional biochar groups have been examined using Fourier Transform Infrared Spectroscopy (FTIR). The morphological characteristics and elemental composition of the rice husk carbon have been investigated by using a Scanning Electron Microscope (SEM).

3 Results and Discussions

3.1 Fourier Transform Infrared Spectroscopy (FTIR)

The research examined the characteristics of untreated rice husk and rice husk carbon at various temperatures and durations, particularly for 3 h (Fig. 1 and Table 1). The raw rice husk exhibited a prominent O-H stretching peak at 3467.39 cm^{-1}, a characteristic feature of lignin, hemicellulose, and cellulose [6]. The strength of the O-H stretching band reduced as the pyrolysis temperatures increased, suggesting that the hydroxyl groups were lost due to heat degradation. The C-H stretching area of the raw rice husk and rice husk carbon exhibited maxima at 2920.2 cm^{-1} and 2927.63 cm^{-1}, respectively.

The inorganic functional group exhibited peaks at 1716.15 cm^{-1} and 1702.7 cm^{-1}, suggesting the existence of $C = O$ stretching in aromatic rings and lignin. The presence of the aromatic group derived from lignin resulted in the occurrence of $C = C$ asymmetric stretching at all temperatures. The peaks associated with $C = O$ and $C = C$ bonds exhibited a significant reduction, indicating the breakdown of aromatic and carbonyl structures. The Si-O-Si stretching peak seen at 797.57 cm^{-1}, 798.09 cm^{-1}, 798.49 cm^{-1}, and 797.68 cm^{-1} suggests an increase in silica content when organic matter decomposes [7]. The heightened Si-O-Si peaks observed suggest a significant degree of silica exposure at a temperature of 450 °C.

Figure 2 and Table 2 show the outcomes of pyrolyzing rice husk carbon at different temperatures for 4 h. The O-H stretching increases between 250 °C and 350 °C and decreases between 450 °C and 550 °C, suggesting alterations in hydrogen-bonded

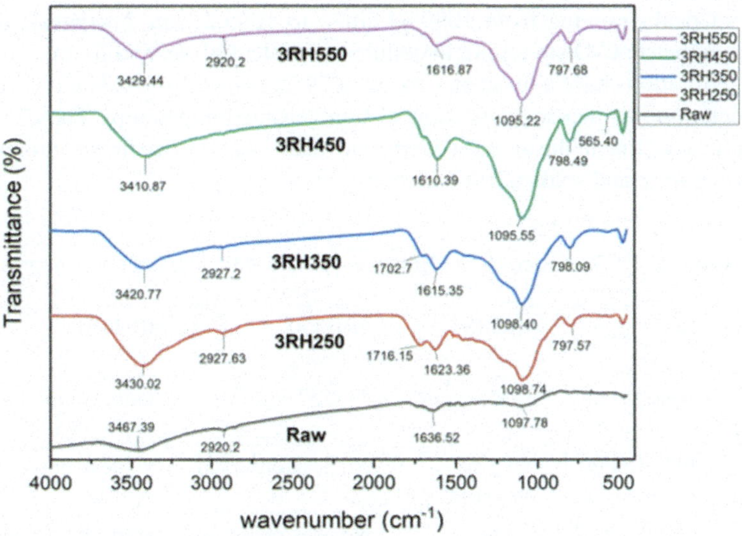

Fig. 1. FTIR spectrum of raw rice husk and rice husk carbon at 3 h with various pyrolysis temperatures

Table 1. FTIR of raw rice husk and rice husk carbon at 3 h with various pyrolysis temperatures

Functional Groups	Raw	3RH250	3RH350	3RH450	3RH550
O-H stretching	3467.39 cm^{-1}	3430.02 cm^{-1}	3420.77 cm^{-1}	3410.87 cm^{-1}	3429.44 cm^{-1}
C-H stretching	2920.2 cm^{-1}	2927.63 cm^{-1}	2927.2 cm^{-1}	No peak observed	2920.2 cm^{-1}
C = O stretching	No peak observed	1716.15 cm^{-1}	1702.7 cm^{-1}	No peak observed	No peak observed
C = C stretching	1636.52 cm^{-1}	1623.36 cm^{-1}	1615.35 cm^{-1}	1610.39 cm^{-1}	1616.87 cm^{-1}
C-O stretching	1097.78 cm^{-1}	1096.74 cm^{-1}	1098.40 cm^{-1}	1095.55 cm^{-1}	1095.22 cm^{-1}
Si-O-Si stretching	No peak observed	797.57 cm^{-1}	798.09 cm^{-1}	m^{-1}	797.68 m^{-1}

hydroxyl groups because of heat breakdown. The O-H stretching is associated with the hydroxyl groups in the rice husk carbon. The presence of peaks at 2922.15 cm^{-1} suggests the stretching of C-H bonds in aliphatic carbon compounds, while peaks at 1716.8 cm^{-1} show the stretching of C = O bonds in aromatic rings and lignin.

At all temperatures, there are observable peaks in the stretching of aromatic C = C bonds. The observation of C-C stretching is seen at both 250 °C and 350 °C. The presence of peaks ranging from 1099.34 cm^{-1} to 1095.23 cm^{-1} suggests the occurrence of symmetric C-O stretching in cellulose, hemicellulose, and lignin. The Si-O-Si stretching vibrations have a frequency range of 795.57 cm^{-1} to 796.74 cm^{-1}, indicating an augmentation in silica concentration with increasing temperature. The FTIR spectra demonstrate a decline in the amount of organic material and a rise in the amount of silica as the temperature and soaking time increase.

Table 2. FTIR of Rice Husk Carbon at 4 Hours with different temperatures

Functional Groups	4RH250	4RH350	4RH450	4RH550
O-H stretching	3436.15 cm^{-1}	3436.65 cm^{-1}	3435.85 cm^{-1}	3433.91 cm^{-1}
C-H stretching	2922.15 cm^{-1}	No peak observed	No peak observed	No peak observed
C = O stretching	1716.8 cm^{-1}	No peak observed	No peak observed	No peak observed
C = C stretching	1631.95 cm^{-1}	1616.33 cm^{-1}	1616.65 cm^{-1}	1610.05 cm^{-1}
C-C stretching	1459.2 cm^{-1}	1452.2 cm^{-1}	No peak observed	No peak observed
C-O stretching	1099.34 cm^{-1}	1098.37 cm^{-1}	1096. 58 cm^{-1}	1095.23 cm^{-1}
Si–O-Si stretching	795.57 cm^{-1}	794.78 cm^{-1}	793.97 cm^{-1}	796.74 cm^{-1}

3.2 Scanning Electron Microscope (SEM)

Figures 3 and 4 show SEM micrograph images of raw rice husk and rice husk carbon at different temperatures and times. The figures clearly show significant pore morphology and dimensions variations between thermally processed rice husks and raw rice husks. The porous structure of raw rice husk and rice husk carbon were seen in the SEM micrographs.

From the result, the different pore structures in the thermally treated rice husk were as apparent as in the raw rice husk, but more pore structures were found in the rice husk carbon at different temperatures. As a result, pore sizes increased from 3 to 4 h. As a result, rice husk heated for 3 h (Fig. 3) has fewer holes or pores and cannot be seen as clearly as rice husk heated for 4 h (Fig. 4). When the temperature rises, the pores become visible and identifiable. As the temperature increases from 250°C to 550°C, the porosity of the rice husk carbon rises significantly. Figures 3 and 4 show a transformation from a dense, intact structure to a highly porous and hollow structure. The samples heated for 4 h have more developed holes than those heated for 3 h.

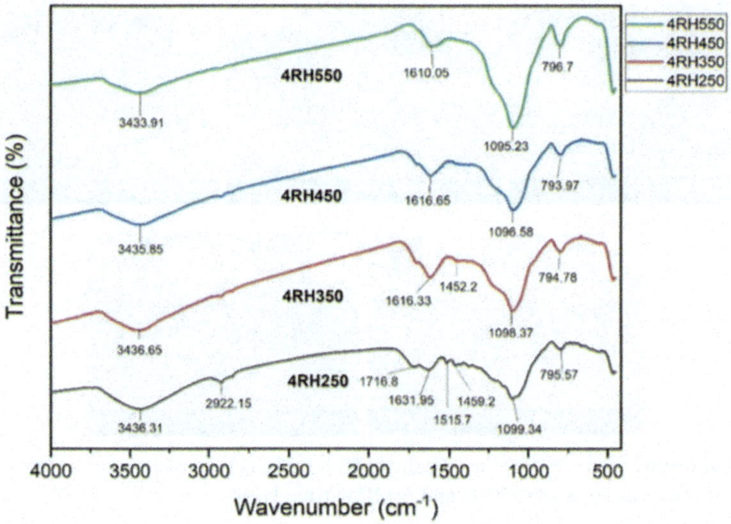

Fig. 2. FTIR Spectrum of Rice Husk Carbon at 4 h with different temperatures

The biochar shown in Figs. 3(a) and 4(a), and raw rice husks exhibited tissue that had not undergone considerable devolatilization, indicating that the pores were not completely established or were in the beginning stage of pore formation. The surface of raw rice husk is compact with little apparent holes. The structure consists mainly of cellulose, hemicellulose, and lignin without substantial thermal treatment. The biochar's morphology changed to a honeycomb-like structure with cylindrical holes connected by large holes at 450 °C and 550 °C. This micrograph reveals that numerous pores have formed over the surface, and the high temperatures have caused cracks and shrinkages on the rice husk carbon at a temperature of 550 °C in 4 h. At 550 °C, the structures of rice husk carbon were very porous, hollow, and well-arranged. Lower temperatures (between 250 to 350 °C) allow the rice husk carbon to keep its natural shape and have fewer holes. At higher temperatures (450 °C and 550 °C), the structure becomes more porous and organized, but many of the volatile substances are eliminated and a structure like a honeycomb emerges. The cellulose-lignin matrix breaks down as raw rice husk becomes biochar [8]. As a result, the pores enlarge and their internal structure becomes more visible.

Analysis showed the presence of carbon (C), oxygen (O), and silicon (Si) at different temperatures and different soaking times (3 and 4 h). According to Table 3, the main rise in the EDS spectrum is for silicon (Si), which means that silica ($SiO2$) is present in large amounts. Compared to rice husk carbon, raw rice husk has a higher percentage of silica.

Carbon levels substantially increase when subjected to a temperature of 550°C, particularly during the conversion into rice husk ash. The carbon content experiences a substantial increase during pyrolysis. The maximum size is achieved after a pyrolysis period of 3 h at a temperature of 450 °C; the minimum size is obtained after a pyrolysis period of 4 h at a temperature of 350 °C. However, at higher temperatures, the carbon

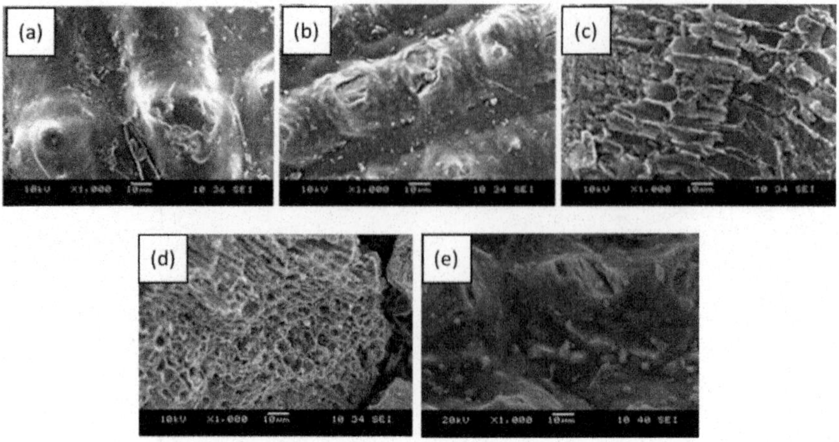

Fig. 3. SEM results with 1000X magnification at 3 h soaking times with various temperatures (a) 250°C (b) 350°C (c) 450°C (d) 550°C and (e) Raw Rice Husk

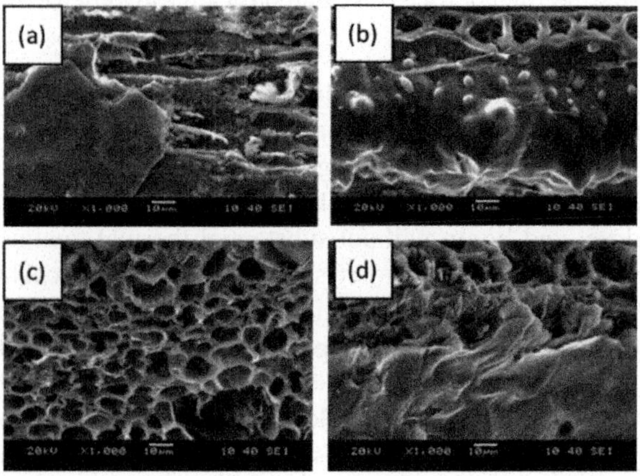

Fig. 4. SEM results with 1000X magnification at 4 h soaking times with various pyrolysis temperatures a) 250°C (b) 350°C (c) 450°C and (d) 550°C

content subsequently falls. The creation of carbon-rich compounds causes an increase in temperature during pyrolysis. In contrast, the reduction at higher temperatures is linked to the generation of rice husk ash. The oxygen concentration diminishes as the temperature rises, suggesting the depletion of oxygen-containing functional groups and volatile chemicals.

Table 3. EDS result for raw rice husk and rice husk carbon produced at various pyrolysis temperatures and soaking times

Element (%)	Raw	250°C		350°C		450°C		550°C	
		3 h	4 h	3 h	4 h	3 h	4 h	3 h	4 h
C	29.69	43.29	35.70	48.98	48.19	50.52	37.15	43.97	45.87
O	37.85	35.00	35.28	30.66	30.81	30.61	36.90	32.34	30.40
Si	32.46	21.71	29.03	20.35	21.00	18.86	25.95	23.69	23.73

4 Conclusion

The research shows that treating rice husk to pyrolysis at different pyrolysis temperatures and times significantly improves its physical and chemical characteristics. FTIR examination demonstrates that higher temperatures lead to the degradation of hydroxyl, aliphatic, carbonyl, and aromatic groups. Meanwhile, the peaks associated with silica grow more pronounced, indicating a rise in the amount of inorganic silica as the organic components degrade. SEM result demonstrates that the porosity of rice husk biochar produced increases with higher pyrolysis temperatures and longer soaking times, resulting in a highly porous structure that resembles a honeycomb. The EDS analysis indicates an increase in carbon content and a decrease in oxygen content, which corresponds to removing oxygen-containing groups and forming silicon-rich ash at elevated temperatures. Typically, raising the pyrolysis temperatures greatly enhances the properties of biochar.

Acknowledgements. The authors thank the Faculty of Chemical Engineering & Technology, Universiti Malaysia Perlis and the Centre of Excellence Geopolymer & Green Technology (CeGeoGTech) for providing the facilities to run this research work.

References

1. Amalina, F., Abd Razak, A.S., Krishnan, S., Sulaiman, H., Zularisam, A.W. Nasrullah, M.: Biochar production techniques utilizing biomass waste-derived materials and environmental applications – A review. J. Hazardous Mater. Adv. **7** (2022)
2. Bushra, B., Remya, N.: Biochar from pyrolysis of rice husk biomass—characteristics, modification, and environmental application. Biomass Conversion and Biorefinery (2020)
3. Potnuri, R., Suriapparao, D., Sankar Rao, C., Yadav, A., Veluru, S., Neelancherry, R.: A review on analysis of biochar produced from microwave-assisted pyrolysis of agricultural waste biomass. J. Anal. Appl. Pyrol. **173** (2023)
4. Hidayat, H., et al.: Analysis of rice husk biochar characteristics under different pyrolysis temperatures. IOP Conf. Ser. **1201**(1), 012095–012095 (2023)
5. Singh Karam, D., et al.: An overview on the preparation of rice husk biochar, factors affecting its properties, and its agriculture application. J. Saudi Soc. Agric. Sci. **21**(3) (2021)

6. Ayala-Armijos, J., et al.: Effect of alkaline and hydrothermal pretreatments in sugars and ethanol production from rice husk waste. Resources **13**, 128 (2024)
7. Almendro-Candel, M.B., Callejas, P., Montealegre, M.Á., Rincón, J.M., Jordán Vidal, M.M.: The emission of volatile components during laboratory vitrification when using fly ash and other waste to obtain ceramic coatings. MDPI (2023)
8. Tsai, W.-T., Lin, Y.-Q., Huang, H.-J.: Valorization of rice husk for the production of porous biochar materials. MDPI (2021)

Transforming Wasted Polystyrene Foam to Conductive Composites by the Incorporation of Graphene-Based Fillers

Lee Yong Shaun[1], Siti Salmi Samsudin[1](✉), Mohd Awaludin Mohamed Bashir[2], Mohd Shukry Abdul Majid[3], and Mohd Ridzuan Mohd Jamir[3]

[1] Faculty of Chemical Engineering Technology, Universiti Malaysia Perlis (UniMAP), 02600 Arau, Perlis, Malaysia
sitisalmi@unimap.edu.my

[2] Unit Latihan & Pendidikan Lanjutan, Politeknik Tuanku Syed Sirajuddin, Pauh Putra, 02600 Arau, Perlis, Malaysia

[3] Faculty of Mechanical Engineering Technology, Universiti Malaysia Perlis (UniMAP), Pauh Putra, 02600 Arau, Perlis, Malaysia

Abstract. This research examines the effectiveness of a conductive polystyrene composites system that incorporates nanofillers which are graphite (Gs) and graphene nanoplatelets (GNPs), as a means of reinforcing polystyrene matrices with enhanced electrical properties. The samples were being investigated on the effect of different filler and filler concentrations. The blending of conductive polystyrene composites was carried out through a straightforward process by solution mixing method. The incorporation of the graphitic-based fillers resulted in diverse effects in the electrical conductivity and mechanical properties of the conductive polystyrene composites. The findings exhibit a correlation between the electrical properties of the conductive polystyrene composites and their respective filler loadings, which an increase in filler loadings led to an increase in electrical properties. The incorporation of 5 wt.% of filler concentration resulted in a significant enhancement of the electrical conductivity of the composites, reaching a value of 4.86×10^{-7} $(\Omega\cdot\text{cm})^{-1}$.

Keywords: Wasted Polystyrene Foam · Conductive Polymer Composites · Sustainable Development

1 Introduction

In recent years, there has been a surge of interest in the development and utilization of conductive polymer composites (CPCs) by combining polymers with electrically conductive materials. These composites exhibit enhanced electrical conductivity and mechanical strength, making them suitable for various applications such as electrolyte, electrostatic discharge (ESD) protection, antistatic and electromagnetic interference (EMI) shielding, supercapacitors, and solar cells. The addition of graphitic-based fillers like carbon

N. R. Abdul Razak et al. (Eds.): ECOOP 1996, SPPHY 418, pp. 327–333, 2025.
https://doi.org/10.1007/978-981-96-2871-1_46

nanotubes and graphene nanoplatelets to the polymer matrix has been shown to significantly improve the conductivity and strength of these composites. Furthermore, the use of renewable resources in the development of conductive polymer composites has been a focus of research, aiming to address the poor mechanical characteristics of polymers derived from renewable resources and provide greater mechanical and electrically conductive capabilities [1]. One of the types of research demonstrated that combining graphite powder with polypropylene resin through melt compounding and compression molding processes resulted in enhanced tensile modulus and improved thermal stability of the composite [2]. Similarly, another research developed thin polystyrene (PS) waste embedded multiwall carbon nanotube (MWCNT) film, showing improved electrochemical properties for flexible dye-sensitized solar cell (DSSC) applications [3]. These studies highlight the potential of conductive polymer composites in various electronic applications, emphasizing their biodegradability and recyclability as promising features for sustainable technological solutions.

2 Experimental Procedure

2.1 Materials

In this study, expanded polystyrene sheets (EPS), known as Styrofoam, were used as the polymer matrix. EPS were taken from recycled Styrofoam that are ready to be disposed of. The first filler that was used in this study is graphite pure powder of graphite (Technical Grade) was provided by BTCorp Generique Nano Pvt. Ltd. (Bengaluru, India). The second filler that was used in this study is graphene nanoplatelets (GNP). Graphene nanoplatelets (GNP) was provided by Skyspring Nanomaterials, Inc. (Houston, USA). The properties of fillers are shown in Table 1.

Table 1. Properties of fillers

Properties	Graphite	GNP
Appearance	Dark grey amorphous powder	Black Powder
Morphology	Single layer hexagonal lattice	Platelet
Content of Carbon	> 92%	> 99.5%
Average diameter/particle size	< 10 μm	< 2 μm
Length	10 μm	-
Thickness Average	10–20 nm	1–5 nm
Density	0.41 g/cm^3	0.2–0.6 g/cm^3
Surface area	52 m^2/g	750 m^2/g

2.2 Sample Preparations

The procedure involved dissolving recycled EPS (r-EPS) in acetone to extract them into melted PS material. Conductive fillers were then added to the melted PS and thoroughly mixed until homogeneous. The mixture was pressed using a hot press machine at 140ºC, with 5 min of preheating and 8 min of compression, followed by 5 min in a cold press. This process was repeated for all sample formulations, which were then cut according to ASTM standards for tensile and electrical conductivity testing. Table 2 details of the formulations of the conductive polystyrene composites.

Table 2. Formulation for conductive polystyrene composites

Type of Sample	Sample Abbreviation	Polystyrene (wt. %)	Graphite (wt. %)	Graphene Nanoplatelets (wt. %)
Control Group	r-EPS 100	100	0	0
r-EPS/G composites	r-EPS/G (1.25)	98.75	1.25	0.00
	r-EPS/G (2.5)	97.50	2.50	0.00
	r-EPS/G (5)	95.00	5.00	0.00
r-EPS/GNP composites	r-EPS/GNP (1.25)	98.75	0.00	1.25
	r-EPS/GNP (2.5)	97.50	0.00	2.50
	r-EPS/GNP (5)	95.00	0.00	5.00

2.3 Evaluation Method

2.3.1 Tensile Test

Tensile testing was conducted by using the Instron 5569 Universal Testing Machine (UTM), following the ASTM D638 standard. The sample dimensions were measured at 110mm (Length) x 15 mm (Width) x 2 mm (Thickness) in a rectangular shape. The gap between the upper jaw and lower jaw of the Instron machine was adjusted and set at 50 mm. The crosshead speed of machine was set to 3 mm/min, at room temperature.

2.3.2 Scanning Electron Microscopy

The morphology of samples was examined using a Scanning Electron Microscope (SEM) (JEOL JSM-6460LA) according to ASTM F1877. After conducting the tensile test, the fracture surface of the samples was cut to 1 mm thickness and platinum-coated using a JFC-1600 Auto Fine Coater (JEOL Ltd. Japan) before SEM imaging at 1000x magnifications.

2.3.3 Electrical Conductivity

The electric conductivity test is according to ASTM D149 (two point probe). The electrical characteristics of the samples were determined using a Fluke 8845A/8846A 6.5-digit precision multimeter with a 5V power source. The samples were produced in a circular form with a thickness of less than 1 mm and a diameter of 10 mm. To ensure good surface contact with the electrodes, a conductive silver paint was applied to the top and bottom surfaces of the samples. Bulk volume conductivity was calculated using Eq. (1).

$$\text{Bulk volume conductivity, } \sigma\ (\Omega.\text{cm})^{-1} = \frac{1}{\rho} \quad (1)$$

3 Results and Discussions

3.1 Tensile Test

The tensile strength of the conductive polystyrene composites was examined to study the mechanical properties as the effects of different filler concentrations in Fig. 1. Adding graphene nanoplatelet (GNP) fillers increases the tensile strength across all concentrations, with 1.25 wt.% GNP increasing it to 5.465 MPa, 2.5 wt.% GNP to 5.711 MPa, and 5 wt.% GNP to 5.625 MPa, the latter decrease likely due to GNP agglomeration. Conversely, graphite fillers reduce tensile strength. 1.25 wt.% of graphite raises it slightly to 5.233 MPa, but 2.5 wt.% graphite fillers and 5 wt.% graphite fillers decrease it to 4.818 MPa and 4.163 MPa respectively due to agglomeration and uneven distribution. This improvement in tensile strength with GNP is attributed to strong interfacial interaction with the r-EPS matrix, which enhances stress transfer. GNP's high aspect ratio and surface area contribute to effective reinforcement, thereby improving the composite's mechanical properties. The highest tensile strength was observed at 2.5 wt.% GNP, indicating that this concentration optimally balances the matrix-filler interaction, where the r-EPS matrix serves as the binder and the GNP as the reinforcing agent. Conversely, graphite fillers demonstrated a less favorable impact, with tensile strength decreasing as the graphite concentration increased. This is likely due to graphite particle agglomeration, leading to stress concentration and a weaker composite structure. The lower aspect ratio of graphite compared to GNP also results in less effective reinforcement and stress transfer, highlighting GNP's superior ability to enhance the mechanical properties of r-EPS composites.

3.2 Fracture Surface Morphologies

Figure 2 shows the SEM micrographs of r-EPS/GNP and r-EPS/G composites with 2.5 wt.% and 5 wt.% filler concentrations. Sem images for 1.25 wt.% are not included, as the fillers are evenly dispersed, resulting in a smooth surface with minimal rough areas and cracks. At 2.5 wt.%, the composite surface remains smooth with visible cracks and undulations, indicating challenges in achieving uniform dispersion at these concentrations, with noticeable GNP agglomerations and graphite aggregations. Despite these challenges, the interfacial interactions appear to improve the composite's properties. For r-EPS/GNP, at 5 wt.%, the surface becomes rough with larger agglomerations, indicating difficulties in achieving consistent distribution at higher concentrations. For r-EPS/G at 5 wt.%, significant agglomeration and entangled morphology are observed, with large graphite clusters densely packed within the composite, disrupting the matrix integrity. Cracks and voids suggest stress concentration areas that could compromise mechanical

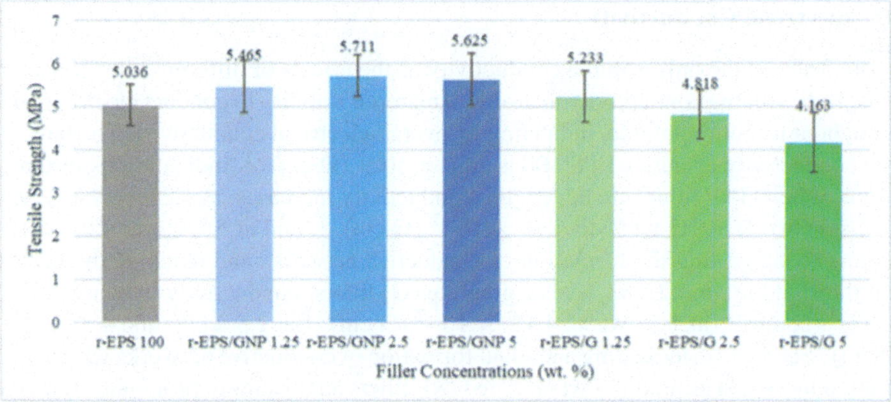

Fig. 1. Tensile strength as the effects of different filler concentrations.

strength. At this magnification, increased surface roughness, irregulaties and prominent filler agglomerates are evident, with some regions showing weak bonding between the fillers and the r-EPS matrix, as seen in Fig. 2(b).

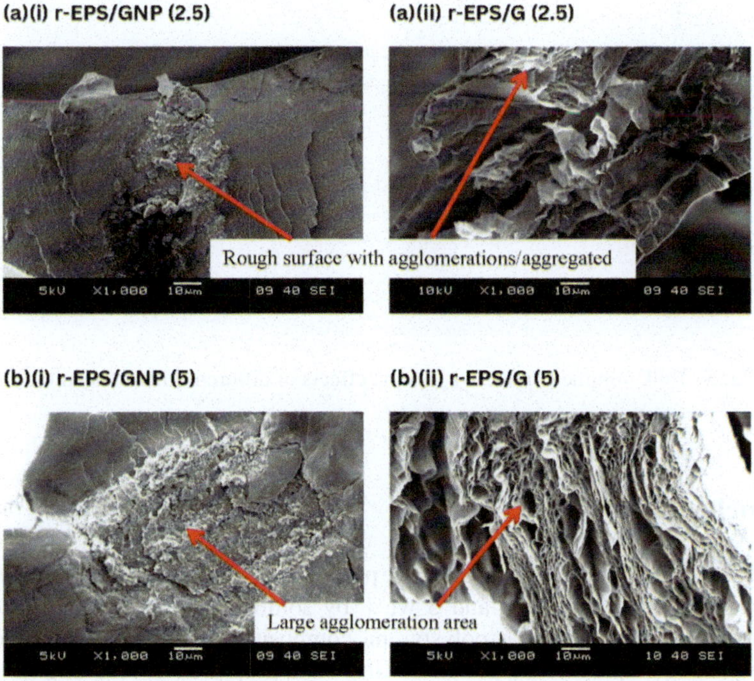

Fig. 2. SEM Micrographs at 1000x magnification of the r-EPS/GNP and r-EPS/G composites at different filler concentrations; (a) 2.5 wt.% (b) 5 wt. %.

3.3 Electrical Conductivity

Figure 3 shows the bulk volume conductivity as the effects of different filler concentrations. Bulk volume conductivity measures a material's ability to conduct electric current throughout its entire volume, indicating its overall electrical conductivity. The bulk volume conductivity of pure r-EPS 100 is 2.60×10^{-7} $(\Omega\cdot cm)^{-1}$, indicating its insulating nature. When GNP fillers are added, the conductivity increases: at 1.25 wt.% to 2.95×10^{-7} $(\Omega\cdot cm)^{-1}$, at 2.5 wt.% to 4.74×10^{-7} $(\Omega\cdot cm)^{-1}$, and at 5 wt.% to 4.86×10^{-7} $(\Omega\cdot cm)^{-1}$, suggesting the formation of conductive networks and reaching the percolation threshold around 2.5 wt.%. For graphite (G) fillers, conductivity increases to 3.26×10^{-7} $(\Omega\cdot cm)^{-1}$ at 1.25 wt.%, 4.12×10^{-7} $(\Omega\cdot cm)^{-1}$ at 2.5 wt.%, and 4.55×10^{-7} $(\Omega\cdot cm)^{-1}$ at 5 wt.%, indicating a gradual formation of conductive networks with the percolation threshold around 2.5 wt.%. The GNP fillers show a more pronounced increase in conductivity between 1.25 and 2.5 wt.%, highlighting their effectiveness in forming conductive networks through electron tunneling and hopping [4]. The conductivity increases at 5 wt.% GNP is less pronounced, indicating saturation. In contrast, graphite fillers show a steady increase in conductivity, with a less pronounced increment at higher concentrations, suggesting nearing saturation but still improving network connectivity.

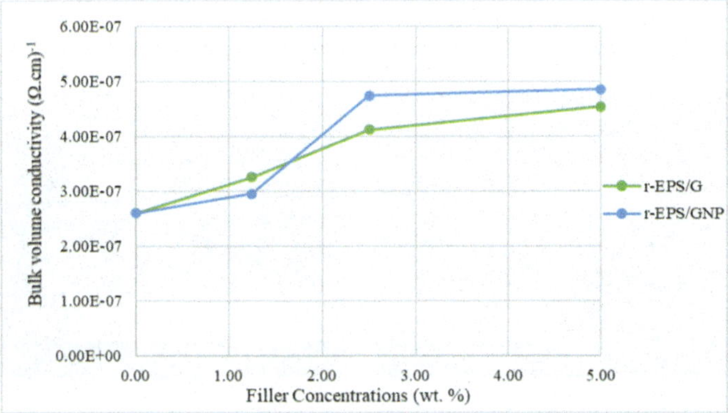

Fig. 3. Bulk volume conductivity as the effects of different filler concentrations.

4 Conclusion

The study successfully prepared conductive polystyrene composites with graphitic-based fillers at 1.25 wt.%, 2.5 wt.%, and 5 wt.% by solution mixing method. The testing and analysis of the conductive polystyrene composites yielded positive results. The incorporation of graphene nanoplatelets (GNP) and graphite (G) resulting in increased electrical properties as the fillers has form conductive pathways for the electricity to travel more effectively in the conductive polystyrene composites. Hence, the research revealed that the electrical conductivity and mechanical properties of the r-EPS/GNP 5 conductive polystyrene composites with GNP fillers at 5 wt. % of GNP filler concentration are the most optimal properties due to better dispersion and higher surface areas in forming conductive pathways.

References

1. Wu, H., Feng, L., Jiang, A., Zhang, B.: Effect of the processing of injection-molded, carbon black-filled polymer composites on resistivity. Polym. J. **43**(11), 930–936 (2011)
2. Tong, D.J.Y., Koay, S.C., Chan, M.Y., Tshai, K.Y., Ong, T.K., Buys, Y.F.: Conductive polymer composites made from polypropylene and recycled graphite scrap. J. Phys. Sci. **32**(3) (2021)
3. Anuar Abdul Muin, M., Sabihah Zakaria, N., Nur Azella Zaini, S.: Optimization of thin polystyrene waste embedded multiwall carbon nanotubes (MWCNT) film for solid electrolyte of flexible dye-sensitized solar cell (DSSC). Mater. Today Proc. **66**(P10) (2022)
4. Mergen, Ö.B., Umut, E., Arda, E., Kara, S.: A comparative study on the AC/DC conductivity, dielectric and optical properties of polystyrene/graphene nanoplatelets (PS/GNP) and multi-walled carbon nanotube (PS/MWCNT) nanocomposites. Polym. Test. **90**, 106682 (2020)

Amphiphilic Graphite Oxide-Based Sponge: A Game Changer for Oil-Water Separation

Tai Le-Thanh[1], Mac Nhu Hiep[1], Tang Kien Ky[1], Ly Minh Dang[1], Le Hoang Tung[1], Vinh-Phuc Ha[2], Vinh-Dat Vuong[1,2,3(✉)], and Thang Van Le[3,4]

[1] Faculty of Materials Technology, Ho Chi Minh City University of Technology (HCMUT), 268 Ly Thuong Kiet Street, District 10, Ho Chi Minh City, Vietnam
vuongvinhdat@hcmut.edu.vn

[2] VNU-HCM Key Laboratory for Materials Technology (MTLab), 268 Ly Thuong Kiet Street, District 10, Ho Chi Minh City, Vietnam

[3] Vietnam National University Ho Chi Minh City, Linh Trung Ward, Thu Duc City, Ho Chi Minh City, Vietnam

[4] International University – Vietnam National University Ho Chi Minh City, Quarter 6, Linh Trung Ward, Thu Duc City, Ho Chi Minh City, Vietnam

Abstract. This study developed a high-performance sponge from graphite oxide nanoparticles. The sponge boasts a large surface area, excellent water absorption, and high porosity due to its network of pores. Analysis confirmed the formation of key bonds in the material, influencing its properties. This novel foam shows promise for applications like oil spill treatment due to its strong attraction to oil over water.

Keywords: Amphiphilic Sponge · Graphite Oxide Nanoparticle · Oil-Water Separation

1 Introduction

Despite scientific advancements, oil remains crucial for global activities (102.7 million barrels produced daily by 2024 according to the IEA [1]). This dependence raises environmental concerns, particularly oil spills [2, 3]. Ocean waves, wind, temperature, and salinity impact oil spills, causing evaporation, surface slicks, dispersion, or sinking [4]. Sustainable development efforts focus on mitigating oil spill impacts through effective cleaning techniques. These techniques fall into two categories: physical/mechanical and chemical [5]. Common methods include skimming [6, 7], dispersants [8–10], solidifiers [11, 12], sorbents [13, 14], and bioremediation [15]. Solidifiers and sorbents react with oil to break it down or form easily managed compounds [11–14]. While convenient and productive, they are limited in effectiveness [15]. Skimming utilizes skimmers and booms to remove floating oil, leveraging oil's properties (flammability and lower density than water) [6, 7]. This method offers a fast response but consumes significant energy and releases CO2 [16]. In contrast, oil-adsorbing materials effectively treat spills by separating oil from the environment [17–21]. This method allows for oil recovery without harming the ecosystem [18, 19, 22].

N. R. Abdul Razak et al. (Eds.): ECOOP 1996, SPPHY 418, pp. 334–339, 2025.
https://doi.org/10.1007/978-981-96-2871-1_47

Physical adsorption for oil spills relies on material properties. Common materials like polyurethane are effective but reusability and production cost are challenges [23–25]. New research explores nanocarbons like graphene for efficient oil adsorption [26]. Their reusability, high stability, and large surface area (up to 3000m2/g) make them promising [26–28]. Studies show surface functionalization of these materials can further enhance oil-water separation [29]. This offers a practical solution for oil spill treatment [29]. Research also reveals these materials can adsorb various oils [30]. Their high adsorption capacity, porous structure, and large pore volume (up to 2630 m2/g) make them excellent for separation, especially in 3D structures [30–33].

Building on the potential of graphite-based oil spill adsorbents, this study proposes a method to synthesize amphiphilic graphite sponges (Fig. 1). The process functionalizes graphite to create graphite oxide (GrO) with enhanced surface area and dual oil-water attraction. GrO serves as a precursor for 3D sponges with cross-linked structures (-CO-NH-, -COO-, -C-C-) formed by cross-linking agents and an expanded porous structure achieved using a foaming agent. This novel material holds promise for improved oil spill treatment efficiency and sustainability.

2 Experimental Procedure

Graphite (95%, 40–150 μm of particle size) was purchased from Xilong Scientific. Sulfuric acid (H2SO4, 98%), acid chlorhidric (HCl, 37%), sucrose ($C_{12}H_{22}O_{11}$, 98%), smmonium carbonate ($(NH_4)_2CO_3$, 99%), potassium permanganate (KMnO4, 99%) and hydrogen peroxide (H_2O_2) were purchased from Xilong Scientific. Ethanol (EtOH, 99%) was purchased from Chemsol. Octane-1,8-diamine ($H_2N\text{-}(CH_2)_8\text{-}NH_2$ 98%) was purchased from Alfa Aesar.

Preparation of graphite oxide, sponge was synthesized from graphite oxide (GrO) via a cross-linking process with sucrose and OMDA as cross-linking agents and $(NH_4)2CO_3$ as the foaming agent. The mixture underwent multiple thermal steps for solidification and foaming, resulting in a highly porous structure. The GrO sponge exhibited excellent oil-water separation efficiency compared to a control sponge made with graphite.

The water adsorption experiments on the sponge samples were conducted to calculate several parameters, including absorption capacity (AC), the ratio of pore volume in the structure (RP), the percentage of open pore volume (RPopen), and the percentage of closed pore volume (RPclose). The water adsorption capacity of the sponge samples was further evaluated by measuring the filtration flux through a fixed sponge volume for the two samples, GrOS-X and GrS-X, using the filtration test.

3 Results and Discussions

3.1 Morphology

Microscopic analysis revealed similar surface profiles for porous samples made with different structural materials (graphite and GrO) at the same reagent ratio (Fig. 1(a)). Samples with a lower weight ratio of structural material to foaming agent (GrOS-4 and GrS-4) exhibited a more uniform surface (Fig. 1(a)). Pore size distribution analysis showed GrOS-4 has smaller pores (10–100 μm^2) compared to GrS-4 (100–1000 μm^2), making GrOS-4 ideal for adsorption (Figs. 1(b) & 1(c)). Furthermore, GrOS-4 displayed the highest adsorption capacity (3.2949 L.kg^{-1}) and porosity (81%) among all samples, suggesting its potential for applications requiring high adsorption efficiency.

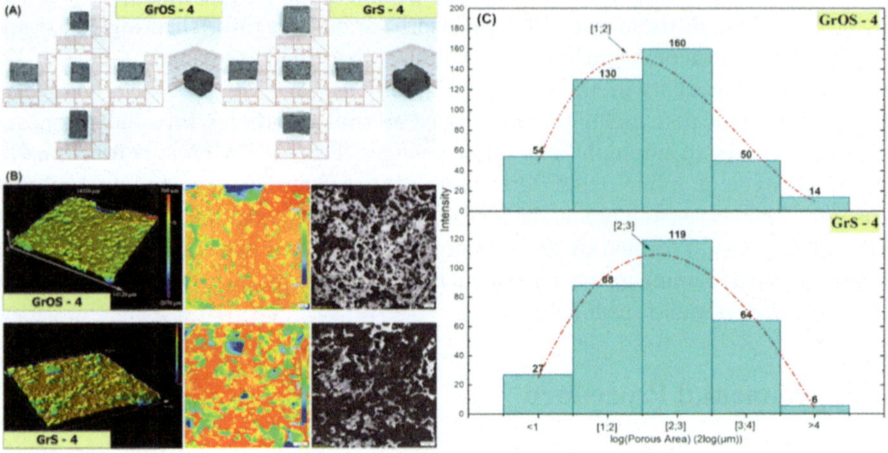

Fig. 1. Morphology of sponge samples GrOS-4 and GrS-4. (a) Appearance of the samples; (b) Microstructure images; (c) Pore area distribution on the surface.

3.2 FTIR Spectroscopy & Raman Spectroscopy

In GrOS samples, the solidification of the sample can be achieved through the formation of chemical cross-links, including peptide bonds (-CONH-), ester bonds (R-COO-R'), and non-planar (sp^3 C-C) bonds between carbon atoms on different graphene sheets. In contrast, the structure of the GrS samples is primarily composed of non-planar (sp^3 C-C) cross-links between graphite sheets. The presence of these bonds in the structure is confirmed through FTIR and Raman spectroscopic analyses (Fig. 2).

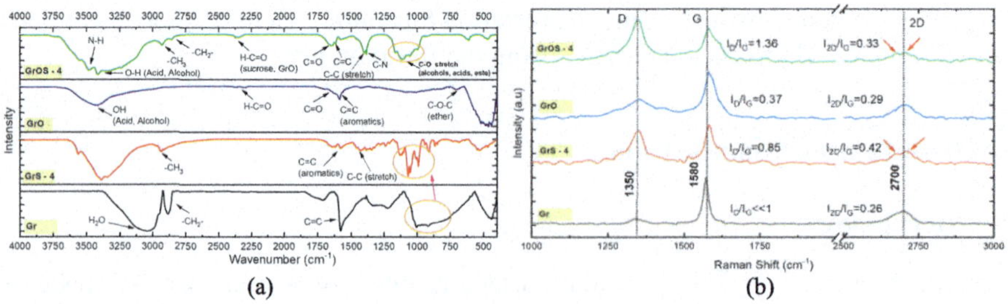

Fig. 2. The (a) FTIR Spectra and (b) The Raman Spectra, of Gr, GrS-4, GrO and GrOS-4.

Raman spectroscopy analysis revealed distinct differences in the structure of graphite (Gr) and graphite oxide (GrO) samples. Gr, with minimal defects, showed a weak D band and high G band intensity. In contrast, GrO displayed a prominent D band indicating a higher defect level. Analysis of the sponge samples (GrS-4 and GrOS-4) confirmed the formation of non-planar cross-links through a significant increase in D band intensity compared to Gr. Additionally, the 2D band splitting suggested the presence of two types of sp^2 C-C bonds. These findings confirm the introduction of defects and cross-links during foaming and functionalization, contributing to the overall structure of the sponge material.

3.3 Amphiphilicity

Modifying graphite with oxygen (oxidation) significantly changes how it interacts with water and oil as shown in Fig. 3. Raw graphite repels both (hydrophobic and oleophilic), while oxidized graphite (GrO) becomes moderately attracted to both (hydrophilic and oleophilic). Interestingly, foams made from these materials retain similar properties. GrS-4 foam, derived from graphite, remains water-repelling and oil-loving, while GrOS-4 foam, made from GrO, keeps its moderate attraction to both liquids. This demonstrates that these graphite oxide foams possess both water and oil-loving characteristics.

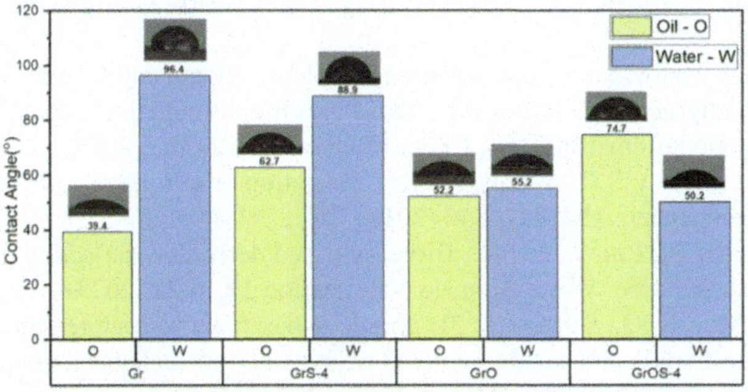

Fig. 3. The contact angle of oil and water on samples of Gr, GrO, GrS-4, and GrOS-4

4 Conclusion

In this study, a highly porous and oil-attracting sponge was synthesized from graphite oxide for separating oil from water. The sponge achieved high porosity with a network of micro-sized pores and maintained its oil-loving properties. This novel material holds promise for large-scale oil spill treatment and industrial wastewater cleanup due to its ability to selectively absorb oil over water.

Acknowledgements. We acknowledge the support of time and facilities from Ho Chi Minh City University of Technology (HCMUT), VNU-HCM for this study.

References

1. U. E. I. Administration: Global oil markets (2024)
2. Khasanova, S., Alieva, E., Shemilkhanova, A.: Environmental pollution: types, causes and consequences. BIO Web Conf. **63**, (2023)
3. Mir, R.A., Mantoo, A., Sofi, Z.A., Bhat, D., Bashir, A., Bashir, S.: Types of Environmental Pollution and Its Effects on the Environment and Society, 1–31 (2023)
4. Doshi, B., Sillanpää, M., Kalliola, S.: A review of Bio-based materials for oil spill treatment. Water Res. **135**, 262–277 (2018)
5. Behnood, R., Anvaripour, B., Jaafarzadeh, N., Farasati, M.: Oil spill sorption using raw and acetylated sugarcane bagasse. J. Cent. South Univ. **23**, 1618–1625 (2016)
6. Fritt-Rasmussen, J.: Review on burn residues from in situ burning of oil spills in relation to Arctic Waters. Water Air Soil Pollut. **226**, 329 (2015)
7. Abdullah, L., Santuso, K., Jamaludin, Z., Zamri, R., Maslan, M.N.: Design and analyses of semi-automated portable oil spill skimmer for water treatment application. Intell. Manuf. Mechatron. 372–384 (2020)
8. Afshar-Mohajer, N., Lam, A., Dora, L., Katz, J., Rule, A.M., Koehler, K.: Impact of dispersant on crude oil content of airborne fine particulate matter emitted from seawater after an oil spill. Chemosphere **256**, 127063 (2020)
9. Liu, Z., Callies, U.: A probabilistic model of decision making regarding the use of chemical dispersants to combat oil spills in the German Bight. Water Res. **169**, 115196 (2020)
10. Liu, Z., Callies, U.: Implications of using chemical dispersants to combat oil spills in the German Bight – depiction by means of a Bayesian network. Environ. Pollut. **248**, 609–620 (2019)
11. Motta, F.L., Stoyanov, S.R., Soares, J.B.P.: Development and application of an amylopectin-graft-poly(methyl acrylate) solidifier for rapid and efficient containment and recovery of heavy oil spills in aqueous environments. Chemosphere **236**, 124352 (2019)
12. Motta, F.L., Stoyanov, S.R., Soares, J.B.P.: Application of solidifiers for oil spill containment: a review. Chemosphere **194**, 837–846 (2018)
13. Zhang, Y., Sam, E., Liu, J., Lv, X.: Biomass-based/derived value-added porous absorbents for oil/water separation. Waste Biomass Valorization **14**, 1–22 (2023)
14. Riaz, M.A., McKay, G., Saleem, J.: 3D graphene-based nanostructured materials as sorbents for cleaning oil spills and for the removal of dyes and miscellaneous pollutants present in water. Environ. Sci. Pollut. Res. **24**(36), 27731–27745 (2017)
15. Graham, L., Hale, C., Maung-Douglass, E., Sempier, S., Swann, L., Wilson, M.: Chem. Dispersants Their Role Oil Spill Response 1–8 (2016)
16. Fingas, M.: An overview of in-situ burning. Oil Spill Sci. Technol. 737–903 (2011)
17. Cojocaru, C., Pricop, L., Samoila, P., Rotaru, R., Harabagiu, V.: Surface hydrophobization of polyester fibers with poly(methylhydro-dimethyl) siloxane copolymers: experimental design for testing of modified nonwoven materials as oil spill sorbents. Polym. Test. **59**, 377–389 (2017)
18. Franco, C.A., Cortés, F.B., Nassar, N.N.: Adsorptive removal of oil spill from oil-in-fresh water emulsions by hydrophobic alumina nanoparticles functionalized with petroleum vacuum residue. J. Colloid Interface Sci. **425**, 168–177 (2014)
19. Alassod, A., Gibril, M., Islam, S.R., Huang, W., Xu, G.: Polypropylene/lignin blend monoliths used as sorbent in oil spill cleanup. Heliyon **6**(9), e04591 (2020)
20. Siddiqa, A., Shahid, A., Gill, R.: Silica decorated CNTs sponge for selective removal of toxic contaminants and oil spills from water. J. Environ. Chem. Eng. **3**(2), 892–897 (2015)
21. Yang, W., Gao, H., Zhao, Y., Bi, K., Li, X.: Facile preparation of nitrogen-doped graphene sponge as a highly efficient oil absorption material. Mater. Lett. **178**, 95–99 (2016)

22. Fan, T., et al.: Robust Graphene@PPS fibrous membrane for harsh environmental oil/water separation and all-weather cleanup of crude oil spill by joule heat and photothermal effect. ACS Appl. Mater. Interfaces **13**(16), 19377–19386 (2021)
23. Wu, D., Wu, W., Yu, Z., Zhang, C., Zhu, H.: Facile preparation and characterization of modified polyurethane sponge for oil absorption. Ind. Eng. Chem. Res. **53**(52), 20139–20144 (2014)
24. Wang, S., Wang, K., Li, Y., Wu, F., Wang, S., Zheng, W.G.: Open-cell polypropylene/polyolefin elastomer blend foams fabricated for reusable oil-sorption materials. J. Appl. Polym. Sci. **133**(33), 43812 (2016)
25. Ying, Q., Lili, Z., Haixiang, S., Peng, L.: PAN/PS elctrospun fibers for oil spill cleanup. IOP Conf Ser Mater Sci Eng **62**, 12032 (2014)
26. Elhenawy, S., Khraisheh, M., AlMomani, F., Hassan, M.K., Al-Ghouti, M.A., Selvaraj, R.: Recent developments and advancements in graphene-based technologies for oil spill cleanup and oil-water separation processes. Nanomaterials **12**(1), 87 (2022)
27. Yuan, J., et al.: Novel high-capacity and reusable carbonaceous sponges for efficient absorption and recovery of oil from water. Appl. Surf. Sci. **487**, 398–408 (2019)
28. Li, L., et al.: Activated carbon prepared from lignite as supercapacitor electrode materials. Electroanalysis **28**(1), 243–248 (2016)
29. Olabintan, A.B., Ahmed, E., Al Abdulgader, H., Saleh, T.A.: Hydrophobic and oleophilic amine-functionalised graphene/polyethylene nanocomposite for oil–water separation. Environ. Technol. Innov. **27**, 102391 (2022)
30. Shiu, R.F., et al.: Superhydrophobic graphene-based sponge as a novel sorbent for crude oil removal under various environmental conditions. Chemosphere **207**, 110–117 (2018)
31. Lee, S., Kim, H., Shim, B.: Graphene: an emerging material for biological tissue engineering. Carbon Lett. **14**(2), 63–75 (2013)
32. Pethsangave, D.A., Wadekar, P.H., Khose, R.V., Some, S.: Super-hydrophobic carrageenan cross-linked graphene sponge for recovery of oil and organic solvent from their water mixtures. Polym. Test. **90**, 106743 (2020)
33. Nguyen, D.D., Tai, N.H., Lee, S.B., Kuo, W.S.: Superhydrophobic and superoleophilic properties of graphene-based sponges fabricated using a facile dip coating method. Energy Environ. Sci. **5**(7), 7908–7912 (2012)

Mechanical Properties of Sea Sand Concrete Incorporating Corrosion Inhibitors and Supplementary Cementitious Materials

Vu Thi Hong Nhung[1,2], Nguyen Ngoc Tri Huynh[1,2], and Nguyen Khanh Son[1,2,3](✉)

[1] Faculty of Materials Technology, Ho Chi Minh City University of Technology (HCMUT), 268 Ly Thuong Kiet Street, District 10, Ho Chi Minh City, Vietnam
ksnguyen@hcmut.edu.vn

[2] Vietnam National University Ho Chi Minh City, Linh Trung Ward, Thu Duc City, Ho Chi Minh City, Vietnam

[3] Centre of Excellence Geopolymer and Green Technology (CEGeoGTech), Universiti Malaysia Perlis (UniMAP), 01000 Kangar, Perlis, Malaysia

Abstract. This study investigates the properties of mortar and concrete using sea sand as fine aggregate, focusing on the effects of fly ash (FA) and granulated blast furnace slag (GBFS) as supplementary cementitious materials (SCM), along with corrosion inhibitors. Sodium nitrite had minimal impact on the flowability of cement mortar, while chromate-based inhibitors significantly reduced it. Compressive strength tests showed that mortar samples without corrosion inhibitors had the highest strength, whereas those with chromate-based inhibitors had significantly lower strength. Surface resistivity measurements indicated that samples with 25% FA and 25% GBFS had significantly higher resistivity, suggesting improved resistance to chloride ion penetration and enhanced durability. These findings demonstrate the effectiveness of FA and slag in improving concrete performance and durability in marine environments. The study highlights the importance of careful admixture and corrosion inhibitor selection in optimizing sea sand concrete properties, promoting sustainable construction practices using locally available resources.

1 Introduction

The construction and infrastructure development sectors are major consumers of raw materials, particularly river sand, in concrete production, leading to significant environmental challenges. In Vietnam, the depletion of natural sand resources and the prevalence of illegal sand mining have become critical issues amid rapid infrastructure development. The extraction and use of river sand present both technical and environmental problems. Technically, erosion reduces grain size, limiting the sand's modulus. Environmentally, unregulated or salvage sand mining causes increased riverbed depth, altered water flow, salinity intrusion, and riverbank erosion. Standards allow blending river sand with crushed stone to mitigate these issues and improve aggregate grading. However, limited extraction sources and high transportation costs directly affect construction progress and costs. This highlights the potential for using locally available, cost-effective materials for infrastructure development in remote coastal and island regions.

N. R. Abdul Razak et al. (Eds.): ECOOP 1996, SPPHY 418, pp. 340–350, 2025.
https://doi.org/10.1007/978-981-96-2871-1_48

The potential of utilizing abundant and inexhaustible sea sand as a substitute for river sand in coastal and island construction projects becomes a pertinent consideration. With substantial reserves, sea sand as a replacement for river sand in cement concrete production can significantly reduce construction costs and address extraction and transportation challenges specific to coastal and island projects. Additionally, this approach helps prevent resource depletion and promotes sustainable development, presenting an innovative and eco-friendly use of natural resources.

However, sea sand contains chloride, which can lead to steel corrosion. Prolonged use of sea sand can cause salt-induced steel corrosion. In chloride-containing environments, the corrosion process involves two stages: the formation of unstable compounds such as $FeCl_3$ and $Fe(OH)_3$ from Cl- ions reacting with the passive film, followed by the decomposition of these unstable compounds to produce Fe^{2+}, which then combines with OH^- ions to form iron hydroxide [1, 2]. This compound readily oxidizes to form red-brown iron hydroxide III, commonly known as rust. Chlorides in sea sand reduce the load-bearing capacity of steel, compromising structural integrity and leading to deterioration.

Globally, the demand for sea sand has increased in recent years due to the scarcity of natural sand. According to the American Concrete Institute (ACI), approximately 30% of the sand used in global cement concrete production is sea sand, and this proportion is expected to rise in the coming years. The UK is a pioneer in researching the application of sea sand as a raw material or fine aggregate in construction. In the 1960s, UK experts extensively studied sea sand usage in concrete construction. Annually, over 21 million tons of sea sand and gravel were extracted around the coasts of England and Wales from 2000 to 2004 [3]. Marine-sourced aggregates accounted for about 17% of the total aggregate used in concrete production in England and Wales. In Japan, sea sand has been used as a raw material in concrete production since 1973. In 2011, of the 70 million tons of aggregate used for ready-mix concrete production in Japan, sea sand accounted for 12.2% [3, 4]. In Vietnam, reports from the Ministry of Construction indicate that approximately 10% of the sand used in concrete production comes from the sea, following specific tap water washing. However, challenges include the need for more specific quality standards and limited technologies for extraction and processing to meet technical and environmental needs.

Several studies on geopolymer concrete indicate that using sea sand can expedite construction and produce concrete as strong as that made with river sand and freshwater. Additionally, using corrosion inhibitors for steel reinforcement in concrete is an effective and practical method. Sodium nitrite prevents corrosion by oxidizing unstable iron (II) oxide to iron (III) oxide [5]. Nitrite ions also strengthen the passive iron (III) oxide film by adsorbing onto the steel surface and covering defects in the iron (II) oxide layer, minimizing chloride ion penetration. Using sodium nitrite does not adversely affect the physical properties of concrete. It effectively inhibits steel reinforcement corrosion when the ratio of chloride to nitrite ions is less than or equal to 2.0 [6, 7].

According to ASTM C1585, chromates are effective corrosion inhibitors for steel reinforcement in chloride-rich environments. However, their use is declining due to toxicity and carcinogenicity concerns. Combining chromate with hexametaphosphate and borate at 5:40:55 by weight percentage results in optimal corrosion inhibition [8]. This combined approach strengthens the oxide films on the steel surface, offering superior protection compared to chromate alone.

In this study, sodium nitrite ($NaNO_2$) and chromates were employed as a corrosion inhibitors, with granulated blast furnace slag (GBFS) and fly ash (FA) used as supplementary cementitious materials in the mortar and concrete mixtures. These materials, by-products of iron production and coal combustion respectively, not only improve concrete durability but also significantly reduce the environmental impact of the concrete industry. Physio-mechanical properties of the mortar and concrete samples were measured according to national standards. Additionally, surface resistivity and chloride ion penetration tests were conducted to examine the effects of the different materials used. In conclusion, we discuss the feasibility of using sea sand in concrete production and recommend further studies to scale up this approach.

2 Methodology

2.1 Materials and Analysis

The cement used in this study consisted of CEM I and CEM IIB (provided by Siam City Cement Vietnam), conforming to the compositions outlined in the EN 197–1 standards. The sea sand samples had much finer grains than river sand. The sea sand collected from the Binh Thuan province coast appeared cleaner and whiter than river sand. The percentage of particles smaller than 75 μm in the sea sand was 0.36%, and the chloride ion content was 0.08%. Grain size distribution tests revealed a significant presence of particles retained on the 0.3 mm (42.7%) and 0.15 mm (52.1%) sieves, indicating fine and uniformly sized particles (Fig. 1). Such particle distribution could lead to sedimentation during mixing and segregation, potentially impacting workability, slump retention, and concrete strength.

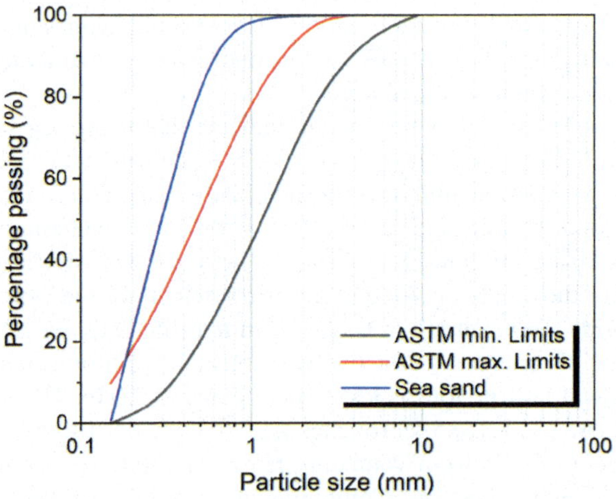

Fig. 1. Particle size distribution curve of sea sand sample collected from the Binh Thuan province coast.

Table 1 presents the various proportional mixtures of mortar samples. In total, as below, 15 sample series were formulated and distinguished by cement types, FA, GBFS components, and inhibitor admixtures.

Table 2 presents the estimated results of the proportional mixtures of concrete samples with an expected design grade of 340, according to the ACI method. Table 3 details the composition of three series of concrete samples with varying inhibitor admixture contents.

3 Results and Discussions

3.1 Results

3.1.1 Flow Testing for Mortar

The initial trials using 1% SikaPlast®-135 VN admixture without any corrosion inhibitor resulted in water separation and sand settling at the bottom of the mix, indicating an inappropriate mix ratio. Optimal results were obtained with a higher 5% SikaPlast®-135 VN admixture dosage by cement weight or with 0.15% Polycarboxylate Ether (PCE). These mixtures produced a stable mortar with desirable flow characteristics, ensuring uniformity and homogeneity. The addition of Sodium Nitrite (Mix 2) had minimal impact on flowability compared to Mix 1, suggesting that Sodium Nitrite can be incorporated without adversely affecting the workability of the mortar. In contrast, the chromate-based corrosion inhibitor (Mix 3) significantly reduced flowability, as evidenced by lower flow values and poor retention, negatively impacting the mortar's workability. This reduction in flowability may be attributed to the interaction between the chromate ions and the cementitious materials, leading to increased viscosity and reduced fluidity. The flow value is a critical measure for practical applications as it affects the ease of mortar application, bonding quality, and the overall finish of the construction work. High flowability ensures better filling of molds and gaps, leading to fewer voids and stronger bonds (Fig. 2).

The results on the flow of mortar suggest that, while optimizing for corrosion resistance, the choice of inhibitors must also consider the impact on flowability to maintain workability standards. Mix 1, with the highest flow value, serves as the control and benchmark for evaluating the effects of corrosion inhibitors. Mix 2 maintains a comparable flow value, making Sodium Nitrite a favorable option when both flowability and corrosion resistance are required. However, the poor performance of Mix 3 highlights the urgent need to explore alternative inhibitors or adjust the admixture composition to mitigate the negative impact on workability. The study emphasizes the importance of balancing admixture composition to achieve desirable workability and corrosion resistance. In practical scenarios, especially in marine environments where chloride-induced corrosion is a concern, using Sodium Nitrite over chromate-based inhibitors is recommended to avoid compromising the flow properties of the mortar.

Table 1. Proportional mixtures for mortar sample preparation.

	W/C	Water (g)	Cement (g)	FA (g)	GBFS (g)	Inhibitor admixture (g)				Superplasticizer (g)
						$NaNO_2$	K_2CrO_4	H_3BO_3	$(NaPO_3)_6$	
CEM I-Mix1	0.38	265	700	–	–	–	–	–	–	3.5
CEM I-Mix2	0.38	265	700	–	–	1.05	–	–	–	3.5
CEM I-Mix3	0.38	265	700	–	–	–	0.04	0.28	0.39	3.5
CEM IIB-Mix1	0.4	286	700	–	–	–	–	–	–	3.5
CEM IIB-Mix2	0.4	286	700	–	–	1.05	–	–	–	3.5
CEM IIB-Mix3	0.4	286	700	–	–	–	0.04	0.28	0.39	3.5
CEM IIB-40%GBFS-Mix1	0.36	252	420	–	280	–	–	–	–	3.5
CEM IIB-40%GBFS -Mix2	0.36	252	420	–	280	1.05	–	–	–	3.5
CEM IIB-40%GBFS -Mix3	0.36	252	420	–	280	–	0.04	0.28	0.39	3.5
CEM IIB-25%FA + 25%GBFS -Mix1	0.33	230	350	175	175	–	–	–	–	3.5
CEM IIB-25%FA + 25%GBFS -Mix2	0.33	230	350	175	175	1.05	–	–	–	3.5
CEM IIB-25%FA + 25%GBFS -Mix3	0.33	230	350	175	175	–	0.04	0.28	0.39	3.5
CEM I-25%FA + 25%GBFS -Mix1	0.33	230	350	175	175	–	–	–	–	3.5
CEM I-25%FA + 25%GBFS -Mix2	0.33	230	350	175	175	1.05	–	–	–	3.5
CEM I-25%FA + 25%GBFS -Mix3	0.33	230	350	175	175	–	0.04	0.28	0.39	3.5

Table 2. Proportional mixtures for concrete samples by batch.

Cement (kg)	FA (kg)	GBFS (kg)	Sand (kg)	Coarse aggregate (kg)	Water (kg)
7.35	3.675	3.675	26.85	34.81	4.37

Table 3. Composition of three concrete sample mixtures with varying inhibitor admixture contents.

			Trial batch 1	Trial batch 2	Trial batch 3
Binder (50% cement + 25% FA + 25% GBFS)		kg	14,70	14,70	14,70
0,5% PCE		g	73,50	73,50	73,50
Sodium nitrite		g	–	20,87	–
0,1% Chromate-based admixture	5% Chromate (K_2CrO_4)	g	–	–	0,74
	40% Boric (H_3BO_3)	g	–	–	5,88
	55% Hexametaphosphate $(NaPO_3)_6$	g	–	–	8,09

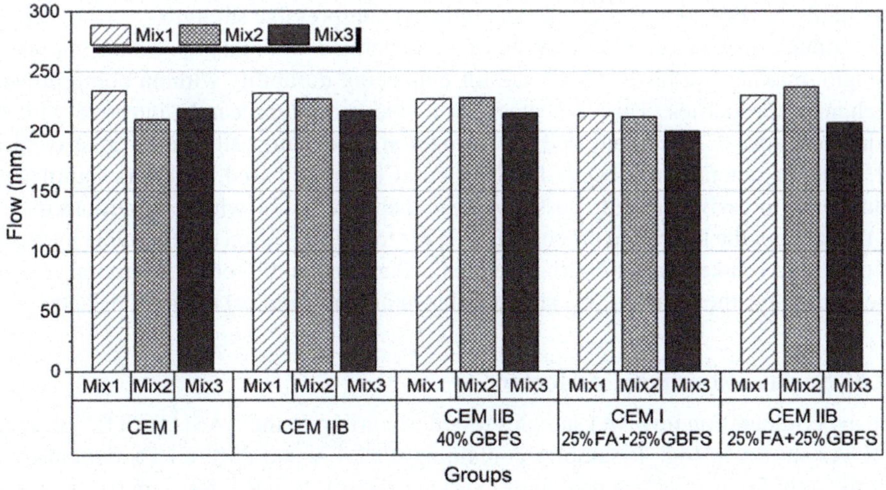

Fig. 2. Result of flow table test for mortar samples.

3.1.2 Compression Strength of Mortar

Figure 3 in the document illustrates the compressive strength of various mortar samples with different compositions and the inclusion of corrosion inhibitors. The control mixes without any corrosion inhibitors showed the highest compressive strength. For instance, CEM I-Mix1 and CEM IIB-Mix1 exhibited superior performance compared to mixes with inhibitors. Including corrosion inhibitors such as sodium nitrite, potassium chromate, and combinations of chromate with hexametaphosphate and borate, negatively impacted the compressive strength. Mix 3, which contained the chromate-based inhibitor, showed a noticeable decrease in compressive strength. Mortar samples incorporating 40% GBFS demonstrated reduced compressive strength after 60 days. This reduction could be due to the interaction of GBFS with other mix components, leading to changes in the hydration process and the microstructure of the mortar. Mixes with 40% GBFS, such as CEM IIB-40%GBFS-Mix1, CEM IIB-40%GBFS-Mix2, and CEM IIB-40%GBFS-Mix3, showed reduced compressive strength after prolonged curing periods (60 days). The W/C was kept consistent across different mixes to ensure comparability. The design aimed for a consistent W/C ratio to maintain uniformity. Mix 2, which included sodium nitrite, showed comparable compressive strength to the control mix without inhibitors, indicating its potential for use without significantly compromising structural integrity. The reduction in compressive strength observed in mortars with 40% GBFS after 60 days can be attributed to several factors. GBFS, as a supplementary cementitious material, participates in the secondary hydration process, which can lead to a delayed strength gain [9–11]. However, over time, the interaction of GBFS with chloride ions (especially in sea sand) may also cause changes in the microstructure, potentially leading to micro-cracks and reduced compressive strength.

Sodium nitrite as a corrosion inhibitor demonstrated minimal impact on compressive strength, making it a favorable choice for enhancing durability without compromising mechanical properties. This indicates that it does not significantly interfere with the hydration process of cement and the formation of calcium silicate hydrate (C-S-H) gel, which is crucial for strength development. Chromate-based inhibitors significantly reduced compressive strength, possibly due to their interaction with cementitious materials that disrupt the hydration process. This leads to the formation of less stable hydration products and weaker mortar microstructures. Alternative inhibitors or admixtures should be explored for the desired corrosion resistance and mechanical strength balance.

3.1.3 Surface Resistivity of Mortar

The experimentation involved samples tested according to the AASHTO TP95 standard [12]. As shown in Fig. 4, samples containing 25% FA and 25% GBFS demonstrated significantly higher surface resistivity. Surface resistivity indicates a material's ability to withstand the movement of ions under an electric field. Higher resistivity signifies enhanced resistance to chloride ion penetration and other aggressive agents through cracks and defects. Thus, combining FA and GBFS effectively improves the durability of concrete when utilizing sea sand by restricting chloride ion ingress. Also, FA and GBFS contribute to the densification of the concrete microstructure by increasing surface resistivity. FA's fine particles fill voids within the concrete matrix, reducing permeability.

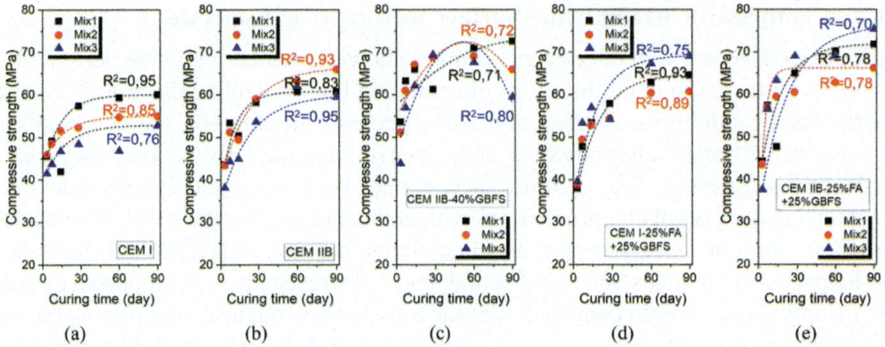

Fig. 3. Compressive strength of mortar samples.

Simultaneously, GBFS, a by-product of steel manufacturing, reacts with calcium hydroxide in concrete to form additional calcium silicate hydrate (C-S-H), further strengthening the concrete and reducing porosity.

The combined effect of these two supplementary cementitious materials results in a more compact and less permeable concrete structure. Besides, the increased surface resistivity observed in the samples suggests that the ion transport mechanisms within the concrete are significantly hindered. This is particularly vital when using sea sand, which naturally contains higher chloride ions. The incorporation of FA and GBFS minimizes the ingress of these detrimental ions, thereby reducing the risk of corrosion of steel reinforcement embedded within the concrete. This reduction in permeability and ion transport enhances the durability and longevity of concrete structures exposed to marine environments. The combination of FA and GBFS improves concrete's mechanical properties and enhances its resistance to chemical attacks. By filling voids and reacting with calcium hydroxide, these materials reduce the pathways for aggressive ions to penetrate the concrete. It is crucial for structures exposed to chloride-rich environments, as it helps maintain the integrity of the concrete and the embedded reinforcement over time.

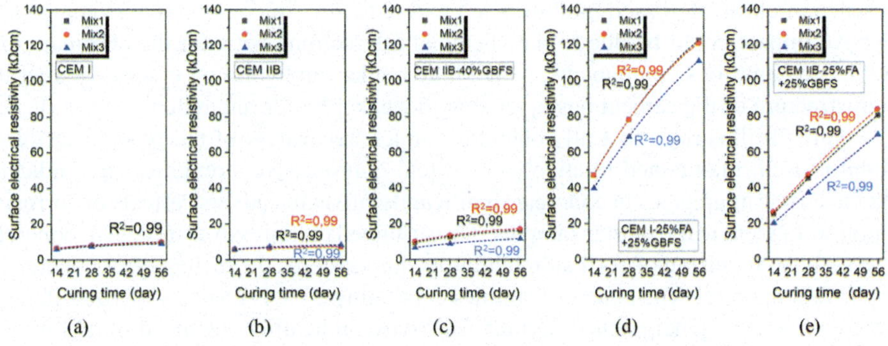

Fig. 4. Surface resistivity of 28-day-old mortar samples.

3.1.4 Compressive Strength and Surface Resistivity of Concrete

The compressive strength of concrete samples showed minimal variation across different mixes with corrosion inhibitors, as illustrated in Fig. 5a. Unlike the mortar samples, where specific inhibitors led to a noticeable decrease in strength, the concrete samples maintained relatively consistent compressive strengths. This consistency can be attributed to coarse aggregates (stones) in concrete. These aggregates play a crucial role in enhancing the overall compressive strength and counteracting the negative impacts of corrosion inhibitors. Coarse aggregates in concrete provide a high-strength framework that helps distribute loads and improve the material's resistance to compressive forces. Their rigidity and strength contribute significantly to the structural integrity of concrete. The mechanical interlocking of coarse aggregates and the surrounding cement paste creates a robust composite material capable of withstanding substantial compressive loads.

Coarse aggregates help mitigate the detrimental effects of corrosion inhibitors on the concrete matrix. While inhibitors may interfere with the hydration process of cement and potentially weaken the cement paste, coarse aggregates offset these effects by contributing to the overall strength of the concrete. While corrosion inhibitors may affect the cement paste and fine aggregates, coarse aggregates preserve the concrete's structural integrity and compressive strength. Concrete samples incorporating 25% FA and 25% GBFS demonstrated high surface resistivity and robust compressive strengths. This indicates that these supplementary cementitious materials contribute positively to the overall performance of the concrete. FA and GBFS enhance concrete's long-term strength and durability by contributing to the pozzolanic reaction, which produces additional calcium silicate hydrate (C-S-H) gel, further strengthening the concrete matrix. The control mixes without corrosion inhibitors (CEM I-Mix1) displayed the highest compressive strength. The absence of additional chemicals was expected to allow the cement hydration process to proceed without interference, resulting in optimal strength development. The addition of Sodium Nitrite showed minimal impact on compressive strength compared to the control. This suggests that Sodium Nitrite can be used as a corrosion inhibitor without compromising the compressive strength of concrete, making it a viable option for enhancing durability without affecting structural performance. The presence of chromate-based inhibitors led to a reduction in compressive strength. This negative impact could be due to the chemical interactions between the chromate ions and the cementitious components, which may hinder the hydration process or alter the microstructure of the cement paste, leading to weaker bonds and reduced strength.

Figure 5b illustrates that all concrete samples, regardless of the type of corrosion inhibitor used, maintained relatively consistent compressive strengths. This indicates that the coarse aggregates in concrete significantly offset the adverse effects of corrosion inhibitors evident in the mortar samples. The sustained compressive strength of concrete, even with the inclusion of corrosion inhibitors, underscores the resilience of concrete as a structural material. The concrete samples containing 25% FA and 25% GBFS demonstrated significantly heightened surface resistivity, indicating enhanced durability and superior resistance to chloride ion penetration (Fig. 5b). The elevated surface resistivity suggests an improved defense against chloride ion ingress and other corrosive agents, which is vital for the long-term performance of concrete structures, especially in marine

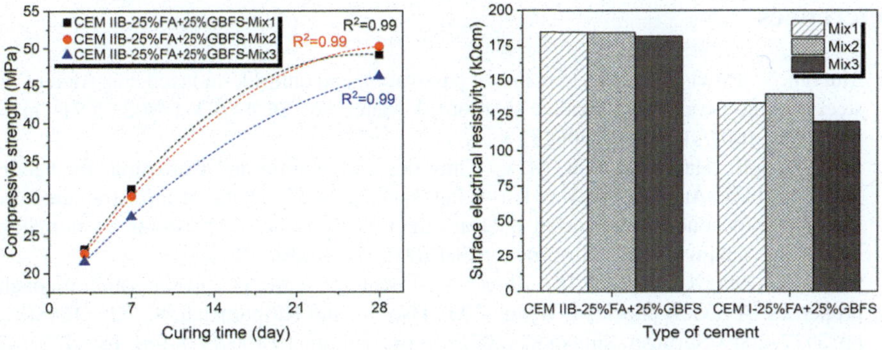

(a) (b)

Fig. 5. Compressive strength of concrete samples (a) and Surface resistivity of 28-day-old concrete samples (b).

environments. Figure 6 emphasizes the efficacy of utilizing FA and GBFS to enhance the durability of concrete produced with sea sand. The increased surface resistivity values in these samples underscore the potential of FA and GBFS to create a more impermeable and resilient concrete matrix. This enhancement is particularly pertinent for applications requiring robust resistance to chloride-induced corrosion, such as in coastal and marine infrastructure.

4 Conclusions

After investigating the characteristics of mortar and concrete with sea sand and various admixtures, including corrosion inhibitors, the study found that optimal flowability and stability of mortar samples were achieved with either 5% SikaPlast®-135 VN or 0.15% Polycarboxylate Ether as superplasticizer admixtures. Incorporating sodium nitrite had no significant impact on flowability, while chromate-based inhibitors reduced it. Compressive strength was highest in Mix 1 (without corrosion inhibitors), and chromate-based inhibitors negatively affected the compressive strength of all mixes. This indicates that sodium nitrite is preferable to chromate-based inhibitors for maintaining compressive strength while protecting against corrosion. Furthermore, mortar samples containing 25% FA and 25% GBFS exhibited the highest surface resistivity, indicating enhanced resistance to chloride ion penetration and improved durability. A combination of FA and GBFS in concrete made with sea sand is recommended to protect the reinforcement from corrosion. Additionally, corrosion inhibitors should be carefully considered to avoid compromising structural integrity. Choosing the suitable admixtures and corrosion inhibitors is crucial for improving concrete performance in marine environments.

Acknowledgments. We acknowledge Ho Chi Minh City University of Technology (HCMUT), VNU-HCM for supporting this study.

References

1. Mundra, S., Provis, J.L.: Mechanisms of passivation and chloride-induced corrosion of mild steel in sulfide-containing alkaline solutions. J. Mater. Sci. **56**, 14783–14802 (2021). https://doi.org/10.1007/s10853-021-06237-x
2. Saouma, V.E., Hariri-Ardebili, M.A.: Chloride, Carbonation and Corrosion. In: Saouma, V.E. and Hariri-Ardebili, M.A. (eds.) Aging, shaking, and cracking of infrastructures: from mechanics to concrete dams and nuclear structures. pp. 191–213. Springer International Publishing, Cham (2021). https://doi.org/10.1007/978-3-030-57434-5_9
3. Vafaei, D., Ma, X., Hassanli, R., Zhuge, Y.: Effects of seawater and sea-sand on concrete properties: A review paper. In: Wang, C.M., Dao, V., and Kitipornchai, S. (eds.) EASEC16. pp. 2037–2049. Springer, Singapore (2021). https://doi.org/10.1007/978-981-15-8079-6_187
4. Nguyen, G.Q., Trinh, K.K.: Current situation of using natural sands and study properties of seasand in some marine areas of Vietnam. Journal of Water Resources & Environment Engineering. **66**, 151–156 (2019)
5. Dong, S., La Plante, E.C., Chen, X., Torabzadegan, M., Balonis, M., Bauchy, M., Sant, G.: Steel corrosion inhibition by calcium nitrate in halide-enriched completion fluid environments. npj Materials Degradation. **2**, 32 (2018)
6. Hill, R.L., Boyd, A.J., Dongell, J.E., Hichborn, G., Neal, R.E., Nmai, C.K., Lamond, J.F., Leming, M.L.: Guide to durable concrete reported by aci committee 201. Guide to Durable Concrete Reported by ACI Committee, 201 (2008)
7. Berke, N.S.: Use of concrete admixtures to provide long-term durability from steel corrosion. Special Publication. **119**, 383–404 (1989)
8. Bahadur, A.: Low chromate corrosion inhibitor with borate and hexametaphosphate. Mater. Trans., JIM **37**, 605–611 (1996)
9. Bentz, D.P., Ferraris, C.F.: Rheology and setting of high volume fly ash mixtures. Cement Concr. Compos. **32**, 265–270 (2010)
10. Papadakis, V.G.: Effect of supplementary cementing materials on concrete resistance against carbonation and chloride ingress. Cem. Concr. Res. **30**, 291–299 (2000)
11. Shi, C., Qian, J.: High performance cementing materials from industrial slags—a review. Resour. Conserv. Recycl. **29**, 195–207 (2000)
12. AASHTO, T.: Standard method of test for surface resistivity indication of concrete's ability to resist chloride ion penetration. AASHTO Provisional Standards, 78–86 (2011)

Research on Using Sandstone as an Active Additive for Portland Cement

Do Quang Minh[1,2] and Le Minh Son[1,2(✉)]

[1] Faculty of Materials Technology, Ho Chi Minh City University of Technology (HCMUT), 268 Ly Thuong Kiet Street, District 10, Ho Chi Minh City, Vietnam
lmson@hcmut.edu.vn

[2] Vietnam National University Ho Chi Minh City, Linh Trung Ward, Thu Duc Dist, Ho Chi Minh City, Vietnam

Abstract. Waste sandstone from the Binh Phuoc quarry has been researched and used as an active mineral additive when mixed with Portland cement. Aspects of the influence of sandstone content (up to 20 wt.%) on the mechanical strength of sandstone blended cement were studied. The results of systematic research according to current standards show that this type of sandstone has the properties of an active mineral additive and the sample mixed with 15% sandstone meets the mechanical strength standards of an active mineral additive after curing days. The results of SEM and XRD microstructure studies have shown that the minerals formed in the sandstone blended cement are equivalent to ordinary Portland cement. This ensures strength development over a longer use of the sandstone blended cement similar to ordinary Portland cement.

Keywords: Waste sandstone · Sandstone blended cement · Ordinary Portland cement · Active mineral additive: Pozzolan activity · The geopolymerization process · Alkaline active substance $Ca(OH)_2$

1 Introduction

Sandstone is a sedimentary rock composed mainly of feldspar and quartz. Its colors vary based on the cement type binding its grains—common colors include bright, grey, green, and red. Although relatively soft and brittle, sandstone has been used historically in construction. However, its modern applicability is unclear, necessitating specific studies for each use [1]. Sandstone types differ based on elemental composition and clay content, and some hold promise for cement concrete production. However, limited research restricts its use as a natural aggregate substitute [2].

The scarcity of natural aggregates prompts exploration of alternatives, including industrial waste from mining, which is a significant sandstone source. Artificial sand from waste sandstone offers a potential river sand replacement [3, 4]. Using local sandstone waste rather than importing high-quality stones can reduce costs, environmental impact, and construction time [5]. Sandstone, due to its softness, is often used as a fine aggregate. Its aggregate size affects concrete flowability, with large aggregates potentially destabilizing the mix [6].

N. R. Abdul Razak et al. (Eds.): ECOOP 1996, SPPHY 418, pp. 351–357, 2025.
https://doi.org/10.1007/978-981-96-2871-1_49

Studies highlight sandstone's suitability as a concrete aggregate, meeting structural, mineral, and chemical standards [7]. Research confirms its viability for low-grade concrete and highway engineering [8], and its resilience to gamma rays and neutron radiation [9]. However, sandstone's high-water absorption and large crushing value make it less desirable for roads [10]. Water absorption can cause dimensional instability and increased concrete shrinkage [11]. Despite these challenges, environmental concerns drive the use of waste materials like sandstone to lower concrete's carbon footprint [12, 13].

This article examines waste sandstone as a mineral additive, assessing its pozzolanic reactivity. It evaluates how replacing 0–20% of cement with sandstone affects compressive and flexural strength over 3, 7, and 28 days, using SEM and XRD for microstructure analysis.

2 Experimental Procedure

Waste sandstone was obtained from a lime mine in Binh Phuoc province, Vietnam. Cement type 1 according to ASTM 150 as Ordinary Portland cement (OPC) was used in this study. The OPC consists of 95% wt. of clinker, 5% wt. of gypsum, and 5% wt. of limestone.

Waste sandstone was ground into granules through a 90 μm sieve. The sandstone blended cement samples were performed by mixing the sandstone with OPC at ratios of 0,15, and 20% wt. According to TCVN 6260: 2020, the additive content is not higher than 20% wt. Compared to clinker. They were denoted by CK0, CK15, and CK20), respectively with the ratio of sandstone/OPC. Sample CK0 is the control OPC sample, to compare properties with samples mixed with sandstone.

The strength activity of sandstone additives is determined according to TCVN 6882: 2016. The standard consistency and setting time of the cement samples are determined according to TCVN 6017:2015. The cement paste samples of standard consistency were formed in a mold with 4×4×16 (cm) sizes. The samples are cured at room temperature for 3, 7, and 28 days, then tested for mechanical strength according to TCVN 6016:1995.

SEM and XRD analyzed the microstructure of the OPC sample and selected sandstone mixed samples to investigate the formed minerals, thereby providing comments on the possibility of applying sandstone additives.

3 Experimental Results

3.1 The Strength Activity Index After Curing 28 days

$$I_R = \frac{R_A}{R_B} \times 100 = 90.67\% > 75\% \quad (1)$$

Therefore, sandstone can be used as an active mineral additive. That means it can partially replace cement according to TCVN 6882:2001.

3.2 Mechanical Intensity

After curing periods of 3 and 7 days, the mechanical strength of the sandstone blended cement samples (CK5 - CK20) was developed similarly to the control sample (CK0). After curing for 28 days, there was a more obvious difference in mechanical durability. According to ASTM 150–20 (Type 1), the compressive strength of samples after curing for 28 days needs to reach 38.5 N/mm^2, which is almost achieved for sandstone blended samples. Sample CK15 had higher compressive strength than the control sample CK0. Thus, although sandstone is considered an active mineral additive due to its high pozzolanic activity, in reality, the impact of sandstone on mechanical strength is not great because it is also affected by other physical factors such as absorption capacity. The water content for standard consistency and setting time of the samples containing hydrophilic sandstone were increased, thus the mechanical strength was reduced (Table 1).

3.3 Microstructure Studies

The samples CK-0 and CK-15 were selected to analyze using SEM and XRD. The results are shown in Figs. 1 and 2.

On the SEM images of sample CK-15, the needle-shaped crystals are observed. That is the characteristic image of ettringite crystals [15]. These crystals are not observed on the SEM images of the sample CK-0. The ettringite crystals demonstrated that sandstone promoted the formation of hydration reactants, such as C-S-H gels, by the reactions of activated silica particles with portlandite and pozzolanic reactions [16, 17].

Figure 2 shows the XRD diffraction of sandstone, CK-0, and CK-15 samples after curing for 28 days. The main minerals in raw sandstone are quartz (Q), calcite (C), kaolinite, and albite. The minerals kaolinite and albite are not observed on the XRD diffraction of samples CK 0 and CK 15. They were reacted in the hydration process of cement. On the XRD diffraction of samples CK0 and CK-15, there are characteristic peaks of Portlandite (P), ettringite (E), and CSH. These are the hydrated minerals of Portland cement.

Table 1. The compressive and flexural strength of the sandstone blended cement samples after 3, 7, and 28 curing days

Sample	Flexural strength (N/mm^2)			Compressive strength (N/mm^2)		
	3 days	7 days	28 days	3 days	7 days	28 days
CK-0	8.550 ± 0.034	12.617 ± 0.118	13.811 ± 0.188	24.716 ± 3.210	32.651 ± 1.318	40.991 ± 0.412
CK-15	8.468 ± 0.128	11.919 ± 0.337	13.830 ± 0.046	24.439 ± 0.541	31.286 ± 1.270	41.462 ± 0.806
CK-20	8.449 ± 0.342	9.567 ± 0.209	11.195 ± 0.758	23.235 ± 1.199	28.364 ± 1.062	32.569 ± 1.774

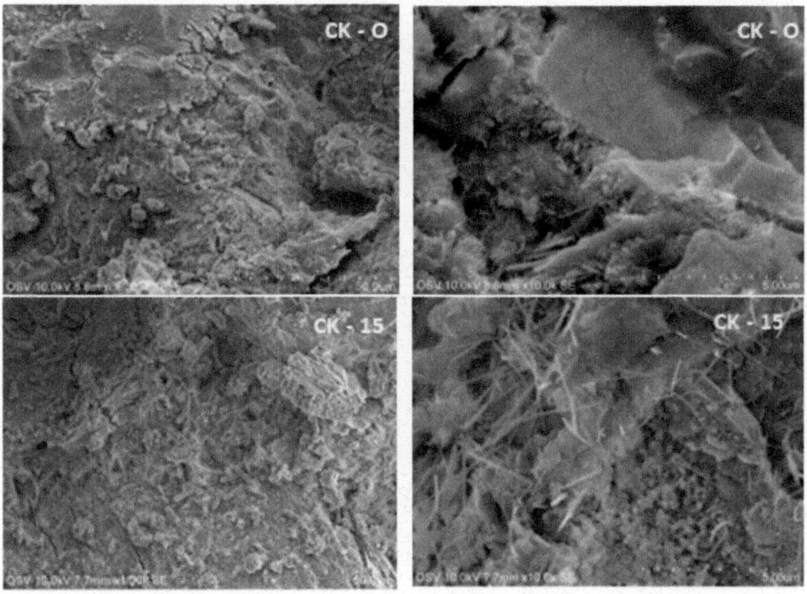

Fig. 1. SEM images of CK-0 and CK-15 after curing 28 days.

4 Conclusions

Binh Phuoc sandstone can be used as an active mineral additive for sandstone blended cement. Compared with OPC, the standard consistency and setting time of the sandstone blended cement increased with the content of blended sandstone. XRD and SEM research results show that hydrated minerals in mixed cement and OPC are equivalent, even the ettringite mineral in sample CK-15 is superior to OPC. This explains that sample CK-15 has flexural and compressive strength that meets Vietnamese Standards.

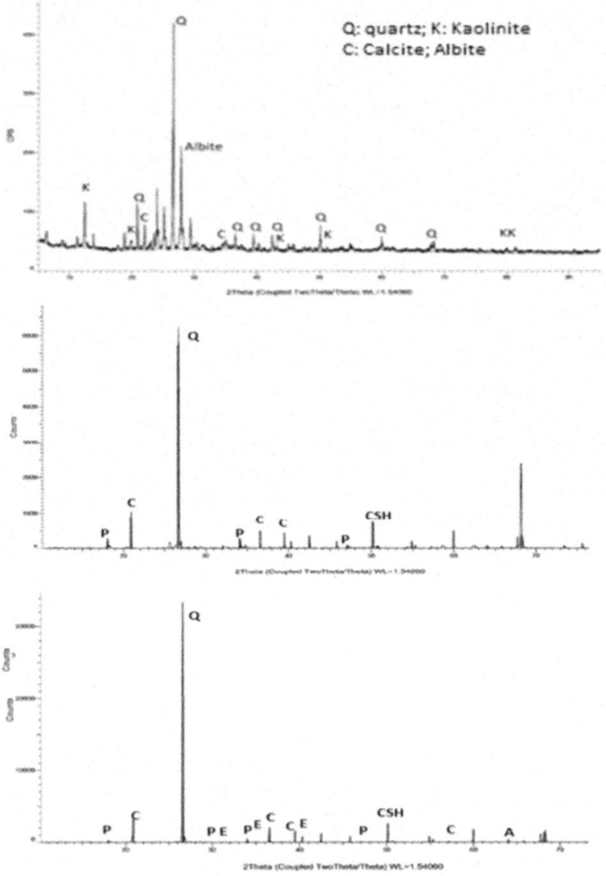

Fig. 2. XRD patterns of CK-0 and SK-15 after curing 28 days.

Acknowledgment. We acknowledge Ho Chi Minh City University of Technology (HCMUT), VNU-HCM for supporting this study.

References

1. Fayong, Y., Haibin, L., Guijuan, Z., Wenbo, L.: Mechanical Performance and Durability Evaluation of Sandstone Concrete. Adv. Mater. Sci. Eng. 13), 1–10 (2020)
2. Kumar, S.: Blessen Skariah Thomas, Vinayak Gupta, Prarthita Basu, Sandeep Shrivastava, "Sandstone wastes as aggregate and its usefulness in cement concrete – A comprehensive review,." Renew. Sustain. Energy Rev. **81**(1), 1147–1153 (2018)
3. Prakash, S., Arun, S., Bhagat, S., Prashant, S.: The impact of sandstone fines on the effectiveness of concrete. In 12th International Scientific Conference "TechSys 2023" – Engineering, Technologies and Systems, Plodiv, Bulgaria (2023)
4. Gayana, B.C., Ram Chandar, K., Reddy, K.R.: Comparative Study on Using Laterite and Sandstone Aggregates on Mechanical Properties of Concrete, in Sustainable Environment and Infrastructure. Lecture Notes in Civil Engineering, 90, Springer, Cham (2020)

5. Qiang, D., Ting, P., Jing, L., Jie, Z., Qingwei, M., Qiang, S.: Mechanical properties of sandstone cement-stabilized Macadam. Appl. Sci. **9**(17), 3460 (2019)
6. Li Jun Zhou: Fang Ning Sun, Qing Hong Zhao.: Particle size distribution of sandstone aggregate influence on recycled aggregate concrete properties. Appl. Mech. Mater. **584–586**, 1504–1508 (2014)
7. Tugrul, A., Murat, Y.: Assessing the quality of sandstones for use as aggregate in concrete. Mag. Concr. Res. **64**(12), 1–12 (2012)
8. Qing, W., Longhui, P.: Studies on use of sandstone ag-gregate for low-grade concrete of wanzhou-dazhou highway. Highw. Eng. **38**(5), 161–164 (2013)
9. Maruyama, Ippei et al.: Radiation-induced alteration of sandstone concrete aggregate. J. Nucl. Mater. **583**, 154547 (2023)
10. Rodgers, M., Hayes, G., Healy, M.G.: Cyclic loading tests on sandstone and limestone shale aggregates used in unbound forest roads. Constr. Build. Mater. **23**, 421–2427 (2009)
11. Mackechnie, J.R.: Shrinkage of concrete containing greywacke sandstone aggregate. Mater. J. **103**(5), 390–396 (2006)
12. Xuan, D.X., Molenaar, A.A.A., Houben, L.J.M.: Compressive and indirect tensile strengths of cement treated mix granulates with recycled masonry and concrete aggregates. J. Mater. Civ. Eng. **21**, 577–585 (2012)
13. El-Shafe, Y.: Waste from stone industry as a cement replacement towards a reduced carbon footprint of concrete. Acad. 3790 (2021)
14. Amalina, A., Elsaid M. M. Zahran, Soon Jian Tan, Nurul Hasan.: Chemical stabilisation of sandstone road aggregate layers: A literature review. In Brunei International Conference on Engineering and Technology 2021 (2023)
15. Hou, P., Kawashima, S., Kong, D., Corr, D.J., Qian, J., Shah, S.P.: Modification effects of colloidal nanoSiO2 on cement hydration and its gel property. Compos. Part B Eng. **45**, 440–448 (2013)
16. Nazari, A., Riahi, S.: Splitting tensile strength of concrete using ground granulated blast furnace slag and SiO_2 nanoparticles as binder. Energy Builds. **43**, 864–872 (2011)
17. Kim, T., Hong, S., Seo, K.-Y., Kang, C.: Characteristics of ordinary portland cement using the new colloidal Nano-Silica mixing method. Appl. Sci. **9**(20), 4358 (2019)

The Feasibility Study of Sustainable Waste Management Framework for Solar Photovoltaic in Malaysia

Najwa Syahirah Mohd Nor Izam[1], Zarina Itam[1](✉), Wong Leong Sing[1], Nazirul Mubin Zahari[1], Nurul Aqilah Razeman[1], Muhammad Imran Najeeb[1], Mohd Hafiz Zawawi[1], and Zuratul Ain Abdul Hamid[2]

[1] Institute of Energy Infrastructure, Universiti Tenaga Nasional, Jalan IKRAM - UNITEN, 43000 Kajang, Selangor, Malaysia
iZarina@uniten.edu.my

[2] Biomaterials Research Niche Group, School of Materials and Mineral Resources Engineering, Universiti Sains Malaysia, 14300 Nibong Tebal, Pulau Pinang, Malaysia

Abstract. This study presents a framework for Sustainable Waste Management of Solar PV Modules in Malaysia, identifying best practices for end-of-life PV waste management with government and stakeholder support. Objectives include defining sustainable elements, developing a conceptual model, and establishing a Waste Management Framework using PLS-SEM. The study used quantitative methods, including data analysis and questionnaire surveys, with SPSS Statistics and SmartPLS for validation. The framework includes four energy parameters of sustainable elements—Economic, Environmental, Social, and Technical as independent variables, and 3 PV waste management methods—Reuse, Recycle, and Disposal as dependent variables. The technical factors were most significant ($\beta = 0.353$), while the social factors were least significant ($\beta = 0.056$). The conceptual model and framework are moderately relevant and acceptable. As a result, the factors that influence the independent variables are the technical factors of sustainable elements, which show the highest values from the framework derived from PLS-SEM using SMARTPLS related to the dependent variables, which are the impacts of Reuse, Recycle, and Disposal on the Sustainable Waste Management Framework for Solar PV Modules in Malaysia.

N. R. Abdul Razak et al. (Eds.): ECOOP 1996, SPPHY 418, pp. 358–365, 2025.
https://doi.org/10.1007/978-981-96-2871-1_50

1 Introduction

The circular economy approach to waste management is crucial as the use of photovoltaic (PV) systems increases, leading to a rise in end-of-life PV modules. The 3Rs—reduce, reuse, and recycle, are essential for sustainable waste management. Reducing the material used in PV production, reusing products for the same function, and recycling materials to high or medium quality are key strategies [1]. Without proper recycling measures, PV systems could significantly contribute to global waste [2].

2 Model Development Using PLS-SEM and SMARTPLS

A conceptual model for solar PV waste management in Malaysia was developed using key energy parameters of sustainable elements identified from preliminary research. The model examines the relationship between these energy parameters and the waste management framework. A survey was conducted with 350 non-expert respondents in Malaysia, yielding 274 valid responses, surpassing the minimum required 138 for PLS-SEM analysis. The model categorizes the parameters into four main clusters—Economic, Environmental, Social, and Technical—as independent variables. The dependent variables include three methods of waste management: Reuse, Recycle, and Disposal. The PLS model links these independent and dependent variables through the following hypotheses, H1-Factors of "Economic" have a significant effect on the Sustainable Waste Management for Solar PV in Malaysia; H2-Factors of "Environmental" have a significant effect on the Sustainable Waste Management for Solar PV in Malaysia; H3-Factors of "Social" have a significant effect on the Sustainable Waste Management for Solar PV in Malaysia; H4-Factors of "Technical" have a significant effect on the Sustainable Waste Management for Solar PV in Malaysia; H5-Sustainable Waste Management for Solar PV in Malaysia have a significant effect on Impacts of "Reused"; H6-Sustainable Waste Management for Solar PV in Malaysia have a significant effect on Impacts of "Recycle"; H7-Sustainable Waste Management for Solar PV in Malaysia have a significant effect on Impacts of "Disposal".

The model is termed Higher-Order Construct (HOC) or Hierarchical Component Model (HCM). Both the Lower-Order Constructs (LOC) and HOC use the reflective-reflective type. Measurement models for LOC and HOC are assessed separately. The HOC is evaluated using the repeated indicator approach. Data analysis with SEM-PLS begins with determining the reliability and validity of each latent variable predictor. This involves using Cronbach's Alpha and internal consistency (composite reliability) to assess the measurement model's validity. Additionally, convergent validity (AVE) and discriminant validity are evaluated. Since the model is a higher-order construct (HOC), additional measurements are necessary to confirm validity.

Cross-loading analysis ensures that each manifest variable has a higher loading with its lower-order components. This is shown in Table 2 (bold numbers), where stronger cross-loadings within their group indicate achieved discriminant validity (Fig. 1 and Tables 1 and 3).

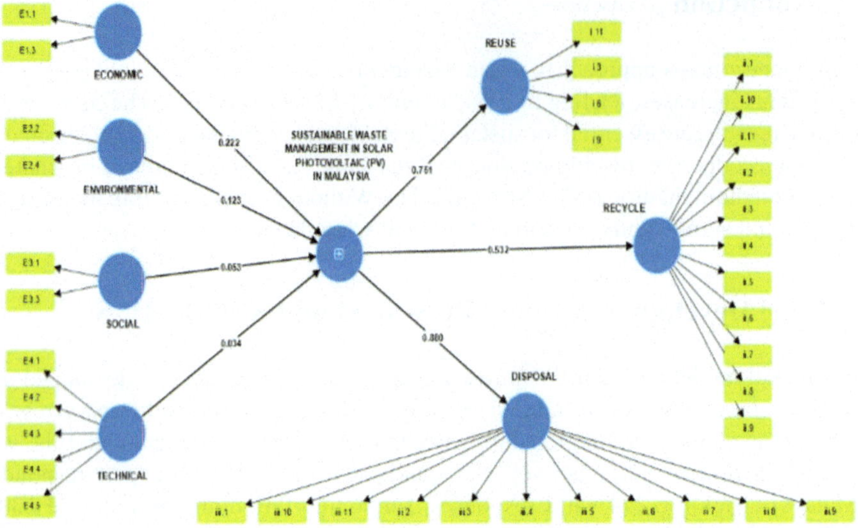

Fig. 1. The developed PLS-SEM for Sustainable Waste Management for Solar PV Modules in Malaysia

Table 1. Internal consistency reliability and convergent reliability (LOC)

	Initial Iteration			Second Iteration		
	Cronbach's alpha, α	Composite reliability (CR)	Average variance extracted (AVE)	Cronbach's alpha, α	Composite reliability (CR)	Average variance extracted (AVE)
REUSE	0.857	0.886	0.423	0.794	0.866	0.618
RECYCLE	0.908	0.922	0.534	0.908	0.918	0.519
DISPOSAL	0.925	0.939	0.595	0.925	0.94	0.596

Changes in the independent variable may have an impact on the output, outcome, or response. On the other hand, an independent variable is one that the researcher manipulates or controls and relates to factors that influence it. In order to see how it affects the dependent variable, it is adjusted or changed. The research study's assumed cause or predictor. It is thought that variations have an impact on the dependent variable. Thus, the hypotheses indicate the importance of sustainable waste management practices in influencing the outcomes of reusing, recycling, and disposing of solar PV waste in Malaysia (Fig. 2).

Table 2. Discriminant validity by using cross-loading analysis

	Impact	Reuse	Recycle	Disposal
Reuse	i.11	**0.765**	0.433	0.33
	i.3	**0.791**	0.24	0.43
	i.6	**0.799**	0.133	0.482
	i.9	**0.789**	0.438	0.275
Recycle	ii.1	0.248	**0.768**	−0.101
	ii.10	0.107	**0.786**	−0.01
	ii.11	0.149	**0.777**	0.002
	ii.2	0.322	**0.721**	−0.199
	ii.3	0.287	**0.772**	−0.159
	ii.4	−0.065	**0.541**	−0.214
	ii.5	−0.029	**0.709**	−0.104
	ii.6	0.216	**0.809**	−0.038
	ii.7	0.208	**0.836**	−0.037
	ii.8	0.493	**0.184**	0.646
	ii.9	0.158	**0.775**	−0.03
Disposal	iii.1	0.312	0.119	**0.848**
	iii.10	0.37	0.176	**0.809**
	iii.11	0.291	0.055	**0.846**
	iii.2	0.377	0.231	**0.821**
	iii.3	0.458	−0.026	**0.747**
	iii.4	0.295	0.375	**0.301**
	iii.5	0.398	0.196	**0.707**
	iii.6	0.312	0.142	**0.869**
	iii.7	0.385	0.043	**0.759**
	iii.8	0.518	0.023	**0.708**
	iii.9	0.408	−0.079	**0.898**

Table 3. Discriminant validity of LOC by using Fornell-Larcker criterion

	Disposal	Recycle	Reuse
Disposal	**0.772**		
Recycle	0.138	**0.720**	
Reuse	0.485	0.392	**0.786**

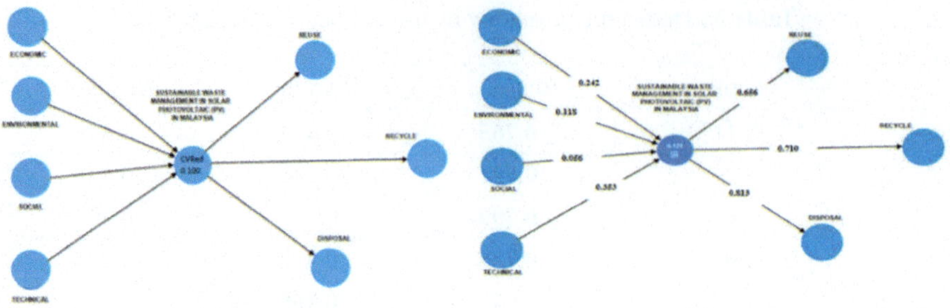

(a) (b)

Fig. 2. (a) Path coefficient in the Structural Model, (b) Value of Q^2 (Cross validated Redundancy)

Meanwhile, for the predictive relevance (Q^2) the R^2 value serves as a measure of predictive accuracy, while the Q^2 value assesses predictive relevance by estimating missing sample values.

This estimation involves systematically omitting portions of the dataset, calculating model parameters, and predicting the excluded values, with the Q^2 indicator gauging the accuracy of these predictions.

3 Results and Discussion

Based on the analysis results, three out of the four primary groups of sustainable energy parameters significantly impact the management of Solar PV waste: economic, environmental, and technical. Notably, the technical group stands out as the most influential, as depicted in Fig. 3, with a path coefficient (β-values) of 0.353. The technical group's pivotal sustainable elements are represented by the following questions: i) In my opinion, Solar PV modules are technology that is easy to use; ii) Solar PV modules can be easily installed in my house and workplace; iii) Solar PV modules are easy to maintain and for long term operation; iv) I can quickly get help when I have trouble with the Solar PV; v) In my opinion, reuse, recycle, and disposal can be a Sustainable Waste Management Framework for Solar PV in Malaysia.

Hence, the findings show that the independent variable for Reused, Recycled, and Disposal of Sustainable Waste Management for Solar PV in Malaysia is significantly influenced by the manifest variables in the technical group of energy parameters of sustainable elements. The simplicity of solar panel installation contributes to this significance, as maintenance is relatively straightforward due to their angled placement, aiding debris removal. Customer perceptions of PV solar projects, as highlighted by H. Abuzaid et al. [3], underscore the importance of technical maintenance for installed PV solar installations. Routine tasks such as panel cleaning, inspecting blockages, and verifying system efficiency are typically part of maintenance inspections conducted by licensed electricians.

However, S. Mujoo et al. suggests that in developing nations, the technical management of solar PV waste is primarily handled by the informal sector, driven by economic necessity despite the health risks associated with handling hazardous chemicals [3]. Aligned with these efforts, the second is to establish a conceptual model of Solar PV Waste Management and last objective is to establish Waste Management Framework for Solar PV in Malaysia using Partial Least Squares-Structural Equation Models (PLS-SEM). A conceptual framework has been established based on the model developed in SmartPLS software to construct a Waste Management Framework for Solar PV Modules in Malaysia using Partial Least Squares-Structural Equation Models (PLS-SEM). Figure 3b shows the Malaysian Sustainable Waste Management Solar PV Framework. The Sustainable Waste Management Framework for Solar PV Modules in Malaysia shown represents the components that work together in this research. The four (4) group sustainable elements influencing solar PV waste management are listed from least to most significant on the left: social, environmental, economic, and technical. Thus, all sustainable elements are related to the most significant and best practices for managing solar PV waste at the end of its lifecycle involving methods of reuse, recycle, and disposal. As a result, all are acceptable in Malaysian Sustainable Waste Management for Solar PV Modules compared to the implementation of PV waste recycling in a few selected countries (including Germany, the United Kingdom, Italy, Switzerland, Norway, the Czech Republic, Japan, the United States, China, South Korea, and India) that have this assessment and life cycle management methods such as laws, policies, guidelines, or regulatory frameworks.

4 Conclusion

The study aims to identify sustainable elements of Solar PV Waste Management in Malaysia, focusing on social, economic, environmental, and technical factors. A questionnaire was developed and distributed to 15 experts, receiving 13 responses, confirming moderate reliability (Cronbach's Alpha of 0.694). Feedback from the pilot test helped identify relevant sustainable elements. The second objective involved establishing a conceptual model through a survey of 350 non-experts, achieving a 78.30% response rate and good reliability (Cronbach's Alpha of 0.899). The model outlines social, technical, economic, and environmental components and indicates that technical variables significantly influence reuse, recycling, and disposal. Collaboration among stakeholders is recommended for effective implementation.

(a)

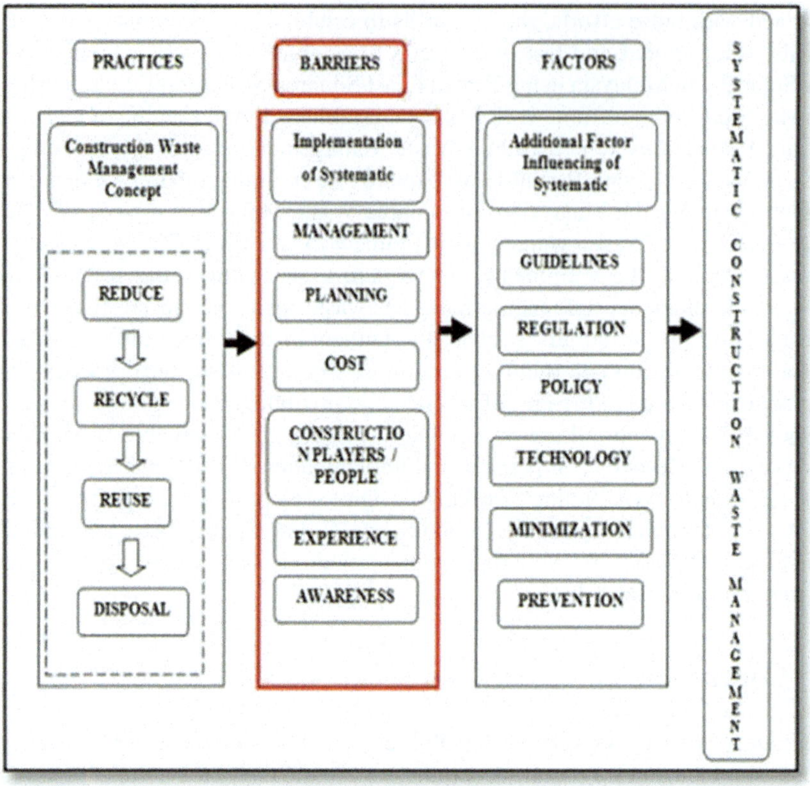

(b)

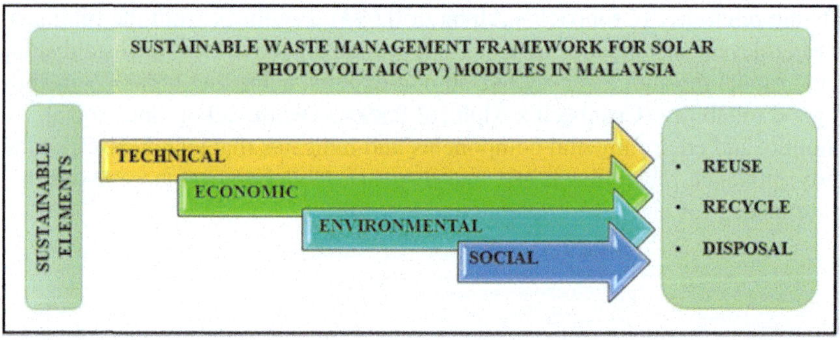

Fig. 3. (a) Conceptual model of systematic construction waste management in the Malaysian construction industry [4], (b) Sustainable Waste Management Solar PV Framework Modules in Malaysia

Acknowledgements. The authors express their gratitude to Universiti Tenaga Nasional (UNITEN), Malaysia, for supporting this project. Funding for the research was received via FRGS 2020-1 Grant No. FRGS/1/2020/TK0/UNITEN/02/18, and publications fee was supported by BOLD HICOE Grant No. J510050002/HICOE05. Special thanks to those who have contributed to this project, directly or indirectly.

References

1. Izam, N.S.M.N., Itam, Z., Sing, W.L., Syamsir, A.: Sustainable development perspectives of solar energy technologies with focus on solar photovoltaic—a review. Energies **15**(8), 2790 (2022Apr 11)
2. Zorpas, A.A.: Strategy development in the framework of waste management. Sci. Total. Environ. **716**, 137088 (2020May)
3. Abuzaid, H., Moeilak, L.A., Alzaatreh, A.: Customers' perception of residential photovoltaic solar projects in the UAE: A structural equation modeling approach. Energy Strateg. Rev. **39**, 100778 (2022Jan)
4. Abd Rahim, M.H.I., Kasim, N.: Conceptual model for systematic construction waste management. MATEC Web Conf. **12**(87), 01008 (2017Dec)

The Impact of Partial Replacement of Sand by Crumb Rubber on Concrete's Characteristic

Nur Liyana Mohd Kamal[1(✉)], Zarina Itam[1], Nazirul Mubin Zahari[1], Salmia Beddu[1], Teh Sabariah Manan[2], and Noorwirdawati Ali[3]

[1] Civil Engineering Department, Universiti Tenaga Nasional, Kajang, Malaysia
yana_kamal@uniten.edu.my

[2] Institute of Tropical Biodiversity and Sustainable Development, Universiti Malaysia, Kuala Terengganu, Terengganu, Malaysia

[3] Jamilus Research Center, Faculty of Civil and Environmental Engineering, Universiti Tun Hussein Onn, Parit Raja, Malaysia

Abstract. This study investigates the potential of using crumb rubber as a partial sand replacement in concrete, addressing environmental concerns related to rubber waste disposal and the depletion of natural aggregates. The research aims to identify optimal ratios of crumb rubber, silica fume, and cenospheres while assessing the impact of crumb rubber on concrete's compressive strength. By integrating crumb rubber (up to 10% sand replacement), silica fume, and cenospheres into concrete mixtures, the study evaluates both workability and compressive strength. The results reveal that while crumb rubber significantly reduces compressive strength, the inclusion of silica fume and cenospheres mitigates this effect. The ideal mix consists of 10% crumb rubber, 20% silica fume, and 10% cenospheres. However, the observed low workability suggests that further adjustments, such as increasing the superplasticizer dosage, may be necessary. Overall, this research demonstrates that crumb rubber can be effectively utilized in concrete as a sustainable alternative to natural aggregates, providing an environmentally friendly solution for rubber waste management.

1 Introduction

The amount of rubber waste generated is growing, particularly from tires that are thrown away, which presents a big problem for traditional trash management techniques. A study shows that about 2% of global waste is made up of end-of-life tires which becomes a hazard to health and environment due to improper disposal (Thudium, 2024). Concurrently, there is worry about resource depletion and environmental damage because traditional concrete production mainly depends on natural aggregates. The necessity for creative solutions that address waste accumulation, and the environmental effect of concrete production is highlighted by these two challenges.

By turning rubber waste into a useful building material, this initiative seeks to address the critical issue of disposing of rubber waste in response to these difficulties. Thiruppathi (2013) conducted a study to evaluate the appropriateness of concrete incorporating recycled tyre aggregates as a concrete material. The study examined a broad range of

N. R. Abdul Razak et al. (Eds.): ECOOP 1996, SPPHY 418, pp. 366–371, 2025.
https://doi.org/10.1007/978-981-96-2871-1_51

physical and mechanical properties of the concrete. According to his observations, the workability levels of concrete mixes with less than 40% rubber substitution by mass nevertheless have high to normal workability levels. The goal of the initiative is to lessen the negative environmental effects of rubber waste by using it to produce green concrete. Furthermore, the goal of the research on rubberised concrete's performance as a sustainable substitute for conventional concrete is to decrease the need for virgin aggregates, which will decrease the environmental effect of construction practices.

2 Materials

In this study, the normal concrete mixture was adopted. The most common type of cement and the main component of concrete, Ordinary Portland Cement (OPC), was utilised.

The experiment utilized river-washed sand with a particle size range of 0–5 mm as the fine aggregate, and washed gravel with size ranges of 5–10 mm and 10–20 mm as the coarse aggregates. A portion of the sand was replaced with shredded rubber from used tires.

Silica fume and cenospheres were also incorporated into the concrete mix. Previous studies, such as Algaifi et al. (2023), have shown that increasing the amount of crumb rubber in the mix tends to decrease the compressive strength of concrete. To counteract this, silica fume was added to improve particle bonding, which is expected to enhance the compressive strength of the rubber-modified concrete.

Additionally, cenospheres from fly ash were included, based on Zanjad et al. (2022), who highlighted their benefits when used in cementitious materials. Cenospheres have been shown to improve the mechanical strength of cement, making it suitable for construction applications. Therefore, in this study, 20% silica fume will be added to all mixes except Mix 1, along with cenospheres to further improve material bonding. The following Table 1 shows the concrete mixtures:

Table 1. Concrete mix design

Mix No.	W/c	Water		Cement		SF (kg/m3)		Cenosphere		FA		CA		Rubber	
		kg/m3	kg	kg/m3	kg	%	kg	%	kg	kg/m3	kg	kg/m3	kg	%	kg
Mix 1	0.55	197	2.1276	358.182	3.8684	-	-	20	0.7737	805.868	8.7034	934.4	10.09152	0	-
Mix 2	0.55	197	2.1276	358.182	3.8684	20	0.774	-	-	805.868	8.7034	934.4	10.09152	0	-
Mix 3	0.55	197	2.1276	358.182	3.8684	20	0.774	10	0.3868	805.868	8.7034	934.4	10.09152	0	-
Mix 4	0.55	197	2.1276	358.182	3.8684	20	0.774	20	0.7737	805.868	8.7034	934.4	10.09152	0	-
Mix 5	0.55	197	2.1276	358.182	3.8684	20	0.774	-	-	725.281	7.833	934.4	10.09152	10	0.87
Mix 6	0.55	197	2.1276	358.182	3.8684	20	0.774	10	0.3868	725.281	7.833	934.4	10.09152	10	0.87
Mix 7	0.55	197	2.1276	358.182	3.8684	20	0.774	20	0.7737	725.281	7.833	934.4	10.09152	10	0.87
Mix 8	0.55	197	2.1276	358.182	3.8684	-	-	20	0.7737	805.868	8.7034	934.4	10.09152	-	-
Mix 9	0.55	197	2.1276	358.182	3.8684	20	0.774	-	-	805.868	8.7034	934.4	10.09152	-	-
Mix 10	0.55	197	2.1276	358.182	3.8684	20	0.774	-	-	725.281	7.833	934.4	10.09152	10	0.87
Total			21.276		38.684		6.189		3.0947		83.552		100.9152		3.481

3 Experimental and Result

3.1 Slump Test

The slump test was conducted to assess the workability of concrete mixes where crumb rubber was used as a partial replacement for sand, following the ASTM C143 standard. The findings revealed that most mixes had slump values close to 0 mm. Specifically, Mixes 3, 5, 6, 7, and 9 registered a slump value of 0 mm, indicating no slump and extremely low workability. Mixes 1, 2, 8, and 10 showed slightly higher slump values, ranging from 9 mm to 16 mm, still within the very low workability range. Mix 4 had the highest slump value at 17 mm, but this too falls under very low workability.

These results suggest that the majority of the concrete mixes were very stiff and dense, which, while beneficial for compressive strength and durability, presents challenges for placement and compaction. Notably, Mixes 5, 6, 7, and 10, all with a 10% replacement of sand by crumb rubber, exhibited very low slump values (0 mm for Mixes 5, 6, and 7; 13 mm for Mix 10). The inclusion of crumb rubber appears to significantly reduce workability, likely because rubber particles do not absorb water and affect the mix's cohesion. This reduced workability may require additional measures such as mechanical compaction or vibration to ensure proper consolidation and to avoid defects like voids.

Mixes containing silica fume (all except Mixes 1 and 8) generally exhibited low slump values. Silica fume, known for its strength-enhancing properties, can also reduce workability by increasing the water demand of the mix. This effect was observed in Mixes 2, 3, 4, 5, 6, 7, 9, and 10, which contained 20% silica fume and showed low slump values.

The addition of cenospheres also affected workability. Mixes 3 and 6, which included 10% cenospheres, and Mixes 1, 4, 7, 8, and 10, which included 20% cenospheres, showed varying slump values. Cenospheres, due to their lightweight and spherical shape, may slightly improve workability, but their overall impact is less significant compared to the reduction caused by crumb rubber and silica fume.

Mixes 8, 9, and 10, which contained a superplasticizer, displayed slightly higher slump values compared to other mixes with similar compositions. For instance, Mix 8 had a slump value of 16 mm, and Mix 10 had a slump value of 13 mm, indicating that the superplasticizer helped improve workability despite the presence of crumb rubber and silica fume.

3.2 Compressive Strength

The main focus of this study was to enhance the compressive strength of rubber-modified concrete. The compression test followed the EN 12390-3 standard, involving the testing of concrete cubes (100 × 100 × 100 mm) after curing periods of 7, 14, and 28 days. The compressive strength results provided valuable insights into the effects of various additives and curing times on concrete performance as indicated in Fig. 1.

Most mixes exhibited an increase in compressive strength from 7 to 14 days, indicating ongoing hydration and strengthening of the concrete matrix. However, strength generally stabilized or slightly decreased by the 28th day. For example, Mix 1, containing 20% cenospheres with no rubber or silica fume, had a compressive strength of 21.9 MPa

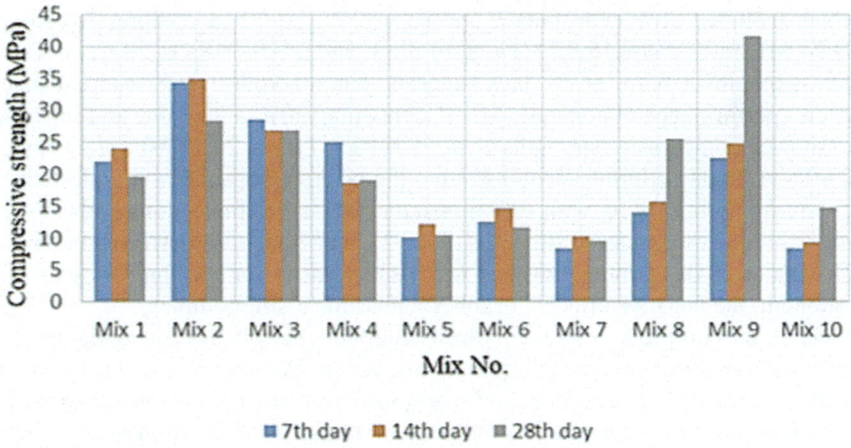

Fig. 1. Compressive strength result.

at 7 days, which increased to 24.03 MPa at 14 days, but then decreased to 19.51 MPa at 28 days. This pattern suggests that while early curing significantly enhances strength, prolonged curing might lead to slight reductions, possibly due to microstructural changes or shrinkage.

The inclusion of silica fume had a notable impact on compressive strength. Mix 2, which contained 20% silica fume without cenospheres or rubber, exhibited high compressive strengths of 34.33 MPa at 7 days, 34.89 MPa at 14 days, and 28.26 MPa at 28 days. Similarly, Mix 9, also with 20% silica fume but no cenospheres or rubber, followed a similar trend, with strengths of 22.58 MPa at 7 days, 24.78 MPa at 14 days, and a significant increase to 41.5 MPa at 28 days, indicating that silica fume can significantly enhance long-term strength under certain conditions.

However, rubber replacement generally had a negative effect on compressive strength. Mix 5, containing 10% rubber and 20% silica fume, showed lower compressive strengths of 10 MPa at 7 days, 12.07 MPa at 14 days, and 10.3 MPa at 28 days. Similarly, Mix 10, with the same composition, displayed compressive strengths of 8.3 MPa at 7 days, 9.16 MPa at 14 days, and 14.71 MPa at 28 days. These results indicate that rubber replacement weakens the concrete matrix, reducing its compressive strength across all curing periods.

The combined effects of cenospheres and rubber replacement yielded mixed results. Mix 6, with 10% cenospheres, 10% rubber, and 20% silica fume, showed compressive strengths of 12.45 MPa at 7 days, increasing to 14.5 MPa at 14 days, but then decreasing to 11.44 MPa at 28 days. This suggests that while the combination of additives may initially boost strength, it may not sustain it over longer curing periods. Mix 7, with 20% cenospheres, 10% rubber, and 20% silica fume, exhibited similar results, with strengths of 8.34 MPa at 7 days, 10.15 MPa at 14 days, and 9.39 MPa at 28 days, highlighting the challenge of maintaining long-term strength with these additives.

Superplasticizer inclusion in Mixes 8, 9, and 10 was another key factor. Superplasticizers are added to improve concrete workability without increasing water content. Mix

8, which included 20% cenospheres and superplasticizer, had compressive strengths of 13.94 MPa at 7 days, 15.52 MPa at 14 days, and 25.59 MPa at 28 days. The significant strength increase at 28 days suggests that superplasticizers can enhance the long-term performance of concrete. Mix 9, containing 20% silica fume and superplasticizer, showed compressive strengths of 22.58 MPa at 7 days, 24.78 MPa at 14 days, and 41.5 MPa at 28 days, further demonstrating the positive impact of superplasticizer on long-term strength. Mix 10, with 10% rubber and 20% silica fume along with superplasticizer, exhibited compressive strengths of 8.3 MPa at 7 days, 9.16 MPa at 14 days, and 14.71 MPa at 28 days, indicating that while superplasticizer can improve workability and strength, the negative effects of rubber replacement still dominate.

Overall, the compressive strength results highlight the complex interactions between curing time, mix composition, and additives in concrete. While curing generally enhances strength, additives like silica fume and superplasticizer can significantly improve long-term performance. However, rubber replacement tends to reduce compressive strength, emphasizing the need for careful consideration when including it in concrete formulations. These findings provide a basis for further research and optimization of concrete mix designs to achieve desired performance characteristics.

4 Conclusion

The results of this study suggest that incorporating cenospheres and silica fume into the concrete mix leads to a drier consistency. Although the use of superplasticizer has been effective in enhancing hydration and improving compressive strength, the mix still tends to be dry with poor workability. This indicates that these materials might be more effective when used as cement replacements rather than as additives. Additionally, increasing the superplasticizer dosage to 1% of the cement weight could potentially improve workability without negatively affecting hydration.

It was also found that cenospheres are less effective than silica fume in boosting the compressive strength of both conventional and rubber-replaced concrete. The introduction of crumb rubber significantly decreases compressive strength, which is consistent with the findings of previous studies by Toutanji (1996), Reda Taha et al. (2008), and Atahan & Yücel (2012).

The study identifies the optimal proportion for rubber-replaced concrete as a 10% replacement of sand with crumb rubber, combined with a 20% addition of silica fume and a 10% addition of cenospheres. This conclusion is in line with Adesina's (2020) research, which demonstrated that using moderate amounts of cenospheres (up to 10%) does not significantly hinder the early strength development of the concrete.

References

1. Adesina, A.: Sustainable application of cenospheres in cementitious materials – Overview of performance. Develop. Built Environ. **4**, 100029 (2020)
2. Algaifi, H.A. et al.: Development of rubberised cementitious material incorporating graphene nanoplatelets and silica fume. Case Stud. Constr. Mater. **19** (2023)

3. Atahan, A.O., Yücel, A.Ö.: Crumb rubber in concrete: Static and dynamic evaluation. Constr. Build. Mater. **36**, 617–622 (2012)
4. Reda Taha, M.M., et al.: Mechanical, fracture, and microstructural investigations of rubber concrete. J. Mater. Civ. Eng. **20**(10), 640–649 (2008)
5. Thiruppathi, R.: Discarded tyre rubber as concrete aggregate: A possible outlet for used tyres. In: 2013 International Conference on Current Trends in Engineering and Technology (ICCTET) [Preprint] (2013)
6. Thudium, M.: Tire waste statistics you need to know, Contec. (2024) https://contec.tech/tire-waste-statistics-need-to-know/
7. Toutanji, H.A.: The use of rubber tire particles in concrete to replace mineral aggregates. Cement Concr. Compos. **18**(2), 135–139 (1996)
8. Zanjad, N., Pawar, S., Nayak, C.: Use of fly ash cenosphere in the construction industry: a review. Mater. Today: Proc. **62**, 2185–2190 (2022)

Evaluating the Influence of Cenosphere Extraction on Physicochemical and Mechanical Performance of Residual Pulverised Fly Ash (RPFA) Incorporated Concrete

Idris Ahmed Jae[1], Amirul Ashraf Norsalam[2], S. M. M. Shaik Ahmad Fadzil[1], Zarina Itam[1,2(✉)], Agusril Syamsir[1,2], Nazirul Mubin Zahari[1,2], Mohd Hafiz Zawawi[1,2], Zakaria Che Muda[3], and Faizatul Shimal Mehamod[4]

[1] Institute of Energy Infrastructure, College of Engineering, IKRAM - UNITEN, Universiti Tenaga Nasional, 43000 Kajang, Selangor, Malaysia
iZarina@uniten.edu.my
[2] Department of Civil Engineering, College of Engineering, IKRAM - UNITEN, Universiti Tenaga Nasional, 43000 Kajang, Selangor, Malaysia
[3] Faculty of Engineering and Quantity Surveying, INTI International University, Negeri Sembilan, Persiaran Perdana BBN, Putra Nilai, 71800 Nilai, Malaysia
[4] Faculty of Science and Marine Environment, Universiti Malaysia Terengganu, 21030 Kuala Nerus, Terengganu, Malaysia

Abstract. Cenospheres are lightweight, hollow spherical artefacts that are extracted from fly ash for specialised industrial applications. However, there is a growing concern that this extraction process could potentially alter the physicochemical and mechanical properties of the residue pulverised fly ash (RPFA), leading to a trade-off between continued extraction and discontinuation or even exploring ways of improving the RPFA. To explore the extent of this impact, this study experimentally investigated the physicochemical and mechanical properties of cement paste and concrete partially replaced with 10% to 40% pulverised fly ash (PFA) and RPFA. Samples from four Malaysian coal-powered plants were analysed. The results revealed an increase in PRFA amorphousness with a surface area between 8% - 37%, a decrease in residual density by 1.5% - 12%, workability by 10% - 35%, compressive strength by 3% - 10% and flexural strength by 13% - 10%.

Keywords: Cenospheres · Residue Pulverised Fly Ash

1 Introduction

Supplementary cementitious materials (SCM) are substances used in combination with cement in concrete mixtures to enhance specific properties. These materials can either be naturally occurring or by-products from manufacturing processes. In whatever form they contribute to the properties of hardened concrete through hydraulic or pozzolanic

N. R. Abdul Razak et al. (Eds.): ECOOP 1996, SPPHY 418, pp. 372–378, 2025.
https://doi.org/10.1007/978-981-96-2871-1_52

activity. Typical examples of SCMs include Pulverised Fuel Ash (PFA), Ground Granulated Blast Furnace Slag (GGBFS), and Silica Fume. Other materials, such as agricultural waste are also continuously being explored for their pozzolanic characteristics to be utilised as SCMs in concrete [1–4]. Generally, the application of SCMs has several advantages including reduced environmental impacts [5], enhanced mechanical and impact resistance [6–8] and improved durability [8, 9].

Over the years, using SCMs in concrete production has gained popularity as a strategy for mitigating the environmental impact posed by cement production as well as reducing construction costs when compared to cement [10]. Many countries with surplus availability of fly ash, including Malaysia, are pushing towards sustainable construction practices through the use of Fly Ash as partial cement replacement, and this trend is gaining traction.

Several studies have explored the influence of incorporating varying amounts of fly ash as a replacement for cement, with dosages ranging from low (<15%) to very high (>50%). While common practice suggests replacing 15% to 35% of cement with fly ash, high dosages of between 40% to 60% have been successfully used in structural applications. In this study, a replacement range within this established norm is maintained.

In the last two decades, research has delved into deriving potentially valuable microspheres from fly ash for various applications. These microspheres, considered as the artefacts of the fly ash include the Cenosphere, Ferrosphere, Plerosphere and Carbonaceous particles and are valued for their unique properties and potential applications in industries such as ceramics, fillers, lightweight materials, composites, polymers, and environmental cleanup.

Cenospheres are the predominant component of fly ash microstructures, followed by Ferrosheres and Plerosheres. All of these are mainly made up of aluminosilicate glasses, mullite, calcium silicates, sulphate, calcite ferrous oxides, and quartz. The silicate content dominates the solid phase, representing about 50%-60%, while alumina and Fe_2O_3 comprise about 19%-65% and 0.7%-6.5% respectively.

However, despite these significant benefits, there seems to be a growing concern about the possible impact of Cenosphere extraction on the residual fly ash. This concern is valid considering the high chemical composition of the Cenosphere as the predominant microstructural component in fly ash including high levels of silica and alumina. There are worries that the extraction process could negatively affect the performance of the residue fly ash. To address these concerns, this study aims to experimentally investigate the influence of Cenosphere extraction on the physicochemical and mechanical properties of the residue fly ash.

2 Experimental Procedure

Four samples of Pulverised Fuel Ash (PFA) from different power plants across Malaysia were collected. The samples were sourced from electrostatic precipitators which capture the range of fly ash produced in coal-fed Peninsular Malaysia power plants.

Cenosphere was extracted separately from the collected samples using the float and sink method. This technique entails dispersing the PFA sample in a bath of water at room temperature. The Cenosphere floats to the water surface due to its lower density compared to the PFA, while the residual PFA (RPFA) sinks to the bottom of the water bath. The floated Cenosphere is collected through decantation, followed by filtration. Collected samples are then spread in thin layers to enhance fast drying. The RPFA is also collected and dried in the same manner.

The elemental chemical composition of PFA and RPFA in the four samples was analysed using X-ray Fluorescence (XRF) following BS EN 12390- 3:2002. The samples were prepared and positioned in the XRF device. Next, the Amorphousness of the samples was determined using an X-ray Diffraction (XRD) test in compliance with ASTM D3906-03. This test examines the crystalline structure and mineralogical composition of the PFA and RPFA samples. Each sample is exposed to an X-ray beam, and the resulting diffracted pattern is recorded by a detector. After that, the Scanning Electron Microscopy (SEM) test was conducted to assess the surface morphology, including the appearance and microstructure of the material at high magnification. The Density test was also conducted using the AccuPyC II model automatic displacement pycnometer available in the civil engineering laboratory at the University Tenaga Nasional, Malaysia.

For concrete mix design, the material utilised in the concrete production consists of Ordinary Portland Cement (OPC) of grade 42.5, crushed aggregate with a maximum size of 20mm, and clean tap water. To examine the influence of PFA and RPFA on both physical and chemical characteristics of concrete, a 10%, 20%, 30% and 40% percentage replacement to the weight of cement was adopted. After that, the compressive strength and flexural strength tests will be conducted.

3 Results and Discussions

3.1 Characterisation of RFA and RPFA

The SEM images of PFA and RPFA analysed at ×1000 and ×5000 magnification are illustrated in Fig. 1. From the PFA images, a combination of solid rough (fly ash) and hollowed spherical particles (Cenosphere) of varying size distributions were predominantly observed in all the samples. In addition, glassy spherical particles were observed in 3 of the samples (A, B and C). Elongated crystalline particles are also effortlessly observed in the samples. The existence of pores in the PFA images is attributed to the variation of particles of different sizes.

For XRD analysis, from the results, minerals such as SiO_2 (Quartz), Al_2O_3 (Diamonite) and Fe_2O_3(Hematite) have been predominantly observed. The diffractogram of the PFAs is observed to be characterised by closely packed peaks compared to the RPFA, which implies the increase in amorphousness of the sample resulting from the extraction of the Cenosphere from the samples. All samples as anticipated are high in Quartz followed by diamonite content.

Based on the level of amorphousness in the samples, in terms of cement replacement utilization, both PFA and RPFA are suitable alternatives since the amorphous components of fly ash contribute to its pozzolanic activity. The amorphous phase of fly ash offers a sizable surface area for this reaction to take place, enhancing the substance's pozzolanic capabilities. To achieve a highly reactive cement replacement material, the crystalline structure of the silica present in the ash is essential. Both samples also contained quartz, indicating that the sandy sample textures were due to silica compounds in the ash samples.

Chemical characterization base on X-ray Fluorescence results indicates only a slight reduction in the oxide composition after Cenosphere extraction with a maximum reduction of 7% SiO_2 from source B, and 4% Al_2O_3 in source D. Reduction in other oxide components is less than 2.5% with samples B and C having the highest and least silica content respectively.

In the density test, a reduction of between 1.5% and 12% is observed in all RPFA samples. With the highest decrease observed in a sample from source A and the least from Source B. Nonetheless, densities of both PFA and RPFA are still within the documented fly ash densities of between 2000 kg/m^3 and 2600kg/m^3. The low decrease in density is attributed to the low quantity of Cenosphere being extracted per sample, which is around 2% for the sample. This is expected and is reflected in the slight variation in densities.

3.2 Influence of Cenosphere Extraction on Mechanical Properties of Concrete

Figure 2 illustrates the visual variation in compressive strength with different content of PFA and RPFA at varying curing ages compared with the control samples. The variation of the compressive strength compared to the control sample can be seen to be least at 10% of PFA and RPFA. In all cases, the variations can be seen to increase with increasing PFA/RPFA content.

Figure 3 compares the flexural strength of concrete samples with cement partially replaced with 10%, 20% and 30% PFA and RPFA. Generally, lower flexural strengths are observed in all samples containing both PFA and RPFA, with about 9% – 13% lower than the control sample observed in PFA and about 13% to 15% in RPFA samples. The variation in strength between PFA and RPFA samples is not significant.

The findings reveal the extent to which Cenosphere extraction can influence the physicochemical and mechanical properties of residual pulverised fly ash. These highlighted cases offer valuable insights for future research, particularly in areas that require improvement for specific intended utilisation. For instance, incorporating fibres or other related means to improve the workability, compressive and flexural strength of the RPFA concrete can lead to enhanced performance.

4 Conclusion

This study delves into the influence of Cenosphere extraction on the physico-chemical and mechanical properties of the residual pulverised fly ash (RPFA). Based on the findings, the following are the specific conclusions are drawn:

Extraction of the Cenosphere leads to increased amorphousness in the residual pulverised fly ash (RPFA), with a consequent increase in the surface area which can influence the strength and durability of the concrete produced.
The chemical composition of RPFA is not significantly altered by Cenosphere extraction, with the maximum reductions observed being 7% for SiO2, 4% for Al_2O_3 and 2.5% for other components.

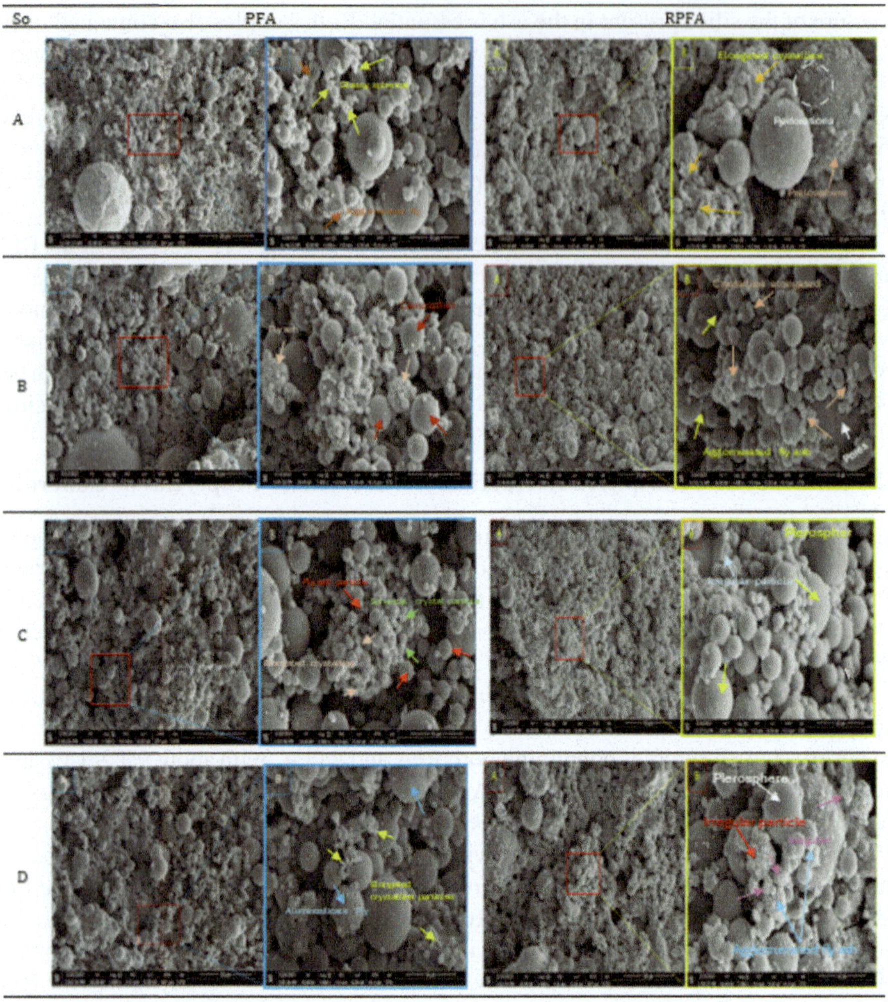

Fig. 1. Illustration of SEM images of PFA and RPFA.

Depending on the source, the extraction of the Cenosphere might influence the change of classification of RPFA.

The densities of both pulverised fly ash (PFA) and the RPFA remained consistent with the documented fly ash densities, ranging from 200kg/m^3 and 2600kg/m^3. However, the extraction of the Cenosphere resulted in a decrease of 1.5% to 12% in RPFA densities.

Concrete containing RPFA experienced a reduction in workability by between 10% to 33% due to its higher surface area. Conversely, concrete containing PFA exhibited increased workability in direct proportion to its content.

The inclusion of PFA and RPFA at 10% to 40% resulted in an increase in surface area by 6% - 41% and 8% - 37% respectively. With the highest increase observed at 30% - 40% of PFA.

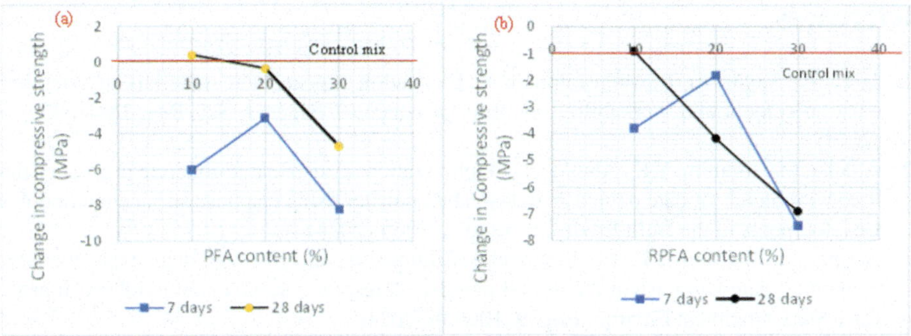

Fig. 2. Variation of compressive strength with different (a) PFA content and (b) RPFA content.

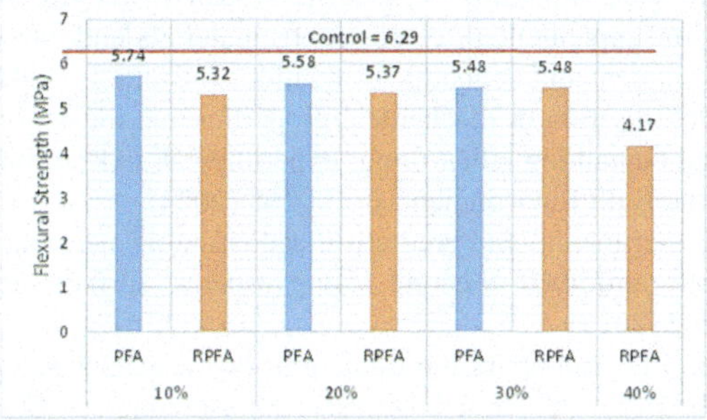

Fig. 3. Average flexural strength at varying content of PFA and RPFA.

The densities of both pulverised fly ash and residue remained consistent with documented fly ash densities, ranging from 200kg/m^3 to 2600kg/m^3. However, the extraction of the Cenosphere resulted in a decrease of 1.5% to 12% in RPFA densities. The compressive strength of samples containing PFA and RPFA does not show a significant difference (3% -10%). However, the compressive strength of the sample with RPFA exhibits slightly lower compressive performance at all percentage replacements, which is attributed to the absence of the Cenosphere. A general decrease in flexural performance ranging from 9% - 13% and 13% - 15% for samples containing PFA and RPFA respectively are observed. This decrease is indicative of the reduced flexural behaviour resulting from Cenosphere extraction.

Acknowledgments. The authors express gratitude to the Innovation & Research Management Centre of Universiti Tenaga Nasional for supporting this research through Project No: J510050002/HICOE05, facilities support from UNITEN Civil Engineering Laboratory, the BOLD Refresh Postdoctoral Fellowships under Grant J510050002-IC-6 BOLDREFRESH2025-Centre of Excellence. Gratitude is also extended to the TNB Seed Fund No. U-TG-RD-20-23 for enabling the need for this research. Special thanks to the IRMC of Universiti Tenaga Nasional for the full cooperation given to make this project achievable.

References

1. Ja'e, I.A., Sulaiman, T., Abdurrahman, A.: Evaluation of pozzolanic materials and their influence on cement and workability retention of concrete. Niger. J. Sci. Res. **18**(4), 329–336 (2019)
2. Adamu, M., Ibrahim, Y.E., Alanazi, H.: Optimization of sustainable concrete properties modified with blends of date palm ash and eggshell powder using response surface methodology. Developments in the Built Environment, p. 100359 (2024)
3. Adamu, M., Ibrahim, Y.E.: Environmental sustainability and cost-benefit analysis of concrete containing date palm ash and eggshell powder: A response surface methodology approach. Case Stud. Chemical Environ. Eng. **9**, 100636 (2024)
4. Adamu, M., Alanazi, H., Ibrahim, Y.E., Abdellatief, M.: Mechanical, microstructural characteristics and sustainability analysis of concrete incorporating date palm ash and eggshell powder as ternary blends cementitious materials. Constr. Build. Mater. **411**, 134753 (2024)
5. Zhu, P., Wang, H., Yan, X., Yang, L., Zhu, L., Liu, H.: Recycled coarse aggregate from parent concrete with supplementary cementitious materials under freeze-thaw environment: recyclability, environment and economic evaluation. J. Build. Eng. **84**, 108699 (2024)
6. Ja'e, I.A. et al.: Optimisation of mechanical properties and impact resistance of basalt fibre reinforced concrete containing silica fume: experimental and response surface assessment. Developments in the Built Environment, p. 100368 (2024)
7. Ja'e, I.A., et al.: Experimental and predictive evaluation of mechanical properties of kenaf-polypropylene fibre-reinforced concrete using response surface methodology. Dev. Built Environ. **16**, 100262 (2023)
8. Wu, R., et al.: Effect of basalt fibers and silica fume on the mechanical properties, stress-strain behavior, and durability of alkali-activated slag-fly ash concrete. Constr. Build. Mater. **418**, 135440 (2024)
9. Ocelić, A., Baričević, A., Smrkić, M.F.: Synergistic integration of waste fibres and supplementary cementitious materials to enhance sustainability of ultra-high-performance concrete (UHPC). Case Stud. Constr. Mater. **20**, e02772 (2024)
10. Pacewska, B., Wilińska, I.: Usage of supplementary cementitious materials: advantages and limitations: Part I. C–S–H, C–A–S–H and other products formed in different binding mixtures. J. Therm. Anal. Calorim. **142**(1), 371–393 (2020)

Effect of Bentonite on Waterproofing of Concrete

Do Quang Minh[1,2], Le Minh Son[1,2](✉), and Nguyen Hoa Duong[3]

[1] Faculty of Materials Technology, Ho Chi Minh City University of Technology (HCMUT), 268 Ly Thuong Kiet Street, District 10, Ho Chi Minh City, Vietnam
{mnh_doquang,lmson}@hcmut.edu.vn

[2] Vietnam National University Ho Chi Minh City, Đại học Bách Khoa - ĐHQG - HCM, 268 Lý Thường Kiệt, P14, Q10, Tp.HCM, Linh Trung Ward, Thu Duc City, Ho Chi Minh City, Vietnam

[3] Ho Chi Minh City Construction College, 190 Vo Van Ngan, Binh Tho Ward, Thu Duc City, Vietnam

Abstract. To reduce CO_2 emissions, bentonite clay is being studied for its potential use as an additive to Portland cement. However, different types of bentonites will also affect the properties of cement mortar and concrete differently. When Binh Thuan bentonite had replaced 1–6% wt. of Portland cement, the results in this study indicated that the mark of cement was guaranteed, meanwhile, the waterproofing ability of concrete increased. The waterproofing ability of concrete has also increased with bentonite content. The waterproofing effect was clearly shown with lower-mark concrete. The reaction of bentonite with alkali from cement hydration to form zeolite faujasite has increased the waterproofing ability of concrete.

Keywords: Bentonite clay · waterproof · additive to Portland cement · carbon footprint · concrete · zeolite faujasite · swelling · geopolymer · pozzolanic reaction

1 Introduction

Bentonite clay (BC), or in short, bentonite is a hydrated clay mineral with a layered structure. The main mineral component of the BC is montmorillonite. Its chemical composition includes silica, alumina, and water. Based on th structure of the mineral montmorillonite, BC can be divided into calcium bentonite (Ca-bent), sodium bentonite (Na-bent), and magnesium bentonite (Mg-bent) [1]. The typical chemical structure of montmorillonite consists of an edge-shared octahedral sheet with aluminum or magnesium hydroxide and two tetrahedral silicate sheets with sodium or calcium cations, which can be chemically replaced by a cationic-exchange reaction [2, 3]. Exchangeable cations in the interlayer region cause osmotic swelling and this swelling could result in significantly lower hydraulic conductivity [4].

BC has excellent swelling properties when combined with water [2, 5, 6], so it can be added to cement mortar to fill small pores in mortar and reduce the movement of

N. R. Abdul Razak et al. (Eds.): ECOOP 1996, SPPHY 418, pp. 379–385, 2025.
https://doi.org/10.1007/978-981-96-2871-1_53

water within the pore structure, thereby providing waterproofing properties for cement mortar. Na-bent has better waterproofing than Ca-ben and Mg-bent. The pore structure and permeability of cement-based materials have a great influence on the durability of the application of these materials.

BC, as a swelling potential clay, was added to mortar at content of 0, 4, and 8% by weight of Portland cement (PC). Test results showed that the addition of 8 wt.% BC could cause a significant increase of 61.48%, 42.09%, and 76.47% for compressive strength, flexural strength, and impermeability pressure, respectively [7].

In another study [8], 0 to 30% of the volume of PC was replaced by BC. The results show that the compressive strength of samples with 5% BC is 94.7% higher and the thermal conductivity of the same sample is 31.2% lower than the control sample. Increasing the amount of BC by more than 15% will significantly reduce the mechanical and thermal properties. However, using BC (0 - 4% by weight) as an additive in cement mortar increased the tensile strength up to a certain percentage. Then the strength decreases as the bentonite ratio increases. The best tensile strength was achieved after 7 days for 3% BC. At this level, the strength increase is 38.7% [9]. Using 1 - 15% by weight of BC to replace cement under different curing conditions [10], the results also show that increasing BC by more than 10% reduced the compressive strength of concrete cylinders. The best possible reason for the increased strength of bentonite is the pozzolanic reaction, which occurs at a slower rate than the hydration of cement, along with the micro filler effect caused by the BC.

BC has been widely used in the construction industry for a long time. The BC is one of the SiO_2-rich pozzolanic clay types that has been employed as a cost-effective and eco-friendly supplementary cementitious material in PC concrete. The current trend is using BC as low-content cement, no clinker, and cement composite, called SCM. The SCM can improve the properties of concrete and reduce pressure on natural resources and CO_2 emissions. However, the use of BCs as the SCM is limited because of their low pozzolanic activity, high hydrophilicity, and dramatic swelling nature. By partially replacing cement with bentonite (10%, 20%, 30%, and 40 wt.%), the study demonstrates the potential to reduce net consumption and cement costs while contributing to reduced energy consumption. Emissions and greenhouse gas emissions despite reduced concrete mechanical durability. Using BC as a partial replacement for cement will result in reducing greenhouse gas emissions because cement production emits a huge volume of greenhouse gases.

Reducing the carbon footprint in cement industry is one of the challenges and most often SCMs are used. The SCMs are soluble silicate, aluminosilicate or calcium aluminosilicate materials used as partial replacements of clinker in cements or partial replacement of cement in concrete mixtures. Clay minerals such as kaolinite, metakaolin, and montmorillonite (MMT) are generally used as SCM due to their potential to reduce the clinker factor, enhance hydration reactions, and reduce CO_2 emission.

In this study, the results of studying the effects of Binh Thuan bentonite on compressive strength and waterproofing were introduced. The results have shown that, with the same rate of bentonite replacing cement from 1 - 6% by weight, but with different concrete mixes, the effects on properties are different. Microstructural investigations using XRD and SEM showed that faujasite zeolite is a mineral of bentonite reaction in alkaline environments that increases the waterproofing ability of the tested concrete.

2 Experimental Method

In this study, Holcim PCB40 cement (PC), and Bentonite from Binh Thuan (Bent-) province, Vietnam were used as starting materials. Two basic concrete mixes of 300 mark (M300) are designed according to the Bolomey – Skramtaev and ACI methods. Moreover, A 200 concrete sample (M200) was designed according to Bolomey – Skramtaev. Part of the cement in the base concrete is replaced with bentonite at a Bent-/PC ratio of 1%, 2%, 3%, 4%, 5%, and 6%wt. corresponding. These samples are denoted as M0, M1,..., and M6 corresponding to the percentage of cement replaced. At the same time, the cement/water ratios and fine aggregate/coarse aggregate content do not change. The test samples were formed in a cylinder with a diameter and height of 150 × 150 mm, Esch sample group was 6 samples and curing time was 28 days.

The compressive strength, workability, and waterproof level of the concrete have been determined according to the current Vietnamese standards (TCVN) 3105:1993. The waterproof level is determined by the maximum water pressure at which 4 out of 6 test pieces of 150 × 150 mm cylinder are not penetrated by water and is denoted by B2, B4, B6, B8, B10, and B12 (numbers of 0,2; 0,4; 0,6,... is the hydrostatic pressure, in MPa. With pressure increasing step by step, each step was 0,2 MPa. The sample holding time at one pressure level was 16 h. Proceed to pressurize until the sample's surface has water flowing through. Then, close the valve and stop testing the transparent dehydrated sample. After that, continue to try the remaining tablets and stop testing when 4/6 tablets have been soaked in water. The chemical composition was analyzed using XRF (X-ray fluorescence), which provided insight into the chemical composition of bentonite. Microstructural changes in concrete were studied using advanced equipment, including an ultra-high resolution SEM (Scanning Electron Microscope) with model JSM-6700F to study the microstructure and a Diffuser. Brucker D8 Advance canopy (X-ray diffraction – XRD)) to study phase changes in concrete. These analytical techniques contribute to a better understanding of the behavior of concrete under different conditions.

3 Results and Discussions

3.1 The Chemical Compositions of Binh Thuan Bentonite

Table 1. The chemical composition of Binh Thuan bentonite

Oxide	SiO2	Al2O3	Fe2O3	CaO	MgO	SO3	K_2O	Na_2O	TiO2	MnO
% wt	50,04	16,98	5,70	4,88	1,48	0,03	2,53	1,49	0,53	0,13

The sodium oxide (Na_2O) content of 1.49% is not outstanding compared to the calcium oxide (CaO) content of 4.88%. Therefore, Binh Thuan Bent- is not Na-bent, but it seems to be a mix of Na-bent and Ca-bent (Table 1).

3.2 Concrete Mixing Compositions

Table 2. Concrete mixing compositions

	M300 (Bolomey-Skramtaev)				M200 (Bolomey-Skramtaev)				M300 (ACI)			
	PC (kg)	Sand (kg)	Stone (kg)	W. (kg)	PC (kg)	Sand (kg)	Stone (kg)	W.(kg))	PC (kg)	Sand (kg)	Stone (kg)	W.(kg)
M0	394	533	1.187	195	300	590	1220	205	479	561	1.110	187
M1	390	533	1.187	195	297	590	1220	205	474	561	1.110	187
M2	386	533	1.187	195	294	590	1220	205	469	561	1.110	187
M3	382	533	1.187	195	291	590	1220	205	464	561	1.110	187
M4	378	533	1.187	195	288	590	1220	205	459	561	1.110	187
M5	374	533	1.187	195	285	590	1220	205	454	561	1.110	187
M6	370	533	1.187	195	282	590	1220	205	449	561	1.110	187

(Bent-/PC ratio 0, 1, 2... 6 wt.% corresponding to M0, M1, M2,... M6).
W. is denoted for Water.

Table 3. The compressive strength of concrete

	M300 (Bolomey-Skramtaev)				M200 (Bolomey-Skramtaev)				M300 (ACI)			
	3 Days	7 Days	14 Days	28 Days	3 Days	7 Days	14 Days	28 Days	3 Days	7 Days	14 Days	28 Days
M0	150	180	305	312	125	146	233	250	205	238	316	362
M1	138	160	245	308	116	132	193	245	166	216	276	320
M2	125	150	220	302	105	124	181	239	164	212	268	315
M3	108	135	212	290	86	110	172	232	157	210	251	313
M4	95	125	208	270	80	103	168	206	152	201	243	311
M5	75	112	200	250	58	88	161	194	146	198	242	308
M6	68	110	195	245	50	78	154	185	141	192	217	303

The concrete Mark expressed as compressive strength after 28 days of standard curing was reduced when cement was replaced by Binh Thuan bentonite. The M0 samples are guaranteed to meet the design requirements of M300. With the Bolomey-Skramtaev design, the M300 had been guaranteed when the BC content was up to 2 wt. % and the M200 had been guaranteed when the bentonite content was up to 4%. Meanwhile, the ACI design had been guaranteed to reach the M300 mark when the BC content was up to 6 wt.% (Table 2).

3.3 The Waterproofing Ability of Concrete

Table 3 shows the results of determining the resistance of concrete samples at different levels. M300 concrete samples (according to ACI) can withstand permeability levels up to B12. Concrete sample M300 (according to ACI) can withstand permeability levels up to B12. In addition, at the B12 level, some M300 samples (according to the Bolomey-Skramtaev design) without bentonite mixture or low bentonite content (1%) are just starting to penetrate. Samples of 2% BC or more did not permeable, although their compressive strength is lower. This proves that BC has increased the waterproofing ability of concrete.

The waterproofing ability of the bent- has been clearly shown in the M200 samples (according to the Bolomey-Skramtaev design). These samples can only withstand permeability level B3. At level B4, samples M0, M1, M2, M3, and M4 were permeated, samples M5, M6 can withstand permeability level B4 (Table 4).

Table 4. The waterproofing ability of concrete

NO	M300	M200	M300 (ACI)
	(Bolomey-Skramtaev)		
	Number of permeated samples at waterproof level		
	B12	B4	B12
M_0	02	06	0
M1	01	05	0
M2	0	03	0
M3	0	02	0
M4	0	01	0
M5	0	0	0
M_6	0	0	0

3.4 XRD

Figure 1 shows XRD patterns of Binh Thuan Bent-, M0, and M 6 with the Bolomey-Skramtaev design.

The XRD analysis results show that limestone minerals ($CaCO_3$) and nontronite $(CaO_{0.5},Na)_{0.3}Fe^{3+}{}_2(Si,Al)_4O_{10}(OH)_2\cdot nH_2O$ are the main minerals of Binh Thuan bentonite. Nontronite is a montmorillonite mineral with a high iron oxide content. This mineral is not present on the XRD pattern of sample M6, proving that it participated in the reaction during the hydration process with cement.

On the XRD pattern of MO, there are characteristic peaks of minerals such as limestone ($CaCO_3$) and portlandite ($Ca(OH)_2$). They are products of the hydration of Portland cement. On the XRD patterns of samples M6, there are peaks typical of minerals such as

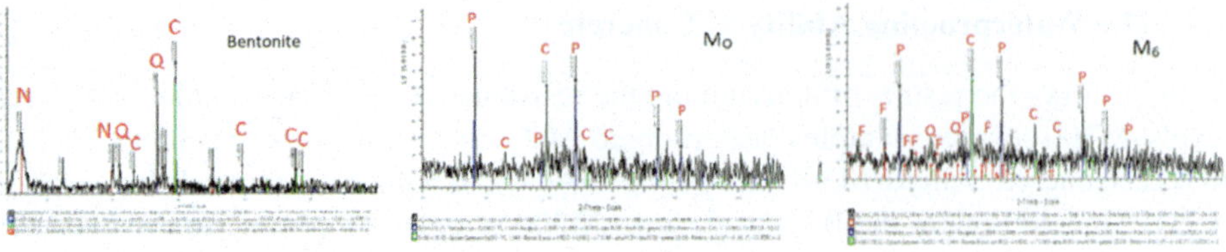

Fig. 1. XRD patterns of Binh Thuan Bent-, M0, and M6

limestone, portlandite, and faujasite ($Na_2Al_2Si_{3.3}O_{10}(6\text{-}7)H_2O$). The faujasite is a type of zeolite formed by the chemical reaction between bentonite and $Ca(OH)_2$ during the hydration of Portland cement.

3.5 SEM

SEM images of samples M0 and M6 are on the Fig. 2. The pyramidal shapes characteristic of faujasite mineral [24] are seen in the image of sample M6 as a product of the reaction of bent- with $Ca(OH)_2$ during PC hydration. Faujasite is the most different in the microstructure of samples with bent- (M6) or not bent- (M6 with the Bolomey-Skramtaev design).

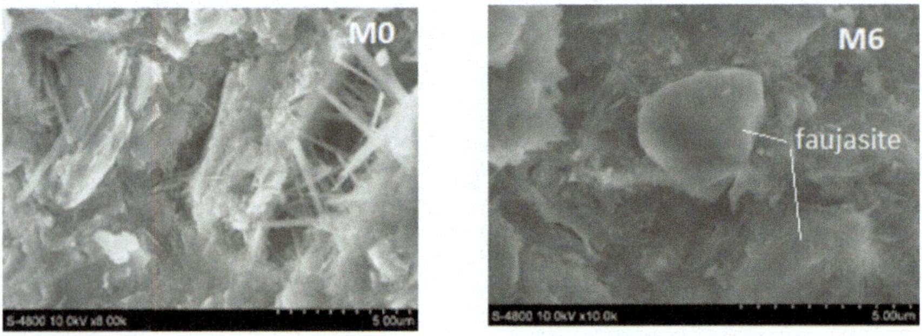

Fig. 2. SEM images of samples M0 and M6.

4 Conclusions

The waterproofing ability of concrete was increased although the compressive strength of concrete was reduced when replacing cement with Binh Thuan bentonite. With the M200 concrete model, the waterproofing ability was quite clear. At water pressure equivalent to waterproofing level B10 (pressure below 1,2 MPa), 4 - 6% wt. Binh Thuan bentonite can replace up to 6% of cement while ensuring the M300 mark (according to ACI) and perfect waterproofing ability. Reducing the amount of cement used with natural ingredients also helps reduce CO_2 emissions. The faujasite crystals in the microstructure of the sample M6 were detected by XRD diffraction and SEM images. This is the product of the puzzolanic reaction of bentonite clay with $Ca(OH)_2$ from the hydration of cement. Thanks to the puzzolanic reaction and faujasite minerals, the waterproofing ability of concrete was increased.

Acknowledgments. We acknowledge Ho Chi Minh City University of Technology (HCMUT), VNU-HCM for supporting this study.

References

1. Köstera, M.H., Williams, L.B., Kudejova, P., Gilg, H.A.: The boron isotope geochemistry of smectites from sodium, magnesium and calcium bentonite deposits. Chem. Geol. **510**(2) (2019)
2. Oren, A.H., Akar, R.Ç.: Swelling and hydraulic conductivity of bentonites permeated with landfill leachates. Appl. Clay Sci. **142**, 81–89, 15 (2017)
3. Anderson, R.L., Ratcliffe, I., Greenwell, H.C., Williams, P.A., Cliffe, S., Coveney, P.V.: Clay swelling—a challenge in the oilfield. Earth Sci. Rev. **98**(3–4), 201–216 (2010)
4. Nehdi, M.: Clay in cement-based materials: critical overview of state-of-the-art. Constr. Build. Mater. **51**, 372–382 (2014)
5. Qiong Wang, Anh Minh Tang, Yu-Jun Cui, Pierre Delage, Behrouz Gatmiri: Experimental study on the swelling behaviour of bentonite/claystone mixture. Eng. Geol. **124**, 59–66 (2012)
6. Saba, S., Barnichon, J.-D., Cui, Y.-J., Tang, A.M., Delage, P.: Microstructure and anisotropic swelling behaviour of compacted bentonite/sand mixture. J. Rock Mech. Geotech. Eng. **6**(2), 126–132 (2014)
7. Hu Yang, et al.: Effects of bentonite on pore structure and permeability of cement mortar. Constr. Build. Mater. **224**, 276–283 (2019)
8. Ebru Gedik, Adem Atmaca: An experimental study investigating the effects of bentonite clay on mechanical and thermal properties of concrete. Constr. Build. Mater. **383**, 131279 (2023)
9. Muwashee, R., Al-jameel, H.A.: Studying the effect of waterproofing admixture on some properties of cement mortar. IOP Conf. Ser.: Mater. Sci. Eng. **737** (2020), 012060IOP (2020)
10. Farhan Mushtaq, S., et al.: Effect of bentonite as partial replacement of cement on residual properties of concrete exposed to elevated temperatures. Sustainability **14**(18), 11580 (2021)

Increasing Bio-Cementation Efficiency Through Microbially Induced Calcium Carbonate Precipitation with Nano-Calcite for Sand, Sandy Soil, and Granular Materials

Nguyen Ngoc Tri Huynh[1,2](✉) and Nguyen Hoang Thien Khoi[2]

[1] Department of Silicate Materials, Faculty of Materials Technology, Ho Chi Minh City University of Technology (HCMUT), 268 Ly Thuong Kiet Street, District 10, Ho Chi Minh City, Vietnam

nnthuynh@hcmut.edu.vn

[2] Vietnam National University Ho Chi Minh City, Linh Trung Ward, Thu Duc City, Ho Chi Minh City, Vietnam

Abstract. The process of Microbial Induced Calcium Carbonate Precipitation (MICP) is an eco-friendly and promising technology for soil stabilization and construction. This study utilized Sporosarcina ureilytica, a non-pathogenic, gram-positive bacterium, to carry out MICP. The bacteria were sourced from Kien Giang province, Vietnam corals, and used in their spore state, demonstrating resilience to high alkalinity and urea concentrations. Nano calcite was employed as a nucleus for crystallization to expedite the MICP process. Almost all organic nutrient sources for biocementation were eliminated in this study. The results showed a significant increase in bulk density, with large soil samples exhibiting a 20% increase from their initial state, indicating improved material density, strength, and stability. Noteworthy changes in grain size distribution were observed, indicating enhanced soil compactness. Thermal analysis revealed a four-step degradation process, highlighting the presence of imperfect and hydrated $CaCO_2$ in MICP-treated samples. Microstructure analysis using an optical microscope and SEM confirmed the impact of MICP on biocementation through grain bridging formation. Moreover, compressive strength and anti-erosion tests substantially improved the treated soil, while anti-corrosion tests indicated enhanced durability. These findings underscore the potential of MICP for sustainable construction and environmental protection, offering a reliable and cost-effective solution for soil stabilization.

1 Introduction

Microbially Induced Calcium Carbonate Precipitation (MICP) is a biochemical process orchestrated by urease-producing bacteria that prompts the generation of carbonate ions ($CO_3{}^{2-}$) and calcium ions (Ca^{2+}) within a nutrient solution [1–4]. The precipitation reactions are influenced by various factors, including the type of bacteria exhibiting urease activity, cell concentration, pH, temperature, reaction period, and nutrient source. Over recent years, significant attention has been directed towards the study of MICP due

N. R. Abdul Razak et al. (Eds.): ECOOP 1996, SPPHY 418, pp. 386–393, 2025.
https://doi.org/10.1007/978-981-96-2871-1_54

to its potential as a sustainable and cost-effective alternative for concrete crack repair [5–10]. Bacillus strains, such as *Bacillus cohnii, Bacillus pasteurii, Bacillus pseudofirmus, and Bacillus subtilis*, have been extensively researched and implemented in MICP applications. It is crucial to carefully consider optimal nutrient sources, such as urea and calcium concentrations, to achieve desired outcomes while minimizing costs. The resulting calcium carbonate products from MICP display diverse morphological forms, including calcite, aragonite, and vaterite, which are known for their exceptional stiffness and durability, particularly in harsh conditions. Based on these concepts and research findings, MICP has been explored for self-healing concrete and soil improvement and stabilization purposes. The method portrays promise in repairing cracks and enhancing the durability of concrete structures.

By utilizing microorganisms capable of creating calcium carbonate, damaged areas can be healed, restoring the structure of the concrete. MICP also aids in improving and strengthening the soil. The introduction of bacteria capable of creating calcium carbonate reduces soil erosion and increases the soil's load-bearing capacity. This approach is environmentally friendly and sustainable. Nevertheless, some limitations include the need for multiple injection pipes to ensure consistent performance in uneven soil and the potential for clogging near the nozzle hole. Despite these limitations, MICP holds promise as a bio-inspired method for improving and stabilizing soil sustainably. Biomineralization technologies are encompassed in aboveground construction materials, near-ground-surface stabilization, and subsurface applications, each with specific implementation requirements. These technologies contribute to innovative building materials, remediation of existing structures, and offer alternatives to traditional soil and dust stabilization methods. Biomineralization aids in rapidly filling cracks in self-healing concrete and facilitating the rapid formation of robust bonds in building materials like bricks. Although biomineralization technologies may incur higher costs, they are reliable and easy to implement. Ureolytic biocementation emerges as a promising greener alternative with lower carbon emissions. Ureolysis-induced calcium carbonate precipitation stands as the most widely employed approach for large-scale applications. The biomineralization process by MICP involves several stages, from urea hydrolysis to the formation of $CaCO_3$ crystals, providing a comprehensive understanding of the process. Furthermore, bacterial biofilm formation, encompassing attachment, colonization, growth, and maturation phases, plays a crucial role in the cementation process. Biofilms comprise intricate communities of bacteria embedded in a self-produced matrix, which aids in the precipitation of $CaCO_3$ and enhances the density of separated particles. The concentration of bacteria significantly influences the biomineralization process, with higher concentrations necessitating increased doses of urea and Ca^{2+}. The addition of a bacterial nutrient, such as yeast extract or glucose, can enhance precipitation even at low initial bacterial concentrations by stimulating bacterial growth and increasing bacterial concentration. Urease activity test results suggest that the presence of this nutrient can impact the activation and strength of bacterial activity, potentially causing a delay in urea decomposition when bacteria are present at low concentrations. Enhancing the capacity and effectiveness of biomineralization through MICP is beneficial for improving biocementation. This, in turn, can facilitate the upscaling of bacterial treatment for larger quantities of sand or soil. However, it is imperative to carefully consider the cost implications associated with enhancing the efficacy of MICP. As effectiveness increases, so do costs. Therefore, optimizing bacterial concentration and nutrient sources is crucial to balance effectiveness and cost-effectiveness. Two valuable suggestions to address this concern are using wasted or recycled nutrient sources and eliminating unnecessary organic nutrient sources, with a primary focus on urea and calcium sources. Repurposing materials that would otherwise go to waste through wasted or recycled nutrient sources can minimize costs. Additionally, eliminating organic nutrient sources and prioritizing essential nutrients can streamline the process and potentially reduce expenses related to unnecessary additives. Various organic carbon sources, including lactose, glucose, corn starch, tapioca, and soybean meal, could be employed for bacterial growth and biomineralization.

MICP is a complex process influenced by various factors, including the specific type of urease-producing bacteria, bacterial density, pH, temperature, reaction time, and nutrient sources. MICP shows potential for soil stabilization, as introducing $CaCO_3$_producing bacteria can help mitigate soil erosion and improve load-bearing capacity, aligning with environmentally friendly and sustainable practices. Despite its promise, MICP encounters limitations, such as requiring multiple treatment cycles to ensure consistent performance on heterogeneous soils and the risk of clogging near injection points. Notwithstanding these challenges, MICP remains a promising approach for sustainable soil stabilization. Biocementation, utilizing biological processes to create cement-like materials, can be applied to surface construction materials, near-surface soil stabilization, and underground applications, offering a reliable and easy-to-implement green alternative with lower carbon emissions. In MICP applications for soil stabilization, forming hydrated ACC can be an initial step that eventually leads to more stable and robust crystalline calcium carbonate. This process can enhance the load-bearing capacity and erosion resistance of treated soils. Over time, hydrated ACC can transform into more stable crystalline forms of calcium carbonate. This transformation can occur naturally and is influenced by temperature, pH, and the presence of ions or organic molecules.MICP presents a solution to improve the physical and mechanical properties of such soils by creating strong bonds between soil particles through $CaCO_3$ precipitation, enhancing soil stiffness and strength. This study aims to explore the feasibility of using nano-calcite in microbial solutions to improve the efficiency of MICP. The research will investigate the transformation of Ca^{2+} into $CaCO_3$ over time and the different polymorphs of $CaCO_3$ that can form. Achieving mineralization efficiency and maintaining microbial activity stability are critical challenges. Using nano-calcite can provide nucleation sites for $CaCO_3$, promoting quicker and more efficient calcite distribution within the soil matrix, thus improving soil quality and performance. Understanding and controlling the factors influencing biocementation can optimize MICP processes to achieve desired soil strength and stability. This research highlights the potential of using microorganisms for sustainable geotechnical and construction applications to create durable materials and stabilize soils using environmentally friendly methods. The application of MICP can reduce settlement, enhance load-bearing capacity, and improve the longevity of structures, particularly in challenging geotechnical conditions.

2 Methodology

Sporosarcina ureilytica was used as the biomineralization agent. It is a gram-positive, aerobic, non-pathogenic bacterial strain that can break down urea. The bacteria were sourced from corals in Kien Giang province, Vietnam, and used in their spore form. These spores can withstand high alkalinity and urea concentrations up to 3M, with a urea hydrolysis rate of 0.76 mM/min. The bacterial density ranged between 108 and 109 colony-forming units per milliliter (CFU/mL). Urea was used as the nutrient source, and calcium chloride served as the calcium source for mineralization. The nano calcite, supplied by Saigon Nanomat Co., Ltd in Vietnam, had a purity of 98% and an average particle size of about 52 nm. Both bacterial solutions, with and without nano calcite, were analyzed to prepare sandy-soil samples. The bacterial solution, a catalyst for MICP and biocementation, was enriched with Nano calcite and urea. Nano calcite was used to create nucleation sites for crystal formation, while urea provided essential nitrogen for the bacteria. Sodium percarbonate (SPC), a powerful oxygen source, was introduced to the solution. This addition not only enhanced the biochemical reactions but also reduced other regular organic nutrient sources for MICP.

The sandy soil test samples were meticulously created using a combination of soil and 10 wt% of sand (fineness modulus of 1.9). The soil used in the mixture is typically rich in organic matter and nutrients, with a neutral to alkaline pH range of 6.5 and an initial moisture content of 18%. The effectiveness of MICP was rigorously assessed through various materials analysis methods. Real-time observation of MICP was conducted using optical microscopy, involving continuous monitoring of the test samples to determine the MICP process in real-time. Thermal analysis of the precipitated products were performed using Differential Scanning Calorimetry combined with Thermogravimetric Analysis (DSC/TG) to qualitatively and quantitatively determine the improvement of MICP aided by nano calcite as a nucleating agent. The TGA/DSC system from Mettler Toledo, a trusted tool in the field, was utilized for this analysis. Phase composition analysis of the materials and products was carried out using X-ray Diffraction (XRD) to compare the initial multi-phase composition of the soil with the soil samples post-MICP treatment. The D2 Phaser from Bruker, a reliable instrument, was employed for this analysis. Microstructural morphology observation was conducted using Scanning Electron Microscopy (SEM) to analyze the $CaCO_3$ structure created during MICP, providing detailed insight into the shape, size, uniformity, and diversity of the precipitated products. After drying, the samples were imaged with the JSM-IT200 system from Jeol, enabling high magnification and better contrast than optical microscopy, ensuring the thoroughness of our analysis.

3 Results and Discussions

The bulk density measurement results of soil samples before and after adding the MICP bacterial solution show a significant increase. Initially, the bulk density of the soil was 203 g/cm^3 for large samples and 123 g/cm^3 for small samples. After adding the MICP bacterial solution, the bulk density increased to 244 g/cm^3 for large and 147 g/cm^3 for small samples. This difference is due to the sample size, leading to uneven MICP distribution, but regardless of sample size, MICP increases bulk density and compactness in soil structure by over 20%. The increase in bulk density, which signifies a rise in material density in the soil, enhances the soil's strength and stability and improves its mechanical properties. This is evident in the enhanced erosion resistance and load-bearing capacity observed after MICP treatment. The significant change in grain size distribution in MICP-treated samples, as visually observed and through particle-size-distribution curves (Figs. 1 and 2), further underscores the potential of MICP technology in construction and environmental protection. The formation of bridges between loose particles leads to increased particle size, resulting in an overall increase in mass density, thus enhancing the compactness of the soil sample structure. The following section will clarify these effects through shape retention and compression-set resistance capabilities.

Observations of sample shapes after mold removal show that untreated control soil samples collapsed immediately, indicating that the particles slid over each other without bonding elements and lacked mold support as shown in Fig. 2 In the case of MICP bacteria (both original and enhanced MICP), the formation of bridges between soil particles occurred, allowing the soil samples to maintain the cylindrical shape of the mold with nearly 100% height retention. However, there are discrepancies in the diameter and height of the soil samples, which can be explained by the fact that the cementation effect has yet to reach complete efficiency as the process has just begun. MICP-treated soil samples show absolute shape retention after mold removal. The results show that moisture directly affects shape retention and the soil sample's strength. The optimal moisture level ensures enough moisture for effective cementation without collapse during mold removal. By observing and analyzing soil samples under different conditions, essential factors necessary to optimize the MICP process for structural shape retention were identified, including bacteria, nano calcite, and appropriate moisture levels. This will help improve the mechanical and physical properties of the soil and open up many practical applications in construction and environmental protection.

Before treatment, particles were dispersed and unconnected. Post-treatment crystallized $CaCO_3$ bridges bonded the particles, forming a solid and cohesive structure. The cementation rate increased rapidly, indicating effective stabilization, although the homogeneity was localized. The microscopic images are quantitatively analyzed using comprehensive image processing methods (including gamma adjustment, grayscale threshold selection, and median processing) to evaluate the biocementation level.

The compression strength tests on soil samples before and after MICP treatment show a significant increase in strength for treated samples. Specifically, the unconfined compression strength of the treated soil samples increased by approximately 50–70% compared to untreated samples. The addition of MICP increased the bulk density and compactness of the soil structure and improved its overall mechanical strength. Figure 3 vividly demonstrates the impact of MICP treatment on the unconfined compression strength of soil samples. After 7 and 14 days of curing, the untreated samples maintained a relatively low compression strength, while the MICP-treated samples exhibited a significant increase in strength. This improvement can be directly attributed to the formation of grain bridges between soil particles, a direct result of the MICP treatment. These bridges have significantly enhanced the soil's structural integrity and its ability to resist deformation under load.

Fig. 1. Change in grain size distribution after biocementation.

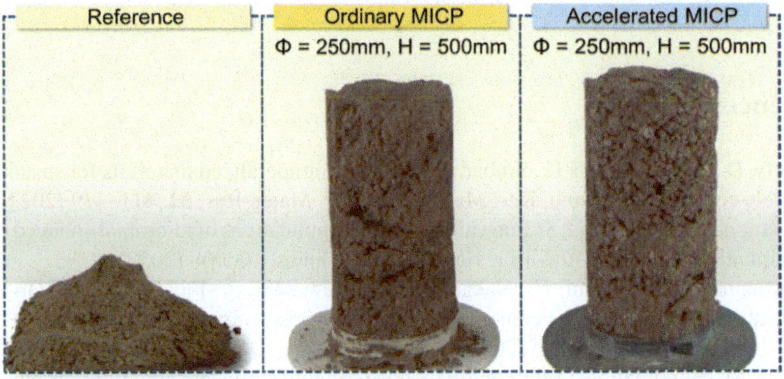

Fig. 2. Sample shapes after mold removal.

4 Conclusions

The results of this study demonstrate that MICP significantly enhances soil's mechanical and physical properties. The formation of $CaCO_3$ bridges between soil particles leads to increased particle size, bulk density, compression strength, and reduced water absorption, thereby improving soil stability and durability. These findings highlight the potential applications of MICP technology in construction and environmental protection, providing new solutions for soil stabilization and addressing environmental challenges such as soil erosion and land degradation. Further research and optimization of the

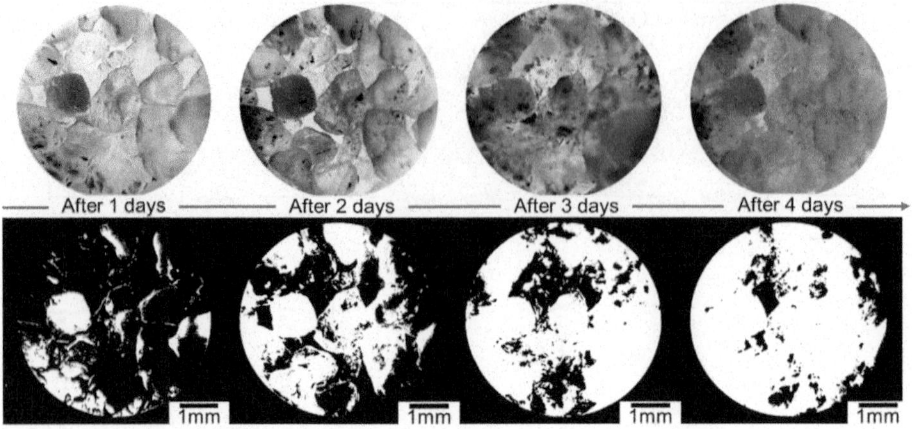

Fig. 3. Optical microscopic observation and binary images of MICP process.

MICP process will help to realize its benefits fully and expand its practical applications in various fields.

Acknowledgments. This research is funded by Vietnam National University HoChiMinh City (VNU-HCM) under grant number C2023-20-12.

References

1. Beatty, D.N., Williams, S.L., Srubar, W.V., III.: Biomineralized materials for sustainable and durable construction. Annu. Rev. Mater. Res.. Rev. Mater. Res. **52**, 411–439 (2022)
2. Y. Wang, C. Konstantinou, S. Tang, and H. Chen, Applications of microbial-induced carbonate precipitation: A state-of-the-art review. Biogeotechnics, 100008 (2023)
3. Gowthaman, S., Koizumi, H., Nakashima, K., Kawasaki, S.: Field experimentation of bio-cementation using low-cost cementation media for preservation of slope surface. Case Stud. Construct. Mater. **18**, e02086 (2023)
4. Seifan, M., Berenjian, A.: Microbially induced calcium carbonate precipitation: a widespread phenomenon in the biological world. Appl. Microbiol. Biotechnol.Microbiol. Biotechnol. **103**, 4693–4708 (2019)
5. De Muynck, W., De Belie, N., Verstraete, W.: Microbial carbonate precipitation in construction materials: a review. Ecol. Eng. **36**, 118–136 (2010)
6. De Rooij, M., Van Tittelboom, K., De Belie, N., Schlangen, E.: Self-Healing Phenomena in Cement-Based Materials: State-of-the-art Report of RILEM Technical Committee 221-SHC: Self-Healing Phenomena in Cement-Based Materials. Springer (2013)
7. Huyen, N.P.H., Huynh, N.N.T.: Bio-cementation by using microbially induced carbonate precipitation technique by using Sporosarcina pasteurii with oxidizing agent supplements. J. Mater. Construct. **11** (2021)
8. Huynh, N.N.T., Imamoto, K., Kiyohara, C.: Eco-friendly technique on nutrient sources and capsulation for bacteria-based self-healing concrete. In: Proceedings of the Japan Concrete Institute, pp. 1288–1293. Japan Concrete Institute, Hiroshima (2020)

9. Huynh, N.N.T., Nhu, N.Q., Son, N.K.: Developing the solution of microbially induced $CaCO_3$ precipitation coating for cement concrete. In: IOP Conference Series: Materials Science and Engineering, p. 062006. IOP Publishing (2018)
10. Huynh, N.N.T., Tu, T.A., Huyen, N.P.H., Son, N.K.: Lab-scale experiments for soil cementing through bio-chemical process. ASEAN Eng. J. **11**, 255–265 (2021)

Surface Coating

Circulating-Infiltrating Reaction: A Promising Method for Large-Scale Manufacturing of Thin Film Composite Membranes with Improved Performance

Vinh-Dat Vuong[1,6](✉), Tran Nhat Minh[1,6], Nguyen Tan Tai[1,6], Tran Van Khai[1,6], Pham Tan Thi[2,6], Dat Quoc Lai[3,6], Mai Thanh Phong[4,6], and Thang Van Le[5,6]

[1] VNU-HCM Key Laboratory for Materials Technology (MTLab), 268 Ly Thuong Kiet Street, District 10, Ho Chi Minh City, Vietnam
vuongvinhdat@hcmut.edu.vn

[2] Faculty of Applied Science, Ho Chi Minh City University of Technology (HCMUT), 268 Ly Thuong Kiet Street, District 10, Ho Chi Minh City, Vietnam

[3] Faculty of Chemical Engineering, Ho Chi Minh City University of Technology (HCMUT), 268 Ly Thuong Kiet Street, District 10, Ho Chi Minh City, Vietnam

[4] Ho Chi Minh City University of Technology (HCMUT), 268 Ly Thuong Kiet Street, Ho Chi Minh City, Vietnam

[5] International University – Vietnam National University Ho Chi Minh City, Quarter 6, Linh Trung Ward, Thu Duc City, Ho Chi Minh City, Vietnam

[6] Vietnam National University Ho Chi Minh City, Linh Trung Ward, Thu Duc City, Ho Chi Minh City, Vietnam

Abstract. Traditional methods for fabricating thin film composite membranes, such as direct polymerization or casting on a polymeric substrate, often have limitations. This study presents a novel approach that overcomes these limitations. By forming chemical bonds between existing carbon nanotube (CNT) precursors and polyamide-based microfilters or fabrics, this method achieves a significant improvement in reaction efficiency, which is confirmed by scanning electron microscopic (SEM) images. The circulating-infiltrating reaction (CIR) ensures a steady flow of CNT dispersion throughout the polyamide matrix, promoting complete contact and preventing clogging. Flow simulations indicate that turbulent and reversed flows achieve uniform CNT coverage on the exposed surface of the filters. Furthermore, ultrasound assistance during the CIR process can further enhance these effects, which can be shown on CNT coverage on back side of the composite membrane. These advantages demonstrate the potential of CIR, with or without ultrasound, for large-scale manufacturing of thin film composite membranes from readily available materials.

Keywords: circulating-infiltrating reaction · polyamide-based filter · carbon nanotubes · anti-clogging · flow simulation

N. R. Abdul Razak et al. (Eds.): ECOOP 1996, SPPHY 418, pp. 397–404, 2025.
https://doi.org/10.1007/978-981-96-2871-1_55

1 Introduction

The burgeoning food, petrochemical, pharmaceutical, and textile industries, coupled with the growing demand for desalination and water treatment, necessitate advanced separation technologies [1–6]. Membrane filtration, renowned for its efficiency and economic viability, has emerged as a promising solution [6–8]. Pressure-driven membranes, including microfiltration, ultrafiltration, nanofiltration, and reverse osmosis, dominate the market [9, 10]. Thin film composite (TFC) membranes, comprising a polyamide selective layer on a support structure, are widely used in nanofiltration and reverse osmosis [11–14]. Despite their high performance, TFC membranes suffer from fouling and chlorination, leading to reduced flux, lifespan, and selectivity [15–18].

To address these challenges, researchers have explored various approaches, including surface treatments and nanomaterial integration [18–20]. Carbon nanotubes (CNTs) offer exceptional potential due to their unique structure, enhancing permeability and selectivity while mitigating fouling [21–28]. However, CNTs' hydrophobicity hinders their dispersion in organic solvents, commonly used in membrane fabrication [25, 29, 30]. Functionalization of CNTs improves their dispersibility and membrane performance, demonstrating enhanced contaminant removal and antifouling properties, especially in desalination [27, 31, 32]. Studies have shown that CNT-incorporated TFC membranes exhibit improved flux and salt rejection [33, 34]. While polymerization is a popular fabrication method, it can lead to non-uniform structures and nanomaterial agglomeration. Given the high cost of nanomaterials, developing efficient and cost-effective fabrication techniques is crucial.

In this study, a novel circulating-infiltrating reaction (CIR) method is introduced for fabricating TFC membranes (Fig. 1a). Nylon 6 fabric is initially modified to increase hydrophilicity by hydrolyzing long polymer chains and regenerating amide bonds with aliphatic diamine. Subsequently, carboxylated multi-walled carbon nanotubes (MWCNT-COOH) are grafted onto the amine-enriched fabric via amidation. The CIR process enhances MWCNT-COOH infiltration, while ultrasound prevents fouling. This approach offers a promising pathway to efficient TFC membrane production.

2 Experimental Procedure

Food-graded nylon 6 fabric (500 mesh/25 μm of nominal aperture) was purchased from JVLab. Octane-1,8-diamine (98%) was purchased from Alfa Aesar. Sodidum hydroxide (96%) was purchased from Xilong Scientific. Ethanol (99%) was purchased from Chemsol. The raw MWCNTs and MWCNT-COOH were produced by VNU-HCM Key Laboratory for Material Technologies (VNU-HCM MTLab).

The treating solution was prepared with a concentration of NaOH and octane-1,8-damine of 1 M and 0.1 M, respectively. The nylon 6 fabric with a diameter of 47 mm was washed in distilled water and incubated at 85 °C for 30 min, followed by sonicating in as-prepared solution at 55 °C for 30 min. Then, the fabric was dried at 85 °C in an oven for 60 min. Finally, the fabric is rinsed with distilled water to remove residual agents and dried again. The amino-enriched fabric is labelled as PA-NH_2. The PA-NH_2 fabric was installed in the CIR. The precursor mixture was prepared by adding 50 mg of

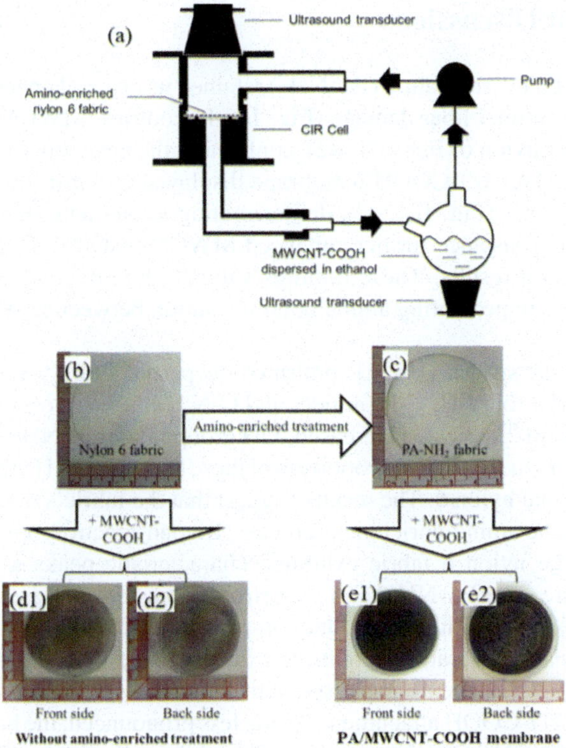

Fig. 1. (a) The CIR process for fabrication of polyamide/MWCNT-COOH membrane. Images of (b) original nylon 6 fabric; (c) PA-NH_2 fabric; (d1, d2) front and back side of PA/MWCNT-COOH without amino-enriched treatment; (e1, e2) front and back side of PA/MWCNT-COOH with amino-enriched treatment.

MWCNT-COOH in 350 mL of ethanol at room temperature. The mixture was sonicated at 60 °C for 15 min, followed by pumping through the CIR cell for 60 min. Ultrasound assistance took place at the surface of PA-NH_2 fabric in the CIR cell and precursor tank to eliminate local clogging of MWCNT-COOH. The membrane was dried and washed with distilled water at the end of the reaction to remove weakly bound MWCNT-COOH on the membrane surface. The as-prepared membrane is labelled as PA/MWCNT-COOH.

The chemical structures of PA-NH_2 and PA-MWCNT-COOH were characterized using Fourier transform infrared (FTIR) and Raman spectroscopies. The FTIR spectra were operated with wavenumber window of 360–4000 cm^{-1}, and resolution of 4 cm^{-1}. The Raman spectra were operated with wavenumber window of 300–3300 cm^{-1} the excitation laser wavelength of 532 nm. Surface morphologies of membranes were observed by field emission scanning electron microscopic (FESEM) and 3D laser scanning microscopic (3D-LSM). The FESEM images were captured with accelerating voltage of 10 kV and SE detector. The 3D-LSM models were scanned with pitch of 5 μm/step. The hydrophilicity of membranes was evaluated by contact angle.

3 Results and Discussions

The amino-enriched nylon 6 fabric (PA-NH_2) retained its original appearance after treatment, indicating minimal fiber damage (Fig. 1c). In contrast, MWCNT-COOH did not adhere to untreated nylon 6 (Fig. 1d1, d2), confirming the necessity of amine groups for bonding. The PA/MWCNT-COOH membrane displayed a uniformly distributed black MWCNT-COOH layer (Fig. 1e1, e2), demonstrating strong adhesion due to chemical bonds. Ultrasound assistance further enhanced MWCNT-COOH distribution, aligning with fluid simulation results. These findings validate the effectiveness of the amino-enrichment process in promoting amide bond formation between MWCNT-COOH and PA-NH_2 fibers.

FTIR and Raman spectra (Fig. 2) confirmed the partial hydrolysis of nylon 6 fabric. Characteristic peaks for N-H, C-H, amide, and C-N bonds were observed. The shift in N-H peak to 3441 cm^{-1} indicated increased -OH groups and free -NH_2 from amino-octyl branches. However, due to the low proportion of these branches, C-H vibrations remained similar to the original nylon 6. The results suggest that the fabric's overall structure was preserved despite the amino-enrichment process. Raman spectroscopy corroborated the FTIR findings. The nylon 6 fabric exhibited characteristic peaks associated with its molecular structure. The PA-NH_2 fabric spectrum indicated increased -CH_2- and $C = O$ stretching, consistent with the introduction of octane-1,8-diamine. Enhanced amide III and -CH_2- twisting vibrations confirmed the presence of abundant -NH_2 groups. The PA/MWCNT-COOH membrane displayed distinct D- and G-band on the front side, confirming MWCNT-COOH attachment. While less pronounced, the back side spectrum also suggested MWCNT-COOH presence, albeit in smaller quantities. These results collectively support the successful grafting of MWCNT-COOH onto the amino-enriched nylon 6 fabric.

Surface morphology analysis by SEM and 3D laser scanning microscopic (Fig. 3) revealed significant changes during the membrane fabrication process. Partial hydrolysis of nylon 6 created nucleation sites for subsequent amine grafting (Fig. 3 c, d). The resulting PA-NH_2 fibers exhibited increased adhesion due to softened surfaces and octamethylene bridges, facilitating MWCNT-COOH cover [34]. A uniform MWCNT-COOH layer formed on the front side (Fig. 3e, f) due to direct precursor exposure. The MWCNT-COOH layer increased fiber radius by approximately 3.22 μm.

Contact angle measurements confirmed the hydrophilic nature of PA-NH_2 fibers compared to nylon 6 (Fig. 3g). However, the MWCNT-COOH layer dramatically increased hydrophobicity on both sides of the PA/MWCNT-COOH membrane. These findings collectively demonstrate the successful integration of MWCNT-COOH onto the nylon 6 fabric through the CIR process.

4 Conclusion

A novel CIR method has enabled the successful fabrication of a double-layered TFC membrane using PA and MWCNT-COOH on food-grade nylon 6 fabric. This approach leverages ultrasound, turbulence, and reverse flows to evenly distribute MWCNT-COOH on the PA-NH_2 substrate, forming a robust structure through chemical bonding

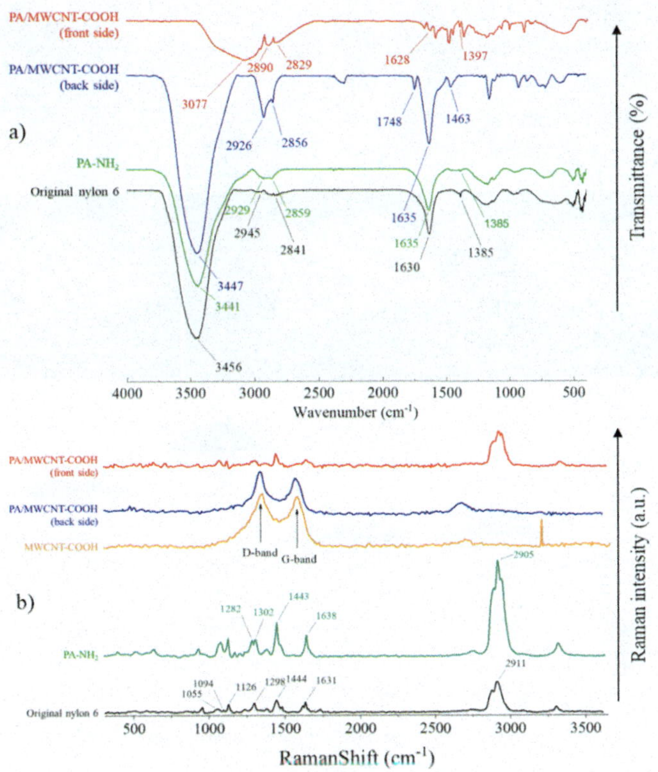

Fig. 2. FTIR (a) and Raman (b) spectra of original nylon 6, $PA-NH_2$ and PA/MWCNT-COOH.

rather than simple adhesion. The resulting TFC membrane outperforms traditional high-temperature methods by creating durable coatings on flexible polyamide substrates. The scalability of the CIR prototype holds promise for large-scale industrial production of TFC membranes.

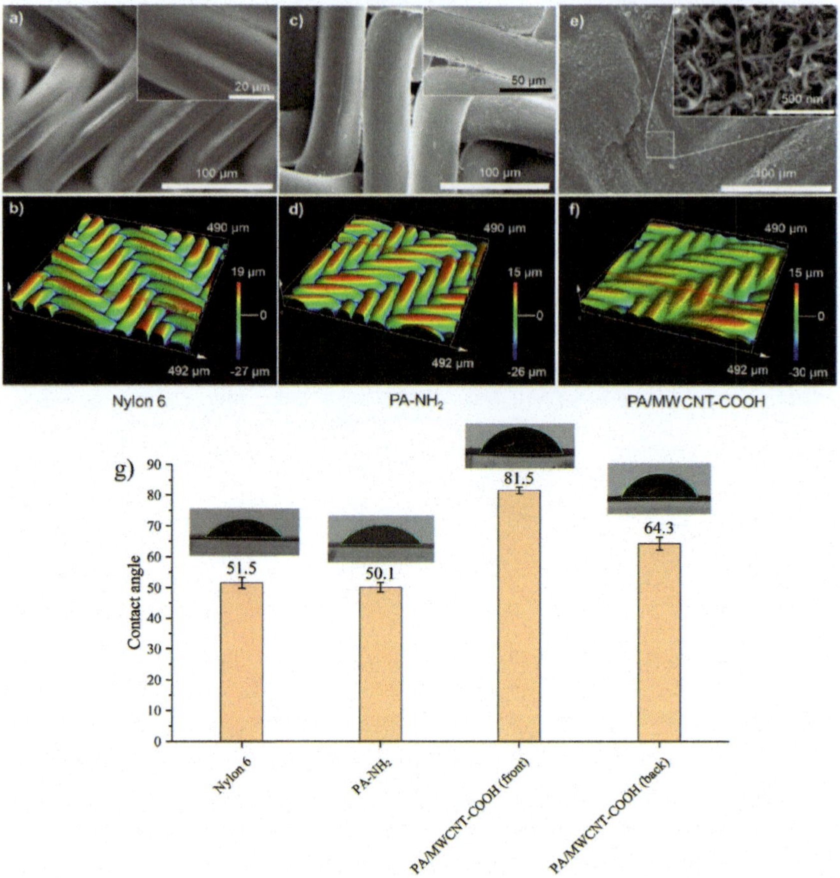

Fig. 3. Surface morphology (a, b) nylon 6, (c, d) PA-NH_2, (e, f) PA/MWCNT-COOH. (g) Contact angle of original nylon 6, PA-NH_2 and PA/MWCNT-COOH membrane.

Acknowledgments. We acknowledge the support of time and facilities from Ho Chi Minh City University of Technology (HCMUT), VNU-HCM for this study.

References

1. Nazir, A., et al.: Membrane separation technology for the recovery of nutraceuticals from food industrial streams. Trends Food Sci. Technol. **86**, 426–438 (2019)
2. Al-Salmi, M., et al.: Application of membrane distillation for the treatment of oil field produced water. Desalination **494**, 114678 (2020)
3. Buonomenna, M.G., Bae, J.: Organic solvent nanofiltration in pharmaceutical industry. Sep. Purif. Rev. **44**(2), 157–182 (2014)
4. Abid, M.F., Zablouk, M.A., Abid-Alameer, A.M.: Experimental study of dye removal from industrial wastewater by membrane technologies of reverse osmosis and nanofiltration. Iranian J. Environ. Health Sci. Eng. **9**(1), https://doi.org/10.1186/1735-2746-9-17 (2012)

5. Ahmad, N.N.R., et al.: Current advances in membrane technologies for saline wastewater treatment: A comprehensive review. Desalination **517**, 115170 (2021)
6. Othman, N.H., et al., A Review on the Use of Membrane Technology Systems in Developing Countries. Membranes (Basel) **12**(1), 30 (2021)
7. Obotey Ezugbe, E., Rathilal, S.: Membrane technologies in wastewater treatment: a review. Membranes (Basel) **10**(5), 89 (2020)
8. Isa, Y.M., Rathilal, S., Aliyu, U.M.: Membrane desalination technologies in water treatment: A review. Water Pract. Technol. **13**(4), 738–752 (2018)
9. Yang, Z., et al., A review on reverse osmosis and nanofiltration membranes for water purification. Polymers (Basel). **11**(8), 1252 (2019)
10. Verma, B., et al.: Pressure-driven membrane process: a review of advanced technique for heavy metals remediation. Processes **9**(5), 752 (2021)
11. Ji, C., et al., Recent advances in high-performance TFC membranes: A review of the functional interlayers. Desalination **500**, 114869 (2021)
12. Asadi Tashvigh, A., Elshof, M.G., Benes, N.E.: Development of Thin-Film Composite Membranes for Nanofiltration at Extreme pH. ACS Appl Polym Mater **3**(11), 5912–5919 (2021)
13. Khoo, Y.S., et al.: Surface modification of PA layer of TFC membranes: Does it effective for performance Improvement? J. Ind. Eng. Chem. **102**, 271–292 (2021)
14. Huang, W., et al.: High performance polyamide TFC reverse osmosis membrane fabricated on co-deposition hydrophilic modified polyethylene substrate. Desalination **538**, 115909 (2022)
15. Liu, C., et al.: Separation, anti-fouling, and chlorine resistance of the polyamide reverse osmosis membrane: From mechanisms to mitigation strategies. Water Res. **195**, 116976 (2021)
16. Liu, M., et al.: Improving fouling resistance and chlorine stability of aromatic polyamide thin-film composite RO membrane by surface grafting of polyvinyl alcohol (PVA). Desalination **367**, 11–20 (2015)
17. Gohil, J.M., Suresh, A.K.: Chlorine attack on reverse osmosis membranes: Mechanisms and mitigation strategies. J. Membr. Sci. **541**, 108–126 (2017)
18. Akther, N., et al.: Recent advances in nanomaterial-modified polyamide thin-film composite membranes for forward osmosis processes. J. Membr. Sci. **584**, 20–45 (2019)
19. Jahangiri, F., et al.: Improvement of performance of polyamide reverse osmosis membranes using dielectric barrier discharge plasma treatment as a novel surface modification method. Polymer Eng. Sci. 59(S1) (2018) https://doi.org/10.1002/pen.25025
20. Zou, L., et al.: Surface hydrophilic modification of RO membranes by plasma polymerization for low organic fouling. J. Membr. Sci. **369**(1–2), 420–428 (2011)
21. Son, M., Novotny, V., Choi, H.: Thin-film nanocomposite membrane with vertically embedded carbon nanotube for forward osmosis. Desalin. Water Treat. **57**(55), 26670–26679 (2016)
22. Gao, S., et al.: Ultrathin Polyamide Nanofiltration Membrane Fabricated on Brush-Painted Single-Walled Carbon Nanotube Network Support for Ion Sieving. ACS Nano **13**(5), 5278–5290 (2019)
23. Baek, Y., et al.: Evaluation of carbon nanotube-polyamide thin-film nanocomposite reverse osmosis membrane: Surface properties, performance characteristics and fouling behavior. J. Ind. Eng. Chem. **56**, 327–334 (2017)
24. Zhao, H., et al.: Improving the performance of polyamide reverse osmosis membrane by incorporation of modified multi-walled carbon nanotubes. J. Membr. Sci. **450**, 249–256 (2014)
25. Kim, H.J., et al.: High-performance reverse osmosis CNT/polyamide nanocomposite membrane by controlled interfacial interactions. ACS Appl. Mater. Interfaces **6**(4), 2819–2829 (2014)
26. Al-Hobaib, A.S., et al.: Modification of thin-film polyamide membrane with multi-walled carbon nanotubes by interfacial polymerization. Appl Water Sci **7**(8), 4341–4350 (2017)

27. Ali, S., et al.: Challenges and opportunities in functional carbon nanotubes for membrane-based water treatment and desalination. Sci. Total. Environ. **646**, 1126–1139 (2019)
28. Rashed, A.O., Esawi, A.M.K., Ramadan, A.R.: Novel Polysulfone/Carbon Nanotube-Polyamide Thin Film Nanocomposite Membranes with Improved Water Flux for Forward Osmosis Desalination. ACS Omega **5**(24), 14427–14436 (2020)
29. Amini, M., Jahanshahi, M., Rahimpour, A.: Synthesis of novel thin film nanocomposite (TFN) forward osmosis membranes using functionalized multi-walled carbon nanotubes. J. Membr. Sci. **435**, 233–241 (2013)
30. Kong, Q., et al.: Fabrication of high performance TFN membrane containing NH(2)-SWCNTs via interfacial regulation. RSC Adv. **10**(42), 25186–25199 (2020)
31. Farid, M.U., Khanzada, N.K., An, A.K.: Understanding fouling dynamics on functionalized CNT-based membranes: Mechanisms and reversibility. Desalination **456**, 74–84 (2019)
32. Khalid, A., et al.: Preparation and properties of nanocomposite polysulfone/multi-walled carbon nanotubes membranes for desalination. Desalination **367**, 134–144 (2015)
33. Lau, W.J., et al.: A review on polyamide thin film nanocomposite (TFN) membranes: History, applications, challenges and approaches. Water Res. **80**, 306–324 (2015)
34. Tai, N.T., et al.: Role of alkane-α, ω-diamine in the nanostructure of carboxylic functionalized multiwalled carbon nanotubes grafted on amino-enriched nylon fabric toward metal ion absorption. Environ. Technol. Innov. **31**, 103165 (2023)

Electrophoretic Deposition and Wet-Transferring Free-Standing (MoS_x@MoO_y)/MWCNT Paper

Vinh-Phuc Ha[1,4], An Thien Nguyen[1,4], Mai Huu Xuan[2,4], Anh Quang Vu[1,4], Vinh-Dat Vuong[1,4(✉)], Khai Van Tran[1,4], and Thang Van Le[3,4]

[1] VNU-HCM Key Laboratory for Materials Technology (MTLab), 268 Ly Thuong Kiet Street, District 10, Ho Chi Minh City, Vietnam
vuongvinhdat@hcmut.edu.vn

[2] Faculty of Applied Science, Ho Chi Minh City University of Technology (HCMUT), 268 Ly Thuong Kiet Street, District 10, Ho Chi Minh City, Vietnam

[3] International University – Vietnam National University Ho Chi Minh City, Quarter 6, Linh Trung Ward, Thu Duc City, Ho Chi Minh City, Vietnam

[4] Vietnam National University Ho Chi Minh City, Linh Trung Ward, Thu Duc City, Ho Chi Minh City, Vietnam

Abstract. Mo-based transition metal dichalcogenide (TMD) materials are being widely researched as a catalyst for hydrogen evolution reaction (HER) due to their near-zero hydrogen adsorption energy and high density of active reaction sides. Only the edges of the MoS_2 layer have high catalytic performance, leading to the need for nanostructured MoS_2. Multiwalled carbon nanotubes (MWCNTs), which have great electrical conductivity and mechanical and chemical stability, are a suitable framework for nanoscale MoS_2 deposition. In this work, we fabricated a free-standing MWCNT paper through repeatedly electrophoretic deposition of MWCNTs-COOH onto indium tin oxide coated polyethylene terephthalate (ITO-PET), with a decoration of shell@core structured MoS_x@MoO_y ($2 < x < 3$ and $2 < y < 3$) nanoparticles through potentiostatic technique, followed by a substrate removal in a neutral medium. The thickness of the MWCNT film increases by approximately 5 μm with each deposition step and shows good electrical conductivity. The XRD pattern shows the hexagonal and orthorhombic structure of MoS_x and MoO_y respectively. Raman spectra and XRD patterns confirm the formation of amorphous MoS_x and MoO_y onto the outer wall of MWCNTs, increase the size of the nanotubes, as well as enhance the mechanical stability of the as prepared thin film.

Keywords: carbon nanotube thin-film · electrophoretic deposition · free-standing nano paper · MoS_x@MoO_y

1 Introduction

Decades of greenhouse gas emissions, primarily from energy production, have accelerated climate change [1–4]. Replacing fossil fuels with sustainable alternatives is urgent. Hydrogen, produced via water electrolysis, is a promising green energy solution [5–8].

N. R. Abdul Razak et al. (Eds.): ECOOP 1996, SPPHY 418, pp. 405–412, 2025.
https://doi.org/10.1007/978-981-96-2871-1_56

However, efficient electrolysis requires robust catalysts that can withstand harsh acidic or basic conditions [9]. While noble metals like platinum are highly effective, their scarcity and cost limit their use [10–16]. Transition metal dichalcogenides (TMDs), such as molybdenum dioxide (MoO_2) and molybdenum disulfide (MoS_2), offer potential as more affordable catalysts due to their conductivity, stability, and hydrogen adsorption properties [10–22]. However, long-term acidic conditions can degrade TMDs. Researchers are exploring combining TMDs with conductive substrates like carbon nanotubes to enhance their performance and durability. This approach aims to create a stable and efficient catalyst for hydrogen production and other applications [23–27].

This work fabricated a porous MWCNT-COOH thin film via electrophoretic deposition (EPD) on an ITO-PET substrate. MWCNT-COOH thin film were attracted to the cathode, forming a dense, horizontally oriented layer. Mo-based TMDs were grafted onto the MWCNTs to enhance adhesion and stability using potentiostatic deposition. The resulting film was detached as a free-standing paper using hydraulic force. This novel approach avoids harmful chemical etching and produces a promising structure for adsorption and energy applications.

2 Experimental

2.1 Materials

Analytical reagent chemicals are included: aluminum nitrate nonahydrate ($Al(NO_3)_3.9H_2O$, 99.99%), sodium sulfide (Na_2S, $\geq$98.0%), ammonium chloride (NH_4Cl, $\geq$99.7%) and isopropyl alcohol (IPA, $\geq$99.7%) was purchased from Xilong. Sodium molybdate (Na_2MoO_4, 99.0%, AR) was purchased from Tianjin Chemical. All the chemicals were used as received without any further treatment. The indium tin oxide coated polyethylene terephthalate (ITO-PET, 125 μm of thickness, 5 Ω/sq) film was purchased from Shenzhen Simihui Electronic Material.

The MWCNTs powder were purified and functionalized through a microwave-assisted chemical functionalization process, as described previously [28, 29]. Briefly, the left-over metal oxide catalyst from MWCNTs synthesis process was purified through heat treatment at 460 °C, and acid treatment with HNO_3 and H_2SO_4 at 60 °C. The carboxyl functional groups were then attached to the outer wall of the MWCNTs to produce MWCNT-COOH. The treated suspension was then centrifuged, filtered, and cleaned to obtain dry powder.

2.2 Preparation of Free-Standing Paper

The designed structure of thin film consists of 3 layers: the ITO-PET flat substrate, the high endurance MWCNT layer, and the MoS_x-MoO_y layer. First, the functionalized MWCNTs layer was deposited on the ITO-PET substrate by EPD method [30], as illustrated in Fig. 1a. The ITO-PET was connected to the negative pole, and stainless steel mesh was connected to the positive pole. Furthermore, a small amount of Al^{3+} (60 µL, 0.05 M) was added to the suspensions to positively charge the MWCNTs-COOH, where 0.06 g MWCNTs-COOH was dispersed into 30 mL IPA solvents through ultrasonication. The assembled acrylic frame was dipped in a well-dispersed suspension. The DC voltage of 40 V was applied for 5 min with 3 repeating cycles.

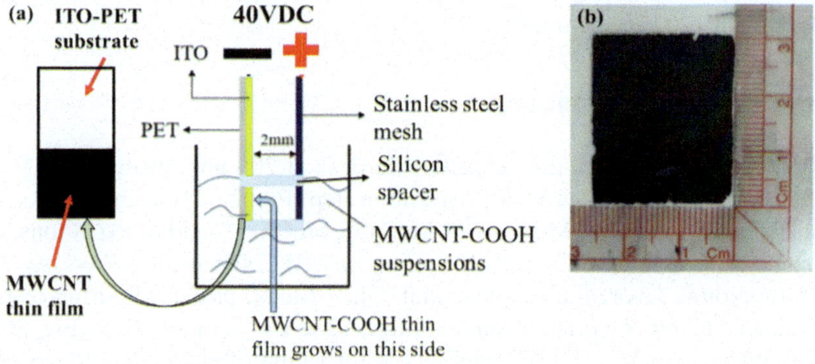

Fig. 1. (a) Schematic of MWCNTs deposition through EPD method. (b) Free-standing (MoS_x@MoO_y)/MWCNT paper.

The Mo-based nanoparticle deposition on MWCNTs was progressed via the potentiostatic method. The $(NH_4)_2MoS_4$ electrolyte was prepared by adding the solution of 0.72 g Na_2S in 20 mL H_2O to the solution of 0.021 g Na_2MoO_4 and 0.321 g NH_4Cl in 30 mL H_2O, followed by constant stirring for 60 min at 90 °C. The deposition cell was dipped in $(NH_4)_2MoS_4$ electrolyte for 15 min at the constant voltage of −1.1 V (vs Ag/AgCl) for 15 min to obtain the thin film. The hydrated thin film was then softly shaken in distilled water to detach from the ITO-PET substrate, then dried to obtain a free-standing (MoS_x@MoO_y)/MWCNT paper (Fig. 1b).

2.3 Characterization

The chemical components of electrolyte for $MoS_x@MoO_y$ deposition were confirmed via ultraviolet-visible spectroscopy (UV-Vis), using Shimadzu UV-1800 with the scanned windows of wavelength from 190 to 1100 nm and step of 1 nm. The thickness of thin film was measured by 3D laser scanning microscope, using Olympus OLS5000, through the profile scan with pitch of 5.00 μm/step. The materials structure thin film was studied by X-ray diffraction (PANalytical instrument) with wavelength of 1.54 Å (Cu-Kα), Raman spectroscopy (Horiba XploraOne) with the excitation laser wavelength of 532 nm and energy-dispersive X-ray (EDX) spectroscopy (Horiba H-7593) with 15kV beam. The morphological features of the samples were observed under Hitachi S-4800 scanning electron microscope (SEM), with the accelerating voltage of 10 kV and SE detector. Transmission electron microscopy (TEM) imaging was carried out using JEOL JEM-1400 with 120 kV operation.

3 Results and Discussions

The UV-Vis spectra showed the characteristic peak at 209 nm, attributed to MoO_4^{2-} ions of electrolyte for $MoS_x@MoO_y$ deposition step (Fig. 2a). The small peaks at 316, 396 and 465 nm attributed to $MoO_2S_2^{2-}$, MoS_4^{2-}, $MoOS_3^{2-}$, and MoS_4^{2-} ions. Cyclic voltammetry reveals that MoO_2 and MoS_2 grew on MWCNTs at −1.1 V (vs. Ag/AgCl). MoS_2 formation is favored at this potential, while MoS_3, produced during oxidation, can be reduced to lower oxidation state of MoS_x ($2 < x < 3$) at −0.72 V (Fig. 2b). The rapid current increase at −1.1 V suggests the formation of conductive MoS_2, lead to transition of x from 3 to 2. Oxidation of MoS_4^{2-} occurs at −0.55 V during the reverse CV sweep. The mechanism of MWCNT-COOH and $MoS_x@MoO_y$ was described in elsewhere [29].

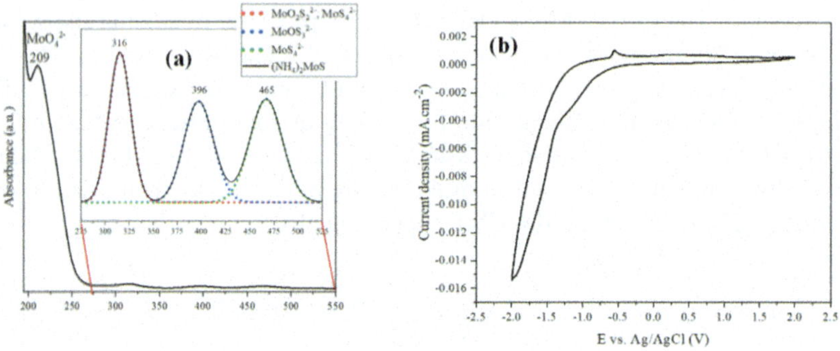

Fig. 2. (a) UV-vis absorption spectra of $(NH_4)_2MoS_4$ solution; (b) cyclic voltammetry curve of MWCNT/ITO-PET electrode in $(NH_4)_2MoS_4$ solution.

Figure 3a illustrated the layered structure of $(MoS_x@MoO_y)$/MWCNT paper attached on ITO-PET substrate. The materials structural analysis were recorded from such layers. EDX spectra (Fig. 3b) confirmed the presence of Mo and S on the deposited

MWCNTs while XRD patterns (Fig. 3c) revealed an amorphous $MoS_x@MoO_y$ phase on the MWCNTs, with characteristic peaks suggesting the presence of MoS_2, MoS_3, and MoO_2. The decrease in the I_{D1}/I_G ratio of Raman spectra (Fig. 3d) after implies a reduction in defect density on the MWCNTs surface due to decoration of $MoS_x@MoO_y$. Raman peaks corresponding to Mo–S and S–S bonds in MoS_2 and Mo–O bonds in MoO_2 further confirm the composition of the deposited layer.

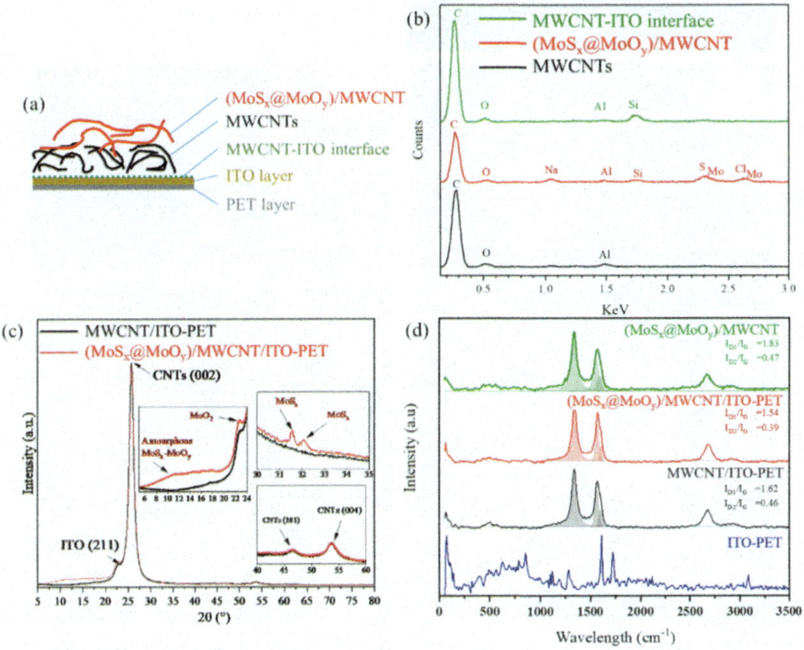

Fig. 3. EDX spectra (a), XRD pattern (b), and Raman spectra (c) of MWCNT/ITO-PET and ($MoS_x@MoO_y$)/MWCNT/ITO-PET.

Thickness of MWCNT layer was measured on surface profile from 3D laser scanning microscopic image (Fig. 4a,b). The average thickness of MWCNT layer from each EPD cycle was about 5 μm. SEM images showed uniformly distributed MWCNTs with sub-100 nm pores (Fig. 4c). $MoS_x@MoO_y$ coating is too thin for SEM observation (Fig. 4d). TEM reveals doubled MWCNT diameter after deposition, confirming $MoS_x@MoO_y$ coating (Fig. 4e,f). $MoS_x@MoO_y$ coated MWCNT exhibit improved agglomeration and mechanical strength compared to bare MWCNTs.

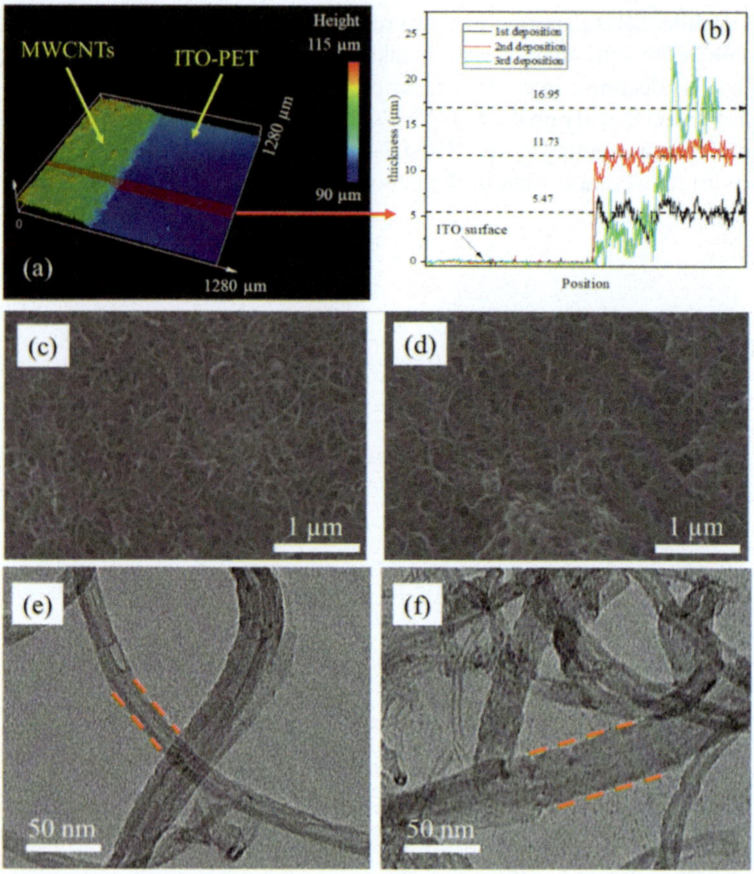

Fig. 4. (a) Surface profile and (b) thickness of MWCNT thin film. SEM images of (c) MWCNT/ITO-PET and (d) (MoS_x-MoO_y)/MWCNT/ITO-PET. TEM images of (e) MWCNT-COOH and (d) (MoS_x-MoO_y)/MWCNT.

4 Conclusion

In summary, the MWCNTs/ITO-PET film is fabricated via the EPD process and decorated with a (MoS_x@MoO_y) layer via static potential deposition. The EPD process can navigate the charged MWCNTs onto ITO surfaces and control the thickness of the MWCNTs layer on the electrode. The static potential technique allows amorphous MoS_x and MoO_y species to form a defect on the sidewall of MWCNTs. The decorated molybdenum layer on the tube wall enhances its adhesion, sufficient for the substrate removal without breaking the film structure. The free-standing MoS_x-MoO_y/MWCNTs film was separated from the ITO-PET substrate in a water instead of etching the PET substrate in a strongly acidic solution, which avoids damaging the as-prepared film. This free-standing paper offers a promising structure for applications in adsorption purposes or energy devices.

Acknowledgments. We acknowledge the support of time and facilities from Ho Chi Minh City University of Technology (HCMUT), VNU-HCM for this study.

References

1. Ritchie, H.: CO_2 emissions dataset: our sources and methods
2. Panchasara, H., Samrat, N.H., Islam, N.: Greenhouse gas emissions trends and mitigation measures in Australian agriculture sector—a review. Agriculture **11**(2), 85 (2021). [Online]. Available: https://www.mdpi.com/2077-0472/11/2/85
3. Yoro, K.O., Daramola, M.O.: Chapter 1—CO_2 emission sources, greenhouse gases, and the global warming effect. In: Rahimpour, M.R., Farsi, M., Makarem, M.A. (eds.) Advances in Carbon Capture. Woodhead Publishing, pp. 3–28 (2020)
4. Candra, O., Chammam, A., Alvarez, J.R.N., Muda, I., Aybar, H.Ş.: The impact of renewable energy sources on the sustainable development of the economy and greenhouse gas emissions. Sustainability **15**(3), 2104 (2023) [Online]. Available: https://www.mdpi.com/2071-1050/15/3/2104
5. Hassan, Q., et al.: Renewable energy-to-green hydrogen: A review of main resources routes, processes and evaluation. Int. J. Hydrogen Energy **48**(46), 17383–17408 (2023). https://doi.org/10.1016/j.ijhydene.2023.01.175
6. Greeley, J., Jaramillo, T.F., Bonde, J., Chorkendorff, I., Nørskov, J.K.J.N.M.: Computational high-throughput screening of electrocatalytic materials for hydrogen evolution, **5**(11), 909–913 (2006)
7. Anantharaj, S., Aravindan, V.J.A.E.M.: Developments and perspectives in 3D transition-metal-based electrocatalysts for neutral and near-neutral water electrolysis, **10**(1), 1902666 (2020)
8. Wang, Y.Z., Yang, M., Ding, Y.M., Li, N.W., Yu, L.J.A.F.M.: Recent advances in complex hollow electrocatalysts for water splitting, **32**(6), 2108681 (2022)
9. Le Goff, A., et al.: From hydrogenases to noble metal–free catalytic nanomaterials for H_2 production and uptake, **326**(5958), 1384–1387 (2009)
10. Pu, D.L.E., Ren, P., Zhou, W., Tang, D., Xiang, B., Wang, Y., Miao, J.:"Ultrathin MoO_2 nanosheets with good thermal stability and high conductivity. AIP Adv. **7**, Art no. 025015 (2017)
11. Syed, R., Ghosh, S.K., Sastry, P.U., Sharma, G., Hubli, R.C., Chakravartty, J.K.: Electrodeposition of thick metallic amorphous molybdenum coating from aqueous electrolyte. Surf. Coatings Technol. **261**, 15–2 (2015). https://doi.org/10.1016/j.surfcoat.2014.11.073
12. Su, H., Pan, X., Li, S., Zhang, H., Zou, R.: Defect-engineered two-dimensional transition metal dichalcogenides towards electrocatalytic hydrogen evolution reaction. Carbon Energy **5**(6), e296 (2023). https://doi.org/10.1002/cey2.296
13. Deng, Q., et al.: Research progress of transition-metal dichalcogenides for the hydrogen evolution reaction. J. Mater. Chem. A **11**(45), 24434–24453 (2023). https://doi.org/10.1039/D3TA04475K
14. Luis-Sunga, M., Regent, L., Pastor, E., García, G.: Non-precious metal graphene-based catalysts for hydrogen evolution reaction, **1**(2), 75–86, 2020 [Online]. Available: https://www.mdpi.com/2673-3293/1/2/8
15. Liu, W., et al.: A durable and pH-universal self-standing MoC–Mo_2C heterojunction electrode for efficient hydrogen evolution reaction, **12**(1), 6776 (2021)
16. Tang, C., Zhong, L., Zhang, B., Wang, H.F., Zhang, Q.J.A.M.: 3D mesoporous van der Waals heterostructures for trifunctional energy electrocatalysis, **30**(5), 1705110 (2018)

17. Chu, S., Majumdar, A.J.N.: Opportunities and challenges for a sustainable energy future, **488**(7411), 294–303 (2012)
18. Yu, W., Porosoff, M.D., Chen, J.G.J.C.R.: Review of Pt-based bimetallic catalysis: from model surfaces to supported catalysts, **112**(11), 5780–5817 (2012)
19. Zhang, R., Tang, Z., Wang, H., Sun, D., Tang, Y., Xie, Z.J.E.A.: The fabrication of hierarchical MoO_2@ MoS_2/rGO composite as high reversible anode material for lithium ion batteries, **364**, 136996 (2020)
20. Benck, J.D., Hellstern, T.R., Kibsgaard, J., Chakthranont, P., Jaramillo, T.F.: Catalyzing the hydrogen evolution reaction (HER) with molybdenum sulfide nanomaterials. ACS Catalysis **4**(11), 3957–3971 (2014). https://doi.org/10.1021/cs500923c
21. He, Z., Que, W.: Molybdenum disulfide nanomaterials: structures, properties, synthesis and recent progress on hydrogen evolution reaction. Appl. Mater. Today **3**, 23–56 (2016). https://doi.org/10.1016/j.apmt.2016.02.001
22. Jaramillo, T.F., Jørgensen, K.P., Bonde, J., Nielsen, J.H., Horch, S., Chorkendorff, I.: Identification of active edge sites for electrochemical H_2 evolution from MoS_2 nanocatalysts. Science **317**(5834), 100–102 (2007). https://doi.org/10.1126/science.1141483
23. Benson, E.E., et al.: Balancing the hydrogen evolution reaction, surface energetics, and stability of metallic MoS_2 nanosheets via covalent functionalization, **140**(1), 441–450 (2018)
24. Dai, Y., Wang, X., Wang, P., Sun, Y., Han, R., Luo, C.: Enhanced hydrogen evolution for MoS_2/CNTs induced by ethylene-glycol via charge transfer and proton concentration in acid solution, **46**(9), 6419–6426 (2021)
25. Liu, Y.-R., et al.: Facile one-pot synthesis of CoS_2-MoS_2/CNTs as efficient electrocatalyst for hydrogen evolution reaction, **384**, 51–57 (2016)
26. Jayabal, S., Saranya, G., Liu, Y., Geng, D., Meng, X.: Unravelling the synergy effects of defect-rich 1T-MoS_2/carbon nanotubes for the hydrogen evolution reaction by experimental and calculational studies, **3**(8), 2100–2110 (2019)
27. Deng et al., J.: High-performance hydrogen evolution electrocatalysis by layer-controlled MoS_2 nanosheets, **4**(66), 34733–34738 (2014)
28. Nguyet, L.V.T.N.T.M., Dong, N.V., Hang, N.T., Anh, L.T., Vinh, C.D.: A facile and effective purification method for multi-walled carbon nanotubes (MWNTs). J. Sci. Technol. **49**(6C), 274–281 (2011)
29. Nguyen, D.-B., et al.: Multi-layered thin film nanocomposite MoS_2@ MoO_2/MWCNP/ITO-PET: electrochemical approaches for synthesis and structural characterizations. Appl. Surf. Sci. **565**, 150508 (2021)

Steel Pile Corrosion in Phu My Port's Brackish Environment: A 7-year Field Study

Nguyen Thi Hong Van[1,4], Vinh-Phuc Ha[1,4], Tuan Anh Le[2,4], Anh Quang Vu[1,4], Vinh-Dat Vuong[1,4(✉)], Nguyen Danh Thao[2,4], and Thang Van Le[3,4]

[1] VNU-HCM Key Laboratory for Materials Technology (MTLab), 268 Ly Thuong Kiet Street, District 10, Ho Chi Minh City, Vietnam
vuongvinhdat@hcmut.edu.vn

[2] Faculty of Civil Engineering, Ho Chi Minh City University of Technology (HCMUT), 268 Ly Thuong Kiet Street, District 10, Ho Chi Minh City, Vietnam

[3] International University – Vietnam National University Ho Chi Minh City, Quarter 6, Linh Trung Ward, Thu Duc City, Ho Chi Minh City, Vietnam

[4] Vietnam National University Ho Chi Minh City, Linh Trung Ward, Thu Duc City, Ho Chi Minh City, Vietnam

Abstract. A long-term field study investigated the corrosion of steel pipe piles (SPPs) used for industrial buildings near Phu My port, Vietnam. This environment is particularly aggressive due to brackish water and pollution from the Thi Vai River. Researchers monitored STK400 steel pipe piles (165 mm diameter, 12 m length) placed in various zones (atmospheric, tidal, submerged, and mud) to assess the impact on corrosion rate and rust formation. The distribution of organisms, particularly barnacles, oysters, and seaweeds, influenced the layering of rust in the tidal and submerged zones. The highest corrosion rates and pitting occurred at the highest water level (HWL) and submerged zones due to constant renewal of corrosive agents at HWL and the influence of organisms at the lowest water level (LWL) and below. Conversely, the tidal and mud zones experienced lower corrosion rates due to cathodic protection from the corroding HWL and submerged zones. This study highlights the complex interplay of environmental factors affecting steel pile corrosion. The findings provide valuable insights for designing and implementing effective anti-corrosion strategies for steel structures in Phu My port's challenging environment.

Keywords: corrosion distribution · brackish environment · tidal zone · field study · mild steel · steel pipe pile

1 Introduction

Corrosion, the degradation of materials, is increasingly impacting society due to the complexity of modern materials and stringent sustainability demands. While traditional metals are affected, ceramics, polymers, and composites are also vulnerable. Addressing corrosion requires material selection, mitigation strategies, and effective monitoring. New materials are crucial for combating this issue [1]. Iron and steel corrosion forms

N. R. Abdul Razak et al. (Eds.): ECOOP 1996, SPPHY 418, pp. 413–418, 2025.
https://doi.org/10.1007/978-981-96-2871-1_57

rust, a protective layer under specific conditions [2, 3]. However, rust degradation can lead to localized corrosion and acid formation [4, 5]. The rust layer generally provides protection by blocking aggressive ions [4]. Studying tidal corrosion is challenging due to the dynamic environment and difficulty in monitoring surface conditions [6, 7]. Numerical models can predict damage and material lifespan, but validation with field data is essential [8]. Steel piles in ports face corrosion from multiple sources, including atmosphere, waves, tides, submersion, and mud [9, 10]. Galvanic coupling between different zones accelerates corrosion, especially in tidal areas [11, 12]. Research on tidal corrosion often focuses on specific locations and experimental conditions [13]. Studies have shown that the entire tidal zone acts as a cathode, with corrosion peaks below the tidal level and in the upper tidal zone [14]. Vietnam has conducted research on steel structure lifespan in corrosive environments. Studies on steel piles in Phu My Port revealed corrosion patterns and factors influencing rust formation [13]. Understanding these patterns is crucial for developing effective corrosion prevention strategies.

Current research on metal corrosion in water primarily centers on steel reinforcement in concrete, assessments of existing hydraulic structures, and laboratory corrosion simulations. These studies often involve small-scale samples and short-term observations, limiting their overall impact. This study investigates the corrosion behavior of STK400 steel, a material commonly used in coastal infrastructure support frames. By employing steel pipe piles at Phu My Port in Ba Ria - Vung Tau province, we aim to comprehensively evaluate the corrosion effects of different atmospheric zones on steel structures.

2 Experimental Procedure

The SPPs were manufactured by JFE corporation. Each pile had an outer diameter of 165 mm, a steel thickness of 9 mm, and a length of 12 m. The SPPs were driven directly into the riverbed and supported upright by the wharf structure (Fig. 1). Field testing of the bare SPPs (without coating) began in January 2014 and continued for seven years. After the testing period, the SPPs were transported to VNU-HCM Key Laboratory for Materials Technology (VNUHCM MTLab) for preservation and detailed analysis. Upon arrival at the laboratory, the steel pipe piles underwent a thorough cleaning process. This process involved:

(i) Removing seaweed, barnacles, and other marine organisms from the outer surface of the piles.
(ii) Removing mud and debris from the interior of the piles.
(iii) Using hand tools to remove thick, brittle rust layers from both the outer and inner surfaces.
(iv) Removing any remaining thin rust layer with a 5% hydrochloric acid solution in ethanol.

Once the piles were cleaned, the steel thickness of the entire SPP surface was measured using an Olympus 38 DL Plus thickness gauge.

Fig. 1. Installation and removal of steel pipe pile on test field at Phu My port.

3 Results and Discussions

The distribution of corrosion according to environmental zones from high to low was shown through the appearance and thickness of the pile after testing. The corrosion products on the outer surface of the SPP after removing environmental agents (seaweed, oysters and barnacles) were shown in Fig. 2. At altitude from about 1.2 m to 3.5 m from pile top, the corrosion products were the substrate for the growth of barnacles and oysters. However, the thickness of the rust layer was not uniform in this tidal zone. The distribution of steel thickness after testing (Fig. 3) showed that the rust layer is very thin in the zone from 2.6 m to 3.5 m.

The remaining steel thickness was plotted on a flat graph instead of a cylindrical one as it should be (the left graph of Fig. 2). The horizontal axis represents the distance (along the vertical axis of the pile) from the top to the point being measured, starting from position 0 m (top of the pile) above sea level, gradually descending to 12 m submerged in mud. The graph on the right shows the variation in the average thickness of the steel pipes at different distances from the top of the pile across various environmental zones.

The atmospheric zone (0 m to 1.1 m), exposed to air, exhibits minimal and uniform corrosion. This is attributed to the less corrosive nature of air compared to submerged environments. The absence of organisms or algae further contributes to the low corrosion rates in this region. Characterized by fluctuating water levels, the tidal zone (1.1 m to 4.1 m) experiences varying corrosion rates. Submersion in seawater, a highly corrosive medium, accelerates corrosion compared to other zones. Electrochemical processes involving anode-cathode pairs drive corrosion in this dynamic environment. Notably, a strong composite anode forms at 1.5 m due to optimal oxygen diffusion in the thin water layer left by receding tides. Conversely, the 3 m position exhibits the lowest composite cathode corrosion. Fully submerged in seawater, this zone (4.1 m to 9 m) continues to

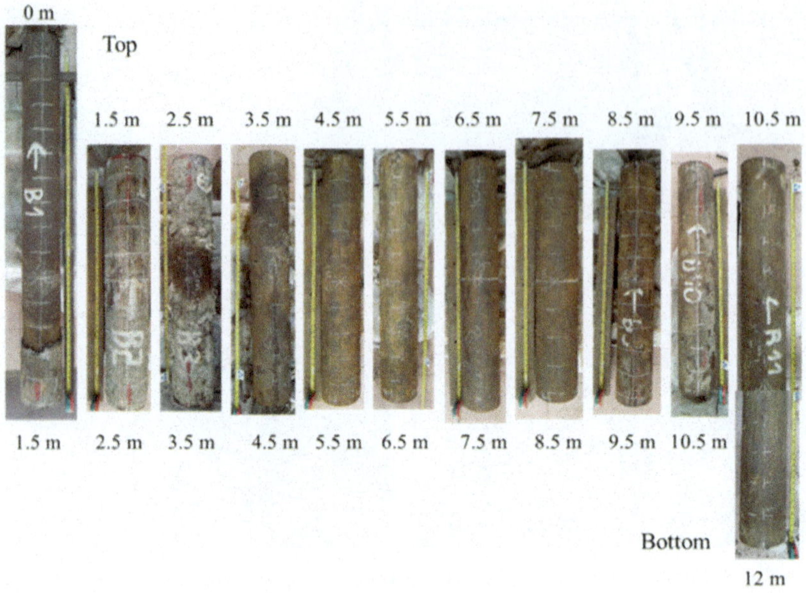

Fig. 2. Steel pipe after removing rust and biofouling.

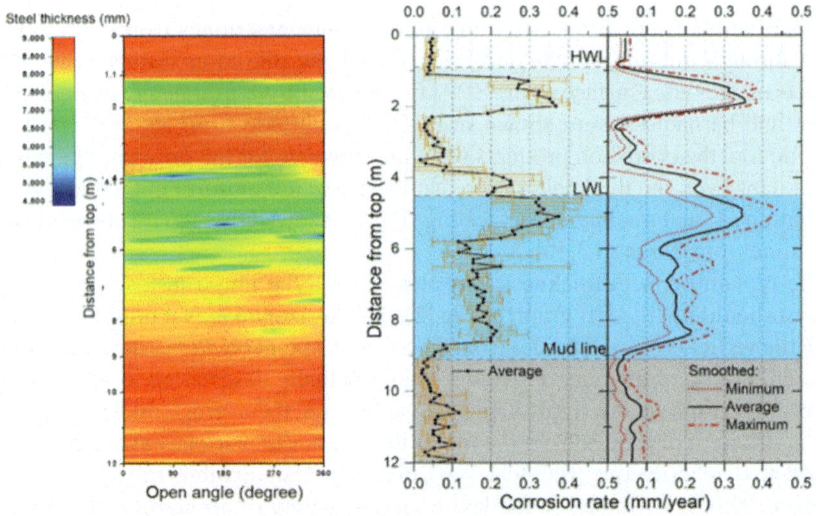

Fig. 3. Distribution chart of steel thickness and corrosion rate.

experience electrochemical corrosion. A strong composite anode exists here, similar to the tidal zone, due to the high oxygen concentration at the top of this layer. However, corrosion rates gradually decrease with depth as oxygen availability diminishes. The zone also displays localized corrosion, primarily caused by chloride ions in seawater. These ions break down the protective oxide layer, creating acidic conditions that accelerate corrosion at specific points. Electrochemical corrosion persists in this mud-filled area (9 m to 12 m) but at a reduced rate. The limited fluid movement prevents the removal of the formed rust layer, hindering the corrosion process.

4 Conclusion

The study analyzed the distribution of the average corrosion rate of STK400 steel according to the height of atmospheric zones (air, tidal, constantly submerged, riverbed mud) after 7 years of testing at the Thi Vai River estuary. The results indicated that severe corrosion appeared at the elevations corresponding to the tidal zone's highest and lowest water levels. The other zones exhibited low corrosion rates, with the primary form being uniform corrosion occurring on the air zone's exterior surface and the pipes' interior surface submerged in mud.

Acknowledgments. We acknowledge the support of time and facilities from Ho Chi Minh City University of Technology (HCMUT), VNU-HCM and JFE Steel Corporation for this study.

References

1. Shi, J., Wu, M., Ming, J.: Long-term corrosion resistance of reinforcing steel in alkali-activated slag mortar after exposure to marine environments. Corros. Sci. **179**, 109175 (2021)
2. Lawal, L., Afolalu, A., Jen, T.-C., Akinlabi, E.: Corrosion control and its application in marine environment - A review. Sol. State Phen. **355**, 61–73 (02/15 2024). https://doi.org/10.4028/p-634sdI
3. Prawoto, Y., Ibrahim, K., wan nik, w.s.: Effect of ph and chloride concentration on the corrosion of duplex stainless steel. Arab. J. Sci. Eng. **34** (12/01 2009)
4. Tran, T., Kannoorpatti, K., Padovan, A., Thennadil, S.: Sulphate-reducing bacteria's response to extreme pH environments and the effect of their activities on microbial corrosion. Appl. Sci. **11**, 2201 (03/03 2021). https://doi.org/10.3390/app11052201
5. Guo, Q., Liu, J., Yu, M., Li, S.-M.: Influence of rust layers on the corrosion behavior of ultra-high strength steel 300M subjected to wet–Dry cyclic environment with chloride and low humidity. Acta Metall. Sinica (Eng. Lett.) **28** (02/01 2015), https://doi.org/10.1007/s40195-014-0174-8
6. Patel, P., Garaniya, V., Abbassi, R., Baalisampang, T., Aryai, V.: Experimental Analysis of Pitting Corrosion in Offshore Structures (pp. 65–87) (2022)
7. Natesan, M., Selvaraj, S., Manickam, T., Venkatachari, G.: Corrosion behavior of metals and alloys in marine-industrial environment (in eng). Sci. Technol. Adv. Mater. **9**(4), 045002 (Dec. 2008). https://doi.org/10.1088/1468-6996/9/4/045002
8. Balangao, J.K.: Corrosion of metals: Factors, types and prevention strategies **14**, 79–87 (01/04 2024)
9. Vashi, R., Kadiya, H.: Corrosion study of metals in marine environment. J. Chem. **6** (10/01 2009). https://doi.org/10.1155/2009/509216
10. Jin, H., Wang, J., Tian, L., Gao, M., Zhao, J., Ren, L.: Recent advances in emerging integrated antifouling and anticorrosion coatings. Mat. Des. **213**, 110307 (2022/01/01/). https://doi.org/10.1016/j.matdes.2021.110307

11. Pham, N.-T., Vuong, V.-D., Tran, T., Van Le, T., Vu, Q.: Numerical simulation of extremely severe corrosion of the STK400 graded steel pipe pile in Southern Vietnam's brackish environment: Effects on macrocell corrosion of seawater flow and tide. Mat. Res. Exp. **9** (05/01 2022). https://doi.org/10.1088/2053-1591/ac6b1c
12. Vuong, V.-D., et al.: Five-year field exposure for visualized corrosion of STK400 graded steel pile in brackish environment of phu my industrial port (South. Vietnam). J. Mat. Eng. Perfor. **31** (11/08 2021). https://doi.org/10.1007/s11665-021-06395-8
13. Zhou, H., et al.: Field test of a reinforced concrete bridge under marine environmental corrosion. Eng. Fail. Anal. **115**, 104669 (06/01 2020). https://doi.org/10.1016/j.engfailanal.2020.104669

Author Index

N. R. Abdul Razak et al. (Eds.): ECOOP 1996, SPPHY 418, pp. 419–421, 2025.
https://doi.org/10.1007/978-981-96-2871-1

The manufacturer's authorised representative in the EU is Springer
Nature Customer Service Centre GmbH, Europaplatz 3, 69115 Heidelberg,
Germany. If you have any concerns regarding our products, please
contact ProductSafety@springernature.com

Printed and bound by CPI Group (UK) Ltd, Croydon, CR0 4YY
01/10/2026
02225905-0003